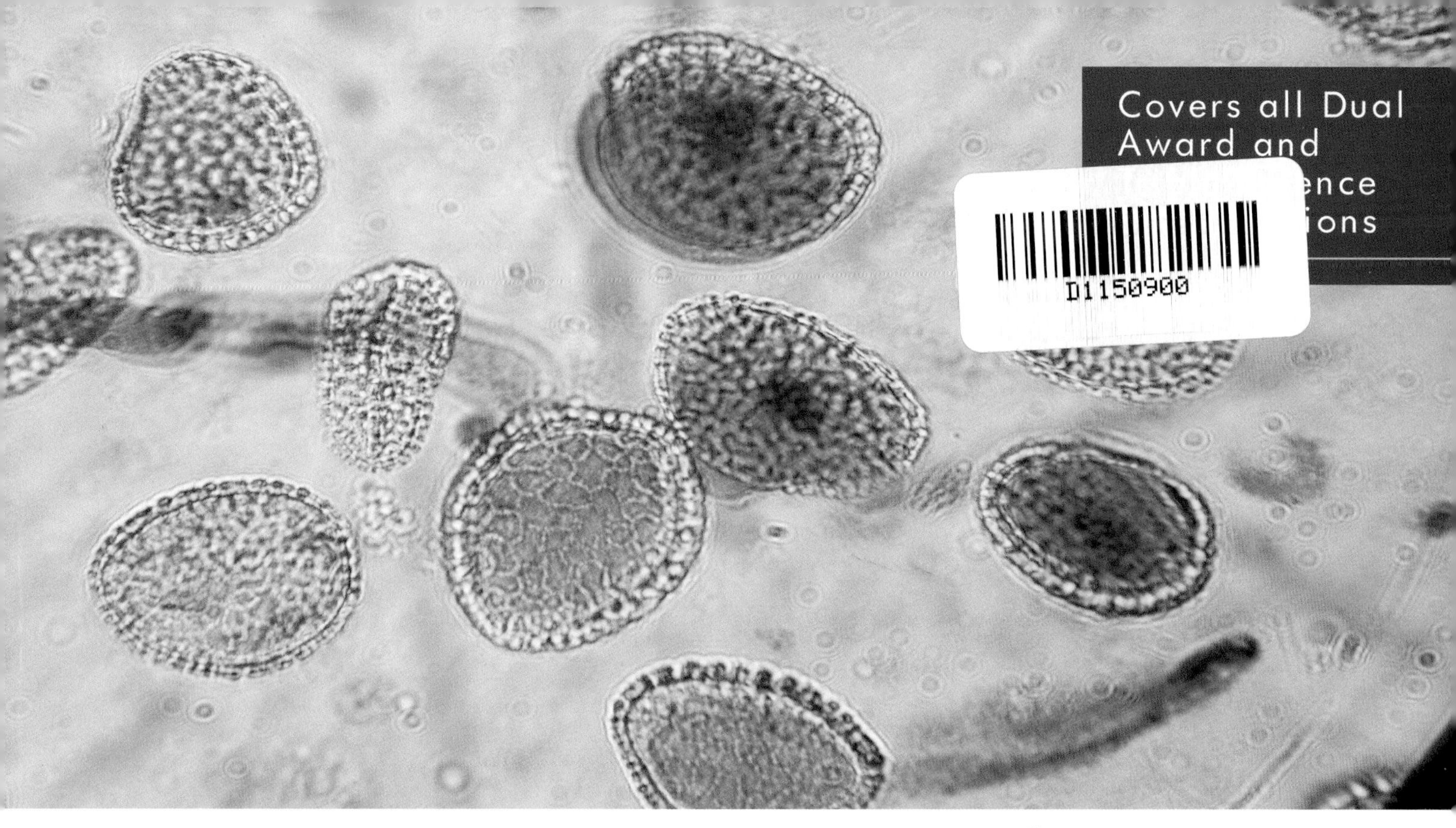

Longman GCSE Biology

PHIL BRADFIELD
AND
STEVE POTTER

Contents

Section F: Microbes – Friend or Foe?

Appendix

Introduction

This book covers everything you would need to know for any GCSE Biology exam set by AQA, Edexcel or OCR. This means that you will not need to know everything in this book – only the parts which are relevant to the GCSE specification that you are following. It is important to find out exactly what your particular examiners are likely to ask you. To do this you will need a copy of your specification and as much other useful information as you can get.

If your teacher hasn't given you a copy of your specification, you can download it from your Awarding Body's website. Find out from your teacher which Awarding Body your school is using. If you are doing Biology as a single subject and are using OCR, find out whether you are doing Extension Block A or B.

The web addresses are:

www.aqa.org.uk

www.edexcel.org.uk

www.ocr.org.uk

Find your way to the GCSE Biology specification you want and download it. Downloading won't take more than a few minutes. The OCR Biology specification contains both Extension Blocks.

While you are on your Awarding Body's website, see what other useful things you can find. You should be able to find examples of exam papers and mark schemes. These are important so that you can see exactly what sort of questions your examiners ask, and how they expect you to answer them. You will also be able to find material written to help teachers. There's no reason why students can't make good use of this as well. Much of it will be free.

To do well in Biology at GCSE you don't need to know everything in this book, but you do need to understand exactly what your examiners want.

About this book

This book has several features to help you with GCSE Biology.

Introduction

Each chapter has a short introduction to help you start thinking about the topic and let you know what is in the chapter.

Section B: Animal Physiology

Chapter 3: Food and Digestion

Food is essential for life. The nutrients obtained from it are used in many different ways by the body. This chapter looks at the different kinds of food, and how the food is broken down by the digestive system and absorbed into the blood, so that it can be carried to all the tissues of the body.

We need food for three main reasons:

- to supply us with a 'fuel' for energy
- to provide materials for growth and repair of tissues
- to help fight disease and keep our bodies healthy.

A balanced diet

The food that we eat is called our **diet**. No matter what you like to eat, if your body is to work properly and stay healthy, your diet must include five groups of food substances – **carbohydrates**, **fats**, **proteins**, **minerals and vitamins** – as well as **water** and **fibre**. Food should provide you with all of these substances, but they must also be present in the *right* amounts. A diet that provides enough of these substances and in the correct proportions to keep you healthy is called a **balanced diet** (Figure 3.1). We will deal with each type of food in turn, to find out about its chemistry and the role that it plays in the body.

Figure 3.1 *A balanced diet contains all the types of food the body needs, in just the right amounts.*

Carbohydrates

The chemical formula for glucose is $C_6H_{12}O_6$. Like all carbohydrates, glucose contains only the elements carbon, hydrogen and oxygen. The 'carbo' part of the name refers to carbon, and the 'hydrate' part refers to the fact that the hydrogen and oxygen atoms are in the ratio two to one, as in water (H_2O).

Carbohydrates only make up about 5% of the mass of the human body, but they have a very important role. They are the body's main 'fuel' for supplying cells with energy. Cells release this energy by oxidising a sugar called **glucose**, in the process called cell respiration (see Chapter 1). Glucose and other sugars are one sort of carbohydrate.

Glucose is found naturally in many sweet-tasting foods, such as fruits and vegetables. Other foods contain different sugars, such as the fruit sugar called **fructose**, and the milk sugar, **lactose**. Ordinary table sugar, the sort

Chapter 3: Food and Digestion

25

Margin boxes

The boxes in the margin give you extra help or information. They might explain something in a little more detail or guide you to linked topics in other parts of the book.

End of chapter checklists

These lists summarise the material in the chapter. They could also help you to make revision notes because they form a list of things that you need to revise. (You need to check your specification to find out exactly what you need to know.)

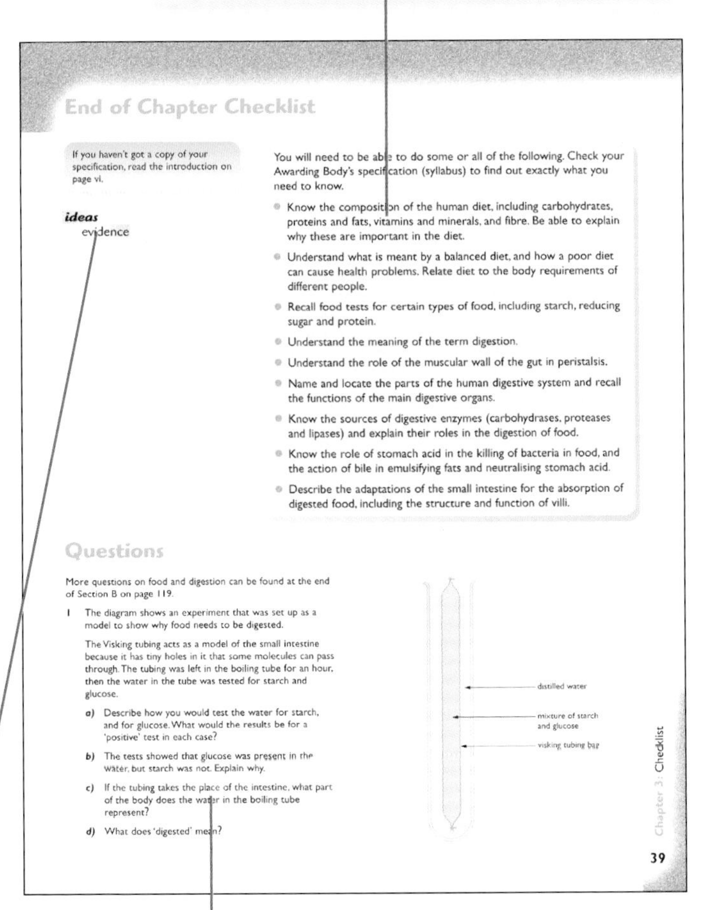

End of Chapter Checklist

If you haven't got a copy of your specification, read the introduction on page vi.

ideas evidence

You will need to be able to do some or all of the following. Check your Awarding Body's specification (syllabus) to find out exactly what you need to know.

- Know the composition of the human diet, including carbohydrates, proteins and fats, vitamins and minerals, and fibre. Be able to explain why these are important in the diet.
- Understand what is meant by a balanced diet, and how a poor diet can cause health problems. Relate diet to the body requirements of different people.
- Recall food tests for certain types of food, including starch, reducing sugar and protein.
- Understand the meaning of the term digestion.
- Understand the role of the muscular wall of the gut in peristalsis.
- Name and locate the parts of the human digestive system and recall the functions of the main digestive organs.
- Know the sources of digestive enzymes (carbohydrases, proteases and lipases) and explain their roles in the digestion of food.
- Know the role of stomach acid in the killing of bacteria in food, and the action of bile in emulsifying fats and neutralising stomach acid.
- Describe the adaptations of the small intestine for the absorption of digested food, including the structure and function of villi.

Questions

More questions on food and digestion can be found at the end of Section B on page 119.

1 The diagram shows an experiment that was set up as a model to show why food needs to be digested.

The Visking tubing acts as a model of the small intestine because it has tiny holes in it that some molecules can pass through. The tubing was left in the boiling tube for an hour, then the water in the tube was tested for starch and glucose.

a) Describe how you would test the water for starch, and for glucose. What would the results be for a 'positive' test in each case?

b) The tests showed that glucose was present in the water, but starch was not. Explain why.

c) If the tubing takes the place of the intestine, what part of the body does the water in the boiling tube represent?

d) What does 'digested' mean?

Chapter 3: Checklist

39

Ideas and Evidence

ideas evidence

This icon means that one of the Awarding Bodies has highlighted this topic as useful for learning about Ideas and Evidence. The Ideas and Evidence parts of the specifications help you to understand how scientific ideas have developed over time and how science relates to our everyday lives.

Questions

There are short questions at the end of each chapter. These help you to test your understanding of the material from the chapter. Some of them may also be research questions – you will need to use the Internet and other books to answer these.

There are also questions at the end of each section. The end of section questions are written in an exam style and cover topics from all the chapters in the section.

Chapter 1: Life Processes

The cells of all living organisms have common features, and the organisms themselves share common processes. In this chapter you will read about these features and look at some of the processes which keep cells alive.

All living organisms are composed of units called **cells**. The simplest organisms are made from single cells (Figure 1.1) but more complex plants and animals, like ourselves, are multicellular, composed of millions of cells. In a **multicellular** organism there are many different types of cells, with different structures. They are specialised so that they can carry out particular functions in the animal or plant. Despite all the differences, there are basic features that are the same in all cells.

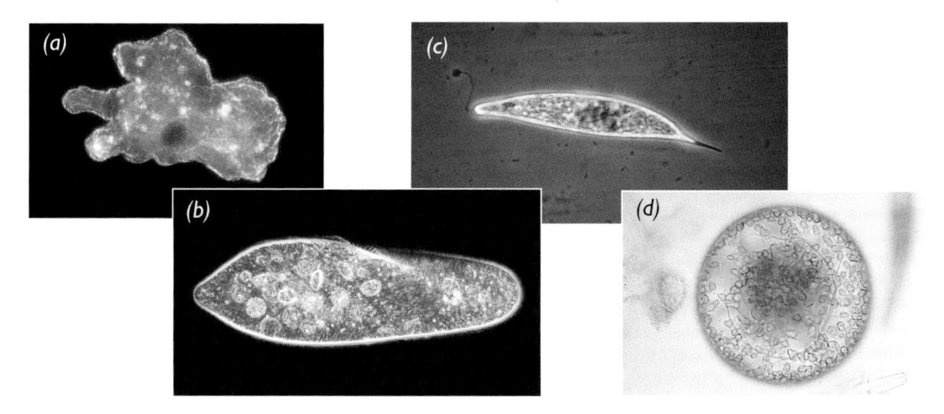

Figure 1.1 *Many simple organisms have 'bodies' made from single cells.*

There are seven life processes which are common to most living things. Organisms:

- **feed** – either by making food, as in plants, or eating other organisms, as animals do
- **excrete** – get rid of toxic waste products
- **move** – by the action of muscles in animals, and slow growth movements in plants
- **grow** – increase in size and mass, using materials from their food
- **respire** – get energy from their food
- **respond to stimuli** – are sensitive to changes in their surroundings
- **reproduce** – produce offspring.

Cell structure

For over 160 years scientists have known that animals and plants are made from cells. All cells contain some common parts, such as the nucleus, cytoplasm and cell membrane. Some cells have structures missing, for instance red blood cells lack a nucleus, which is unusual. The first chapter in a biology textbook usually shows diagrams of 'typical' plant and animal cells. In fact there is really no such thing as a 'typical' cell. Humans, for example, are composed of hundreds of different kinds of cells from nerve cells to blood cells, skin cells to liver cells. What we really mean by a 'typical' cell is a general diagram that shows all the features that you might find in most cells, without them being too specialised. Figure 1.2 shows the features you would expect to see in many animal and plant cells. However not all these are present in all cells – the parts of a plant which are not green do not have chloroplasts, for example.

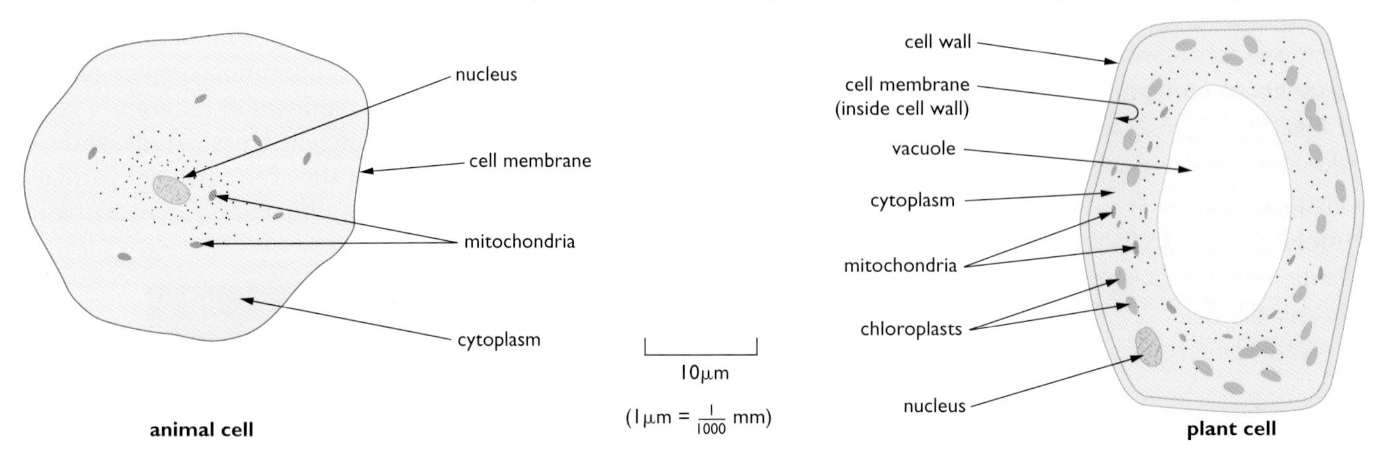

Figure 1.2 *The structure of a 'typical' animal and plant cell.*

The living material that makes up a cell is called **cytoplasm**. It has a texture rather like sloppy jelly, in other words somewhere between a solid and a liquid. Unlike a jelly, it is not made of one substance but is a complex material made of many different structures. You can't see many of these structures under an ordinary light microscope. An electron microscope has a much higher magnification, and can show the details of these structures, which are called **organelles** (Figure 1.3).

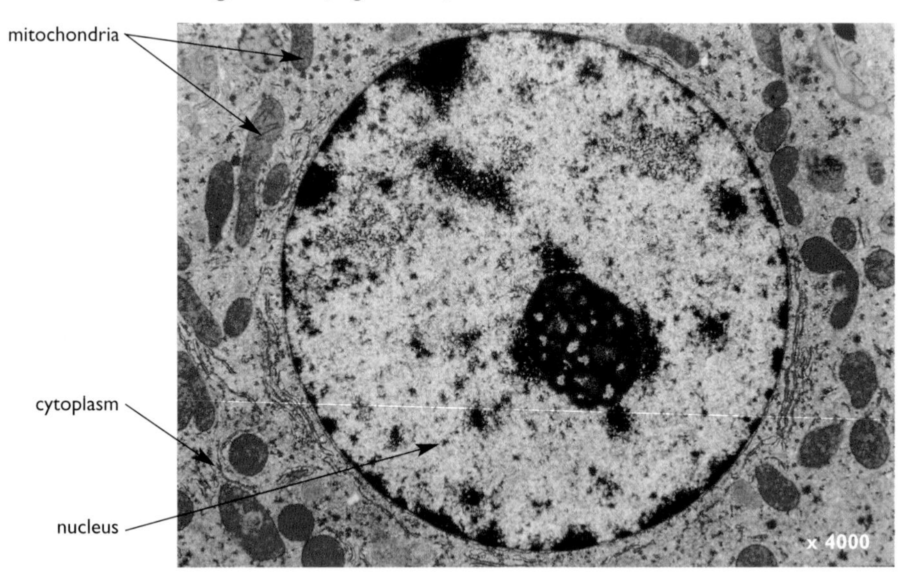

Figure 1.3 *The organelles in a cell can be seen using an electron microscope.*

The largest organelle in the cell is the **nucleus**. Nearly all cells have a nucleus. The few types that don't are usually dead (e.g. the xylem vessels in a stem, Chapter 11) or don't live for very long (e.g. mature red blood cells, Chapter 5). The nucleus controls the activities of the cell. It contains **chromosomes** (46 in human cells) which carry the genetic material, or **genes**. You will find out much more about genes and inheritance later in the book. Genes control the activities in the cell by determining which proteins the cell can make. One very important group of proteins found in cells is **enzymes** (see page 3). Enzymes control chemical reactions that go on in the cytoplasm.

All cells are surrounded by a **cell surface membrane** (often simply called the cell membrane). This is a thin layer like a 'skin' on the surface of the cell. It forms a boundary between the cytoplasm of the cell and the outside. However, it is not a complete barrier. Some chemicals can pass into the cell and others can pass out (the membrane is **permeable** to them). In fact the cell membrane *controls* which substances pass in either direction. We say that it is **selectively** permeable.

One organelle that is found in the cytoplasm of all living cells is the **mitochondrion** (plural **mitochondria**). There are many mitochondria in cells that need a lot of energy, such as muscle or nerve cells. This gives us a clue to the role of mitochondria. They carry out some of the reactions of **respiration** (see page 5) to release energy that the cell can use. In fact most of the energy from respiration is released in the mitochondria.

All of the structures we have seen so far are found in both animal and plant cells. However, some structures are only ever found in plant cells. There are three in particular – the cell wall, a permanent vacuole and chloroplasts.

The **cell wall** is a layer of non-living material that is found outside the cell membrane of plant cells. It is made mainly of a carbohydrate called **cellulose**, although other chemicals may be added to the wall in some cells. Cellulose is a tough material that helps the cell keep its shape. This is why plant cells have a fairly fixed shape. Animal cells, which lack a cell wall, tend to be more variable in shape. Plant cells absorb water, producing internal pressure which pushes against other cells of the plant, giving them support. Without a cell wall to withstand these pressures, this method of support would be impossible. The cell wall has large holes in it, so it is not a barrier to water or dissolved substances. In other words it is **freely permeable**.

Mature (fully grown) plant cells often have a large central space surrounded by a membrane, called a **vacuole**. This vacuole is a permanent feature of the cell. It is filled with a watery liquid called **cell sap**, a store of dissolved sugars, mineral ions and other solutes. Animal cells can have small vacuoles, but they are only temporary structures.

Cells of the green parts of plants, especially the leaves, have another very important organelle, the **chloroplast**. Chloroplasts absorb light energy to make food in the process of photosynthesis (see Chapter 10). The chloroplasts are green because they contain a green pigment called **chlorophyll**. Cells from the parts of a plant that are not green, such as the flowers, roots and woody stems, have no chloroplasts.

Figure 1.4 shows some animal and plant cells seen through the light microscope.

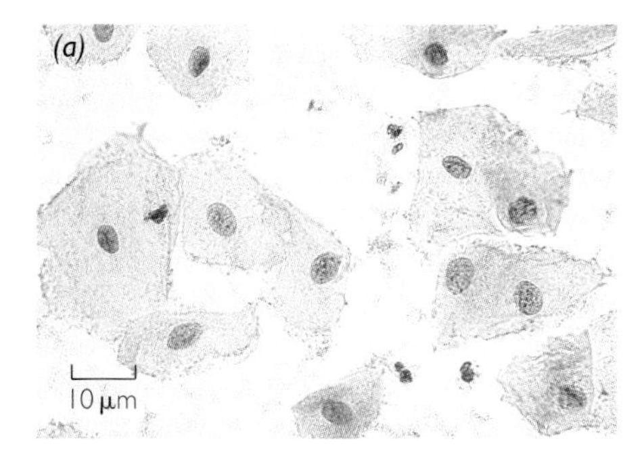

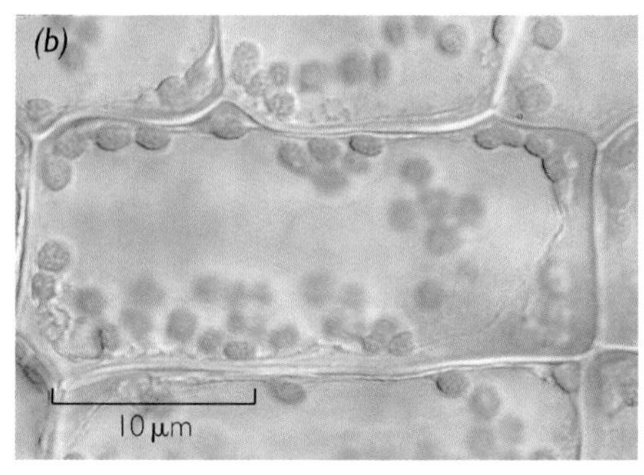

Figure 1.4 *(a) Cells from the lining of a human cheek. (b) Cells from the photosynthetic tissue of a leaf.*

> Nearly all cells have cytoplasm, a nucleus, a cell membrane and mitochondria. As well as these, plant cells have a cell wall and a permanent vacuole, and plant cells which photosynthesise have chloroplasts.

Enzymes: controlling reactions in the cell

The chemical reactions that go on in a cell are controlled by a group of proteins called enzymes. Enzymes are *biological catalysts*. A catalyst is a chemical which speeds up a reaction without being used up. It takes part in the reaction, but afterwards is unchanged and free to catalyse more reactions. Cells contain hundreds of different enzymes, each catalysing a different reaction. This is how the activities of a cell are controlled – the nucleus contains the genes, which control the production of enzymes, which catalyse reactions in the cytoplasm:

genes → proteins (enzymes) → catalyse reactions

Everything a cell does depends on which enzymes it can make, which in turn depends on which genes in its nucleus are working.

You have probably heard of the enzymes involved in digestion of food. They are secreted by the intestine onto the food to break it down. They are called **extracellular** enzymes, which means 'outside cells'. However, most enzymes stay *inside* cells – they are **intracellular**. You will read about digestive enzymes in Chapter 3.

What hasn't been mentioned is why enzymes are needed at all. This is because the temperatures inside organisms are low (e.g. the human body temperature is about 37°C) and without catalysts, most of the reactions that happen in cells would be far too slow to allow life to go on. Only when enzymes are present to speed them up do the reactions take place quickly enough.

It is possible for there to be thousands of different sorts of enzymes because they are made of proteins, and protein molecules have an enormous range of structures and shapes (see Chapter 3). The molecule that an enzyme acts on is called its **substrate**. Each enzyme has a small area on its surface called the **active site**. The substrate attaches to the active site of the enzyme. The reaction then takes place and products are formed. When the substrate joins up with the active site, it lowers the energy needed for the reaction to happen, allowing the products to be formed more easily.

The substrate fits into the active site of the enzyme rather like a key fitting into a lock. That is why this is called the 'lock and key' model of enzyme action (Figure 1.5).

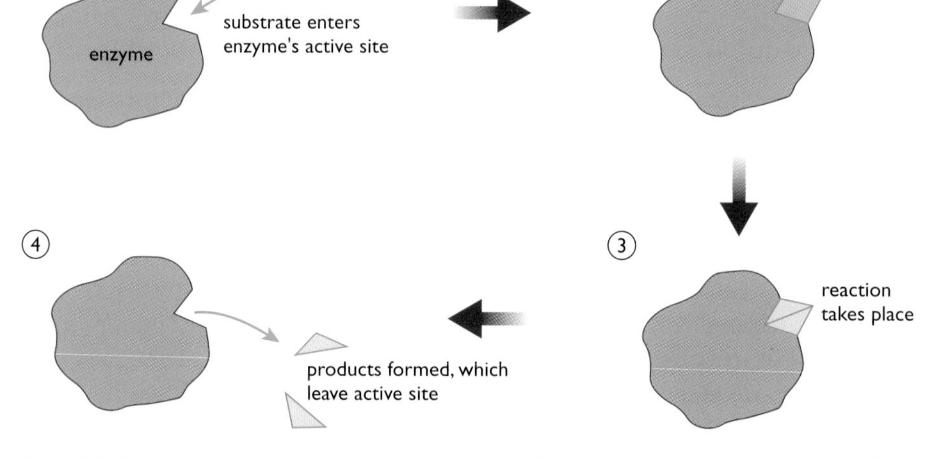

Figure 1.5 *Enzymes catalyse reactions at their active site. This acts like a 'lock' to the substrate 'key'. The substrate fits into the active site, and products are formed. This happens more easily than without the enzyme – so enzymes act as catalysts.*

Notice how, after it has catalysed the reaction once, the enzyme is free to act on more substrate molecules.

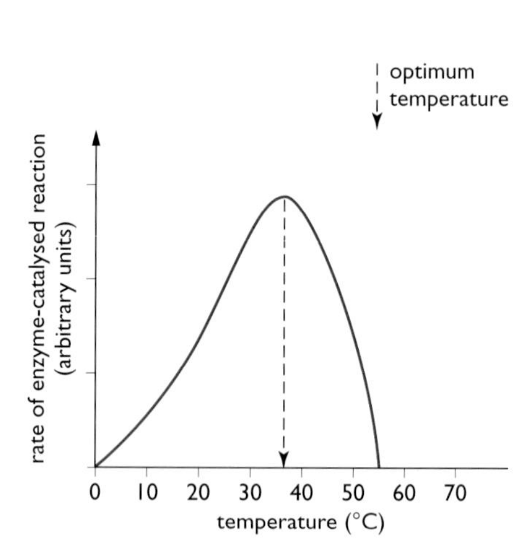

Figure 1.6 *Effect of temperature on the action of an enzyme.*

'Optimum' temperature means the 'best' temperature, in other words the temperature at which the reaction takes place most rapidly.

Factors affecting enzymes

Temperature affects the action of enzymes. This is easiest to see as a graph, where we plot the rate of the reaction controlled by an enzyme against the temperature (Figure 1.6).

Enzymes in the human body have evolved to work best at about body temperature (37°C). The graph shows this, because the peak on the curve happens at about this temperature. In this case 37°C is called the **optimum temperature** for the enzyme.

As the enzyme is heated up to the optimum temperature, increasing temperature speeds up the rate of reaction. This is because higher temperatures give the molecules of enzyme and substrate more energy, so they collide more often. More collisions mean that the reaction will take place more frequently. However, above the optimum temperature another factor comes into play. Enzymes are made of protein, and proteins are broken down by heat. From 40°C upwards, the heat destroys the enzyme. We say that it is **denatured**. You can see the effect of denaturing when you boil an egg. The egg white is made of protein, and turns from a clear runny liquid into a white solid as the heat denatures the protein.

Not all enzymes have an optimum temperature near 37°C, just those of animals such as mammals and birds, which all have body temperatures close to this value. Enzymes have evolved to work best at the normal body temperature of the organism. Bacteria that always live at an average temperature of 10°C will probably have enzymes with an optimum temperature of 10°C.

Temperature is not the only factor that affects an enzyme's activity. The rate of reaction may also be increased by raising the concentration of the enzyme or the substrate. The pH of the surroundings is also important. The pH inside cells is around neutral (pH 7) and not surprisingly, most enzymes have evolved to work best at this pH. At extremes of pH either side of neutral, the enzyme activity decreases, as shown by Figure 1.7. The pH at which the enzyme works best is called the **optimum pH** for that enzyme. Either side of the optimum, the pH affects the structure of the enzyme molecule, and changes the shape of its active site so that the substrate will not fit into it so well.

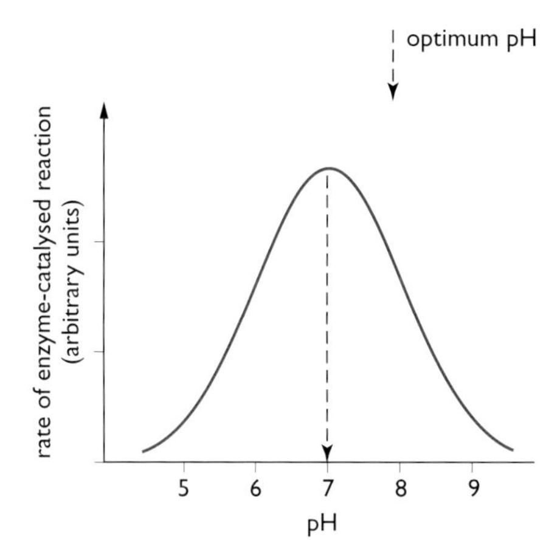

Figure 1.7 *Most enzymes work best at a neutral pH.*

Although most enzymes work best at a neutral pH, a few have an optimum below or above pH 7. The stomach produces hydrochloric acid, which makes its contents very acidic (see Chapter 3). Most enzymes stop working at a low pH like this, but the stomach makes an enzyme called pepsin which has an optimum pH of about 2, so that it is adapted to work well in these unusually acidic surroundings.

How the cell gets its energy

To be able to carry out all the processes needed for life, a cell needs a source of energy. It gets this by breaking down food molecules to release the stored chemical energy that they contain. This process is called **cell respiration**. Many people think of respiration as meaning 'breathing', but although there are links between the two processes, the biological meaning of respiration is very different.

The process of respiration happens in all the cells of our body. Oxygen is used to oxidise food, and carbon dioxide (and water) are released as waste products. The main food oxidised is glucose (a sugar). Glucose contains stored chemical energy that can be converted into other forms of energy that the cell can use. It is rather like burning a fuel to get the energy out of it, except that burning releases all its energy as heat, whereas respiration releases some heat energy, but most is trapped as energy in other chemicals. This chemical energy can be used for a variety of purposes, such as:

- contraction of muscle cells, producing movement
- active transport of molecules and ions (see page 6)
- building large molecules, such as proteins
- cell division.

In respiration, carbon passes from glucose out into the atmosphere as carbon dioxide. The carbon can be traced through the pathway using radioactive C^{14}.

The energy released as heat is also used to maintain a steady body temperature in animals such as mammals and birds (see Chapter 9).

The overall reaction for respiration is:

$$\text{glucose} + \text{oxygen} \rightarrow \text{carbon dioxide} + \text{water} \quad (+\text{ energy})$$
$$C_6H_{12}O_6 + 6O_2 \rightarrow 6CO_2 + 6H_2O \quad (+\text{ energy})$$

This is called **aerobic** respiration, because it uses oxygen. It is not just carried out by human cells, but by all animals and plants and many other organisms. It is important to realise that the equation on page 5 is just a *summary* of the process. It actually takes place gradually, as a sequence of small steps which release the energy of the glucose in small amounts. Each step in the process is catalysed by a different enzyme. The later steps in the process are the aerobic ones, and these release the most energy. They happen in the cell's mitochondria.

There are some situations where cells can respire *without* using oxygen. This is called **anaerobic** respiration. In anaerobic respiration, glucose is not completely broken down, and less energy is released. However, the advantage of anaerobic respiration is that it can occur in situations where oxygen is in short supply. Two important examples of this are in yeast cells and muscle cells.

Yeasts are single-celled fungi. They are used in commercial processes such as making wine and beer, and baking bread. When yeast cells are prevented from getting enough oxygen, they stop respiring aerobically, and start to respire anaerobically instead. The glucose is partly broken down into ethanol (alcohol) and carbon dioxide:

glucose → ethanol + carbon dioxide (+ some energy)

$$C_6H_{12}O_6 \rightarrow 2C_2H_5OH + 2CO_2$$

This process is looked at in more detail in Chapter 22. The ethanol from this respiration is the alcohol in wine and beer. The carbon dioxide is the gas that makes bread rise when it is baked. Think about the properties of ethanol – it makes a good fuel and will burn to produce a lot of heat, so it still has a lot of 'stored' chemical energy in it.

Muscle cells can also respire anaerobically when they are short of oxygen. If muscles are overworked, the blood cannot reach them fast enough to deliver enough oxygen for aerobic respiration. This happens when a person does a 'burst' activity, such as a sprint, or quickly lifting a heavy weight. This time the glucose is broken down into a substance called **lactic acid**:

glucose → lactic acid (+ some energy)

$$C_6H_{12}O_6 \rightarrow 2C_3H_6O_3$$

Anaerobic respiration provides enough energy to keep the overworked muscles going for a short period, but continuing the 'burst' activity makes lactic acid build up in the bloodstream, producing muscle cramps. The person then has to rest, to oxidise the lactic acid fully. This uses oxygen. The volume of oxygen needed to completely oxidise the lactic acid that builds up in the body during anaerobic respiration is called the **oxygen debt**. You can read more about the workings of muscles in Chapter 8.

Anaerobic respiration has two main disadvantages over aerobic respiration. It converts much less of the energy stored in food into a form of chemical energy that cells can use. It also produces toxic waste products, such as lactic acid or ethanol.

Movements of materials in and out of cells

Cell respiration shows the need for cells to be able to take in certain substances from their surroundings, such as glucose and oxygen, and get rid of others, such as carbon dioxide and water. As you have seen, the cell surface membrane is selective about which chemicals can pass in and out. There are three main ways that molecules and ions can move through the membrane. They are diffusion, active transport and osmosis.

Many substances can pass through the membrane by **diffusion**. Diffusion happens when a substance is more concentrated in one place than another. For example, if the cell is making carbon dioxide by respiration, the concentration of carbon dioxide inside the cell will be higher than outside. This difference in concentration is called a **concentration gradient**. The molecules of carbon dioxide are constantly moving about because of their kinetic energy. The cell membrane is permeable to carbon dioxide, so they can move in either direction through it. Because there is a higher concentration of carbon dioxide molecules inside the cell than outside, over time more molecules will move from inside the cell to outside than move in the other direction. We say that there is a *net* movement of the molecules from inside to outside (Figure 1.8).

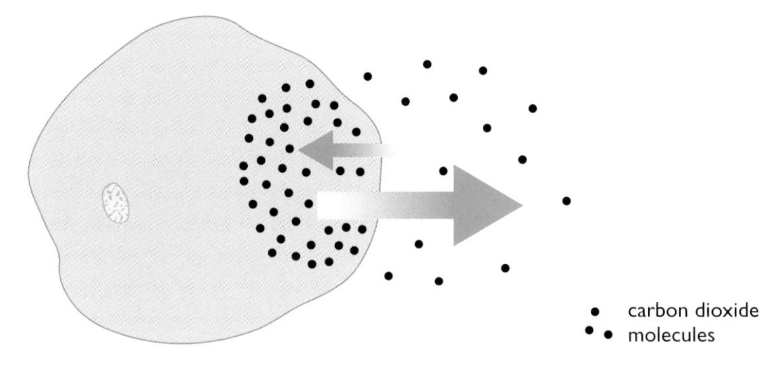

Figure 1.8 *Carbon dioxide is produced by respiration, so its concentration builds up inside the cell. Although the carbon dioxide molecules diffuse in both directions across the cell membrane, the overall (net) movement is out of the cell, down the concentration gradient.*

Diffusion is the net movement of particles (molecules or ions) from a region of high concentration to a region of low concentration, i.e. down a concentration gradient.

The opposite happens with oxygen. Respiration uses up oxygen, so there is a concentration gradient of oxygen from outside to inside the cell. There is therefore a net movement of oxygen *into* the cell by diffusion.

Diffusion happens because of the kinetic energy of the particles. It does not need an 'extra' source of energy from respiration. However, sometimes a cell needs to take in a substance when there is very little of that substance outside the cell, in other words *against* a concentration gradient. It can do this by another process, called **active transport**. The cell uses energy from respiration to take up the particles, rather like a pump uses energy to move a liquid from one place to another. In fact biologists usually speak of the cell 'pumping' ions or molecules in or out. The pumps are large protein molecules located in the cell membrane. An example of a place where this happens is in the human small intestine, where some glucose in the gut is absorbed into the cells lining the intestine by active transport. The roots of plants also take up certain mineral ions in this way (Chapter 11).

Active transport is the movement of particles against a concentration gradient, using energy from respiration.

Earlier in this chapter we called the cell membrane 'selectively' permeable. This term is sometimes used when describing osmosis. It means that the membrane has control over which molecules it lets through (e.g. by active transport). 'Partially' permeable just means that small molecules such as water and gases can pass through, while larger molecules cannot. Strictly, the two words are not interchangeable, but they are often used this way in biology books.

Water moves across cell membranes by a special sort of diffusion, called **osmosis**. Osmosis happens when the total concentrations of all dissolved substances inside and outside the cell are different. Water will move across the membrane from the more dilute solution to the more concentrated one. Notice that this is still obeying the rules of diffusion – the water moves from where there is a higher concentration of *water* molecules to a lower concentration of *water* molecules. Osmosis can only happen if the membrane is permeable to water but not to some other solutes. We say that it is **partially** permeable.

Osmosis is important for moving water from cell to cell, for example in plant roots. You can read about osmosis in much more detail in Chapter 11.

Osmosis in cells is the net movement of water from a dilute solution to a more concentrated solution across the partially permeable cell membrane.

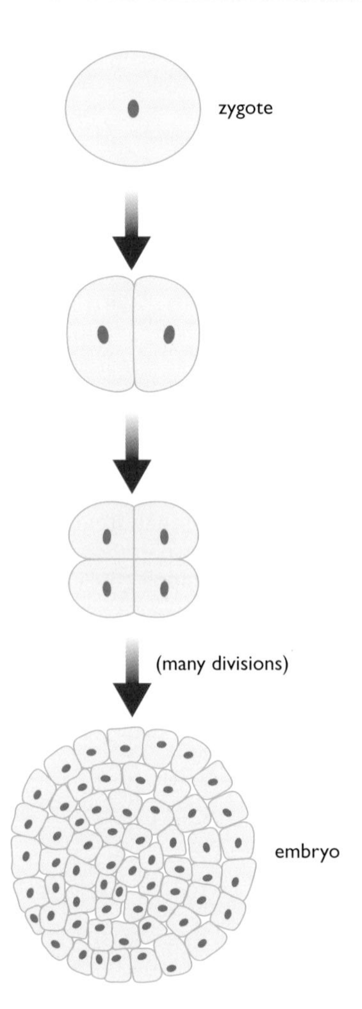

Figure 1.9 *Animals and plants grow by cell division.*

All cells exchange substances with their surroundings, but some parts of animals or plants are specially adapted for the exchange of materials because they have a very large surface area in proportion to their volume. In animals, two examples are the alveoli of the lungs (Chapter 2) and the villi of the small intestine (Chapter 3). Diffusion is a slow process, and organs that rely on diffusion need a large surface over which it can take place. The alveoli (air sacs) allow exchange of oxygen and carbon dioxide to take place between the air and the blood, during breathing. The villi of the small intestine provide a large surface area for the absorption of digested food. In plants, exchange surfaces are also adapted by having a large surface area, such as the spongy mesophyll of the leaf (Chapter 10) or the root hairs (Chapter 11).

Cell division and differentiation

Multicellular organisms like animals and plants begin life as a single fertilised egg cell, called a **zygote**. This divides into two cells, then four, then eight and so on, until the adult body contains countless millions of cells (Figure 1.9).

This type of cell division is called **mitosis** and is under the control of the genes. You can read a full account of mitosis in Chapter 17, but it is worthwhile considering an outline of the process now. First of all the chromosomes in the nucleus are copied, then the nucleus splits into two, so that the genetic information is shared equally between the two 'daughter' cells. The cytoplasm then divides (or in plant cells a new cell wall develops) forming two smaller cells. These then take in food substances to supply energy and building materials so that they can grow to full size. The process is repeated, but as the developing **embryo** grows, cells become specialised to carry out particular roles. This specialisation is also under the control of the genes, and is called **differentiation**. Different kinds of cells develop depending on where they are located in the embryo, for example a nerve cell in the spinal cord, or an epidermal cell in the outer layer of the skin (Figure 1.10). Throughout this book you will read about cells that have a structure adapted for a particular function.

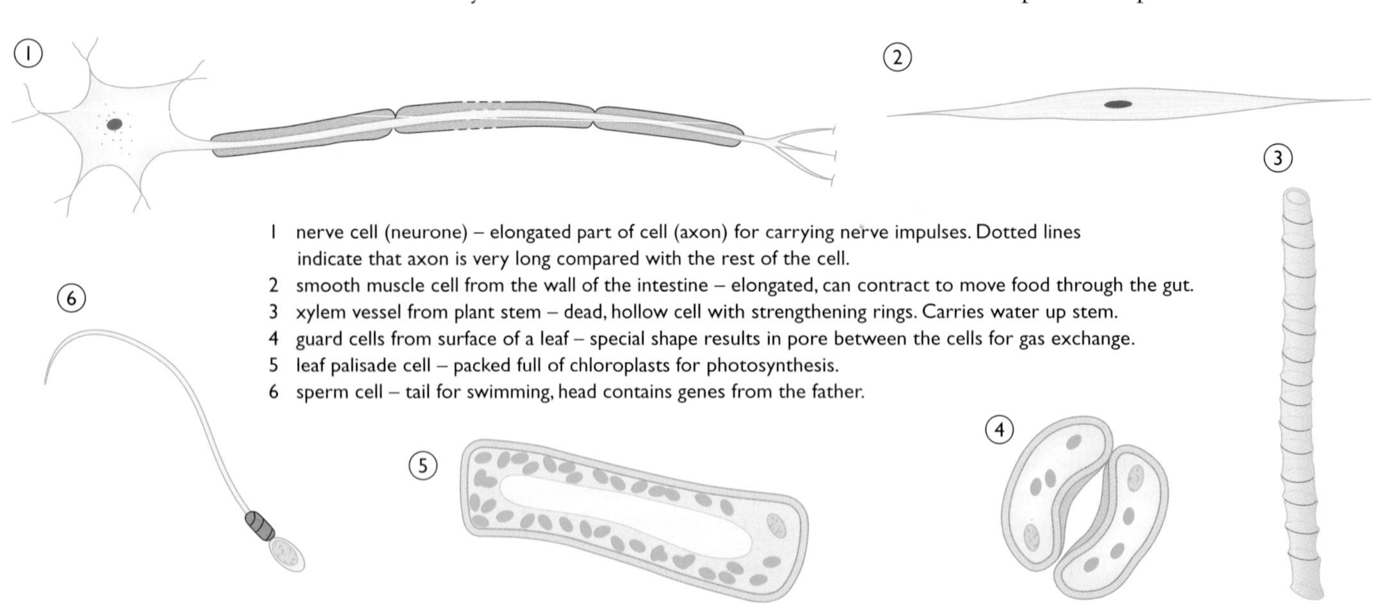

Figure 1.10 *Some cells with very specialised functions. They are not drawn to the same scale.*

What is hard to understand about this process is that through mitosis all the cells of the body have the *same* genes. How is it that some genes are 'switched on' and others are 'switched off' to produce different cells? The answer to this question is very complicated, and scientists are only just beginning to work it out.

Cells, tissues and organs

Cells with a similar function are grouped together as **tissues**. For example the muscle of your arm contains millions of similar muscle cells, all specialised for one function – contraction to move the arm bones. This is muscle tissue. However, a muscle also contains other tissues, such as blood, nervous tissue and epithelium (lining tissue). A collection of several tissues carrying out a particular function is called an **organ**. The main organs of the human body are shown in Figure 1.11. Plants also have tissues and organs. Leaves, roots, stems and flowers are all plant organs.

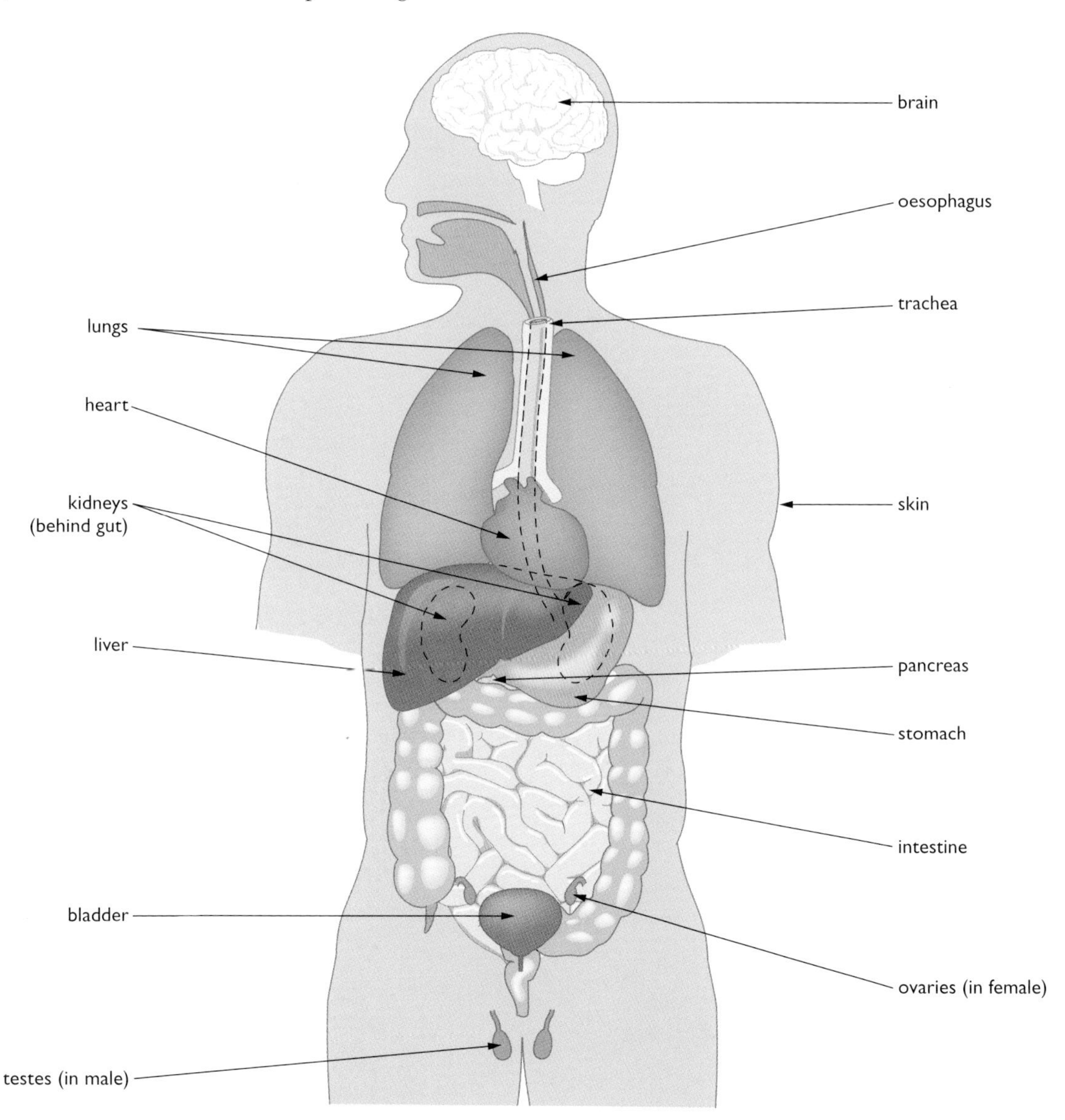

Figure 1.11 *Some of the main organs of the human body.*

In animals, jobs are usually carried out by several different organs working together. This is called an **organ system**. For example, the digestive system consists of the gut, along with glands such as the pancreas and gall bladder. The function of the whole system is to digest food and absorb the digested products into the blood. There are seven main systems in the human body:

- the **digestive** system
- the **respiratory** system: the lungs, which exchange oxygen and carbon dioxide
- the **circulatory** system: the heart and blood vessels, which transport materials around the body
- the **excretory** system, including the kidneys, which filter toxic waste materials from the blood
- the **nervous** system, consisting of the brain, spinal cord and nerves, which coordinate the body's actions
- the **endocrine** system: glands secreting hormones, which act as chemical messengers
- the **reproductive** system, producing sperm in males and eggs in females, and allowing the development of the embryo.

End of Chapter Checklist

If you haven't got a copy of your specification, read the introduction on page vi.

You will need to be able to do some or all of the following. Check your Awarding Body's specification (syllabus) to find out exactly what you need to know.

- Recall the structure of animal and plant cells, and the roles of the nucleus, cytoplasm, cell membrane, cell wall, chloroplasts, permanent vacuole and mitochondria.
- Relate the structure of different types of cells to their function in a tissue, an organ or a whole organism. (You will have to read the other relevant chapters of the book for this.)
- Know that chemical reactions inside cells are controlled by enzymes. Understand the 'lock and key' model of enzyme action. Be able to describe the effects of temperature and pH on enzymes.
- Know that cells obtain their energy by respiration. Recall a word and balanced symbol equation for aerobic respiration. Understand the processes of anaerobic respiration in yeast and muscle cells, including word equations for these.
- Understand the ways by which substances can enter or leave cells, by diffusion, active transport and osmosis. Recall examples of organs that are adapted for exchange of materials by having a large surface area.
- Know that body cells divide to produce more cells during growth and that cells become specialised to perform different functions.
- Appreciate that life processes are achieved by the coordinated action of cells acting together in tissues, organs and organ systems.

Questions

More questions on life processes can be found at the end of Section A on page 13.

1 ***a)*** Draw a diagram of a plant cell. Label all of the parts. Alongside each label write the function of that part.

b) Write down *three* differences between the cell you have drawn and a 'typical' animal cell.

2 Write a short description of the nature and function of enzymes. It would be easier if you used a word processor. Include in your description:

- a definition of an enzyme
- a description of the 'lock and key' model of enzyme action
- an explanation of the difference between intracellular and extracellular enzymes.

Your description should be about a page in length, including a labelled diagram.

3 The graph shows the effect of temperature on an enzyme. The enzyme was extracted from a microorganism that lives in hot mineral springs near a volcano.

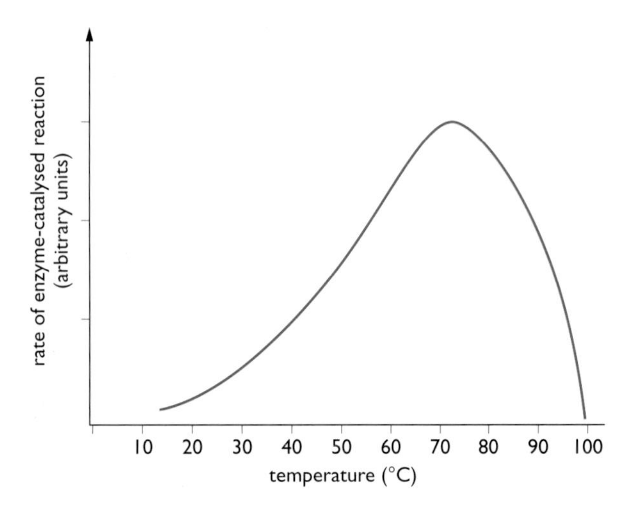

a) What is the optimum temperature of this enzyme?

b) Explain why the activity of the enzyme is greater at 60°C than at 30°C.

c) The optimum temperature of enzymes in the human body is about 37°C. Explain why this enzyme is different.

d) What happens to the enzyme at 90°C?

4 Explain the differences between diffusion and active transport.

5 The nerve cell called a **motor neurone** (page 65) and a **palisade** cell of a leaf (page 125) are both very specialised cells. Read about each of these and explain very briefly (three or four lines) how each is adapted to its function.

6 The diagram shows a cell from the lining of a human kidney tubule. A major role of the cell is to absorb salt from the fluid passing along the tubule and pass it into the blood, as shown by the arrow on the diagram.

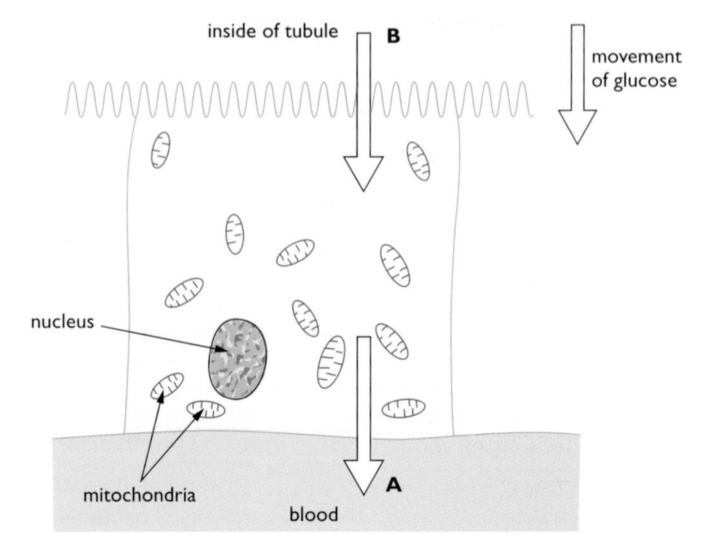

a) What is the function of the mitochondria?

b) The tubule cell contains a large number of mitochlondria. They are needed for the cell to transport glucose across the cell membrane into the blood at 'A'. Suggest the method that the cell uses to do this and explain your answer.

c) The mitochondria are *not* needed to transport the glucose into the cell from the tubule at 'B'. Name the process by which the ions move across the membrane at 'B' and explain your answer.

d) The surface membrane of the tubule cell at 'B' is greatly folded. Explain how this adaptation helps the cell to carry out its function.

End of Section Questions

1 These three organelles are found in cells: nucleus, chloroplast and mitochondrion.

a) Which of the above organelles would be found in:

i) a cell from a human muscle? *(1 mark)*

ii) a palisade cell from a leaf? *(1 mark)*

iii) a cell from the root of a plant? *(1 mark)*

b) Explain fully why the answers to *ii)* and *iii)* above are different. *(1 mark)*

c) What is the function of each organelle? *(3 marks)*

Total 7 marks

2 In multicellular organisms, cells are organised into tissues, organs and organ systems.

a) The diagram shows a section through an artery and a capillary.

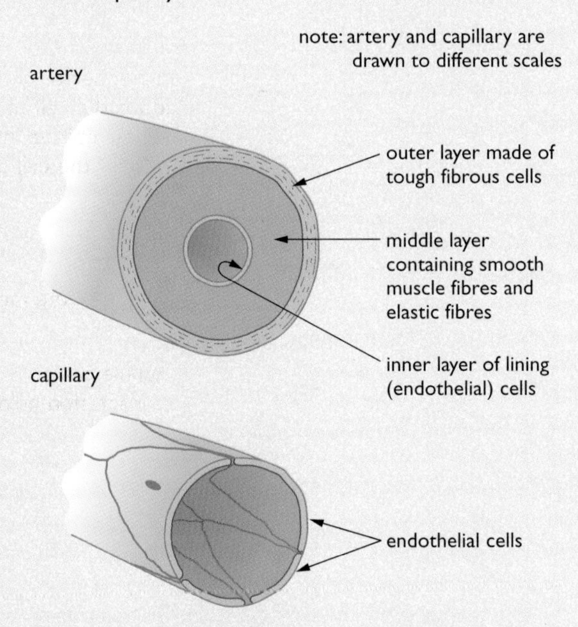

Explain why an artery can be considered to be an organ whereas a capillary cannot. *(2 marks)*

b) Organ systems contain two or more organs whose functions are linked. The digestive system is one human organ system. (See chapter 3.)

i) What does the digestive system do? *(2 marks)*

ii) Name three organs in the human digestive system. Explain what each organ does as part of the digestive system. *(6 marks)*

iii) Name two other human organ systems and, for each system, name two organs that are part of the system. *(6 marks)*

Total 16 marks

3 Catalase is an enzyme found in many plant and animal cells. It catalyses the breakdown of hydrogen peroxide into water and oxygen.

$$\text{hydrogen peroxide} \xrightarrow{\text{catalase}} \text{water} + \text{oxygen}$$

a) In an investigation into the action of catalase in potato, 20 g potato tissue was put into a small beaker containing hydrogen peroxide weighing 80 g in total. The temperature was maintained at 20°C throughout the investigation. As soon as the potato was added, the mass of the beaker and its contents was recorded until there was no further change in mass. The results are shown in the graph.

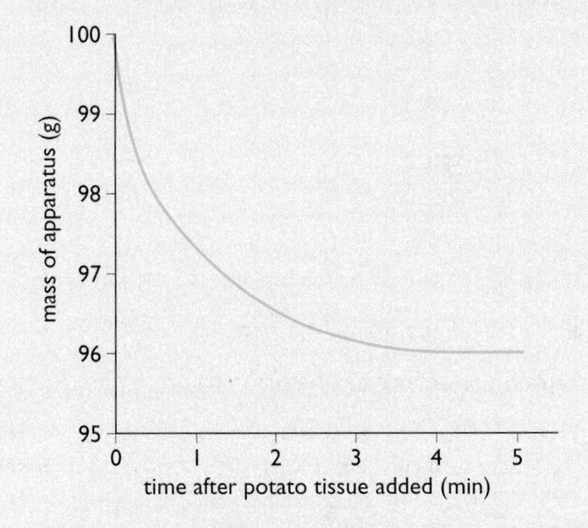

i) How much oxygen was formed in this investigation? Explain your answer. *(2 marks)*

ii) Estimate the time by which half this mass of oxygen had been formed. *(2 marks)*

iii) Explain, in terms of collisions between enzyme and substrate molecules, why the rate of reaction changes during the course of the investigation. *(2 marks)*

b) The students repeated the investigation at 30°C. What difference, if any, would you expect in:

i) the mass of oxygen formed?

ii) the time taken to form this mass of oxygen?

Explain your answers. *(4 marks)*

Total 10 marks

4 Different particles move across cell membranes using different processes.

a) The table shows some ways in which active transport, osmosis and diffusion are similar and some ways in which they are different. Copy and complete the table with ticks and crosses. *(12 marks)*

Feature	Active transport	Osmosis	Diffusion
particles must have kinetic energy			
requires energy from respiration			
particles move down a concentration gradient			
process needs special carriers in the membrane			

b) The graph shows the results of an investigation into the rate of diffusion of sodium ions across the membranes of potato cells.

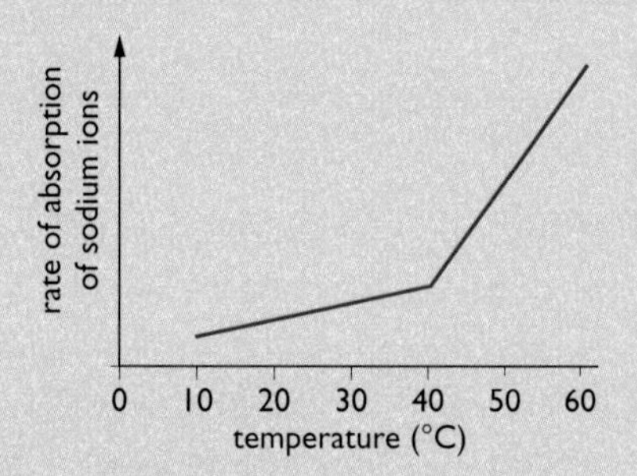

i) Explain the increase in the rate of diffusion up to 40°C. *(2 marks)*

ii) Suggest why the rate of increase is much steeper at temperatures above 40°C. *(2 marks)*

Total 16 marks

5 Cells in the wall of the small intestine divide by mitosis to replace cells lost as food passes through.

a) Chromosomes contain DNA. The graph shows the changes in the DNA content of a cell in the wall of the small intestine as it divides by mitosis.

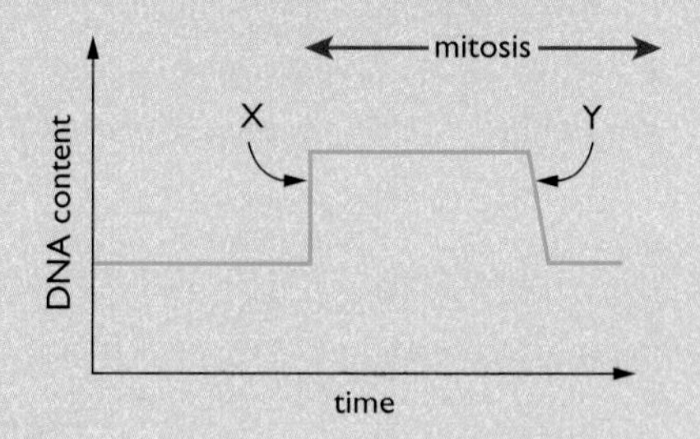

i) Why is it essential that the DNA content is doubled (X) before mitosis commences? *(2 marks)*

ii) What do you think happens to the cell at point Y? *(1 mark)*

b) The diagram shows a cell in the wall of a villus in the small intestine. Some of the processes involved in the absorption of glucose are also shown.

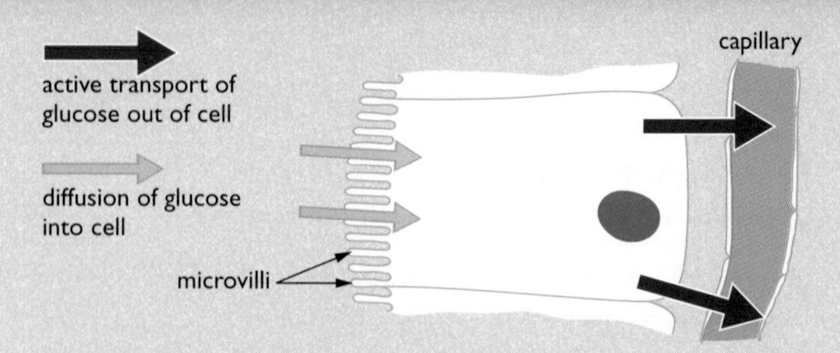

i) What is the importance of the small intestine having villi? *(1 mark)*

ii) Suggest how the microvilli adapt this cell to its function of absorbing glucose. *(1 mark)*

iii) Suggest how the active transport of glucose out of the cell and into the blood stream helps with the absorption of glucose from the small intestine. *(2 marks)*

Total 7 marks

6 A respirometer is used to measure the rate of respiration. The diagram shows a simple respirometer. The sodium hydroxide solution in the apparatus absorbs carbon dioxide. Some results from the investigation are also shown.

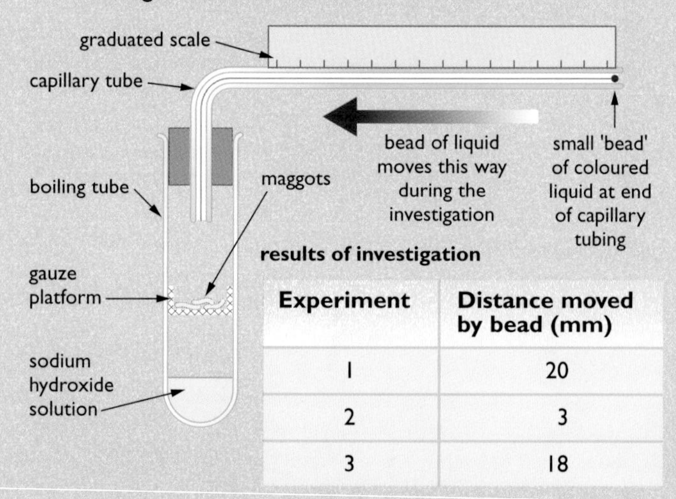

results of investigation

Experiment	Distance moved by bead (mm)
1	20
2	3
3	18

a) Assume that the maggots in the apparatus respire aerobically.

i) Write the symbol equation for aerobic respiration. *(4 marks)*

ii) From the equation, what can you assume about the amount of oxygen taken in and carbon dioxide given off by the maggots? Explain your answer. *(3 marks)*

iii) Result 2 is significantly different from the other two results. Suggest a reason for this. *(2 marks)*

iv) How would the results be different if the organisms under investigation respired anaerobically? *(2 marks)*

Total 11 marks

Chapter 2: Breathing

Breathing is the mechanism whereby air is drawn into and out of the lungs and gas exchange between the air and the blood takes place. This chapter looks at breathing mechanisms in humans and other animals. It also deals with some ways that smoking can damage the lungs and stop these vital organs working properly.

Cells get their energy by oxidising foods such as glucose. This process is called cell respiration (see Chapter 1). If the cells are to respire aerobically, they need a continuous supply of oxygen from the blood. In addition, the waste product of respiration, carbon dioxide, needs to be removed from the body. In humans these gases are exchanged between the blood and the air in the lungs.

Breathing and respiration

It is important to understand the difference between breathing and respiration. Breathing consists of **ventilation**, the mechanism that moves air in and out of the lungs, and **gas exchange** between the air and the blood. Respiration is the oxidation reaction described above. It would probably be better to call the lungs and associated organs the 'breathing system'. However, they are usually called the 'respiratory system', which can be confusing!

The structure of the respiratory system

The lungs are enclosed in the chest or **thorax** by the ribcage and a muscular sheet of tissue called the **diaphragm** (Figure 2.1). As you will see, the actions of these two structures bring about the movements of air into and out of the lungs. Joining each rib to the next are two sets of muscles called **intercostal muscles** ('costals' are rib bones). If you eat meat, you will have seen intercostal muscle attached to the long bones of 'spare ribs'. The diaphragm separates the contents of the thorax from the abdomen. It is not flat, but a shallow dome shape, with a fibrous middle part forming the 'roof' of the dome, and muscular edges forming the walls.

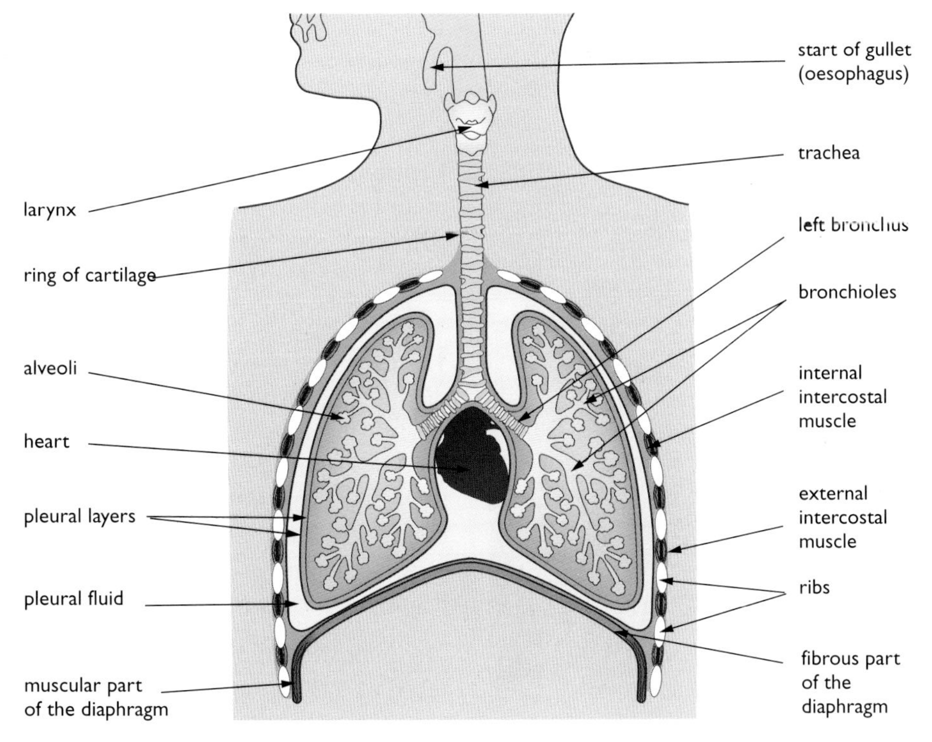

Figure 2.1 *The human respiratory system.*

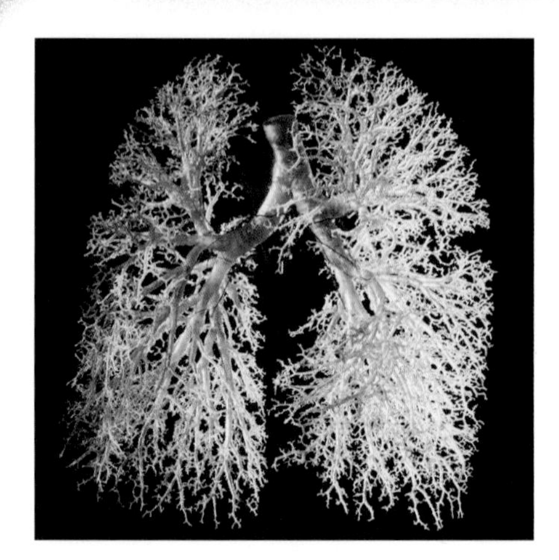

Figure 2.2 *This cast of the human lungs was made by injecting a pair of lungs with a liquid plastic. The plastic was allowed to set, then the lung tissue was dissolved away with acid.*

In the bronchi, the cartilage forms complete, circular rings. In the trachea, the rings are incomplete, and shaped like a letter 'C'. The open part of the ring is at the back of the trachea next to the oesophagus (gullet) passes through the thorax, where it lies behind. When food passes along the oesophagus by peristalsis (see Chapter 3) the gaps in the rings allow the lumps of food to pass through more easily, without the peristaltic wave 'catching' on the rings (Figure 2.3).

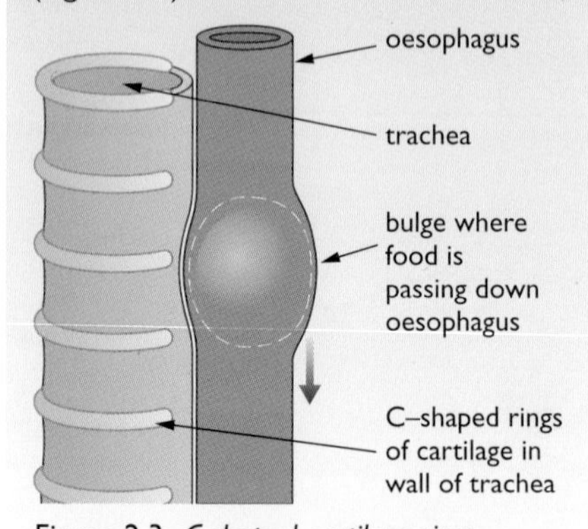

Figure 2.3 *C-shaped cartilage rings in the trachea.*

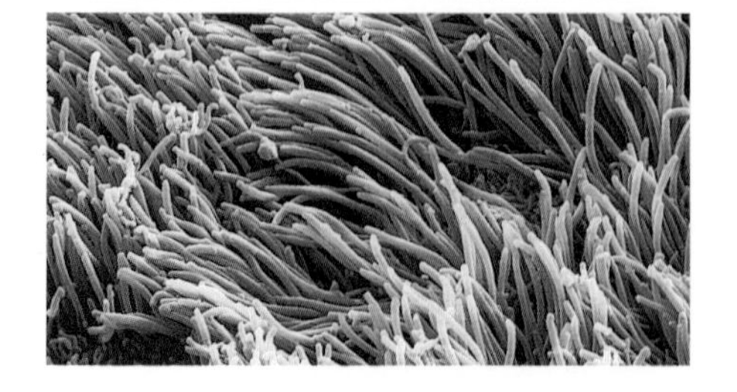

Figure 2.4 *This electron microscope picture shows cilia from the lining of the trachea.*

The air passages of the lungs form a highly branching network (Figure 2.2). This is why it is sometimes called the **bronchial tree**.

When we breathe in, air enters our nose or mouth and passes down the windpipe or **trachea**. The trachea splits into two tubes called the **bronchi**, one leading to each lung. Each **bronchus** divides into smaller and smaller tubes called **bronchioles**, eventually ending at microscopic air sacs, called **alveoli**. It is here that gas exchange with the blood takes place.

The walls of trachea and bronchi contain rings of gristle or **cartilage**. These support the airways and keep them open when we breathe in. They are rather like the rings in a vacuum cleaner hose – without them the hose would squash flat when the cleaner sucks air in.

The inside of the thorax is separated from the lungs by two thin, moist membranes called the **pleural layers**. They make up a continuous envelope around the lungs, forming an airtight seal. Between the two layers is a space called the **pleural cavity**, filled with a thin layer of liquid called **pleural fluid**. This acts as lubrication, so that the surfaces of the lungs don't stick to the inside of the chest wall when we breathe.

Keeping the airways clean

The trachea and larger airways are lined with a layer of cells that have an important role in keeping the airways clean. Some cells in this lining secrete a sticky liquid called **mucus**, which traps particles of dirt or bacteria that are breathed in. Other cells are covered with tiny hair-like structures called **cilia** (Figure 2.4). The cilia beat backwards and forwards, sweeping the mucus and trapped particles out towards the mouth. In this way, dirt and bacteria are prevented from entering the lungs, where they might cause an infection. As you will see, one of the effects of smoking is that it destroys the cilia and stops this protection mechanism from working properly.

Ventilation of the lungs

Ventilation means moving air in and out of the lungs. This requires a difference in air pressure – the air moves from a place where the pressure is high to one where it is low. Ventilation depends on the fact that the thorax is an airtight cavity. When we breathe, we change the volume of our thorax, which alters the pressure inside it. This causes air to move in or out of the lungs.

There are two movements that bring about ventilation, those of the ribs and the diaphragm. If you put your hands on your chest and breathe in deeply, you can feel your ribs move upwards and outwards. They are moved by the intercostal muscles (Figure 2.5). The outer (external) intercostals contract, pulling the ribs up. At the same time the muscles of the diaphragm contract, pulling the diaphragm down into a more flattened shape (Figure 2.6a). Both these movements increase the volume of the chest and cause a slight drop in pressure inside the thorax compared with the air outside. Air then enters the lungs.

The opposite happens when you breathe out deeply. The external intercostals relax, and the internal intercostals contract, pulling the ribs down and in. At the same time, the diaphragm muscles relax and the diaphragm goes back to its normal dome shape. The volume of the thorax decreases, and the pressure in the thorax is raised slightly above atmospheric pressure. This time the difference in pressure forces air out of the lungs (Figure 2.6b). Exhalation is helped by the fact that the lungs are elastic, so that they tend to empty like a balloon.

During normal (shallow) breathing, the elasticity of the lungs and the weight of the ribs acting downwards is enough to cause exhalation. The external intercostals are only really used for deep (forced) breathing out, for instance when we are exercising.

It is important that you remember the changes in volume and pressure during ventilation. If you have trouble understanding these, think of what happens when you use a bicycle pump. If you push the pump handle, the air in the pump is squashed, its pressure rises and it is forced out of the pump. If you pull on the handle, the air pressure inside the pump falls a little, and air is drawn in from outside. This is similar to what happens in the lungs. In exams, students sometimes talk about the lungs *forcing* the air in and out – they don't!

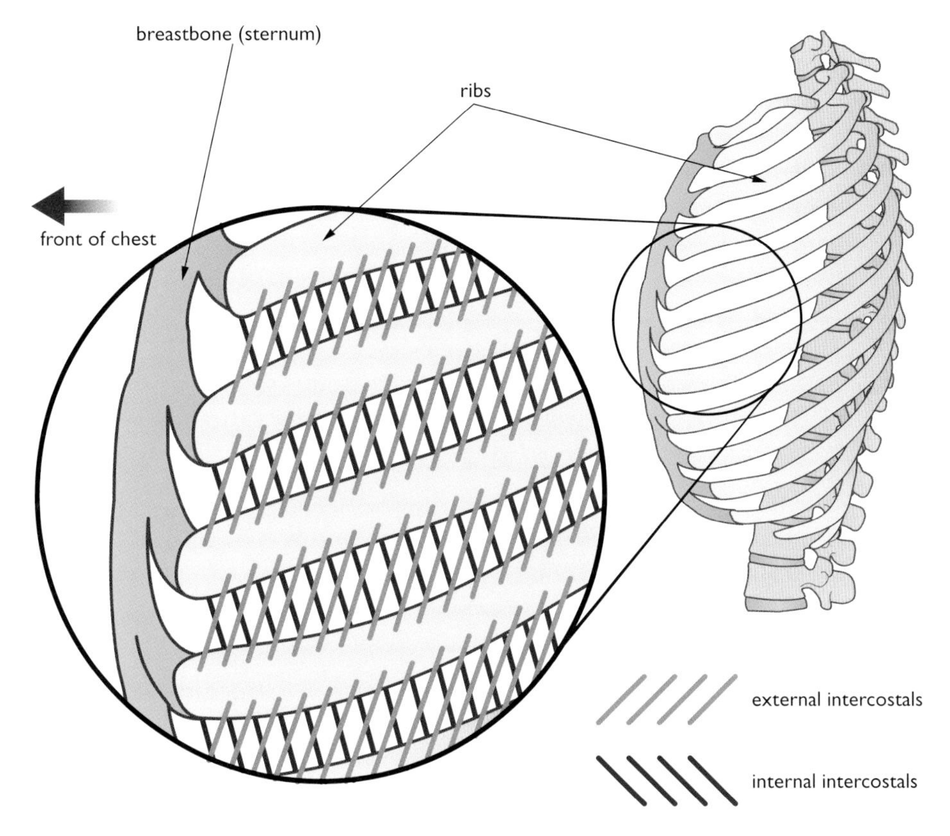

Figure 2.5 *X-ray of side view of the chest wall, showing the ribs. The diagram shows how the two sets of intercostal muscles run between the ribs. When the external intercostals contract, they move the ribs upwards. When the internal intercostals contract, the ribs are moved downwards.*

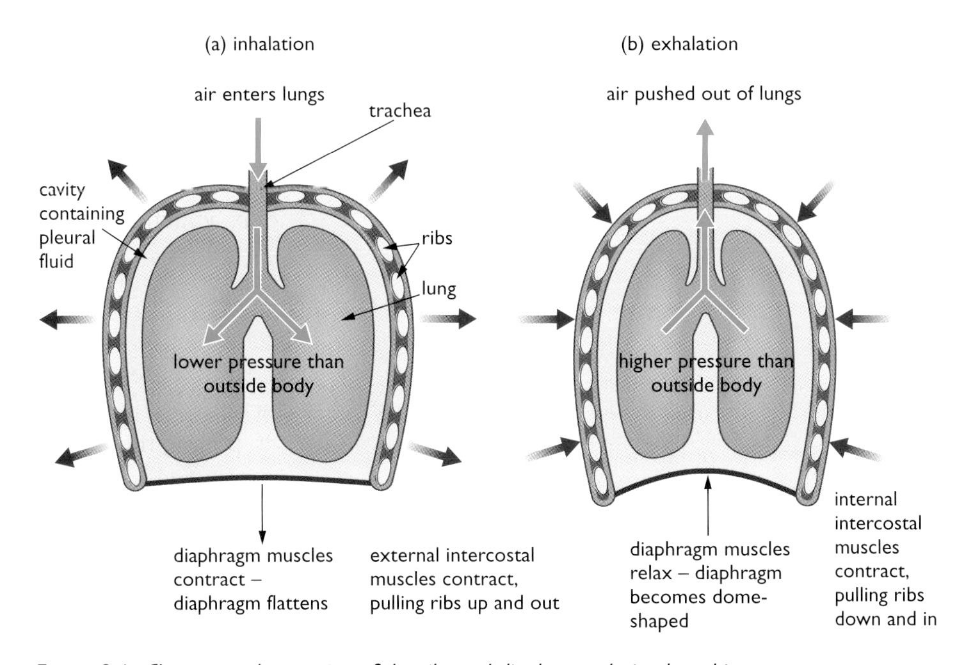

Figure 2.6 *Changes in the position of the ribs and diaphragm during breathing. (a) Breathing in (inhalation) (b) Breathing out (exhalation).*

Be careful when interpreting percentages! The *percentage* of a gas in a mixture can vary, even if the actual *amount* of the gas stays the same. This is easiest to understand from an example. Imagine you have a bottle containing a mixture of 20% oxygen and 80% nitrogen. If you used a chemical to absorb all the oxygen in the bottle, the nitrogen left would now be 100% of the gas in the bottle, despite the fact that the *amount* of nitrogen would still be the same. That is why the percentage of nitrogen in inhaled and exhaled air is slightly different.

Gas exchange in the alveoli

You can tell what is happening during gas exchange if you compare the amounts of different gases in atmospheric air with the air breathed out (Table 2.1).

Gas	Atmospheric air	Exhaled air
nitrogen	78	79
oxygen	21	16
carbon dioxide	0.04	4
other gases (mainly argon)	1	1

Table 2.1: *Approximate percentage volume of gases in atmospheric (inhaled) and exhaled air.*

Exhaled air is also warmer than atmospheric air, and is saturated with water vapour. The amount of water vapour in the atmosphere varies, depending on weather conditions.

Clearly, the lungs are absorbing oxygen into the blood and removing carbon dioxide from it. This happens in the alveoli. To do this efficiently, the alveoli must have a structure which brings the air and blood very close together, over a very large surface area. There are enormous numbers of alveoli. It has been calculated that the two lungs contain about 700 000 000 of these tiny air sacs, giving a total surface area of 60 m^2. That's bigger than the floor area of an average classroom! Viewed through a high-powered microscope, the alveoli look rather like bunches of grapes, and are covered with tiny blood capillaries (Figure 2.7).

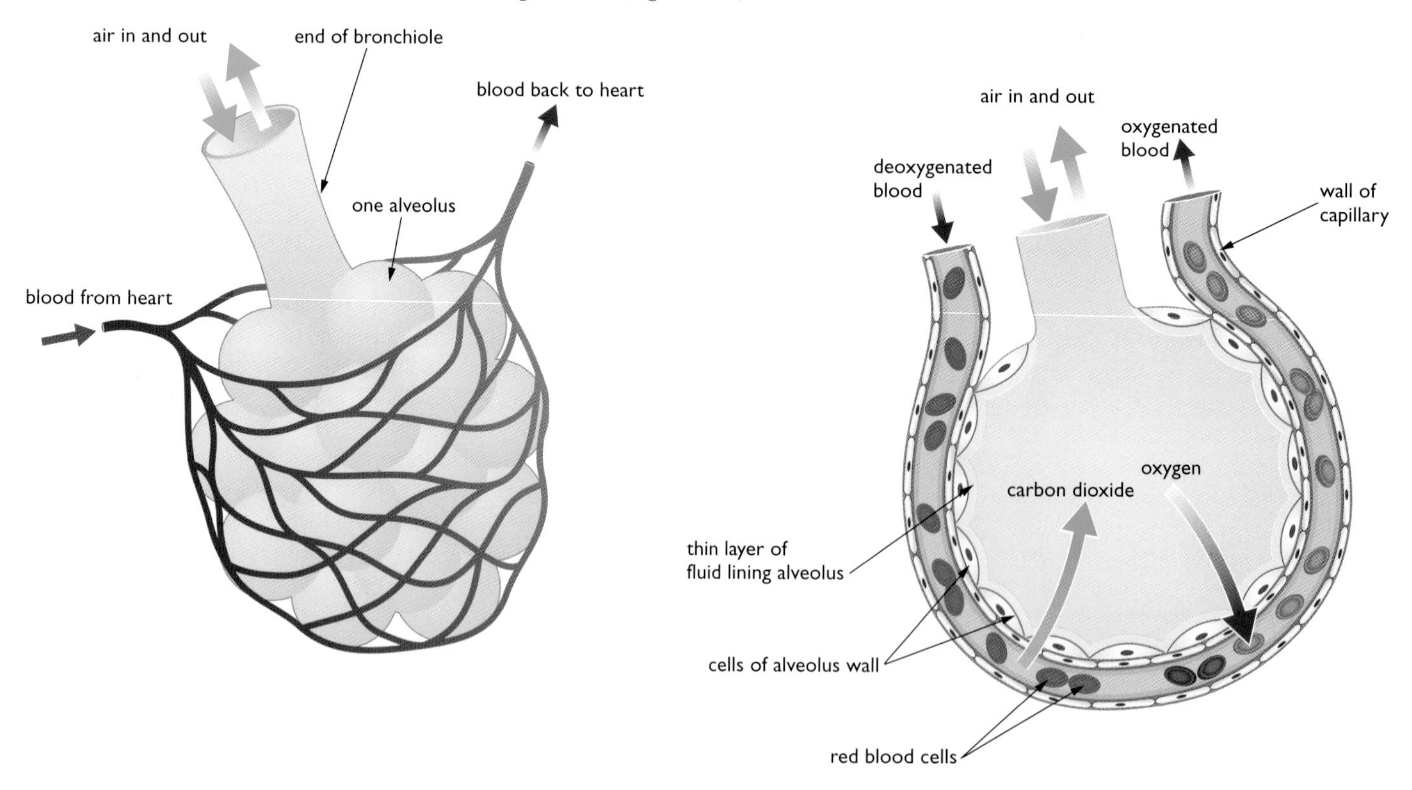

Figure 2.7 *An alveolus and its surrounding capillary network. Diffusion of oxygen and carbon dioxide takes place between the air in the alveolus and the blood in the capillaries.*

Blood is pumped from the heart to the lungs and passes through the capillaries surrounding the alveoli. The blood has come from the respiring tissues of the body, where it has given up some of its oxygen to the cells, and gained carbon dioxide. Around the lungs, the blood is separated from the air inside each alveolus by only two cell layers; the cells making up the wall of the alveolus, and the capillary wall itself. This is a distance of less than a thousandth of a millimetre.

Because the air in the alveolus has a higher concentration of oxygen than the blood entering the capillary network, oxygen diffuses from the air, across the wall of the alveolus and into the blood. At the same time there is more carbon dioxide in the blood than there is in the air in the lungs. This means that there is a diffusion gradient for carbon dioxide in the other direction, so carbon dioxide diffuses the other way, out of the blood and into the alveolus. The result is that the blood which leaves the capillaries and flows back to the heart has gained oxygen and lost carbon dioxide. The heart then pumps the blood around the body again, to supply the respiring cells (see Chapter 5).

The thin layer of fluid lining the inside of the alveoli comes from the blood. The capillaries and cells of the alveolar wall are 'leaky' and the blood pressure pushes fluid out from the blood plasma into the alveolus. Oxygen dissolves in this moist surface before it passes through the alveolar wall into the blood.

The effects of smoking on the lungs and associated tissues

If the lungs are to be able to exchange gases properly, the air passages need to be clear, the alveoli to be free from dirt particles and bacteria, and they must have as big a surface area as possible in contact with the blood. There is one habit that can upset all of these conditions – smoking.

Links between smoking and diseases of the lungs are now a proven fact. Smoking is associated with lung cancer, bronchitis and emphysema. It is also a major contributing factor to other problems, such as coronary heart disease and ulcers of the stomach and duodenum (part of the intestine). Pregnant women who smoke are more likely to give birth to underweight babies. We need to deal with some of these effects in more detail.

Effects of smoke on the lining of the air passages

You saw above how the lungs are kept free of particles of dirt and bacteria by the action of mucus and cilia. In the trachea and bronchi of a smoker, the cilia are destroyed by the chemicals in cigarette smoke. Compare Figure 2.4 with the same photo taken of the lining of a smoker's airways (Figure 2.8).

The reduced numbers of cilia mean that the mucus is not swept away from the lungs, but remains to clog the air passages. This is made worse by the fact that the smoke irritates the lining of the airways, stimulating the cells to secrete more mucus. The clogging mucus is the source of 'smoker's cough'. Irritation of the bronchial tree, along with infections from bacteria in the mucus can cause the lung disease **bronchitis.** Bronchitis blocks normal air flow, so the sufferer has difficulty breathing properly.

Figure 2.8 *This electron micrograph shows cilia from the trachea of a smoker. Notice the reduced numbers of cilia compared with a normal trachea.*

Emphysema

Emphysema is another lung disease that kills about 20 000 people in Britain every year. Smoke damages the walls of the alveoli, which break down and fuse together again, forming enlarged, irregular air spaces (Figure 2.9).

This greatly reduces the surface area for gas exchange, which becomes very inefficient. The blood of a person with emphysema carries less oxygen. In serious cases, this leads to the sufferer being unable to carry out even mild exercise, such as walking. Emphysema patients often have to have a supply of oxygen nearby at all times (Figure 2.10). There is no cure for emphysema, and usually the sufferer dies after a long and distressing illness.

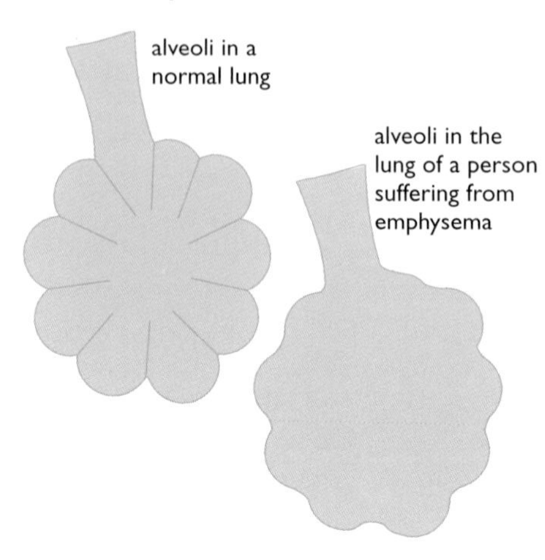

Figure 2.9 *The alveoli of a person suffering from emphysema have a greatly reduced surface area and inefficient gas exchange.*

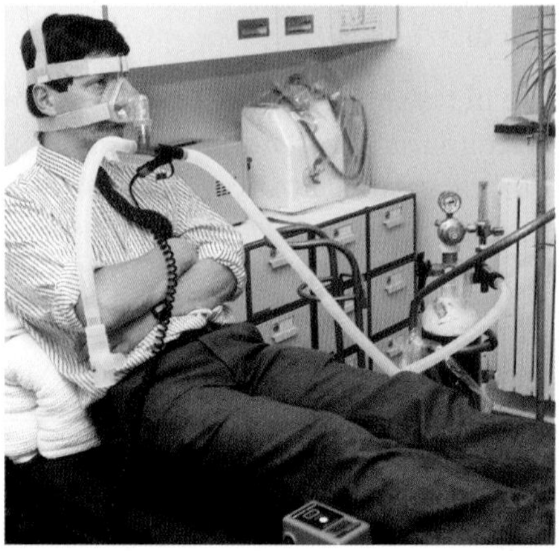

Figure 2.10 *This man suffers from emphysema and has to breathe from an oxygen cylinder to stay alive.*

ideas
evidence

Lung cancer

Evidence of the link between smoking and lung cancer first appeared in the 1950s. In one study, a number of patients in hospital were given a series of questions about their lifestyles. They were asked about their work, hobbies, housing and so on, including a question about how many cigarettes they smoked. The same questionnaire was given to two groups of patients. The first group were all suffering from lung cancer. The second, **control** group were in hospital with various other illnesses, but not lung cancer. To make it a fair comparison, the control patients were matched with the lung cancer patients for sex, age and so on.

When the results were compared, one difference stood out (Table 2.2). A greater proportion of the lung cancer patients were smokers than in the control patients. There seemed to be a connection between smoking and getting lung cancer.

	Percentage of patients who were non-smokers	**Percentage of patients who smoked more than 15 cigarettes a day**
lung cancer patients	0.5	25
control patients (with illnesses other than lung cancer)	4.5	13

Table 2.2: *Comparison of the smoking habits of lung cancer patients and other patients.*

Although the results didn't prove that smoking caused lung cancer, there was a statistically significant link between smoking and the disease: this is called a 'correlation'.

Over 20 similar investigations in nine countries have revealed the same findings. In 1962 a report called 'Smoking and health' was published by the Royal College of Physicians of London, which warned the public about the dangers of smoking. Not surprisingly, the first people to take the findings seriously were doctors, many of whom stopped smoking. This was reflected in their death rates from lung cancer. In ten years, while deaths among the general male population had risen by 7%, the deaths of male doctors from the disease had *fallen* by 38%.

People often talk about 'yellow nicotine stains'. In fact it is the *tar* that stains a smoker's fingers and teeth. Nicotine is a colourless, odourless chemical.

Cigarette smoke contains a strongly addictive drug – **nicotine**. It also contains at least 17 chemicals that are known to cause cancer. These chemicals are called **carcinogens**, and are contained in the **tar** that collects in a smoker's lungs. Cancer happens when cells mutate and start to divide uncontrollably, forming a **tumour** (Figure 2.11). If a lung cancer patient is lucky, he or she may have the tumour removed by an operation before the cancer cells spread to other tissues of the body. Unfortunately tumours in the lungs usually cause no pain, so they are not discovered until it is too late – it may be inoperable, or tumours may have developed elsewhere.

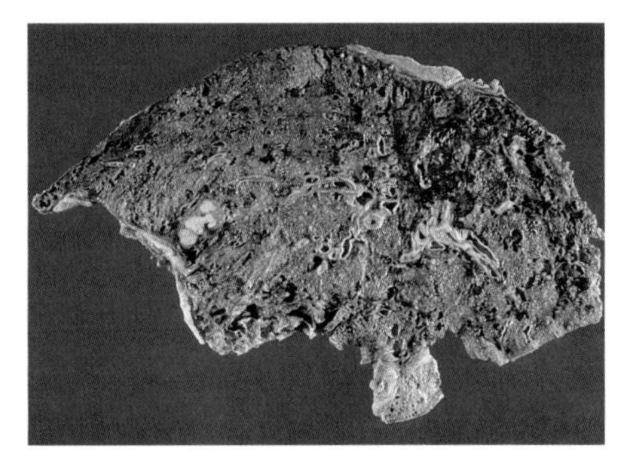

Figure 2.11 *This lung is from a patient with lung cancer.*

If you smoke you are not *bound* to get lung cancer, but the risks that you will get it are much greater. In fact, the more cigarettes you smoke, the more the risk that you will get the disease increases (Figure 2.12).

The obvious thing to do is not to start smoking. However, if you are a smoker, giving up the habit soon improves your chance of survival (Figure 2.13). After a few years the likelihood of your dying from a smoking-related disease is almost back to the level of a non-smoker.

Studies have shown that the type of cigarette smoked makes very little difference to the smoker's risk of getting lung cancer. Filtered and 'low tar' cigarettes only reduce the risk slightly.

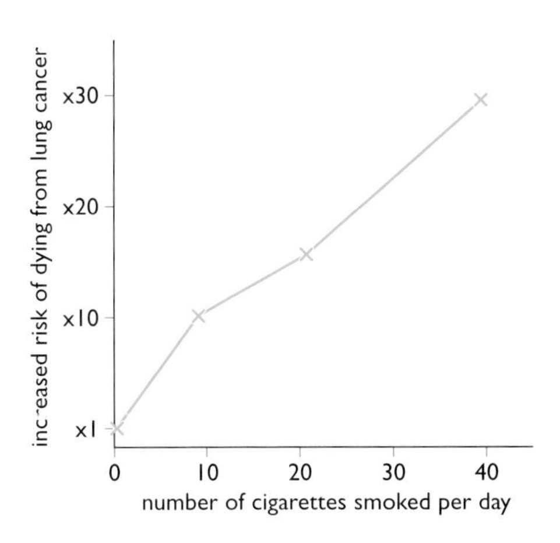

Figure 2.12 The more cigarettes a person smokes, the more likely it is they will die of lung cancer. For example, smoking 20 cigarettes a day increases the risk by about 15 times.

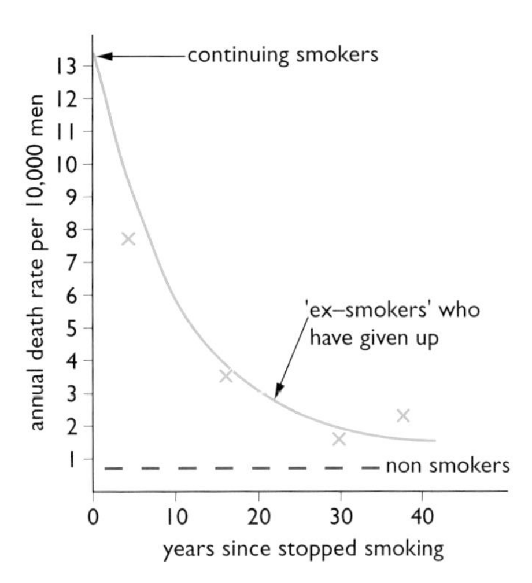

Figure 2.13 *Death rates from lung cancer for smokers, non-smokers and ex-smokers.*

Carbon monoxide in smoke

One of the harmful chemicals in cigarette smoke is the poisonous gas **carbon monoxide**. When this gas is breathed in with the smoke, it enters the bloodstream and interferes with the ability of the blood to carry oxygen. Oxygen is carried around in the blood in the red blood cells, attached to a chemical called **haemoglobin** (see Chapter 5). Carbon monoxide can combine with the haemoglobin much more tightly than oxygen can,

Figure 2.14 *Smoking during pregnancy affects the growth and development of the baby.*

forming a compound called **carboxyhaemoglobin**. The haemoglobin will combine with carbon monoxide in preference to oxygen. When this happens, the blood carries much less oxygen around the body. Carbon monoxide from smoking is also a major cause of heart disease (Chapter 5).

If a pregnant woman smokes, she will be depriving her unborn fetus of oxygen (Figure 2.14). This has an effect on its growth and development, and leads to the mass of the baby at birth being lower, on average, than the mass of babies born to non-smokers.

Some smoking statistics

- 13 million people in the UK smoke cigarettes, 28% of men and 26% of women.
- Smoking is highest among the 20–24 age group: 42% of men and 39% of women.
- More than 80% of smokers take up the habit as teenagers. In the UK about 450 children start smoking every day.
- Among 15 year-olds, 21% of boys and 25% of girls are regular smokers.
- In the UK, smoking kills around 120 000 people every year, about 37 times more people than are killed in road traffic accidents.
- Every year several hundred lung cancer deaths are due to passive smoking (breathing other people's smoke).
- In 1998, the UK government received £10 090 million from tax on tobacco (in 1997 the cost of treating smoking-related diseases was £1 700 million).

Source: Action on Smoking and Health (ASH) fact sheet (January 2001)

Figure 2.15 *Increasing numbers of young women are smokers. Many turn to nicotine patches to help them give up.*

Giving up smoking

Most smokers admit that they would like to find a way to give up the habit. The trouble is that the nicotine in tobacco is a very addictive drug, and causes withdrawal symptoms when people stop smoking. These include cravings for a cigarette, restlessness and a tendency to put on weight (nicotine depresses the appetite).

There are various ways that smokers can be helped to 'kick the habit'. One of the most successful methods is the use of nicotine patches (Figure 2.15) or nicotine chewing gum. These provide the smoker who is trying to give up with a source of nicotine, without the harmful tar of cigarettes. The nicotine is absorbed through the skin (with patches) or through the mouth (from gum) and reduces the craving for a cigarette. Gradually the 'ex-smoker' reduces the nicotine dose until they are weaned off the habit.

There are several other ways that people use to help them give up smoking, including the use of drugs that reduce withdrawal symptoms, acupuncture and even hypnotism.

You could carry out an Internet search to find out about the different methods people use to help them give up smoking. Which methods have the highest success rate?

End of Chapter Checklist

If you haven't got a copy of your specification, read the introduction on page vi.

You will need to be able to do some or all of the following. Check your Awarding Body's specification (syllabus) to find out exactly what you need to know.

- Locate and name the parts of the human respiratory system and describe how these parts are related to their functions.
- Describe how mucus-secreting cells and cilia help to keep the lungs clean and avoid infection.
- Describe how actions of the intercostal muscles, ribs and diaphragm change the volume of the thorax, altering the pressure and causing ventilation of the lungs.
- Recall the approximate percentage volumes of gases in inhaled and exhaled air.
- Describe the structure and function of the alveoli in gas exchange.
- Explain how cigarette smoke causes excess mucus production and destruction of cilia, leading to bronchitis, and can destroy alveoli, reducing gas exchange and causing emphysema.
- Interpret evidence linking smoking and lung cancer. Describe how tar in smoke contains carcinogens which can cause lung cells to mutate to form cancer cells.
- Know that carbon monoxide in smoke reduces the oxygen-carrying capacity of the blood, and that in pregnant women this deprives the fetus of oxygen and leads to low birth mass.
- Describe some methods people can use to help them give up smoking.

Questions

More questions on breathing can be found at the end of Section B on page 119.

1 Copy and complete the table, which shows what happens in the thorax during ventilation of the lungs. Two boxes have been completed for you.

	Action during inhalation	**Action during exhalation**
external intercostal muscles	contract	
internal intercostal muscles		
ribs		move down and in
diaphragm		
volume of thorax		
pressure in thorax		
volume of air in lungs		

2 A student wrote the following about the lungs:

> When we breathe in, our lungs inflate, sucking air in and pushing the ribs up and out, and forcing the diaphragm down. This is called respiration. In the air sacs of the lungs the air enters the blood. The blood then takes the air around the body, where it is used by the cells. The blood returns to the lungs to be cleaned. When we breathe out, our lungs deflate, pulling the diaphragm up and the ribs down. The stale air is pushed out of the lungs.

The student did not have a good understanding of the workings of the lungs. Re-write her description, using correct biological words and ideas.

3 Sometimes, people injured in an accident such as a car crash suffer from a *pneumothorax*. This is an injury where the chest wall is punctured, allowing air to enter the pleural cavity (see Figure 2.1). A patient was brought to the casualty department of a hospital, suffering from a pneumothorax on the left side of his chest. His left lung had collapsed, but he was able to breathe normally with his right lung.

a) Explain why a pneumothorax caused the left lung to collapse.

b) Explain why the right lung was not affected.

c) If a patient's lung is injured or infected, a surgeon can sometimes 'rest' it by performing an operation called an *artificial pneumothorax*. What do you think might be involved in this operation?

4 Briefly explain the importance of the following:

a) The trachea wall contains C-shaped rings of cartilage.

b) The distance between the air in an alveolus and the blood in an alveolar capillary is less than 1/1000th of a millimetre.

c) The lining of the trachea contains mucus-secreting cells and cells with cilia.

d) Smokers have a lower concentration of oxygen in their blood than non-smokers.

e) Nicotine patches and nicotine chewing gum can help someone give up smoking.

f) The lungs have a surface area of about 60 m^2 and a good blood supply.

5 Explain the differences between the lung diseases bronchitis and emphysema.

6 A long-term investigation was carried out into the link between smoking and lung cancer. The smoking habits of male doctors aged 35 or over were determined while they were still alive, then the number and causes of deaths among them were monitored over a number of years. (Note that this survey was carried out in the 1950s – very few doctors smoke these days!). The results are shown in the graph.

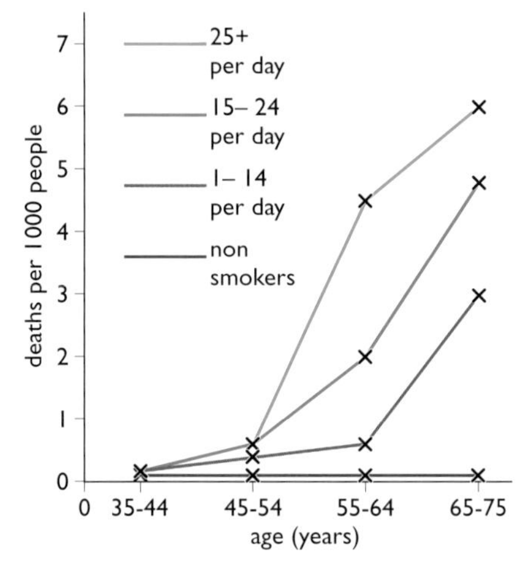

a) Write a paragraph to explain what the researchers found out from the investigation.

b) How many deaths from lung cancer would be expected for men aged 55 who smoked 25 cigarettes a day up until their death? How many deaths from lung cancer would be expected for men in the same age group smoking 10 a day?

c) Table 2.2 (page 20) shows the findings of another study linking lung cancer with smoking. Which do you think is the more convincing evidence of the link, this investigation or the findings illustrated in Table 2.2?

7 Design and make a hard-hitting leaflet explaining the link between smoking and lung cancer. It should be aimed at encouraging an adult smoker to give up the habit. You could use a suitable computer software package to produce your design. Include some smoking statistics, perhaps from an Internet search. However don't use too many, or they may put the person off reading the leaflet!

Chapter 3: Food and Digestion

Food is essential for life. The nutrients obtained from it are used in many different ways by the body. This chapter looks at the different kinds of food, and how the food is broken down by the digestive system and absorbed into the blood, so that it can be carried to all the tissues of the body.

We need food for three main reasons:

- to supply us with a 'fuel' for energy
- to provide materials for growth and repair of tissues
- to help fight disease and keep our bodies healthy.

A balanced diet

The food that we eat is called our **diet**. No matter what you like to eat, if your body is to work properly and stay healthy, your diet must include five groups of food substances – **carbohydrates**, **fats**, **proteins**, **minerals and vitamins** – as well as **water** and **fibre**. Food should provide you with all of these substances, but they must also be present in the *right* amounts. A diet that provides enough of these substances and in the correct proportions to keep you healthy is called a **balanced diet** (Figure 3.1). We will deal with each type of food in turn, to find out about its chemistry and the role that it plays in the body.

Figure 3.1 *A balanced diet contains all the types of food the body needs, in just the right amounts.*

Carbohydrates

The chemical formula for glucose is $C_6H_{12}O_6$. Like all carbohydrates, glucose contains only the elements carbon, hydrogen and oxygen. The 'carbo' part of the name refers to carbon, and the 'hydrate' part refers to the fact that the hydrogen and oxygen atoms are in the ratio two to one, as in water (H_2O).

Carbohydrates only make up about 5% of the mass of the human body, but they have a very important role. They are the body's main 'fuel' for supplying cells with energy. Cells release this energy by oxidising a sugar called **glucose**, in the process called cell respiration (see Chapter 1). Glucose and other sugars are one sort of carbohydrate.

Glucose is found naturally in many sweet-tasting foods, such as fruits and vegetables. Other foods contain different sugars, such as the fruit sugar called **fructose**, and the milk sugar, **lactose**. Ordinary table sugar, the sort

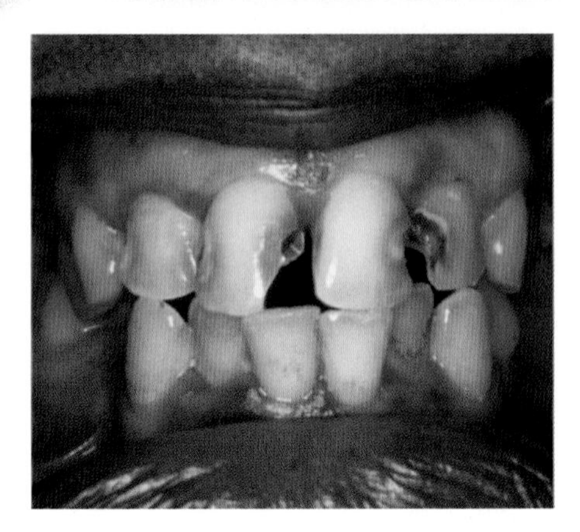

Figure 3.2 *A bad case of tooth decay. One of the causes was too much sugar in the person's diet.*

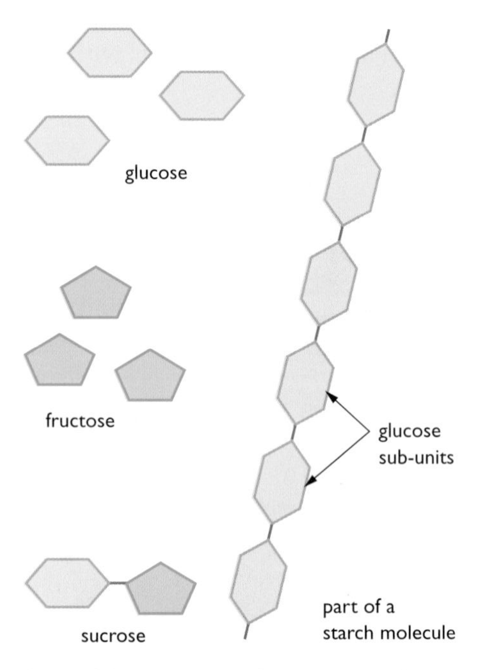

Figure 3.3 *Glucose and fructose are 'single sugar' molecules. A molecule of glucose joined to a molecule of fructose forms the 'double sugar' called sucrose. Starch is a polymer of many glucose sub-units.*

'Single' sugars such as glucose and fructose are called **monosaccharides**. Sucrose molecules are made of two monosaccharides (glucose and fructose) joined together, so sucrose is called a **disaccharide**. Lactose is also a disaccharide of glucose joined to a monosaccharide called galactose. Polymers of sugars, such as starch and glycogen, are called **polysaccharides**.

some people put in their tea or coffee, is called **sucrose**. Sucrose is the main sugar that is transported through plant stems. This is why we can extract it from sugar cane, which is the stem of a large grass-like plant. Sugars have two physical properties that you will probably know: they all taste sweet, and they are all soluble in water.

We can get all the sugar we need from natural foods such as fruits and vegetables, and from the digestion of starch. Many 'processed' foods contain large amounts of *added* sugar. For example, a typical can of cola can contain up to 27 g, or seven teaspoonfuls! There is hidden sugar in many other foods. A tin of baked beans contains about 10 g of added sugar. This is on top of all the food that we eat with a more obvious sugar content, such as cakes, biscuits and sweets. One of the health problems resulting from all this sugar in our diet is **tooth decay**. Bacteria in the mouth feed on sugar, breaking it down and making acids that dissolve the tooth enamel. Once the enamel is penetrated, the acid breaks down the softer dentine underneath, and eventually a cavity is formed in the tooth (Figure 3.2). Bacteria can then enter this cavity and enlarge it until the decay reaches the nerves at the centre of the tooth. Then you feel the pain!

In fact, we get most of the carbohydrate in our diet not from sugars, but from **starch**. Starch is a large, *insoluble* molecule. Because it does not dissolve, it is found as a storage carbohydrate in many plants, such as potato, rice, wheat and millet. The 'staple diets' of people from around the world are starchy foods like rice, potatoes, bread and pasta. Starch is made up of long chains of hundreds of glucose molecules joined together. It is called a **polymer** of glucose (Figure 3.3).

Starch is only found in plant tissues, but animal cells sometimes contain a very similar carbohydrate called **glycogen**. This is also a polymer of glucose, and is found in tissues such as liver and muscle, where it acts as a store of energy for these organs.

As you will see, large carbohydrates such as starch and glycogen have to be broken down into simple sugars during digestion, so that they can be absorbed into the blood.

Another carbohydrate that is a polymer of glucose is **cellulose**, the material that makes up plant cell walls. Humans are *not* able to digest cellulose, because our gut doesn't make the enzyme needed to break down the cellulose molecule. This means that we are not able to use cellulose as a source of energy. However, it still has a vitally important function in our diet. It forms **dietary fibre** or **roughage**, which gives the muscles of the gut something to push against as the food is moved through the intestine. This keeps the gut contents moving, avoiding constipation and helping to prevent serious diseases of the intestine, such as colitis and bowel cancer.

Fats

Fats contain the same three elements as carbohydrates – carbon, hydrogen and oxygen, but the proportion of oxygen in a fat is much lower than in a carbohydrate. For example, beef and lamb both contain a fat called tristearin, which has the formula $C_{51}H_{98}O_6$. This fat, like other animal fats, is a solid at room temperature, but melts if you warm it up. On the other hand, plant fats are usually liquid at room temperature, and are called **oils**.

Meat, butter, cheese, milk, eggs and oily fish are all rich in animal fats, as well as foods fried in fat or oil, such as chips. Plant oils include many types used for cooking, such as olive oil, corn oil and rapeseed oil, as well as products made from oils, such as margarine (Figure 3.4).

Fats make up about 10% of our body's mass. They form an essential part of the structure of all cells, and fat is deposited in certain parts of the body as a long-term store of energy, for example under the skin and around the heart and kidneys. The fat layer under the skin acts as insulation, reducing heat loss through the surface of the body. Fat around organs such as the kidneys also helps to protect them from mechanical damage.

Figure 3.4 *These foods are all rich in fats.*

The chemical 'building blocks' of fats are two types of molecule called **glycerol** and **fatty acids**. Glycerol is an oily liquid. It is also known as glycerine, and is used in many types of cosmetics. In fats, a molecule of glycerol is joined to three fatty acid molecules. There are a large number of different fatty acid molecules, which gives us the many different kinds of fat found in food (Figure 3.5).

Although fats are an essential part of our diet, too much fat is unhealthy, especially a type called **saturated** fat, and a fat-like compound called **cholesterol**. These substances have been linked to heart disease (see Chapter 5).

Proteins

Proteins make up about 18% of the mass of the body. This is the second largest fraction after water. All cells contain protein, so we need it for growth and repair of tissues. Many compounds in the body are made from protein, including enzymes.

Most foods contain some protein, but certain foods such as meat, fish, cheese and eggs are particularly rich in it. You will notice that these foods are animal products. Plant material generally contains less protein, but some foods, especially beans, peas and nuts, are richer in protein than others.

Fats and oils are both known as **lipids**.

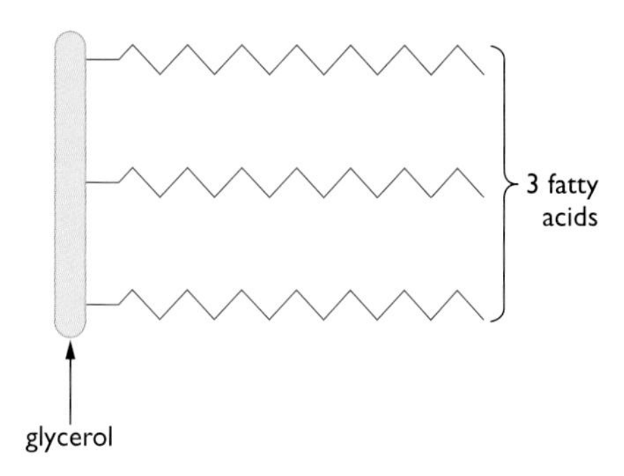

Figure 3.5 *Fats are made up of a molecule of glycerol joined to three fatty acids. The many different fatty acids form the variable part of the molecule.*

Saturated fats are more common in food from animal sources, such as meat and dairy products. 'Saturated' is a word used in chemistry, which means that the fatty acids of the fats contain no double bonds. Other fats are **unsaturated**, which means that their fatty acids contain double bonds. These are more common in plant oils. There is evidence that unsaturated fats are healthier for us than saturated ones.

Cholesterol is a substance that the body gets from food such as eggs and meat, but we also make cholesterol in our liver. It is an essential part of all cells, but too much cholesterol causes heart disease.

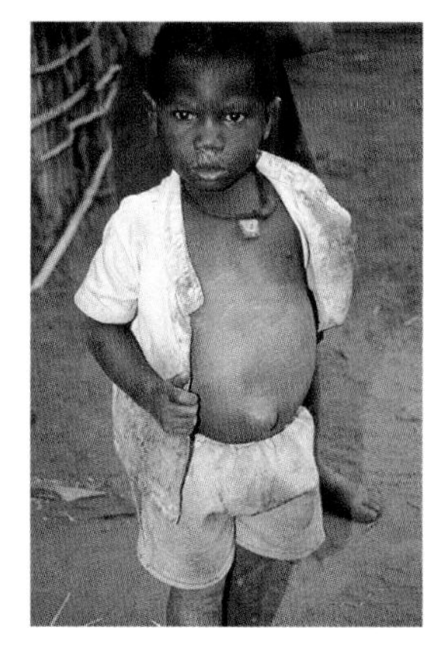

Figure 3.6 *This child is suffering from a lack of protein in his diet, a disease called kwashiorkor. His swollen belly is not due to a full stomach, but is caused by fluid collecting in the tissues. Other symptoms include loss of weight, poor muscle growth, general weakness and flaky skin.*

Humans can make about half of the 20 amino acids that they need, but the other 10 have to be taken in as part of the diet. These 10 are called **essential amino acids.** There are higher amounts of essential amino acids in meat, fish, eggs and dairy products. If you are a vegetarian, you can still get all the essential amino acids you need, as long as you eat a varied diet that includes a range of different plant materials.

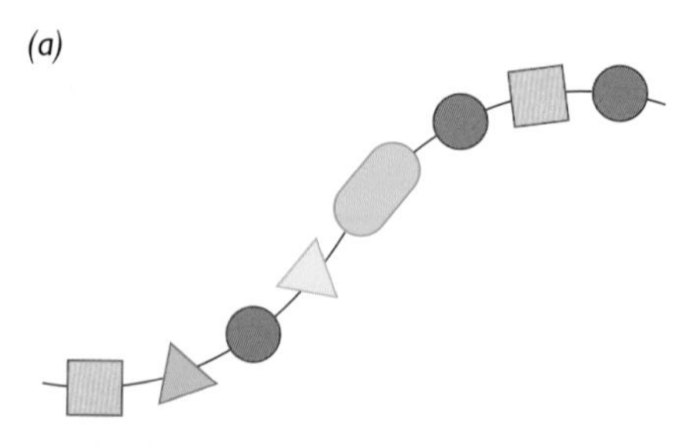

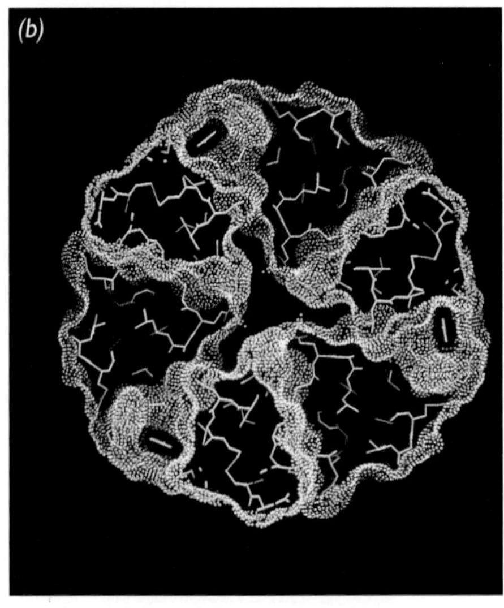

Figure 3.7 *(a) A chain of amino acids forming part of a protein molecule. Each shape represents a different amino acid. (b) A computer model of the protein insulin. This substance, like all proteins, is made of a long chain of amino acids arranged in a specific order.*

However, we don't need much protein in our diet to stay healthy. Doctors recommend a maximum daily intake of about 70 g. In more economically developed countries, people often eat far more protein than they need, whereas in many poorer countries a protein-deficiency disease called **kwashiorkor** is common (Figure 3.6).

Like starch, proteins are also polymers, but whereas starch is made from a single molecular building block (glucose), proteins are made from 20 different sub-units called **amino acids**. All amino acids contain four chemical elements: carbon, hydrogen and oxygen (as in carbohydrates and fats) along with nitrogen. Two amino acids also contain sulphur. The amino acids are linked together in long chains, which are usually folded up or twisted into spirals, with cross-links holding the chains together (Figure 3.7).

The *shape* of a protein is very important in allowing it to carry out its function, and the *order* of amino acids in the protein decides its shape. Because there are 20 different amino acids, and they can be arranged in any order, the number of different protein structures that can be made is enormous. As a result, there are thousands of different kinds of proteins in organisms, from structural proteins such as collagen and keratin in skin and nails, to proteins with more specific functions, such as enzymes and haemoglobin.

Minerals

All the foods you have read about so far are made from just five chemical elements: carbon, hydrogen, oxygen, nitrogen and sulphur. Our bodies contain many other elements which we get from our food. Some are present in large amounts in the body, for example calcium, which is used for making teeth and bones. Others are present in much smaller amounts, but still have essential jobs to do. For instance our bodies contain about 3 g of iron, but without it our blood would not be able to carry oxygen. Table 3.1 shows just a few of these elements and the reasons they are needed. They are called **minerals** or **mineral elements**.

Mineral	Approximate mass in an adult body (g)	Location or role in body	Examples of foods rich in minerals
calcium	1 000	making teeth and bones	dairy products, fish, bread, vegetables
phosphorus	650	making teeth and bones; part of many chemicals, e.g. DNA	most foods
sodium	100	in body fluids, e.g. blood	common salt, most foods
chlorine	100	in body fluids, e.g. blood	common salt, most foods
magnesium	30	making bones; found inside cells	green vegetables
iron	3	part of haemoglobin in red blood cells, helps carry oxygen	red meat, liver, eggs, some vegetables, e.g. spinach

Table 3.1: *Some examples of minerals needed by the body.*

If a person doesn't get enough of a mineral from their diet, they will show the symptoms of a **mineral deficiency disease**. For example, a one-year-old child needs to consume about 0.6 g (600 mg) of calcium every day, to make the bones grow properly and harden. Anything less than this over a prolonged period could result in poor bone development. The bones will become deformed, resulting in a disease called **rickets** (Figure 3.8). Rickets can also be caused by lack of vitamin D in the diet (see below).

Similarly, 16-year-olds need about 12 mg of iron in their daily food intake. If they don't get this amount, they can't make enough haemoglobin for their red blood cells (see Chapter 5). This causes a condition called **anaemia**. People who are anaemic become tired and lack energy, because their blood doesn't carry enough oxygen.

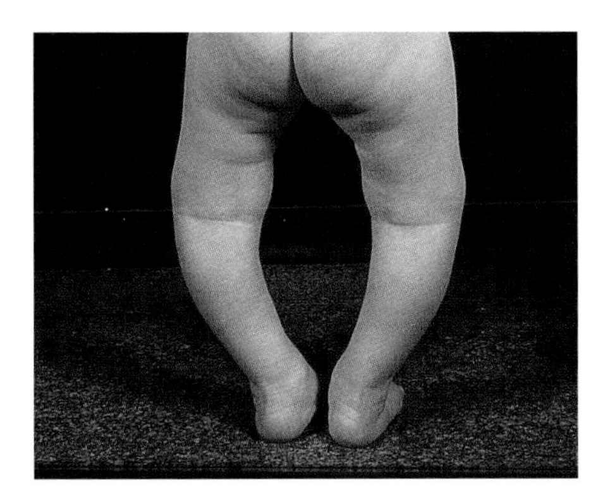

Figure 3.8 *The legs of this child show the symptoms of rickets. This is due to lack of calcium or a lack of vitamin D in the diet, leading to poor bone growth. The bones stay soft and can't support the weight of the body, so they become deformed.*

Vitamins

During the early part of the twentieth century, experiments were carried out that identified another class of food substances. When young laboratory rats were fed a diet of pure carbohydrate, fat and protein, they all became ill and died. If they were fed on the same pure foods with a little added milk, they grew normally. The milk contained chemicals that the rats needed in small amounts to stay healthy. These chemicals are called **vitamins**. The results of one of these experiments are shown in Figure 3.9.

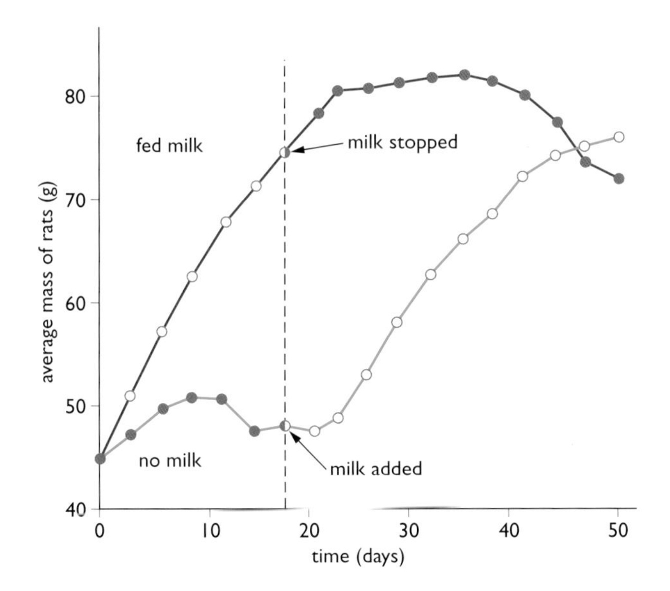

Figure 3.9 *Rats were fed on a diet of pure carbohydrate, fat and protein, with and without added milk. Vitamins in the milk had a dramatic effect on their growth.*

At first, the chemical nature of vitamins was not known, and they were given letters to distinguish between them, such as vitamin A, vitamin B and so on. Each was identified by the effect a lack of the vitamin, or **vitamin deficiency**, had on the body. For example, **vitamin D** is needed for growing bones to take up calcium salts. A deficiency of this vitamin can result in rickets (Figure 3.8), just as a lack of calcium can.

We now know the chemical structure of the vitamins and the exact ways in which they work in the body. As with vitamin D, each has a particular function. **Vitamin A** is needed to make a light-sensitive chemical in the retina of the eye (see Chapter 6). A lack of this vitamin causes **night blindness**,

The cure for scurvy was discovered as long ago as 1753. Sailors on long voyages often got scurvy because they ate very little fresh fruit and vegetables (the main source of vitamin C). A ship's doctor called James Lind wrote an account of how the disease could quickly be cured by eating fresh oranges and lemons. The famous explorer Captain Cook, on his world voyages in 1772 and 1775, kept his sailors healthy by making sure that they ate fresh fruit. By 1804, all British sailors were made to drink lime juice to prevent scurvy. This is how they came to be called 'limeys', a word that was later used by Americans for all British people.

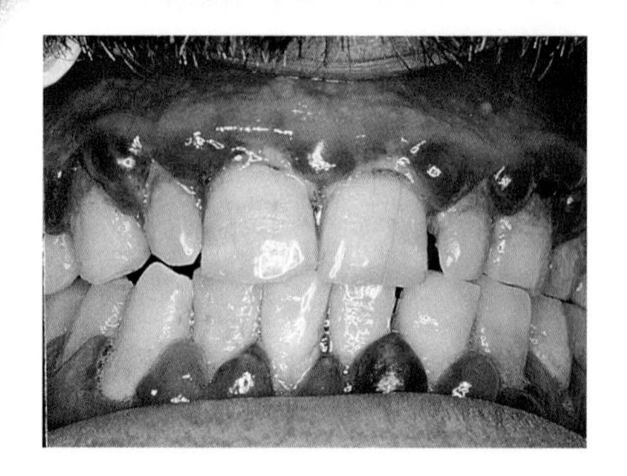

Figure 3.10 *Vitamin C helps lining cells such as those in the mouth and gums stick to each other. Lack of vitamin C causes scurvy, where the mouth and gums become damaged and bleed.*

where the person finds it difficult to see in dim light. **Vitamin C** is needed to make fibres of a material called connective tissue. This acts as a 'glue', bonding cells together in a tissue. It is found in the walls of blood vessels and in the skin and lining surfaces of the body. Vitamin C deficiency leads to a disease called **scurvy**, where wounds fail to heal, and bleeding occurs in various places in the body. This is especially noticeable in the gums (Figure 3.10).

Vitamin B is not a single substance, but a collection of many different substances called the vitamin B group. It includes **vitamins B1 (thiamine), B2 (riboflavin)** and **B3 (niacin)**. These compounds have roles in helping with the process of cell respiration. A different deficiency disease is produced if any of them is lacking from the diet. For example, lack of vitamin B1 results in weakening of the muscles and paralysis, a disease called **beri-beri**.

The main vitamins, their role in the body and some foods which are good sources of each are summarised in Table 3.2.

Notice that the amounts of vitamins that we need are very small, but we cannot stay healthy without them.

Vitamin	Recommended daily amount in diet[1]	Use in the body	Effect of deficiency	Some foods that are a good source of the vitamin
A	0.8 mg	making a chemical in the retina; also protects the surface of the eye	night blindness, damaged cornea of eye	fish liver oils, liver, butter, margarine, carrots
B1	1.4 mg	helps with cell respiration	beri-beri	yeast extract, cereals
B2	1.6 mg	helps with cell respiration	poor growth, dry skin	green vegetables, eggs, fish
B3	18 mg	helps with cell respiration	pellagra (dry red skin, poor growth, and digestive disorders)	liver, meat, fish.
C	60 mg	sticks together cells lining surfaces such as the mouth	scurvy	fresh fruit and vegetables
D	5 μg	helps bones absorb calcium and phosphate	rickets, poor teeth	fish liver oils; also made in skin in sunlight

[1]Figures are the European Union's recommended daily intake for an adult (1993). 'mg' stands for milligram (a thousandth of a gram) and 'μg' for microgram (a millionth of a gram).

Table 3.2: *Summary of the main vitamins.*

Food tests

You can carry out some simple chemical tests to find out if a food contains starch, glucose, protein or fat. In the following descriptions we will use pure food samples to try out the tests, but you can do them on normal foods too. Unless the food is already a liquid, like milk, you may need to cut it up into small pieces and grind these with a pestle and mortar, then shake it up with some water in a test tube. This is done to extract the components of the food and dissolve any soluble substances, such as sugars.

Test for starch

Place a little starch powder in a spotting tile. Add a couple of drops of dilute **iodine** solution. The iodine reacts with the starch, forming a very dark blue, or '**blue-black**' colour (Figure 3.11a). Starch is insoluble, but this test will work on a solid sample of food, such as a potato, or a suspension of starch in water.

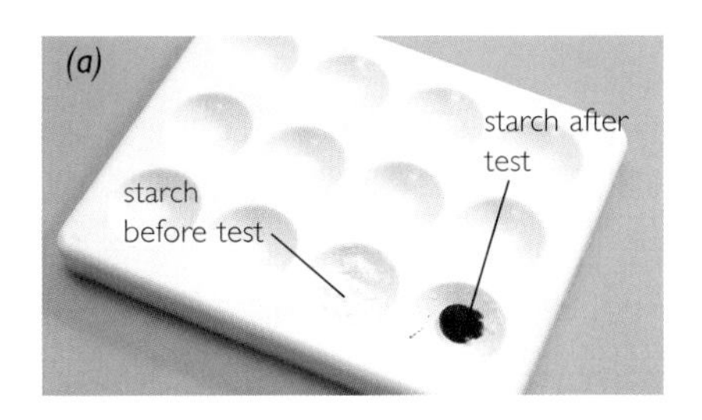

Test for glucose

Glucose is called a **reducing sugar**. This is because the test for glucose involves chemically reducing an alkaline solution of copper sulphate to copper(I) oxide.

Place a small spatula measure of glucose in a test tube and add a little water (about 2 cm deep). Now add several drops of **Benedict's solution**, which contains alkaline copper sulphate. You must add enough to colour the mixture blue. Half fill a beaker with water and heat it on a tripod and gauze. Place the test tube in the beaker and allow the water to boil (using a beaker as a water bath is safer than heating the tube directly in the Bunsen burner). Continue boiling the water for a few seconds. The clear blue solution will gradually change colour, forming a cloudy orange or 'brick red' precipitate of copper(I) oxide (Figure 3.11b).

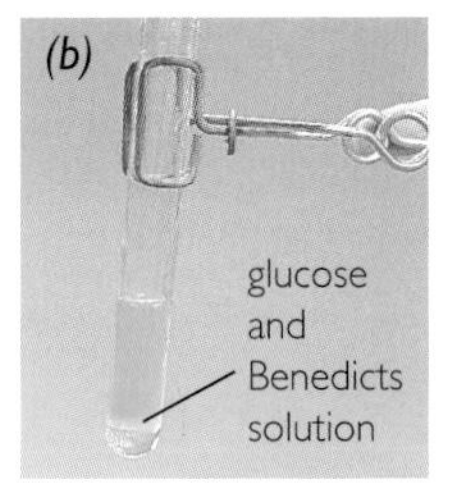

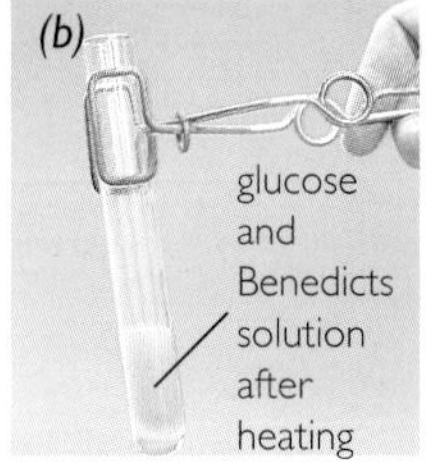

All other 'single' sugars, such as fructose, are reducing sugars, as well as some 'double' sugars, such as the milk sugar, lactose. However, ordinary table sugar (sucrose) is not. Try boiling some sucrose with Benedict's solution – it will stay a clear blue colour.

Test for protein

The test for protein is sometimes called the 'biuret' test, after the coloured compound that is formed.

Add a little powdered egg white (albumen) to 2 cm depth of water in a test tube. Shake to mix the powder with the water. Now add an equal volume of dilute (5%) **potassium hydroxide** solution and mix. Finally, add two drops of 1% **copper sulphate** solution. A mauve colour develops (Figure 3.11c).

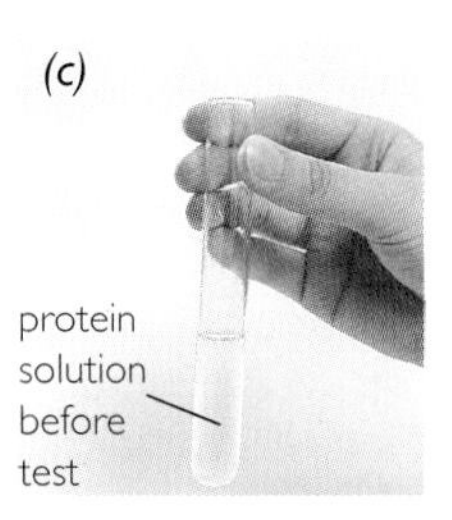

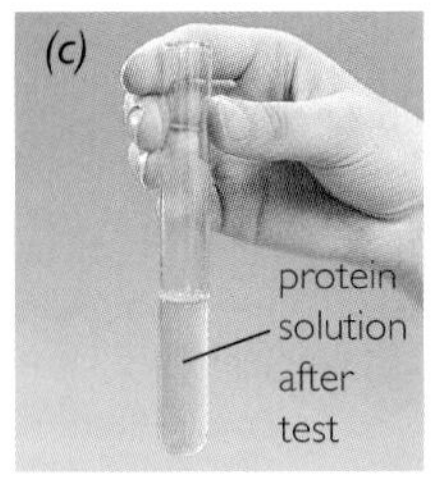

Test for fat

Fats and oils are insoluble in water, but will dissolve in ethanol (alcohol). The test for fats uses this fact.

Use a pipette to place one drop of olive oil in the bottom of a test tube. Add about 2 cm (depth) of **ethanol** and shake to dissolve the oil. Pour the mixture slowly into a test tube that is about three-quarters full with cold water. A white cloudy layer forms on the top of the water (Figure 3.11d). The white layer is caused by the ethanol dissolving in the water and leaving the fat behind as a suspension of tiny droplets, called an **emulsion**.

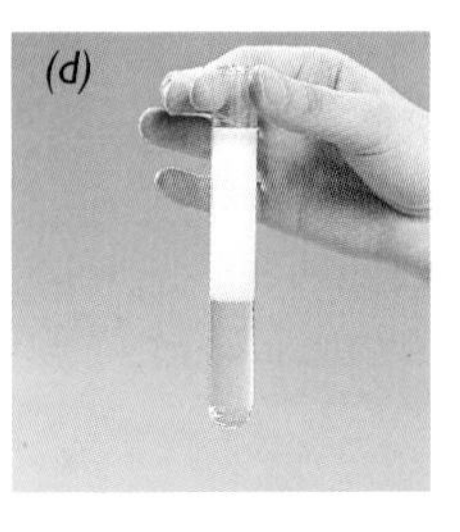

Figure 3.11 *Results of tests for (a) starch, (b) glucose, (c) protein and (d) fat.*

If sucrose is first broken down into its component monosaccharides (glucose and fructose), it will then give a positive Benedict's test. You can do this by boiling a sucrose solution with a few drops of dilute hydrochloric acid. The acid breaks the bond between the two monosaccharides. You must then let the solution cool and add a few drops of an alkali such as sodium hydrogencarbonate to neutralise the acid. The solution will now reduce the copper sulphate in the Benedict's test.

Energy from food

Some foods contain more energy than others. It depends on the proportions of carbohydrate, fat and protein that they contain. Their energy content is measured in **kilojoules (kJ)**. If a gram of carbohydrate is fully oxidised, it produces about 17 kJ, whereas a gram of fat yields over twice as much as this (39 kJ). Protein can produce about 18 kJ. If you look on a food label, it usually shows the energy content of the food, along with the amounts of different nutrients that it contains (Figure 3.12).

Typical values	Amount per 100 g	Amount per serving (207 g)
Energy	313 kJ / 76 kcal	646 kJ / 155 kcal
Protein	4.7 g	9.7 g
Carbohydrate (of which sugars)	13.6 g (6.0 g)	28.2 g (12.4 g)
Fat (of which saturates)	0.2 g (Trace)	0.4 g (0.1 g)
Fibre	3.7 g	7.7 g
Sodium	0.5 g	1.0 g

Figure 3.12 *Food packaging is labelled with the proportions of different food types that it contains, along with its energy content. The energy in units called kilocalories (kcal) is also shown, but scientists no longer use this old-fashioned unit.*

Foods with a high percentage of fat, such as butter or nuts, contain a large amount of energy. Others, like fruits and vegetables, which are mainly composed of water, have a much lower energy content (Table 3.3).

Food	kJ per 100 g	Food	kJ per 100 g
margarine	3200	fried beefburger	1100
butter	3120	white bread	1060
peanuts	2400	chips	990
samosa	2400	grilled beef steak	930
chocolate	2300	fried cod	850
Cheddar cheese	1700	roast chicken	770
grilled bacon	1670	boiled potatoes	340
table sugar	1650	milk	270
grilled pork sausages	1550	baked beans	270
cornflakes	1530	yoghurt	200
rice	1500	boiled cabbage	60
spaghetti	1450	lettuce	40

Table 3.3: *Energy content of some common foods.*

Even while you are asleep you need a supply of energy, for keeping warm, for your heartbeat, to allow messages to be sent through your nerves, and for other body functions. However, the energy you need at other times depends on the physical work that you do. The total amount of energy that a person needs to keep healthy depends on their age and body size, and also on the amount of activity they do. Table 3.4 shows some examples of how much energy is needed each day by people of different age, sex and occupations.

Food scientists measure the amount of energy in a sample of food by burning it in a calorimeter (Figure 3.13). The calorimeter is filled with oxygen, to make sure that the food will burn easily. A heating filament carrying an electrical current ignites the food. The energy given out by the burning food is measured by using it to heat up water flowing through a coil in the calorimeter.

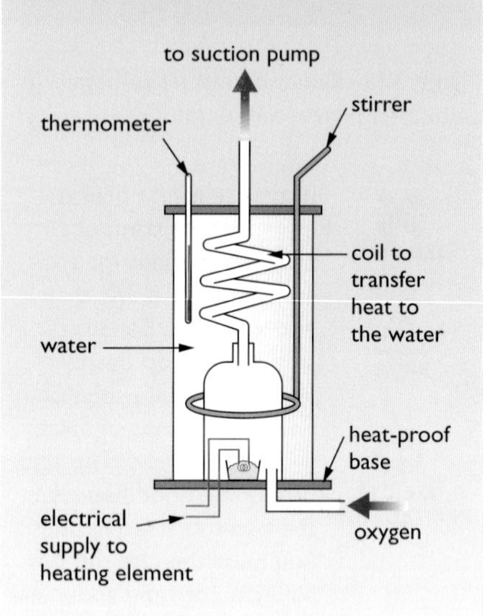

Figure 3.13 *A food calorimeter.*

If you have samples of food that will easily burn in air, you can measure the energy in them by a similar method, using the heat from the burning food to warm up water in a test tube (see question 6 at the end of this chapter).

Age/sex/occupation of person	Energy needed per day (kJ)
newborn baby	2000
child aged 2	5000
child aged 6	7500
girl aged 12–14	9000
boy aged 12–14	11 000
girl aged 15–17	9000
boy aged 15–17	12 000
female office worker	9500
male office worker	10 500
heavy manual worker	15 000
pregnant woman	10 000
breast-feeding woman	11 300

Table 3.4: *The daily energy needs of different types of people.*

Remember that these are approximate figures, and they are averages. Generally, the greater a person's weight, the more energy that person needs. This is why men, with a greater average body mass, need more energy than women. The energy needs of a pregnant woman are increased, mainly because of the extra weight that she has to carry. A heavy manual worker such as a labourer needs extra energy for increased muscle activity.

It is not only the recommended energy requirements that vary with age, sex and pregnancy, but also the *content* of the diet. For instance, during pregnancy a woman may need extra iron or calcium in her diet, for the growth of the fetus. In younger women, the blood loss during menstruation (periods) can result in anaemia, producing a need for extra iron in the diet.

Digestion

Food, such as a piece of bread, contains carbohydrates, fats and proteins, but they are not the same carbohydrates, fats and proteins as in our tissues. The components of the bread must first be broken down into their 'building blocks' before they can be absorbed through the wall of the gut. This process is called **digestion**. The digested molecules – sugars, fatty acids, glycerol and amino acids – along with minerals, vitamins and water, can then be carried around the body in the blood. When they reach the tissues they are reassembled into the molecules that make up our cells.

Digestion is speeded up by **enzymes**, which are biological catalysts (see Chapter 1). Although most enzymes stay inside cells, the digestive enzymes are made by the tissues and glands in the gut, and pass out of cells onto the gut contents, where they act on the food. This **chemical** digestion is helped by **mechanical** digestion. This is the physical breakdown of food. The most obvious place where this happens is in the mouth, where the teeth bite and chew the food, cutting it into smaller pieces that have a larger surface area. This means that enzymes can act on the food more quickly. Other parts of

This is a good definition of digestion, useful for exam answers:
'Digestion is the chemical and mechanical breakdown of food. It converts large insoluble molecules into small soluble molecules, which can be absorbed into the blood.'

the gut also help with mechanical digestion. For example, muscles in the wall of the stomach contract to churn up the food while it is being chemically digested.

Muscles are also responsible for moving the food along the gut. The walls of the intestine contain two layers of muscles. One layer has fibres running in rings around the gut. This is the **circular** muscle layer. The other has fibres running down the length of the gut, and is called the **longitudinal** muscle layer. Together these two layers act to push the food along. When the circular muscles contract and the longitudinal muscles relax, the gut is made narrower. When the opposite happens, i.e. the longitudinal muscles contract and the circular muscles relax, the gut becomes wider. Waves of muscle contraction like this pass along the gut, pushing the food along, rather like squeezing toothpaste from a tube (Figure 3.14). This is called **peristalsis**. It means that movement of food in the gut doesn't depend on gravity – we can still eat standing on our heads!

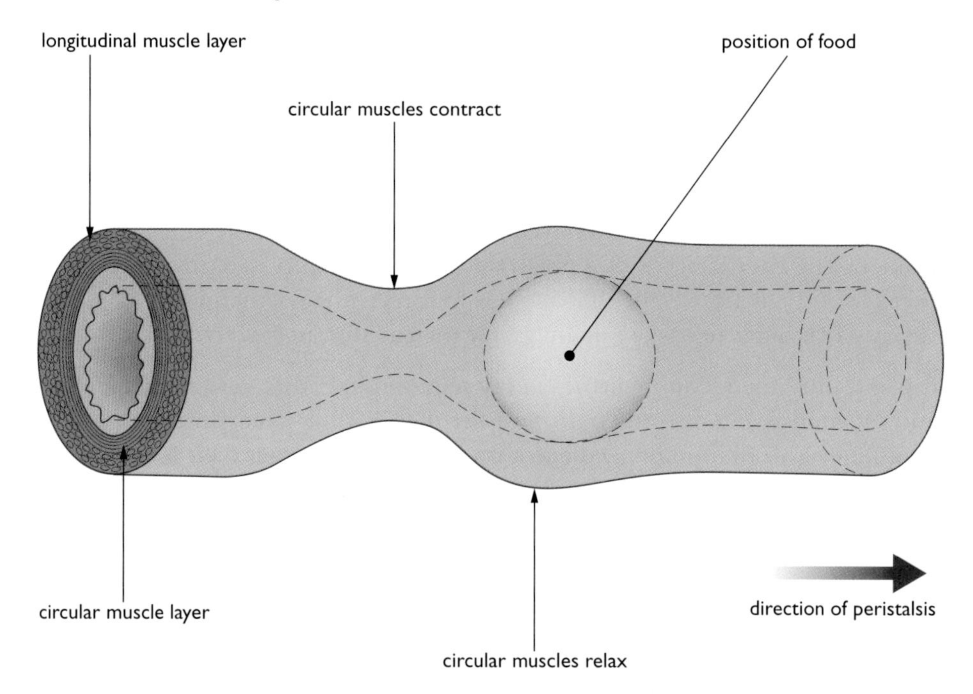

Figure 3.14 *Peristalsis: contraction of circular muscles behind the food narrows the gut, pushing the food along. When the circular muscles are contracted, the longitudinal ones are relaxed, and* vice versa.

Figure 3.15 shows a simplified diagram of the human digestive system. It is simplified so that you can see the order of the organs along the gut. The real gut is much longer than this, and coiled up so that it fills the whole space of the abdomen. Overall, its length in an adult is about 8 m. This gives plenty of time for the food to be broken down and absorbed as it passes through.

The mouth, stomach and the first part of the small intestine (called the **duodenum**) all break down the food using enzymes, either made in the gut wall itself, or by glands such as the **pancreas**. Digestion continues in the last part of the small intestine (the **ileum**) and it is here that the digested food is absorbed. The last part of the gut, the large intestine, is mainly concerned with absorbing water out of the remains, and storing the waste products (**faeces**) before they are removed from the body.

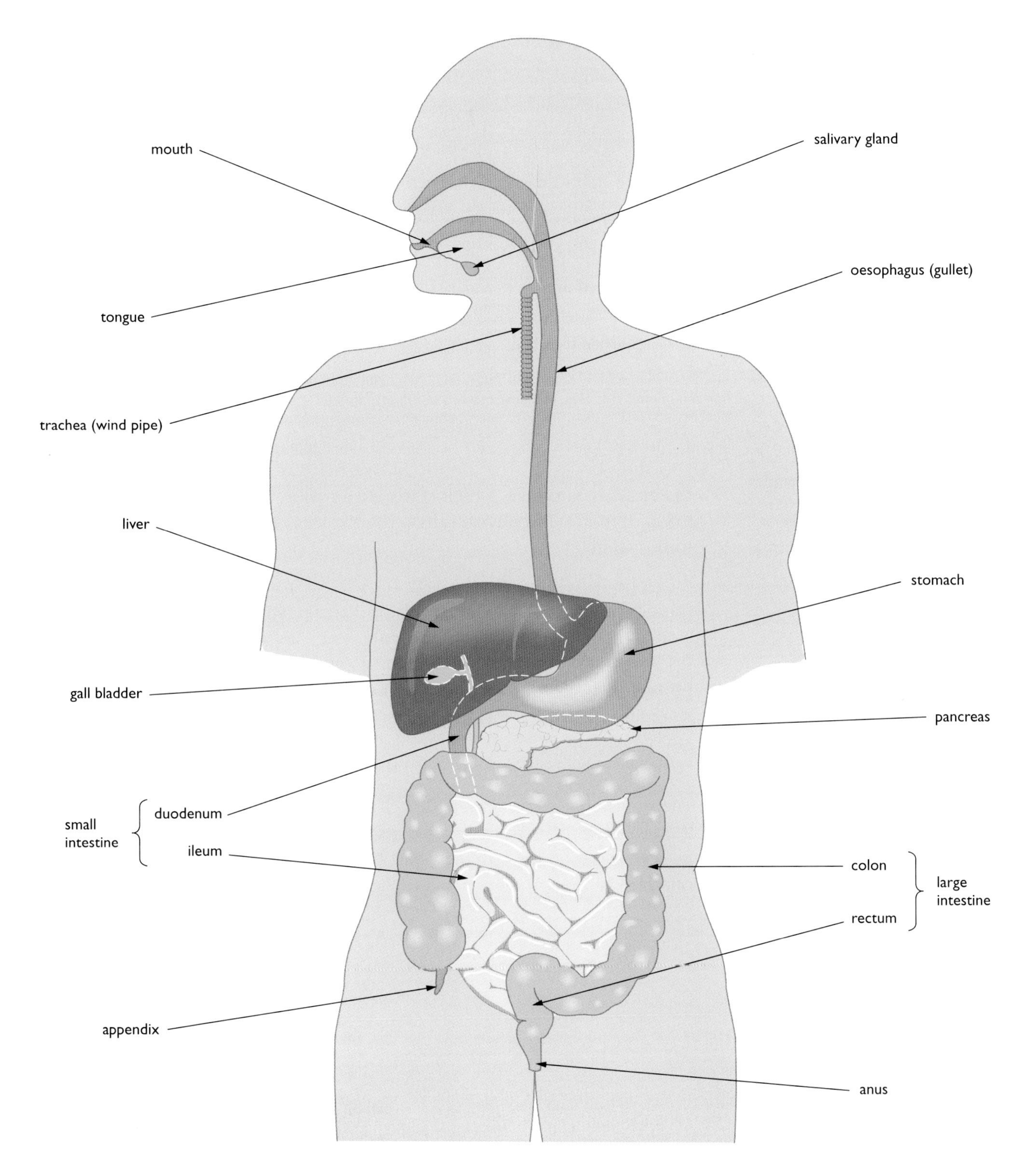

Figure 3.15 *The human digestive system.*

The three main classes of food are broken down by three classes of enzymes. Carbohydrates are digested by enzymes called **carbohydrases**. Proteins are acted upon by **proteases**, and enzymes called **lipases** break down fats (this word comes from the fact that fats and oils are also called lipids). Some of the places in the gut where these enzymes are made are shown in Table 3.5.

Class of enzyme	Examples	Digestive action	Source of enzyme	Where it acts in the gut
carbohydrases	amylase	starch → maltose[1]	salivary glands	mouth
		starch → maltose	pancreas	small intestine
		maltose → **glucose**	wall of small intestine	small intestine
proteases	pepsin	proteins → peptides[2]	stomach wall	stomach
	trypsin	proteins → peptides	pancreas	small intestine
	peptidases	peptides → **amino acids**	wall of small intestine	small intestine
lipases	lipase	fats → **glycerol** and **fatty acids**	pancreas	small intestine

[1]Maltose is a disaccharide made of two glucose molecules joined together.
[2]Peptides are short chains of amino acids.

Table 3.5: *Some of the enzymes that digest food in the human gut. The substances shown in red are the end products of digestion that can be absorbed from the gut into the blood.*

Amylase digests starch into maltose. In this reaction we say that starch is the **substrate** and maltose is the **product**.

Digestion begins in the mouth. **Saliva** helps moisten the food and contains the enzyme **amylase**, which starts the breakdown of starch. The chewed lump of food, mixed with saliva, then passes along the **oesophagus** (gullet) to the stomach.

The food is held in the stomach for several hours, while initial digestion of protein takes place. The stomach wall secretes **hydrochloric acid**, so the stomach contents are strongly acidic. This has a very important function. It kills bacteria that are taken into the gut along with the food, helping to protect us from food poisoning. The protease enzyme that is made in the stomach, called **pepsin**, has to be able to work in these acidic conditions, and has an optimum pH value of about 2. This is unusually low – most enzymes work best at near neutral conditions (see Chapter 1).

The semi-digested food is held back in the stomach by a ring of muscle at the outlet of the stomach, called a **sphincter** muscle. When this relaxes, it releases the food into the first part of the small intestine, called the **duodenum** (Figure 3.16).

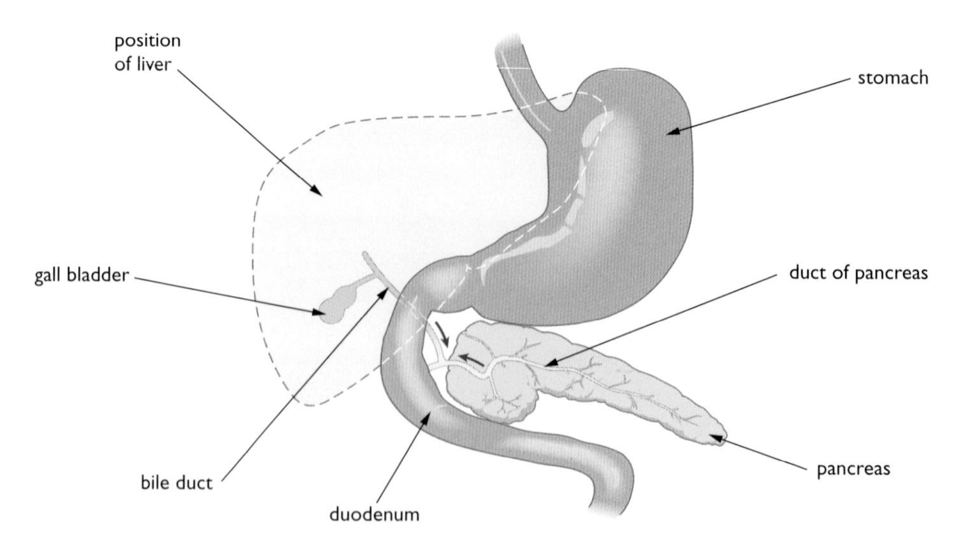

Figure 3.16 *The first part of the small intestine, the duodenum, receives digestive juices from the liver and pancreas through tubes called ducts.*

Several digestive enzymes are added to the food in the duodenum. These are made by the **pancreas**, and digest starch, proteins and fats (Table 3.5). As well as this, the **liver** makes a digestive juice called **bile**. Bile is a green liquid that is stored in the **gall bladder** and passes down the **bile duct** onto the food. Bile does not contain enzymes, but has another important function. It turns any large fat globules in the food into an emulsion of tiny droplets (Figure 3.17). This increases the surface area of the fat, so that **lipase** enzymes can break it down more easily.

Figure 3.17 *Bile turns fats into an emulsion of tiny droplets for easier digestion.*

Bile and pancreatic juice have another function. They are both alkaline. The mixture of semi-digested food and enzymes coming from the stomach is acidic, and needs to be neutralised by addition of alkali before it continues on its way through the gut.

As the food continues along the intestine, more enzymes are added, until the parts of the food that can be digested have been fully broken down into soluble end products, which can be absorbed. This is the role of the last part of the small intestine, the **ileum**.

Absorption in the ileum

The ileum is highly adapted to absorb the digested food. The lining of the ileum has a very large surface area, which means that it can quickly and efficiently absorb the soluble products of digestion into the blood. The length of the intestine helps to provide a large surface area, and this is aided by folds in its lining, but the greatest increase in area is due to tiny projections from the lining, called **villi** (Figure 3.18). The singular of villi is 'villus'. Each villus is only about 1–2 mm long, but there are millions of them, so that the total area of the lining is thought to be about 300 m^2. This provides a massive area in contact with the digested food. As well as this, high-powered microscopy has revealed that the surface cells of each villus themselves have hundreds of minute projections, called **microvilli**, which increase the surface area for absorption even more (Figure 3.19).

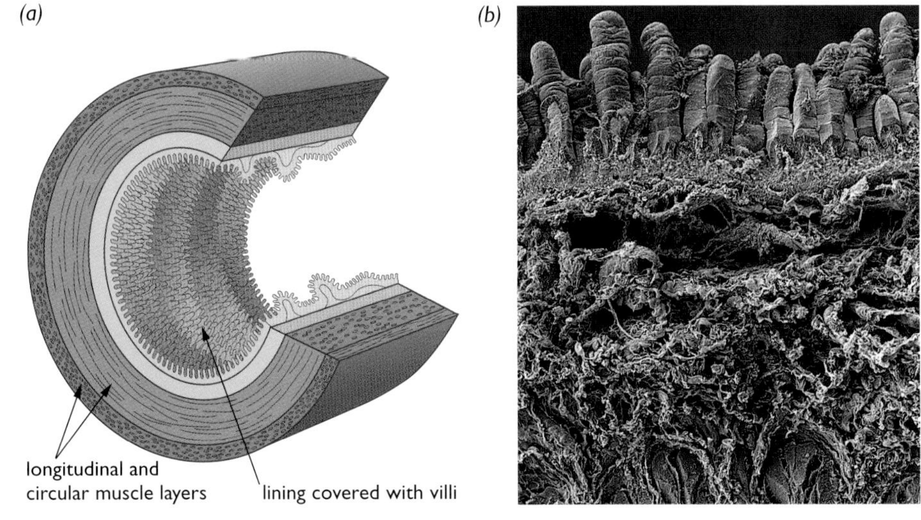

Figure 3.18 *(a) The inside lining of the ileum is adapted to absorb digested food by the presence of millions of tiny villi. (b) A section through the lining, showing the villi.*

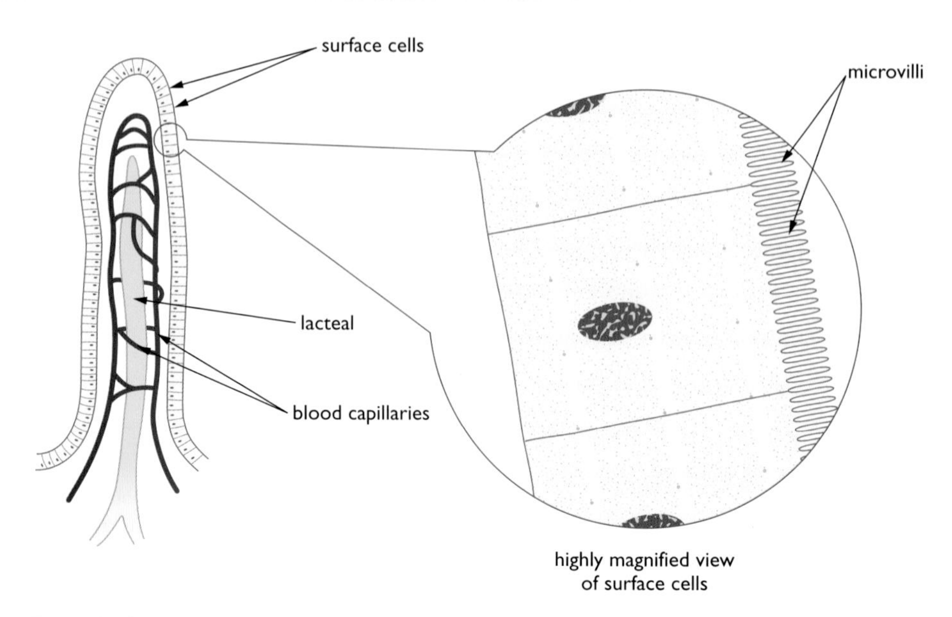

Figure 3.19 *Each villus contains blood vessels and a lacteal, which absorb the products of digestion. The surface cells of the villus are covered with microvilli, which further increase the surface area for absorption.*

Each villus contains a network of blood capillaries. Most of the digested food enters these blood vessels, but the products of fat digestion, as well as tiny fat droplets, enter a tube in the middle of the villus, called a **lacteal**. The lacteals form part of the body's **lymphatic** system, which transports a liquid called lymph. This **lymph** eventually drains into the blood system too.

The blood vessels from the ileum join up to form a large blood vessel called the **hepatic portal vein**, which leads to the liver (see Chapter 5). The liver acts rather like a food processing works, breaking some molecules down, and building up and storing others. For example, glucose from carbohydrate digestion is converted into **glycogen** and stored in the liver. Later, the glycogen can be converted back into glucose when the body needs it (see Chapter 7).

The digested food molecules are distributed around the body by the blood system (see Chapter 5). The soluble food molecules are absorbed from the blood into cells of tissues, and are used to build new parts of cells. This is called **assimilation**.

Removal of faeces by the body is sometimes incorrectly called excretion. Excretion is a word that should only apply to materials that are the waste products of cells of the body. Faeces are not – they consist of waste which has passed through the gut without entering the cells. The correct name for this process is **egestion**.

The large intestine – elimination of waste

By the time that the contents of the gut have reached the end of the small intestine, most of the digested food, as well as most of the water, has been absorbed. The waste material consists mainly of cellulose (fibre) and other indigestible remains, water, dead and living bacteria and cells lost from the lining of the gut. The function of the first part of the large intestine, called the **colon**, is to absorb most of the remaining water from the contents, leaving a semi-solid waste material called **faeces**. This is stored in the **rectum**, until expelled out of the body through the **anus**.

End of Chapter Checklist

If you haven't got a copy of your specification, read the introduction on page vi.

You will need to be able to do some or all of the following. Check your Awarding Body's specification (syllabus) to find out exactly what you need to know.

- Know the composition of the human diet, including carbohydrates, proteins and fats, vitamins and minerals, and fibre. Be able to explain why these are important in the diet.
- Understand what is meant by a balanced diet, and how a poor diet can cause health problems. Relate diet to the body requirements of different people.
- Recall food tests for certain types of food, including starch, reducing sugar and protein.
- Understand the meaning of the term digestion.
- Understand the role of the muscular wall of the gut in peristalsis.
- Name and locate the parts of the human digestive system and recall the functions of the main digestive organs.
- Know the sources of digestive enzymes (carbohydrases, proteases and lipases) and explain their roles in the digestion of food.
- Know the role of stomach acid in the killing of bacteria in food, and the action of bile in emulsifying fats and neutralising stomach acid.
- Describe the adaptations of the small intestine for the absorption of digested food, including the structure and function of villi.

Questions

More questions on food and digestion can be found at the end of Section B on page 119.

1 The diagram shows an experiment that was set up as a model to show why food needs to be digested.

The Visking tubing acts as a model of the small intestine because it has tiny holes in it that some molecules can pass through. The tubing was left in the boiling tube for an hour, then the water in the tube was tested for starch and glucose.

a) Describe how you would test the water for starch, and for glucose. What would the results be for a 'positive' test in each case?

b) The tests showed that glucose was present in the water, but starch was not. Explain why.

c) If the tubing takes the place of the intestine, what part of the body does the water in the boiling tube represent?

d) What does 'digested' mean?

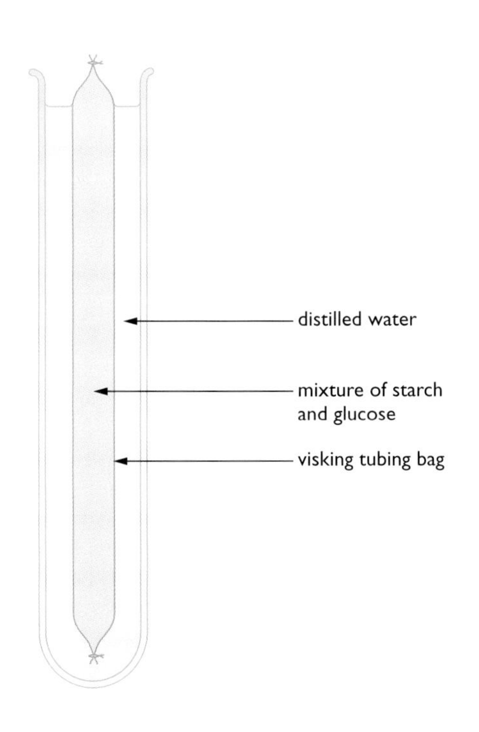

2 (Hint: page 4, Chapter 1 will help with this question.) A student carried out an experiment to find out the best conditions for the enzyme pepsin to digest protein. For the protein, she used egg white powder, which forms a cloudy white suspension in water. The table below shows how the four tubes were set up.

Tube	Contents
A	5 cm^3 egg white suspension, 2 cm^3 pepsin, 3 drops of dilute acid. Tube kept at 37 °C
B	5 cm^3 egg white suspension, 2 cm^3 distilled water, 3 drops of dilute acid. Tube kept at 37 °C
C	5 cm^3 egg white suspension, 2 cm^3 pepsin, 3 drops of dilute acid. Tube kept at 20 °C
D	5 cm^3 egg white suspension, 2 cm^3 pepsin, 3 drops of dilute alkali. Tube kept at 37 °C

The tubes were left for 2 hours and the results were then observed. Tubes B, C and D were still cloudy. Tube A had gone clear.

a) Three tubes were kept at 37 °C. Why was this temperature chosen?

b) Explain what had happened to the protein in tube A.

c) Why did tube D stay cloudy?

d) Tube B is called a **control**. Explain what this means.

e) Tube C was left for another 3 hours. Gradually it started to clear. Explain why digestion of the protein happened more slowly in this tube.

f) The lining of the stomach secretes hydrochloric acid. Explain the function of this.

g) When the stomach contents pass into the duodenum, they are still acidic. How are they neutralised?

3 Copy and complete the following table of digestive enzymes:

Enzyme	Food on which it acts	Products
amylase		
trypsin		
		fatty acids and glycerol

4 Describe four adaptations of the small intestine (ileum) that allow it to absorb digested food efficiently.

5 Bread is made mainly of starch, protein and fat. Imagine a piece of bread about to start its journey through the human gut. Describe what happens to the bread as it passes through the mouth, stomach, duodenum, ileum and colon. Explain how the bread is moved along the gut. Your description should be illustrated by two or three simplified diagrams. It would be easier to word process your account, leaving room for illustrations, or you might obtain these from websites or a CD-ROM.

6 The diagram shows a method that can be used to measure the energy content of some types of food. A student placed 20 cm^3 of water in a boiling tube and measured its temperature. He weighed a small piece of pasta, and then held it in a Bunsen burner flame until it caught alight. He then used the burning pasta to heat the boiling tube of water, until the pasta had finished burning. Finally, he measured the temperature of the water at the end of the experiment.

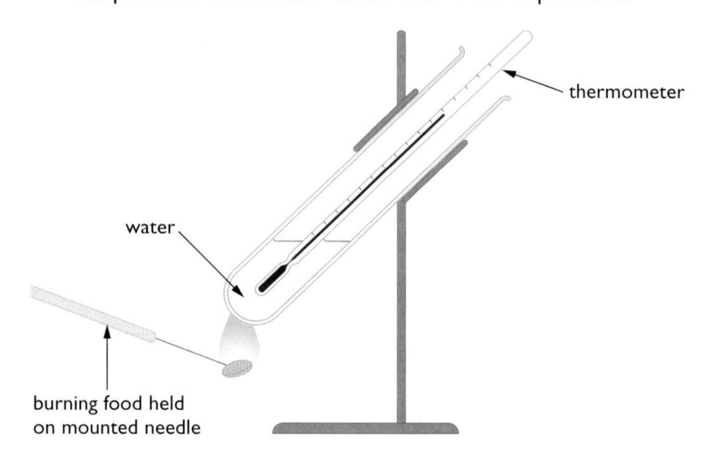

To answer the questions that follow, use the following information:

- The density of water is 1 g/cm^3.
- The pasta weighed 0.22 g.
- The water temperature at the start was 21 °C and at the end was 39 °C.
- The heat energy supplied to the water can be found from the formula:

energy (in joules) = mass of water × temperature change × 4.2

a) Calculate the energy supplied to the water in the boiling tube in joules (J). Convert this to kilojoules (kJ) by dividing by 1000.

b) Calculate the energy released from the pasta as kilojoules per gram of pasta (kJ/g).

c) The correct figure for the energy content of pasta is 14.5 kJ/g. The student's result is an underestimate. Write down three reasons why he may have got a lower than expected result. (Hint: think about how the design of the apparatus might introduce errors.)

d) Suggest one way the apparatus could be modified to reduce these errors.

e) The energy in a peanut was measured using the method described above. The peanut was found to contain about twice as much energy per gram as the pasta. Explain why this is the case.

Although you can do this experiment in a school laboratory, you must not use any kinds of nuts. This is because a small number of people are allergic to nuts and can be severely affected by the fumes from a burning nut. It is OK to use pasta, chocolate or any other foods that you can set fire to in a Bunsen burner flame.

Chapter 4: Patterns of Feeding

This chapter describes different ways of obtaining and digesting food. It considers some mechanisms that invertebrate animals use to feed. It also compares the adaptations of the teeth in humans and carnivores for dealing with solid food, and looks at ways in which herbivores can digest cellulose.

Filter feeding

Animals that **filter feed** are always **aquatic** (live in water). They feed on microscopic particles of food floating in the water. Filter feeding means that the animal sets up feeding currents in the water, so that the food particles are sucked into its body. Inside the body, the animal has some kind of sieve or filter that traps the particles and passes them to the gut. **Mussels** are filter feeders (Figure 4.1).

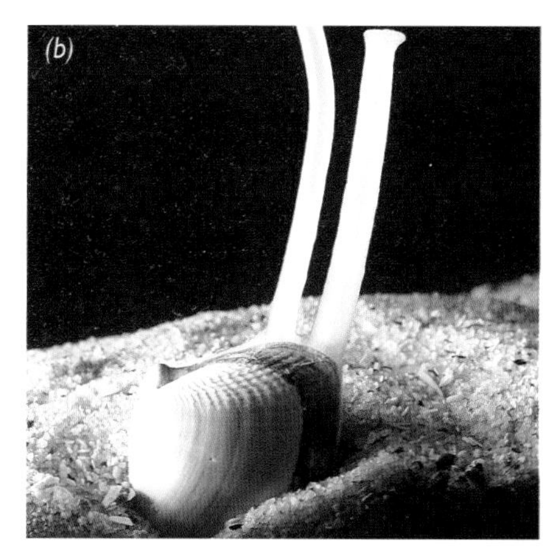

Figure 4.1 *(a) A freshwater mussel. (b) A marine (seawater) mussel. You can see the siphons in the marine mussel that allow water in and out of the space between the shells.*

Molluscs are a group of animals that include snails, slugs, oysters, clams, cockles, squid and octopus. Many (but not all) of the group have a shell made of calcium carbonate, protecting the soft body parts. Mussels belong to a sub-division of the group. They are molluscs called **bivalves**, which have two shells.

Mussels are a type of mollusc that live in water. Both freshwater and marine species feed in a similar way, by using their **gills** as a filter. The food particles consist of tiny pieces of dead organic matter and living microscopic organisms called **plankton**.

The gills in a mussel are inside the cavity protected by the animal's two shells. Each gill is covered in tiny hair-like **cilia**, which beat backwards and forwards, drawing water in through a tube called the **inhalent siphon**. As the water passes over the gills, particles of food get trapped in a sticky **mucus** on the gills. The water then passes out through an **exhalent siphon**. Other cilia pass the food particles into the animal's mouth. These cilia can actually sort out the smaller food particles from larger grains of sand, so that only the food enters the mussel's gut.

Sucking insects

Insects are a very successful group of animals, and there are millions of different species living on Earth. One of the reasons for their success is that they have adapted to feeding on a wide range of materials, from leaves and plant sap to blood and flesh. There are even insects that can feed on materials that at first sight look indigestible, such as tree bark or animal bones.

Many groups of insects have evolved sucking mouthparts, which allow them to feed on liquid food. We will look at just four examples: mosquitoes, aphids (greenflies), butterflies and houseflies.

Parasites are organisms that live in or on another organism and get their food from it. The second organism is called the **host**. Mosquitoes are parasites of a number of species of mammals and birds, and several species of mosquito are parasites of humans.

Mosquitoes

The mouthparts of mosquitoes are modified to form a sucking tube called a **proboscis**. Male mosquitoes and virgin females (those that have not yet mated) only feed on plant sap, but after mating, the females need to feed on blood to help their eggs develop. The end of the proboscis has sharp needle-like structures called **stylets** that pierce the skin of the host animal (Figure 4.2). The stylets help the proboscis to penetrate down through the skin until it reaches a blood vessel (Figure 4.3). Blood then passes up the proboscis into the mosquito's gut, helped by the contraction of muscles in the insect's throat. While the insect is feeding on the blood, it secretes saliva down the proboscis. The saliva contains a substance that stops the blood clotting, so that it flows freely.

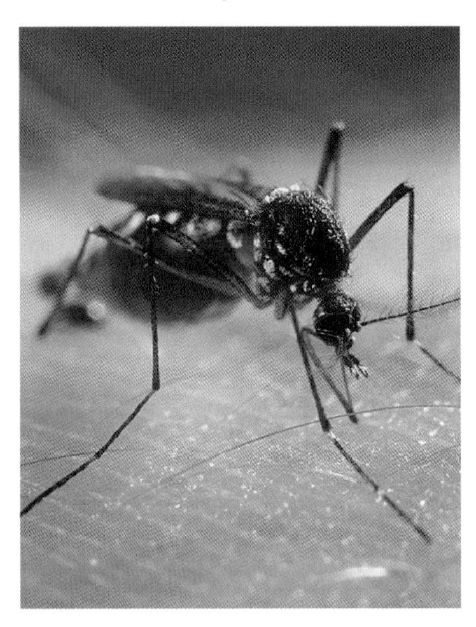

Figure 4.2 *A mosquito using its proboscis to penetrate human skin for a blood meal. Mosquitoes like this can spread a number of parasites that cause human diseases, including malaria.*

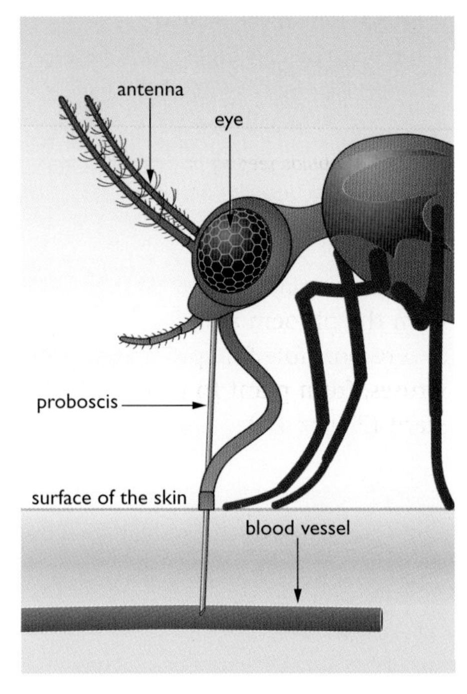

Figure 4.3 *Head and mouthparts of a feeding mosquito.*

The saliva of the mosquito may contain organisms that cause diseases, such as **malaria**. This disease is caused by a single-celled organism called *Plasmodium*. *Plasmodium* is a parasite; it feeds and reproduces inside human blood cells, destroying them and producing a severe fever. The mosquito is responsible for passing the malaria parasite from one person to another. We call it a **vector** of the disease. Malaria will be dealt with in more detail in Chapter 24.

Aphids

Aphids, more commonly known as greenfly and blackfly, are also sucking insects with a proboscis modified for piercing, but they only feed on plant sap (Figure 4.4). The proboscis has sharp stylets that allow the insect to penetrate a stem or leaf and push its proboscis down until it reaches the tubes carrying sap, in a tissue called the **phloem** (see Chapter 11). The sap, containing sugars and other solutes, passes up through the hollow proboscis into the aphid's gut, helped by muscular contractions of the gullet (Figure 4.5).

Figure 4.4 *Aphids feeding on plant sap. Both adults and young aphids (nymphs) feed in this way.*

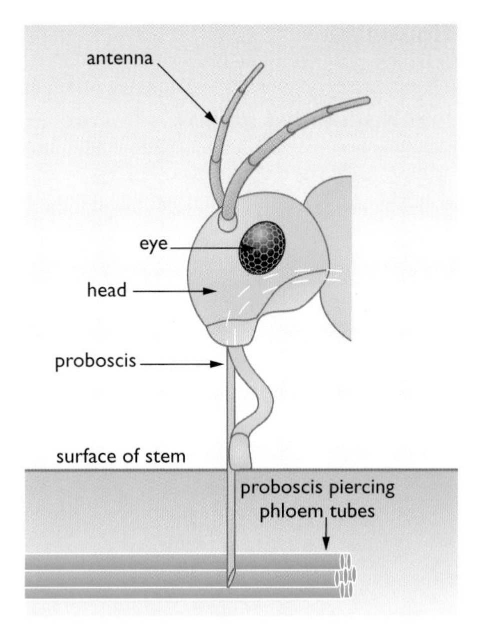

Figure 4.5 *Aphid proboscis penetrate into phloem tubes to reach plant sap.*

Aphids can cause serious damage to their host plants. They remove nutrients from the phloem and stunt the growth of shoots and young leaves. They are also responsible for spreading many disease-causing organisms, such as viruses, from plant to plant. A bad infestation with aphids can easily kill a plant (Figure 4.6).

Figure 4.6 *This broad bean plant has a heavy infestation of blackfly, a species of aphid.*

Figure 4.7 *The proboscis of a butterfly is able to extend to a length of several centimetres to reach the nectaries deep in the flower.*

Butterflies

A butterfly (or moth) has a much longer proboscis than mosquitoes or aphids. When it is not in use, the proboscis is coiled up under the head of the insect, but it contains muscles and nerves that allow it to be uncoiled to a length of several centimetres (Figure 4.7). With the proboscis uncoiled, the butterfly can probe down into the depths of a flower to reach the **nectaries**. These are the source of **nectar**, a sugar-rich liquid that some plants make to attract pollinating insects. The proboscis has a central hollow tube that the butterfly uses to suck up the nectar.

Houseflies

The housefly is also a sucking insect, feeding on liquid food through a highly modified proboscis (Figure 4.8).

Figure 4.8 *The housefly normally carries its proboscis bent under its head when it is not feeding. Here, a combination of blood pressure and muscular action has been used to extend the proboscis onto the food.*

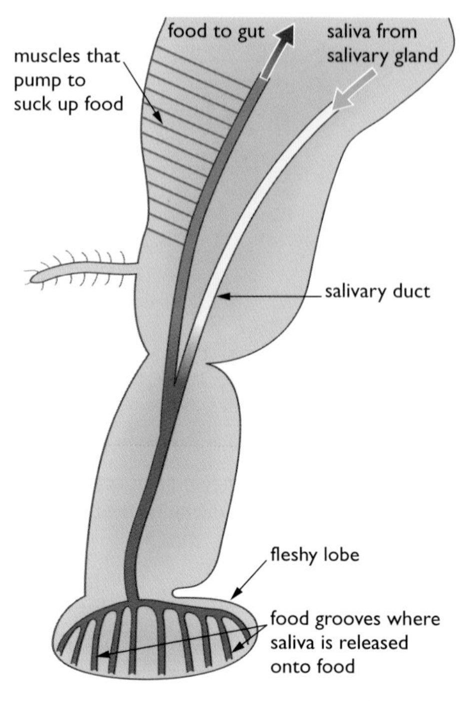

Figure 4.9 *Diagram of a section through the proboscis of a housefly.*

The end of the extended proboscis has two fleshy lobes covered with grooves, which are pressed onto the surface of the food. Saliva is secreted down a duct and squirted out into the grooves, where it softens and dissolves the food. This is an example of **external digestion** (digestion outside the body). The semi-liquid food is then sucked up the proboscis into the fly's gut (Figure 4.9).

Flies feed from many different sources, including waste organic material, faeces, animal dung and human food. They may feed from animal dung one minute, and an uncovered sandwich the next! You won't be surprised to find that they are responsible for carrying many organisms that cause human diseases, from bacteria to parasitic worms.

Mammalian teeth

Human digestion (see Chapter 3) is helped by the mechanical breakdown of food, much of which is achieved in the mouth by the action of the teeth. Teeth are used to bite off pieces of food and to chew it into smaller pieces before it is swallowed. At the same time the food is mixed with saliva, which lubricates it, allowing it to be swallowed more easily. Saliva also contains the enzyme amylase, which starts the breakdown of starch.

In mammals, only the lower jaw is movable, while the upper jaw is fused to the skull. Chewing is brought about by the cheek muscles, which move the lower jaw up and down, and allow some side-to-side movement.

As with most mammals, humans have four types of teeth, called incisors, canines, premolars and molars (Figure 4.10).

The arrangement of an animal's teeth is called its **dentition.**

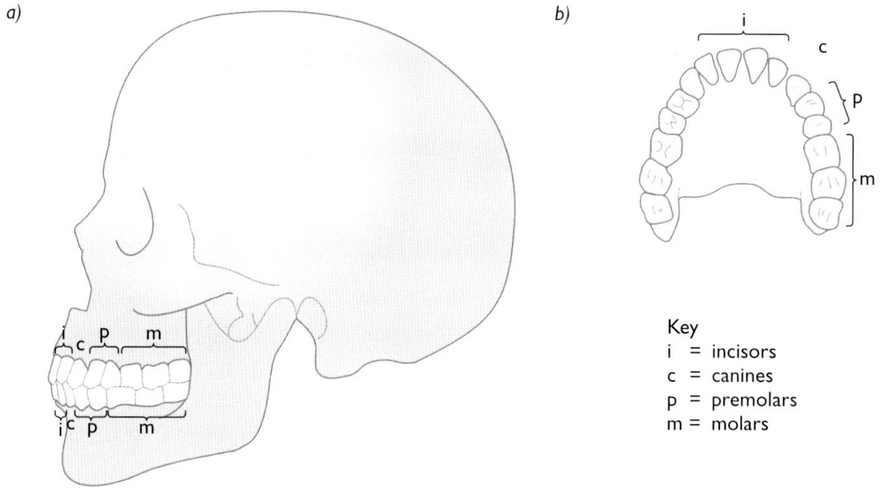

a) b)

Key
i = incisors
c = canines
p = premolars
m = molars

Figure 4.10 *(a) Side view of human skull and teeth. (b) Upper teeth shown from below.*

At the front of the mouth are the **incisors**. They are relatively sharp, chisel-shaped teeth, used for biting off pieces of food. In each jaw behind the incisors is a pair of **canines** (one on each side of the mouth). In humans the canines are similar in shape to the incisors, and have the same biting function. Behind the canines are the **cheek teeth** (**molars** and **premolars**). The crowns of the cheek teeth each have a flatter top surface, and the top and bottom sets of cheek teeth meet crown to crown, so that they can be used for chewing or crushing food. Human teeth are adapted for dealing with a wide range of food types, from meat and fish to plant roots, stems and leaves.

We can see how teeth can be adapted to a more selective diet if we look at the dentition of a dog, which is a carnivore.

A **carnivore** eats mainly meat. A **herbivore** eats plant material (leaves, stems, roots, etc.). An animal that eats a mixture of animal and plant material is called an **omnivore.** Humans are omnivores.

Feeding in a carnivore

Dogs have teeth and jaws that are adapted for a carnivorous diet (Figure 4.11).

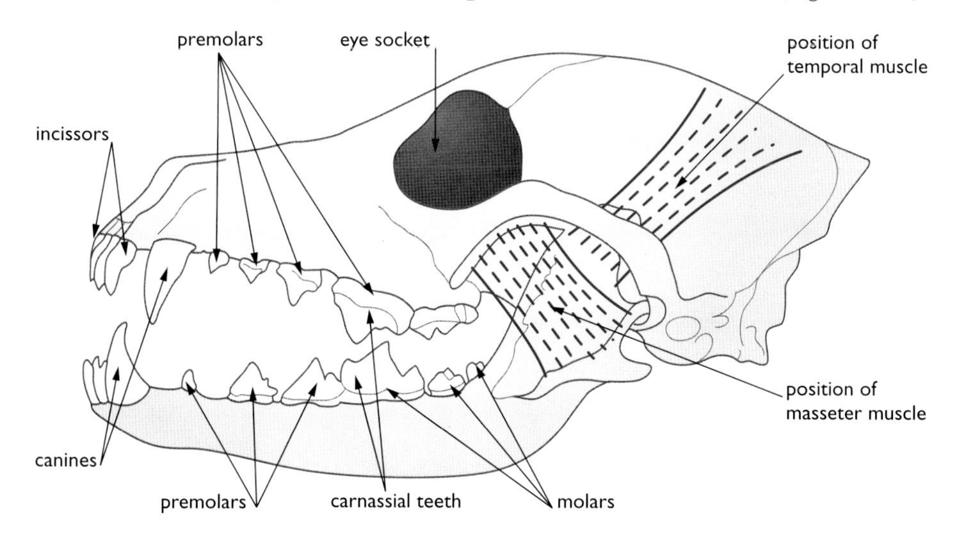

Figure 4.11 *The skull of a dog, side view.*

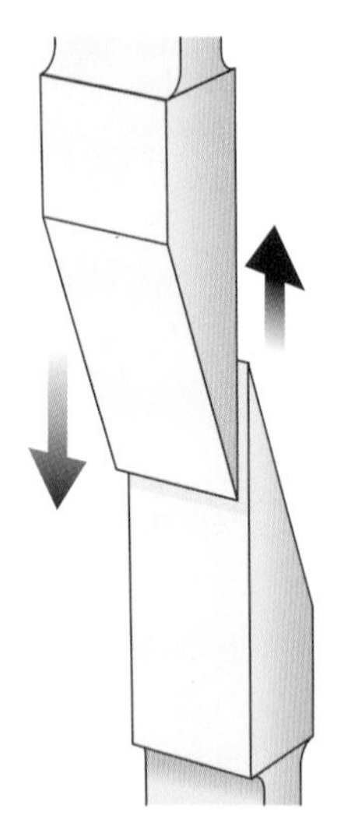

Figure 4.12 *The carnassial teeth of a carnivore act like the blades of scissors, overlapping so that they shear through the meat and bones.*

Figure 4.13 *This wolf is using its cheek teeth to slice through the meat.*

Dogs have the same types of teeth as humans, but they have a very different structure and arrangement. The incisors are small and pointed, and meet together so that they can be used to get a grip on the meat and pull it apart. The canines are large and pointed, and near the front of the long jaw. They stab the prey, stopping it from escaping, as well as helping to kill it. They are also used to help tear the meat.

The cheek teeth of a dog, like other carnivores are adapted for shearing through the meat and bones. They include a pair of very large cheek teeth in each jaw, called the **carnassial** teeth. These have very sharp edges, which overlap like the blades of a pair of scissors (Figure 4.12).

Two sets of muscles help to close the lower jaw of a dog. The **temporal** muscles are attached to the ridge on the top of the skull, and the **masseter** muscles to the bones below the eye sockets. Their positions are shown in Figure 4.11. The shape of the jaw pivot at the back of the skull allows very little side-to-side movement of the jaw – all the effort goes into an up-and-down movement, which gives a strong scissor action. Because the carnassial teeth are near the back of the mouth, near the pivot, this means that they can exert a large force on the food (Figure 4.13).

Of course, it is not only the dentition of a carnivore that is adapted to its diet. A carnivore's whole body shows adaptations. If you look at Figure 4.11, you can see that the dogs's eye sockets are at the front of the skull. This gives it good forward vision and enables it to judge distances well – needed for a predator to catch its prey. The legs of many predators are muscular and powerful for chasing the prey, and claws are a useful aid for holding on to it. Inside the body, the gut is adapted for a meat diet. Meat contains much protein, which is fairly easy to digest. The gut of a carnivore tends to be shorter than that of an omnivore such as a human, or a herbivore such as a rabbit or cow (Figure 4.14). In the next section you will see why the herbivore's gut needs to be very long.

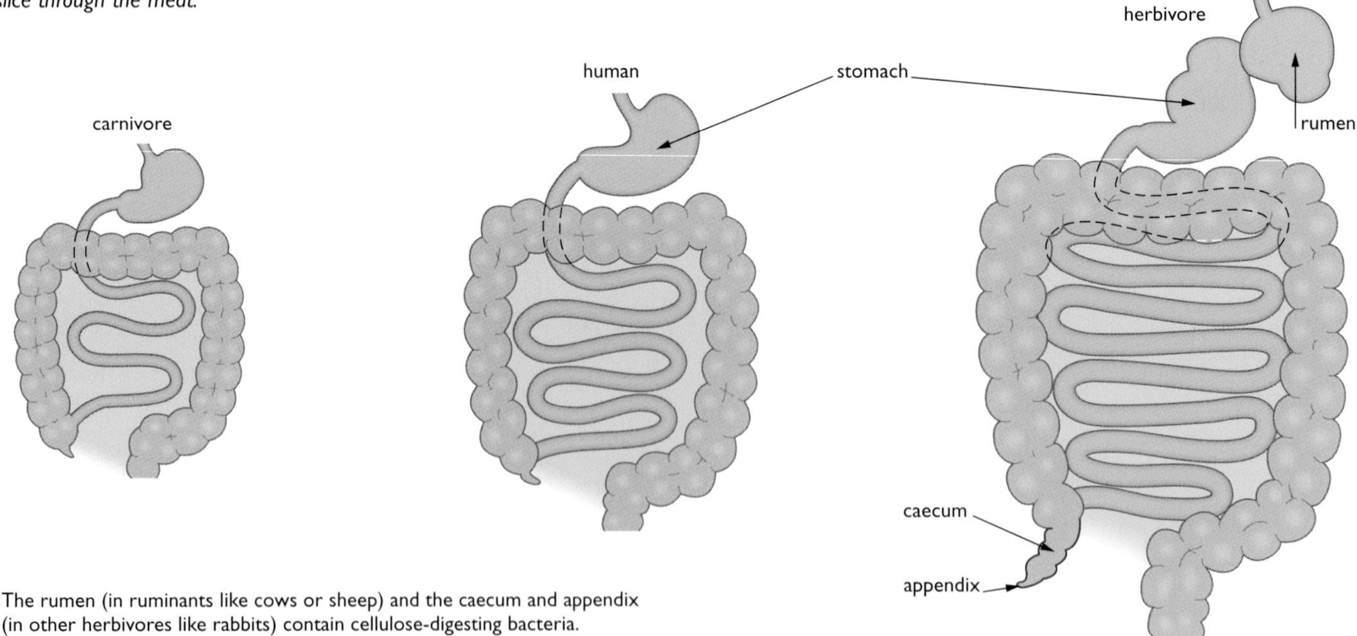

Figure 4.14 *The length of an animal's gut is related to its diet.*

Digestion of cellulose

Much of a herbivore's food consists of **cellulose**, the material that makes up plant cell walls. Humans (and many other omnivores and carnivores) cannot digest cellulose, because we do not produce the enzyme needed for this (see Chapter 3). In humans, cellulose passes through the gut as roughage, or dietary fibre. In carnivores the diet contains little cellulose. Cellulose is made of chains of glucose molecules, and would be an important source of energy if an animal could break it down.

This digestion of cellulose does happen in herbivores. Herbivorous mammals often have guts that contain bacteria and other microorganisms. These bacteria produce enzymes, including **cellulase** that break down the cellulose into glucose. The bacteria live in specialised parts of the herbivore gut, which is one reason why the gut is long. A very large mass of plant material needs to be digested in order to extract enough energy from it, so the herbivore's gut needs to be long in order to hold this amount, and give the bacteria enough time to act on the food.

The bacteria benefit from living in the gut of the animal too. They receive a constant supply of cellulose and other nutrients, and are protected inside a warm, moist environment. This relationship, where two organisms both benefit from living together, is called **mutualism**. You will meet other examples of mutualistic relationships later on in this book.

Mutualism is one type of **symbiosis**. Symbiosis is where two organisms of different species (called **symbionts**) live in a very close association with each other. There are three types of symbiosis:

- parasitism, where one symbiont benefits from the relationship, but the other is harmed
- commensalism, where one symbiont benefits and the other is unaffected
- mutualism, where both symbionts benefit.

Ruminants

In cows and sheep, the bacteria are in an organ called the **rumen**, located between the oesophagus and the stomach. This is why these animals are called **ruminants**. Ruminants eat their food twice. First of all, the chewed grass and other plant material passes into the rumen, where the bacteria start the work of digestion. Later on the partly digested food is **regurgitated** (brought back up to the mouth) for further chewing. This is known as 'chewing the cud'. The chewed cud then passes down into the true stomach for further digestion.

The bacteria in the rumen of cows and sheep also produce a large volume of the waste gases methane and carbon dioxide. Both of these are 'greenhouse gases', and their production by ruminants is thought to have a significant effect on global warming (see Chapter 15).

The caecum and appendix

Other herbivores such as rabbits have cellulose-digesting bacteria in an enlarged caecum and appendix. These organs are located in a 'side-branch' of the intestine, at the point where the ileum enters the large intestine (Figure 4.15).

Any food that enters the rabbit's caecum and appendix has already passed through the small intestine, so the products of digestion have already been absorbed into the blood. However, the bacteria in the caecum and appendix can break down the cellulose into more useful sugars. So that they get the benefit of these nutrients, rabbits eat their own faeces from this first process, so that these nutrients can be absorbed on their second passage through the body.

Because carnivores eat little or no cellulose, they have no need for cellulose-digesting bacteria, so their guts do not have specialised regions to contain them.

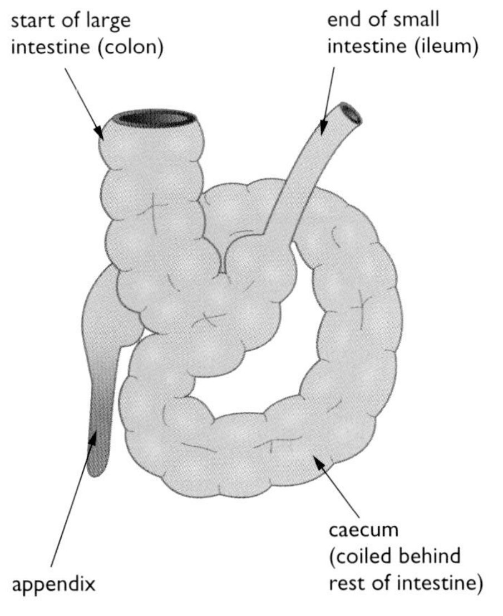

Figure 4.15 *The enlarged caecum and appendix of a rabbit.*

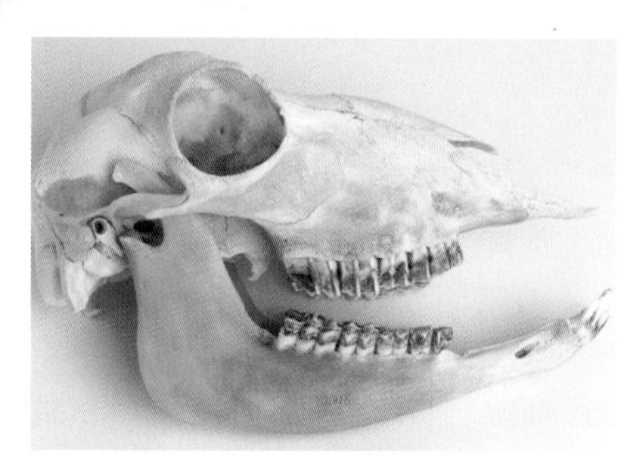

Figure 4.16 *The skull of a sheep.*

Herbivore dentition

The skulls and teeth of herbivores are also adapted for their function (Figure 4.16). Plant material is difficult to break down, and needs thorough chewing and grinding. A herbivore like the sheep has molars which meet crown to crown, with surface ridges of enamel, forming a specialised grinding surface. A sheep's canines are small, and missing completely from the upper jaw. The pivot of its lower jaw is 'loose', allowing side-to-side movements for chewing the cud. The eyes of a sheep are situated in sockets on the side of the skull, giving it good all-round vision to spot predators.

End of Chapter Checklist

If you haven't got a copy of your specification, read the introduction on page vi.

You will need to be able to do some or all of the following. Check your Awarding Body's specification (syllabus) to find out exactly what you need to know.

- Describe the method of filter feeding as shown by the example of the mussel.
- Describe how adult mosquitoes are adapted for feeding on blood.
- Explain how the mosquito is involved in the transmission of the malaria parasite.
- Describe how other sucking insects (aphids, butterflies and houseflies) feed.
- Relate the shapes of human teeth to their function.
- Explain how the teeth and jaws of a dog are adapted for a carnivorous diet.
- Know that mammals do not produce an enzyme that digests cellulose, and that herbivores have specialised regions of their gut that contain cellulose-digesting bacteria.
- Explain the role of the rumen in sheep and cows, and the caecum and appendix in rabbits, in the digestion of cellulose.
- Understand the meaning of the terms parasitism and mutualism.

Questions

More questions on patterns of feeding can be found at the end of Section B on page 119.

1 Explain the meaning of 'symbiosis'. Explain the difference between parasitism and mutualism, giving one example of each.

2 The cockle is a marine mollusc with two shells. It lives buried in the sand along the seashore. You suspect that it may live by filter feeding.

   a) What *external* features of the animal would support your idea that the cockle was a filter feeder?

   b) What would you look for *inside* the shells?

3 Explain how the feeding method of ***a)*** a mosquito and ***b)*** a housefly can make them both harmful to humans.

4 The photograph shows the skull of a cat. Write down three ways you can tell from the photograph that the cat is a carnivore.

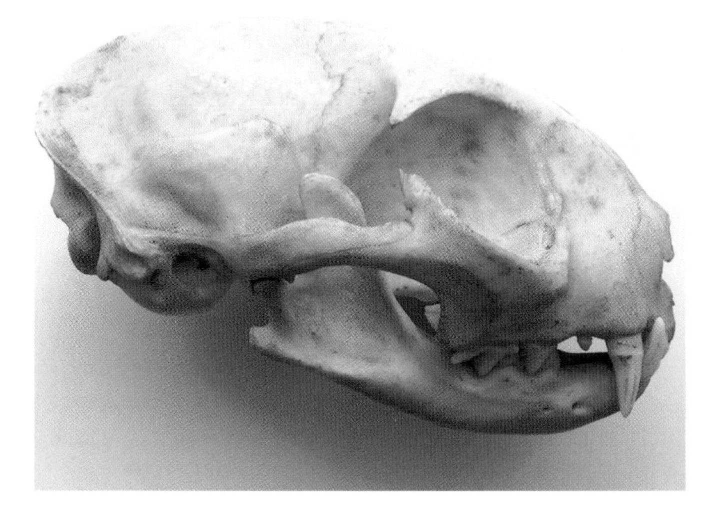

5 The table below compares the digestive system of a dog and a sheep. Copy the table and complete it with a tick (✔) if the feature is present and a cross (✗) if it is absent.

Feature	Dog	Sheep
large canine teeth		
rumen		
gut wall able to secrete cellulase enzymes		
mutualistic relationship with cellulose-digesting bacteria		
intestine longer than in humans		

6 The diagram shows the skull of a sheep.

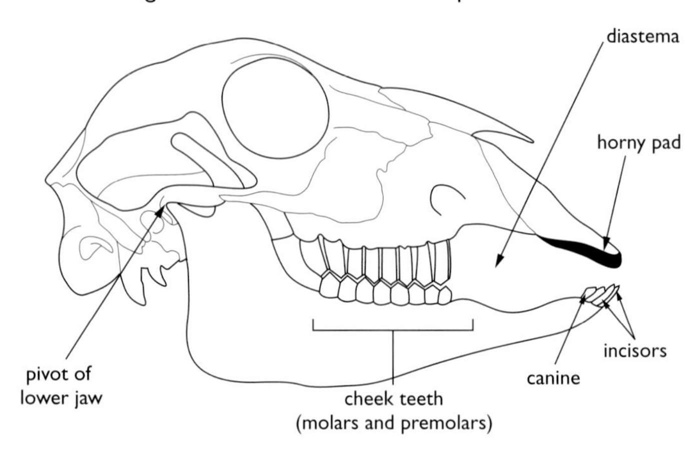

a) Briefly describe three features shown in the diagram which indicate that the sheep is not a carnivore.

b) Explain the importance of the following points, writing a short paragraph for each:

i) The molars and premolars of the sheep have a grinding surface of enamel ridges and the jaw pivot is 'loose'.

ii) A sheep is a ruminant.

iii) A sheep has a much longer gut than a dog.

c) The incisor teeth of a sheep are broad and spade-like. There are no incisors or canines in the upper jaw. Suggest the functions of the horny pad and the space behind the front teeth (diastema).

d) Compared with a dog, would you expect the sheep to be able to exert a large force on the food with its cheek teeth? Explain your answer, using information from Figure 4.12 and the figure in question 6.

Chapter 5: Blood and Circulation

Large, multicellular animals need a circulatory system to transport substances to and from the cells of the body. This chapter looks at the structure and function of the circulatory systems of humans and other animals, the composition of mammalian blood, and disorders associated with the heart and circulation.

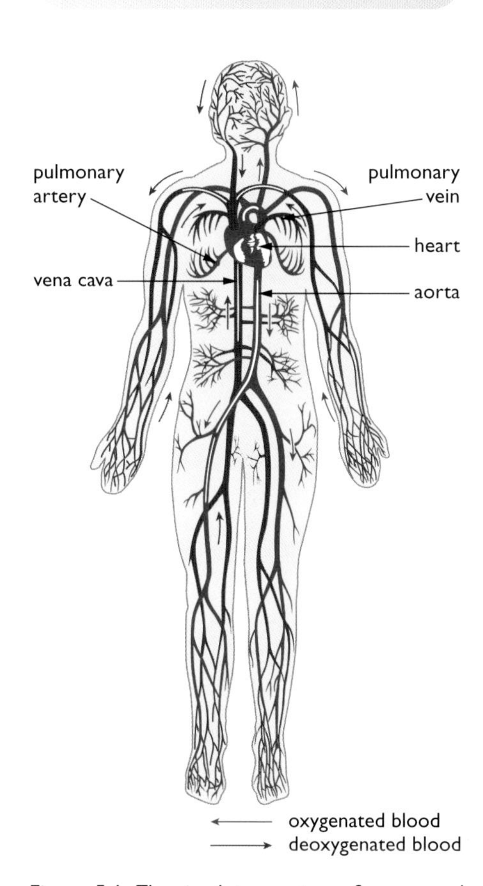

Figure 5.1 *The circulatory system of a mammal.*

The need for circulatory systems

Figure 5.1 shows the circulatory system of a mammal.

Blood is pumped round and round a closed circuit made up of the heart and blood vessels. As it travels around, it collects materials from some places and unloads them in others. In mammals, blood transports:

- oxygen from the lungs to all other parts of the body
- carbon dioxide from all parts of the body to the lungs
- nutrients from the gut to all parts of the body
- urea from the liver to the kidneys.

Hormones, antibodies and many other substances are also transported by the blood as well as distributing heat around the body.

Single-celled organisms, like the ones shown in Figure 5.2, do not have circulatory systems.

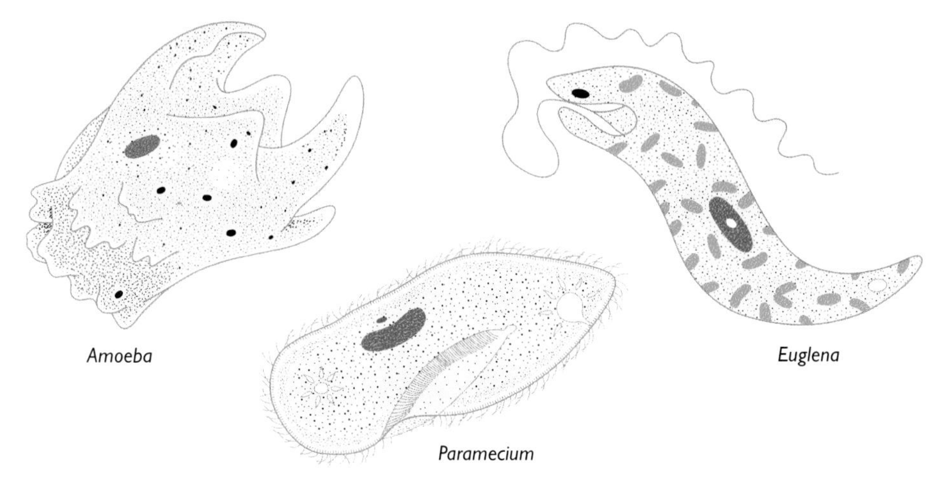

Figure 5.2 *Unicellular organisms do not have circulatory systems.*

Circulatory systems carry materials around the very small 'bodies' of these one-celled organisms. Materials can easily move around the cell without a special system. There is no need for lungs or gills to obtain oxygen from the environment either. One-celled organisms obtain oxygen through the cell surface membrane. The rest of the cell then uses the oxygen. The area of the cell surface determines how much oxygen the organism can get (the supply rate), and the volume of the cell determines how much oxygen the organism uses (the demand rate).

The ratio of supply to demand can be written as: $\frac{\text{surface area}}{\text{volume}}$

This is called the **surface area to volume ratio** (s.a. : vol.) and it is affected by the size of an organism. Single-celled organisms have a high surface area to volume ratio. Their cell surface membrane has a large enough area to supply all the oxygen that their volume demands. In larger animals, the surface area to volume ratio is lower.

Let's pretend that the organism is the shape of a cube!

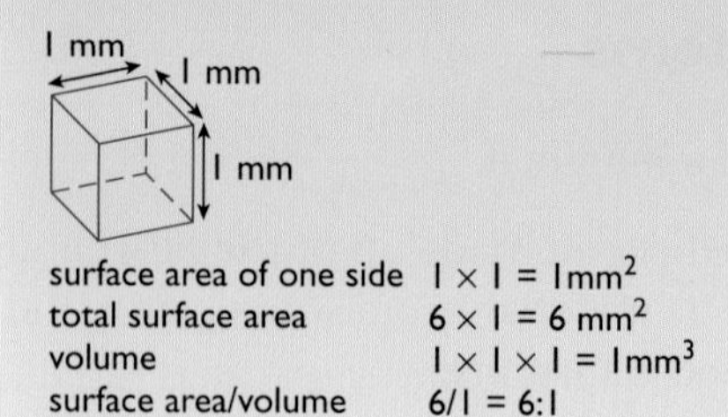

surface area of one side	$1 \times 1 = 1 mm^2$
total surface area	$6 \times 1 = 6 mm^2$
volume	$1 \times 1 \times 1 = 1 mm^3$
surface area/volume	$6/1 = 6:1$

A larger organism has a lower surface area to volume ratio.

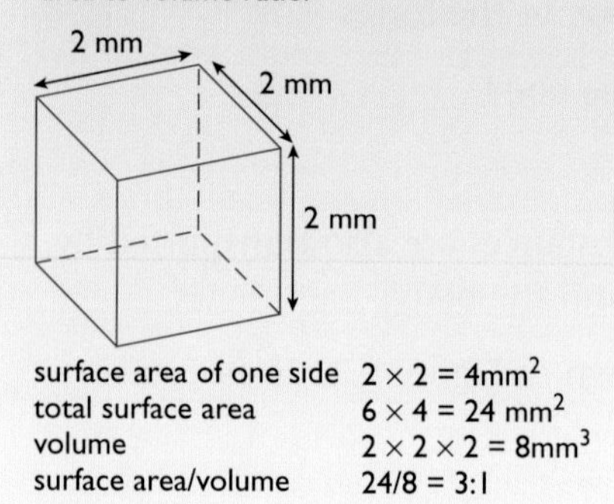

surface area of one side	$2 \times 2 = 4 mm^2$
total surface area	$6 \times 4 = 24 mm^2$
volume	$2 \times 2 \times 2 = 8 mm^3$
surface area/volume	$24/8 = 3:1$

Figure 5.3 *An illustration of surface area to volume ratio.*

The bigger cube has a smaller surface area to volume ratio. It would be less well able to obtain all the oxygen it needs through its surface.

Large animals cannot get all the oxygen they need through their surface (even if the body surface would allow it to pass through) – there just isn't enough surface to supply all that volume. To overcome this problem, large organisms have evolved special gas exchange organs and circulatory systems. The gills of fish and the lungs of mammals are linked to a circulatory system that carries oxygen to all parts of the body. The same idea applies to obtaining nutrients – the gut obtains nutrients from food and the circulatory system distributes the nutrients around the body.

The circulatory systems of different animals

One of the main functions of a circulatory system in animals is to transport oxygen. Blood is pumped to a gas exchange organ to load oxygen. It is then pumped to other parts of the body where it unloads the oxygen. There are two main types of circulatory systems in animals:

- In **single circulatory systems** the blood is pumped from the heart to the gas exchange organ and then directly to the rest of the body.
- In **double circulatory systems** the blood is pumped from the heart to the gas exchange organ, back to the heart and then to the rest of the body.

Figure 5.4 shows the difference between these systems.

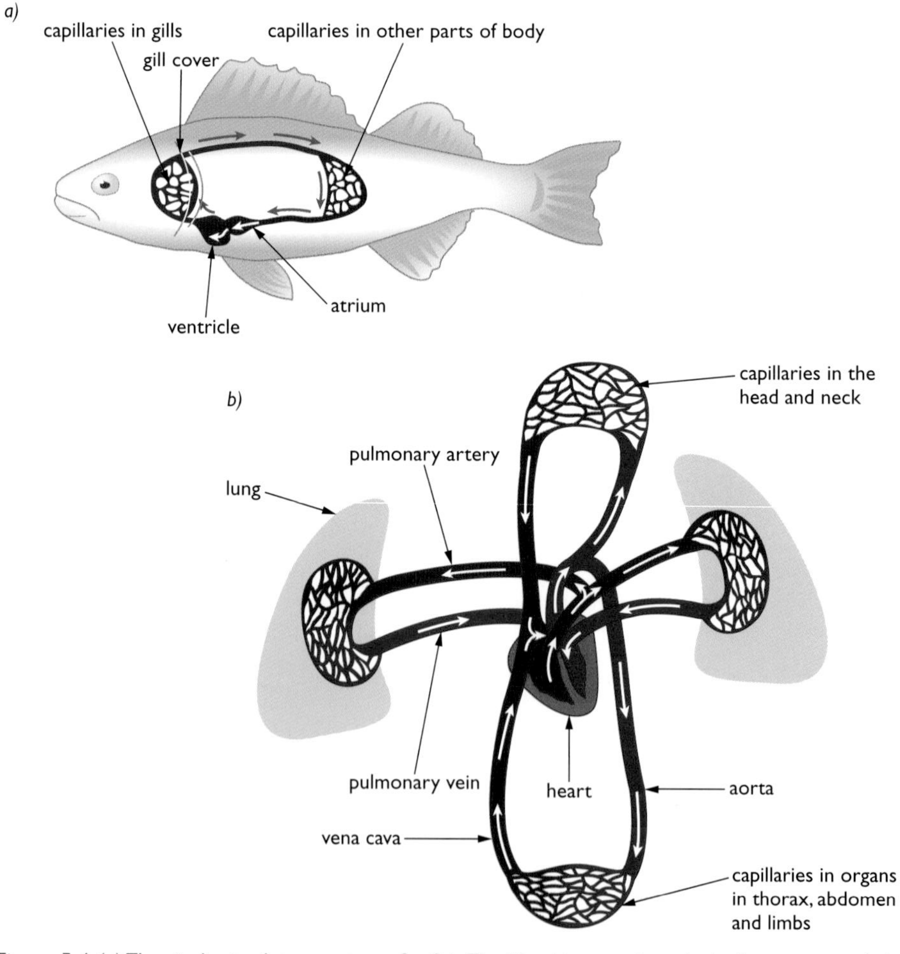

Figure 5.4 *(a) The single circulatory system of a fish. The blood passes through the heart once only in a complete circuit of the body. (b) The double circulatory system of a human (and other mammals). The blood passes through the heart twice in one complete circuit of the body.*

There are two distinct parts to a double circulation:

- the **pulmonary** circulation, in which blood is circulated through the lungs where it is oxygenated
- the **systemic** circulation, in which blood is circulated through all other parts of the body where it unloads its oxygen.

Pulmonary means concerning the lungs.

A double circulatory system is more efficient than a single circulatory system. The heart pumps the blood twice, so higher pressures can be maintained. The blood travels more quickly to organs. In the single circulatory system of a fish, blood loses pressure as it passes through the gills. It then travels relatively slowly to the other organs.

The human circulatory system comprises:

- the **heart** – this is a pump
- **blood vessels** – these carry the blood around the body; **arteries** carry blood away from the heart and towards other organs, **veins** carry blood towards the heart and away from other organs, **capillaries** carry blood through organs
- **blood** – the transport medium.

Figure 5.5 shows the main blood vessels in the human circulatory system.

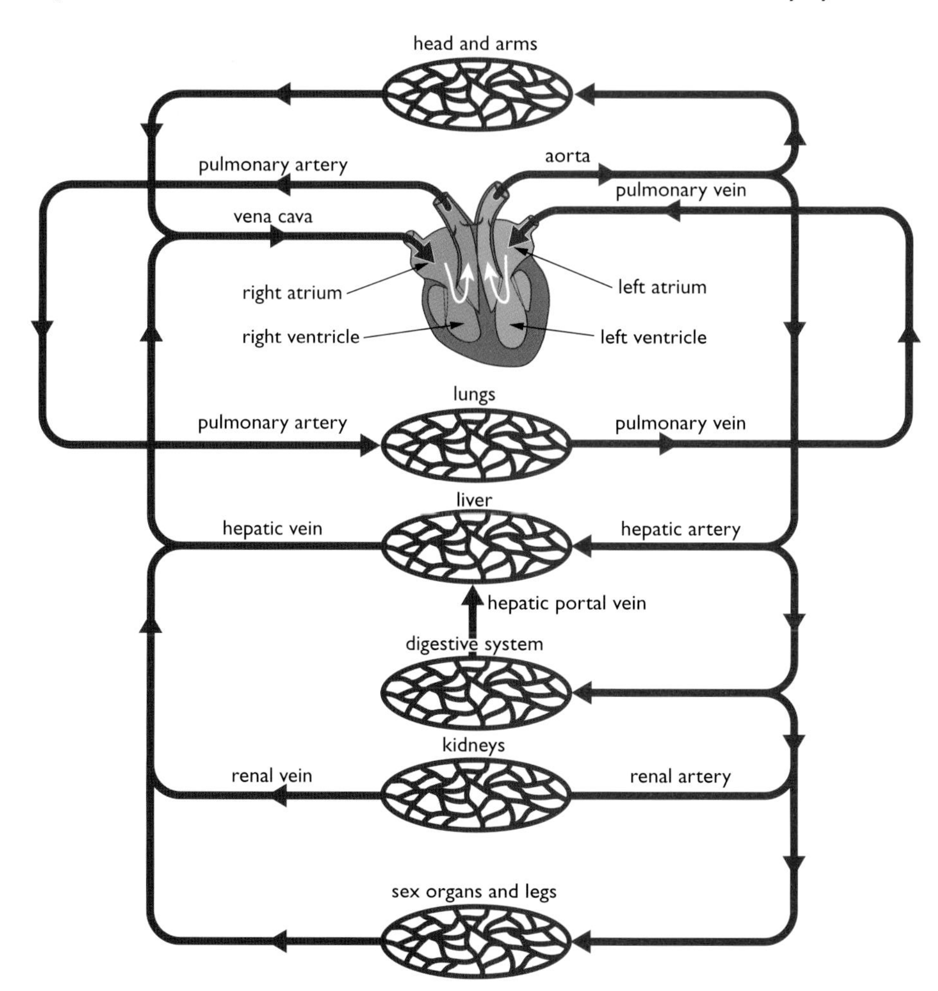

Figure 5.5 *The main components of the human circulatory system.*

Cardiac means 'to do with the heart'.

Cardiac muscle is unlike any other muscle in our bodies. It never gets fatigued ('tired') like skeletal muscle. On average, cardiac muscle fibres contract and then relax again about 70 times a minute. In a lifetime of 70 years, this special muscle will contract over two billion times – and never take a rest!

The structure and function of the human heart

The human heart is a pump. It pumps blood around the body at different speeds and at different pressures according to the body's needs. It can do this because the wall of the heart is made from **cardiac muscle**.

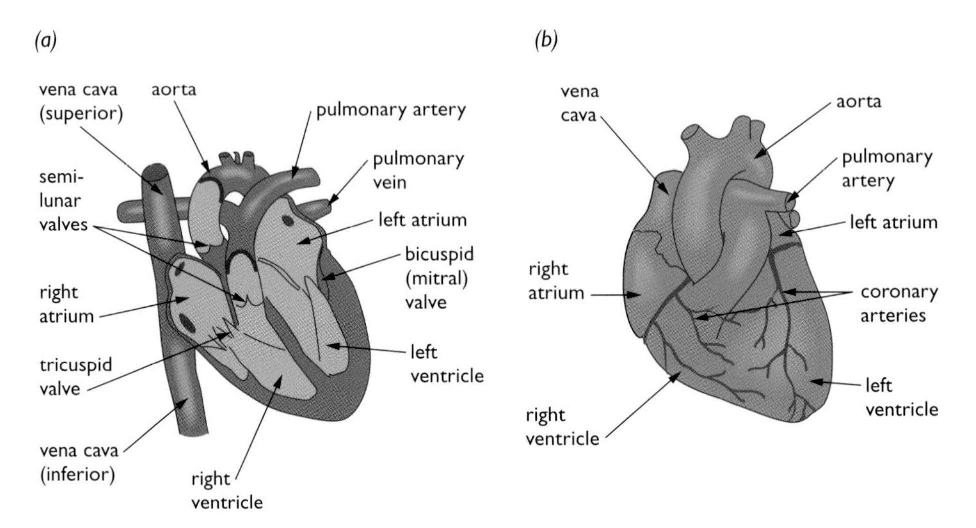

Figure 5.6 *The human heart (a) vertical section, (b) external view.*

The bicuspid (mitral) and tricuspid valves are both sometimes called **atrio-ventricular** valves, as each controls the passage of blood from an atrium to a ventricle.

Blood is moved through the heart by a series of contractions and relaxations of cardiac muscle in the walls of the four chambers. These events form the **cardiac cycle**. The main stages are illustrated in Figure 5.7.

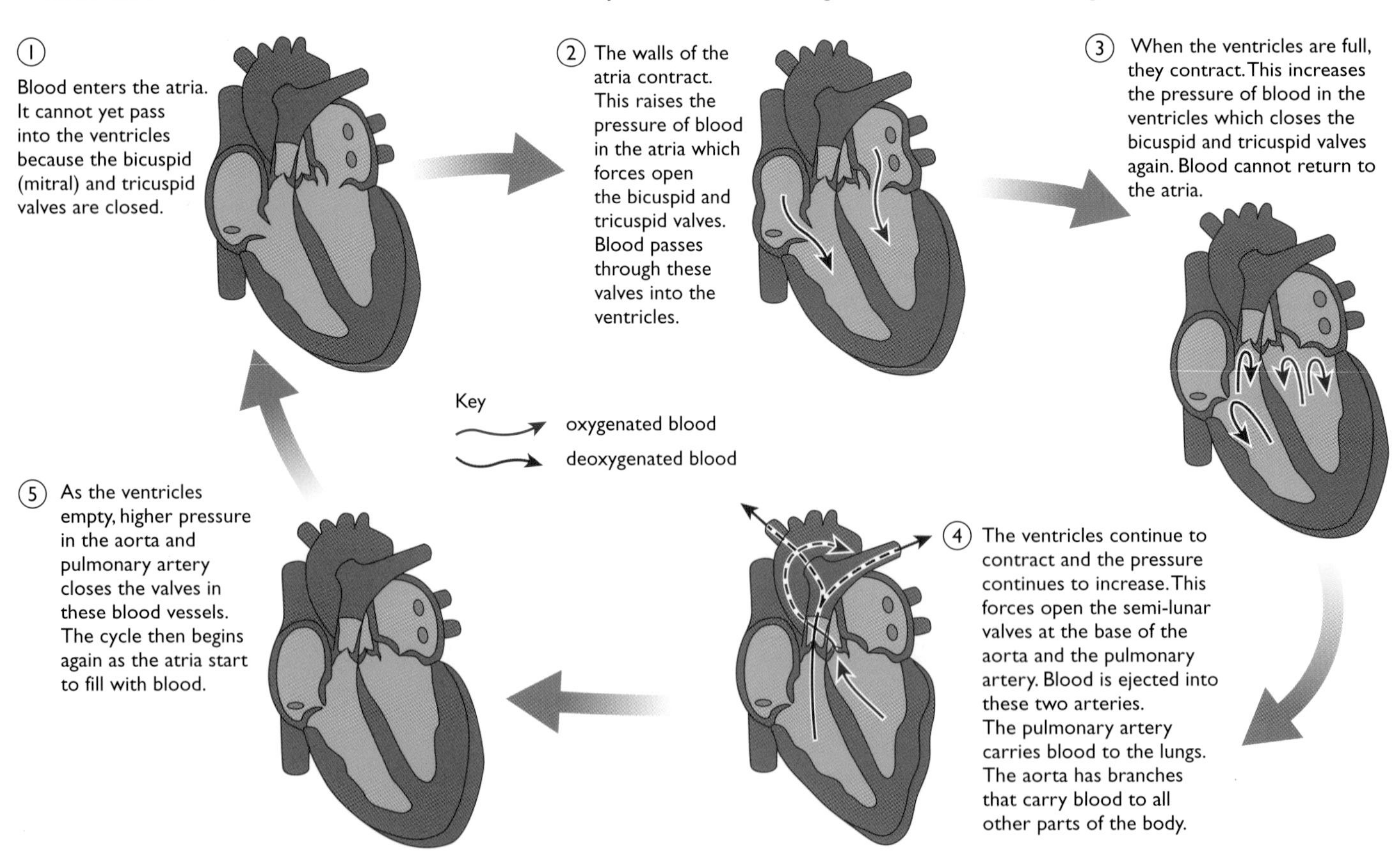

Figure 5.7 *The cardiac cycle.*

When a chamber of the heart is contracting, we say it is in **systole**. When it is relaxing, we say it is in **diastole**.

The structure of the heart is adapted to its function in several ways:

- It is divided into a left side and a right side by the **septum**. The right ventricle pumps blood only to the lungs while the left ventricle pumps blood to all other parts of the body. This requires much more pressure, which is why the wall of the left ventricle is much thicker than that of the right ventricle.
- Valves ensure that blood can flow only in one direction through the heart.
- The walls of the atria are thin. They can be stretched to receive blood as it returns to the heart but can contract with enough force to push blood through the bicuspid and tricuspid valves into the ventricles.
- The walls of the heart are made of cardiac muscle which can contract and then relax continuously, without becoming fatigued.
- The cardiac muscle has its own blood supply – the **coronary circulation**. Blood reaches the muscle via **coronary arteries**. These carry blood to capillaries that supply the heart muscle with oxygen and nutrients. Blood is returned to the right atrium via **coronary veins**.

Heart rate

Normally the heart beats about 70 times a minute, but this can change according to circumstances. When we exercise, muscles must release more energy. They need an increased supply of oxygen for aerobic respiration (see Chapter 1). To deliver the extra oxygen, both the number of beats per minute (heart rate) and the volume of blood pumped with each beat (called stroke volume) increase.

When we are stressed (angry or afraid) – our heart rate again increases. The increased output supplies extra blood to the muscles, enabling them to release extra energy through aerobic respiration. This allows us to fight or run away and is called the 'fight or flight' response. It is triggered by secretion of the hormone adrenaline from the adrenal glands (see Chapter 7).

When we sleep, our heart rate decreases as all our organs are working more slowly. They need to release less energy and so need less oxygen.

These changes in the heart rate are brought about by nerve impulses from a part of the brain called the **medulla** (Figure 5.9). When we start to exercise, our muscles produce more carbon dioxide in aerobic respiration. Sensors in the aorta and the carotid artery (the artery leading to the head) detect this increase. They send nerve impulses to the medulla. The medulla responds by sending nerve impulses along the accelerator nerve. When carbon dioxide production returns to normal, the medulla receives fewer impulses. It responds by sending nerve impulses along a decelerator nerve.

The accelerator nerve increases the heart rate. It also causes the heart to beat with more force and so increases blood pressure. The decelerator nerve decreases the heart rate. It also reduces the force of the contractions. Blood pressure then returns to normal.

Atria is the plural of **atrium**.

Coronary heart disease
The coronary arteries are among the narrowest in the body. They are easily blocked by a build-up of fatty substances (including **cholesterol**) in their walls. This can cut off the blood supply to an area of cardiac muscle. The affected muscle is unable to contract and a heart attack results.

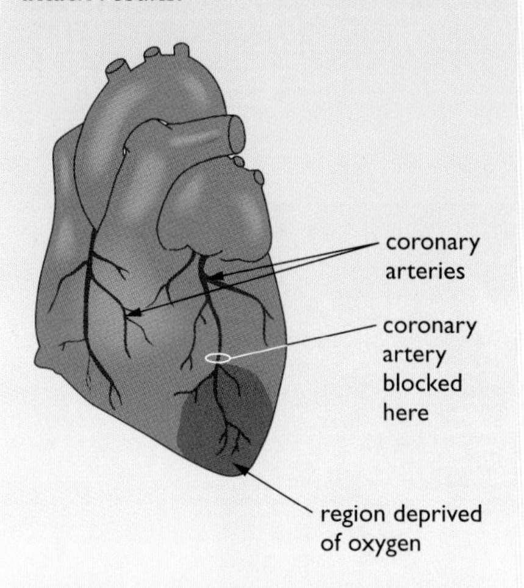

Figure 5.8 *A blockage of a coronary artery cuts off the blood supply to part of the heart muscle.*

A number of factors make coronary heart disease more likely:

- Heredity – some people inherit a tendency to coronary heart disease.
- High blood pressure – puts more strain on the heart.
- Diet – eating more saturated fat is likely to raise cholesterol levels.
- Smoking – raises blood pressure and makes blood clots more likely to form.
- Stress – raises blood pressure.
- Lack of exercise – regular exercise helps to reduce blood pressure and strengthens the heart.

Butterflies in your stomach? Have you noticed a 'hollow' feeling in your stomach when you are anxious? There are no butterflies involved! It happens because blood that would normally flow to your stomach and intestines has been diverted to the muscles to allow the 'fight or flight' response.

The precise region of the medulla that controls heart functions is called the **cardiac centre**.

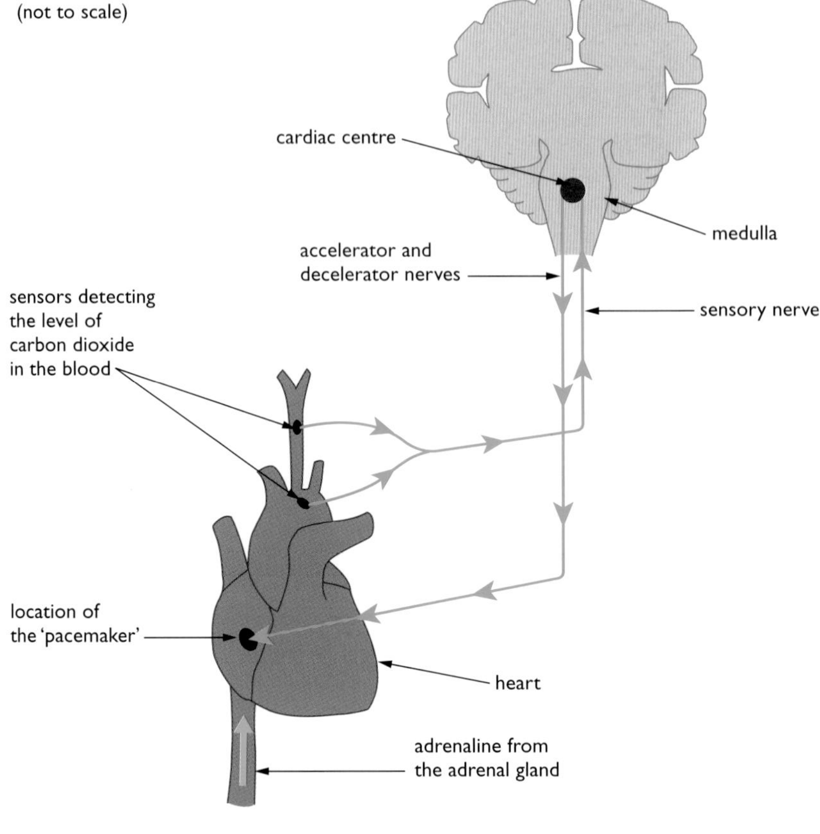

Figure 5.9 *How the heart rate is controlled.*

These controls are both examples of **reflex actions** (see Chapter 6).

Figure 5.10 *As we exercise, the circulatory system works harder.*

ideas evidence

A sustained high blood pressure when *not* exercising can be dangerous in many ways:

- it is a risk factor in the development of coronary heart disease (see page 55)
- it can damage the capillaries in the glomeruli (in the kidney), making filtration of the blood less effective and possibly leading to renal (kidney) failure
- it can cause delicate blood vessels in the brain to burst, causing a **stroke**.

Sustained high blood pressure can be a result of stress or an inappropriate diet. Some people develop high blood pressure as a result of too much salt (sodium chloride) in their diet. Other people do not seem to be affected in this way.

Exercise and the circulatory system

As soon as we start to exercise, our muscle cells respire faster to release more energy and produce more carbon dioxide. This triggers the reflex action to increase heart rate and blood pressure. More intense exercise results in a greater increase in heart rate and blood pressure.

During a period of exercise, the heart rate and blood pressure increase to a maximum to deliver the extra oxygen needed by the muscles. They remain at this level during the period of exercise, then begin to decrease to the pre-exercise levels as soon as the exercise ends. Figure 5.11 shows these changes.

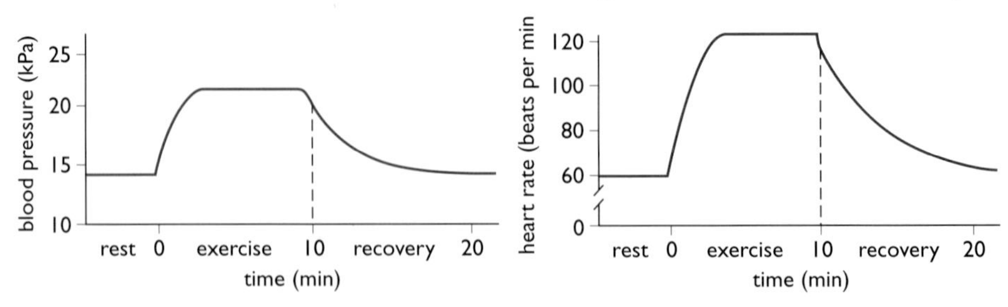

Figure 5.11 *How heart rate and blood pressure change during exercise.*

The period when heart rate and blood pressure are returning to normal following exercise is called the **recovery period**. They do not drop straight away to pre-exercise levels, but decrease gradually during this period. This is because, during exercise, lactic acid is formed in the muscles by anaerobic respiration (see Chapter 1). To get rid of lactic acid it must be oxidised, so as long as there is any lactic acid left in the muscles, extra oxygen will be needed to remove it. The heart must beat faster and stronger to deliver this extra oxygen. As the amount of lactic acid drops, so do the heart rate and blood pressure. When all the lactic acid has been oxidised, both return to pre-exercise levels. Figure 5.12 shows the changes in the heart rate and level of lactic acid in the blood following exercise.

The length of recovery time is a rough measure of how healthy your heart is. If the heart can pump enough oxygen to the muscles during exercise, only a little lactic acid will be formed and the recovery period will be short.

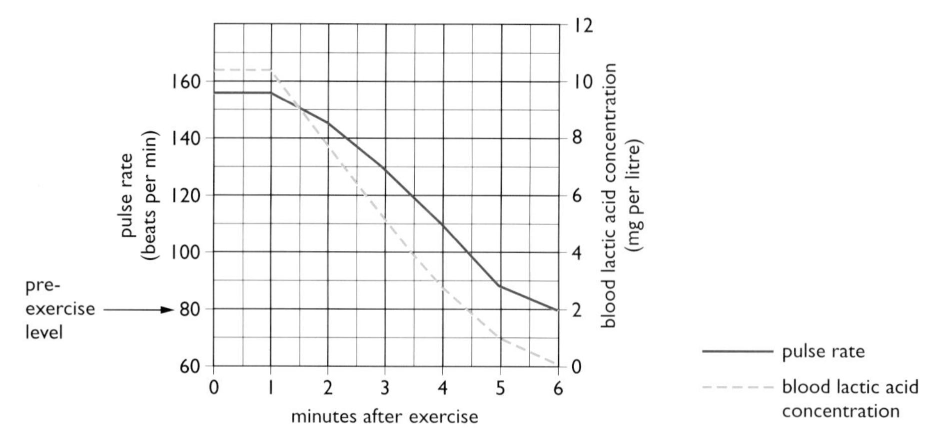

Figure 5.12 *The changes in pulse rate and lactic acid levels in the period following exercise.*

Arteries, veins and capillaries

Arteries carry blood from the heart to organs of the body. This blood (**arterial blood**) has been pumped out by the ventricles and puts a lot of pressure on the walls of the arteries. They must be able to 'give' under the pressure and allow their walls to stretch. They must also have the ability to recoil (pull back into shape) and help to push the blood along.

Veins carry blood from organs back towards the heart. The pressure of this blood (**venous blood**) is much lower than that in the arteries. It puts very little pressure on the walls of the veins. Veins must be able to allow the blood to pass through easily and prevent it from flowing in the wrong direction. Figure 5.13 shows the structure of a typical artery and a typical vein with the same diameter.

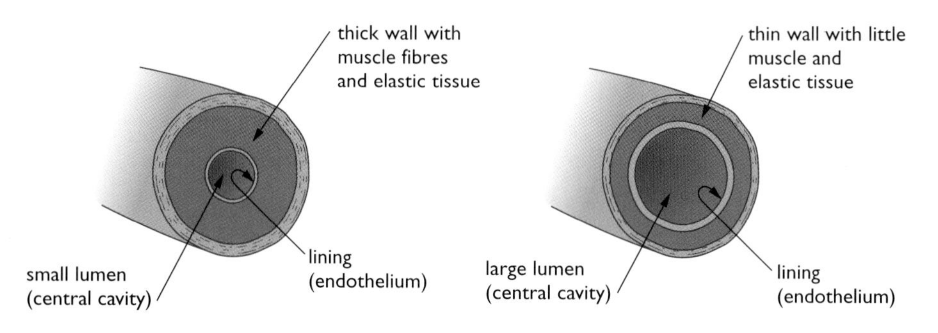

Figure 5.13 *The structure of (a) an artery and (b) a vein as seen in cross section.*

Arterioles are small arteries. They carry blood into organs from arteries. Their structure is similar to the larger arteries, but they have a larger proportion of muscle fibres in their walls. They are also **innervated** (have nerve endings in their walls) and so can be made to dilate (become wider) or constrict (become narrower) to allow more or less blood into the organ.

If *all* the arterioles constrict, it is harder for blood to pass through them – there is more resistance. This increases blood pressure. Prolonged stress can cause arterioles to constrict and so increase blood pressure.

All arteries carry **oxygenated blood** (blood containing a lot of oxygen) except the pulmonary artery and the umbilical artery of an unborn baby. All veins carry **deoxygenated blood** (blood containing less oxygen) except the pulmonary vein and umbilical vein.

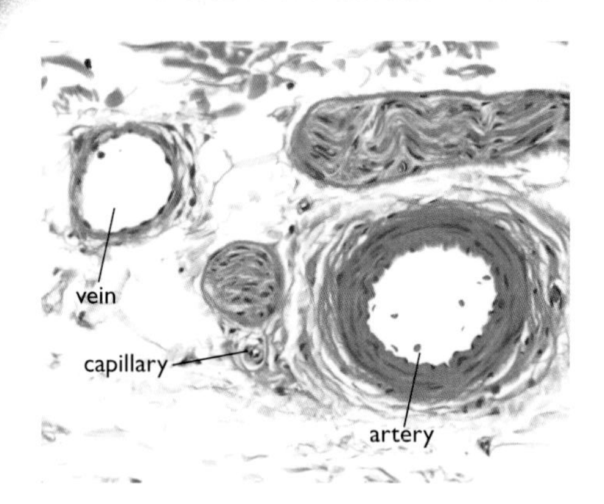

Figure 5.16 *Photograph of a section through an artery, vein and capillary.*

Veins also have valves called 'watch-pocket valves' which prevent the backflow of blood. The action of these valves is explained in Figure 5.14.

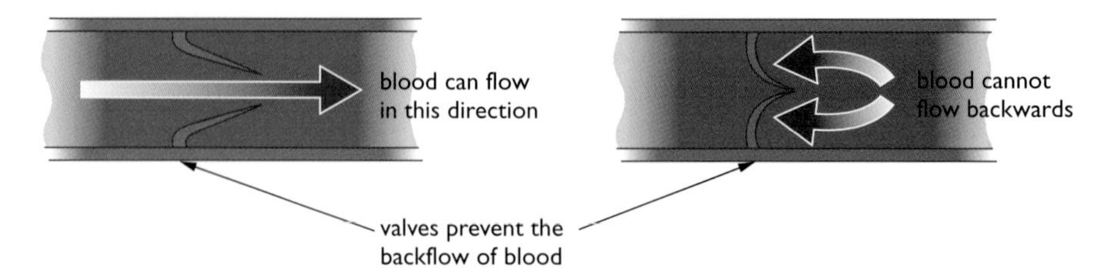

Figure 5.14 *The action of watch-pocket valves in veins.*

Capillaries carry blood through organs, bringing the blood close to every cell in the organ. Substances are transferred between the blood in the capillary and the cells. To do this, capillaries must be small enough to 'fit' between cells, and allow materials to pass through their walls easily. Figure 5.15 shows the structure of a capillary and how exchange of substances takes place between the capillary and nearby cells.

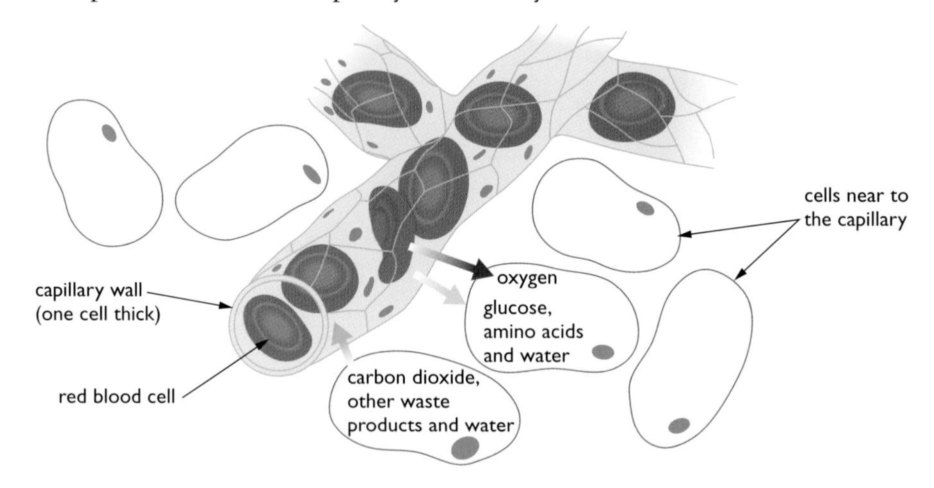

Figure 5.15 *How capillaries exchange materials with cells.*

ideas evidence

We owe much of our knowledge of the circulation of blood to William Harvey. Harvey was a doctor in the seventeenth century. He studied the action of the heart in many different animals and used these studies as a basis for studying the human heart and circulation. He was able to demonstrate that blood actually circulated around the body. Previously, many doctors believed that the heart pumped blood out, but that it was 'used up' in the tissues and remade in the liver. These beliefs were based on the incorrect ideas of a man called **Galen** who had lived fifteen hundred years earlier. Harvey relied only on scientific method. He observed the action of the heart in other animals. From these observations he suggested that the blood circulated around the human body. He performed a famous experiment to show that veins carry blood back towards the heart (Figure 5.18).

Bandaging the arm makes the veins and their valves stand out. Pushing a finger along a vein from one valve in a direction away from the heart causes that part of the vein to empty of blood. The blood cannot flow 'backwards' into the empty vein because the valve prevents it from doing so. Only on removing the finger does the empty portion refill with blood, showing that blood flows in one direction only in veins – towards the heart. You could carry out an Internet search to find out more about Harvey.

Figure 5.17 *William Harvey (1587–1657).*

Figure 5.18 *A drawing from Harvey's book* 'The circulation of the blood' *showing the action of valves in veins.*

The composition of blood

Blood is a lot more than just a red liquid flowing through your arteries and veins! In fact, blood is a complex tissue. Figure 5.19 illustrates the main types of cells found in blood.

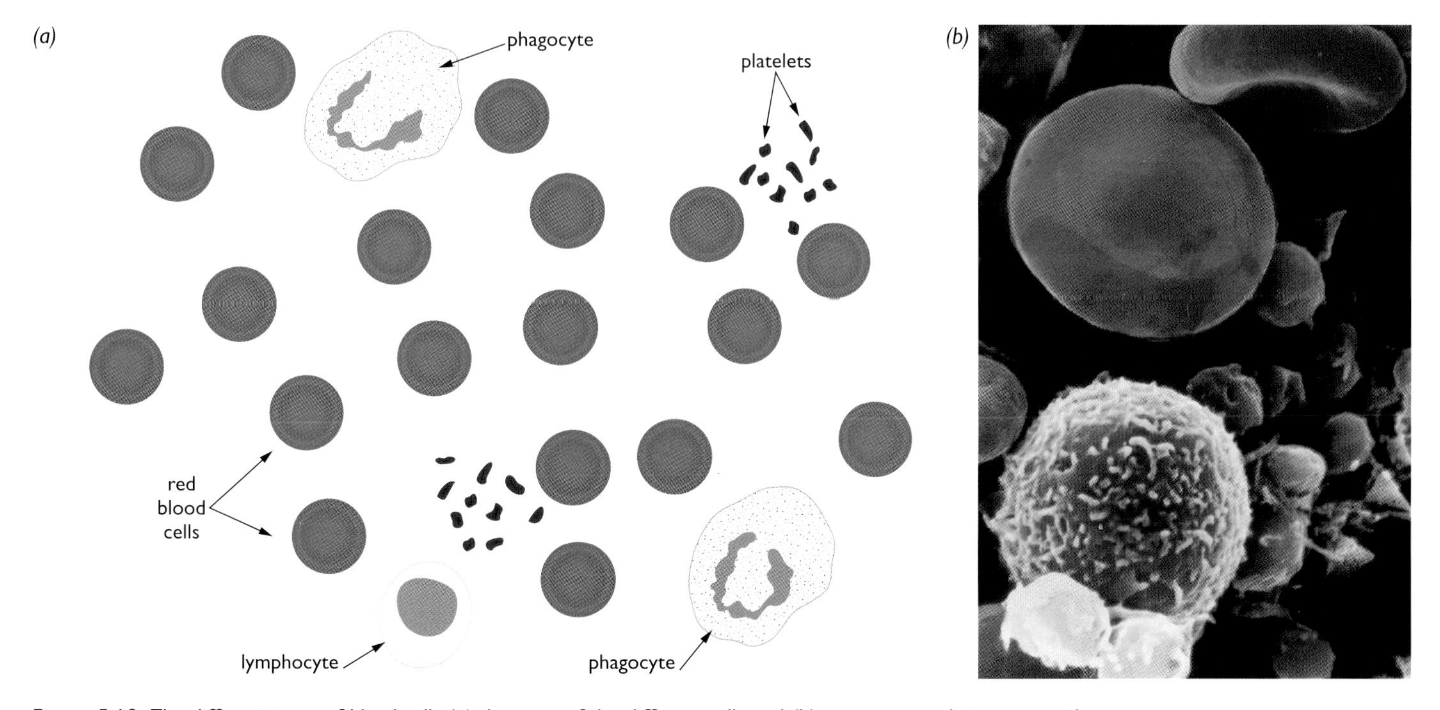

Figure 5.19 *The different types of blood cells (a) drawings of the different cells and (b) as seen in a photomicrograph.*

The different parts of blood have different functions. These are described in Table 5.1.

Component of blood	Description of component	Function of component
plasma	liquid part of blood: mainly water	carries the blood cells around the body; carries dissolved nutrients, hormones, carbon dioxide and urea; also distributes heat around the body
red blood cells	biconcave, disc-like cells with no nucleus; millions in each mm^3 of blood	transport of oxygen – contain mainly haemoglobin, which loads oxygen in the lungs and unloads it in other regions of the body
white blood cells:		
lymphocytes	about the same size as red cells with a large spherical nucleus	produce antibodies to destroy microorganisms – some lymphocytes persist in our blood after infection and give us immunity to specific diseases
phagocytes	much larger cells with a large spherical or lobed nucleus	engulf bacteria and other microorganisms that have infected our bodies
platelets	the smallest cells – are really fragments of other cells	release chemicals to make blood clot when we cut ourselves

Table 5.1: *Functions of the different components of blood.*

The red blood cells are highly specialised cells made in the bone marrow. They have a limited life span of about 100 days after which time they are destroyed in the spleen. They have only one function – to transport oxygen. Several features enable them to carry out this function very efficiently.

Red blood cells contain **haemoglobin.** This is an iron-containing protein that associates (combines) with oxygen to form **oxyhaemoglobin** when there is a high concentration of oxygen in the surroundings. We say that the red blood cell is *loading* oxygen. When the concentration of oxygen is low, oxyhaemoglobin turns back into haemoglobin and the red blood cell *unloads* its oxygen.

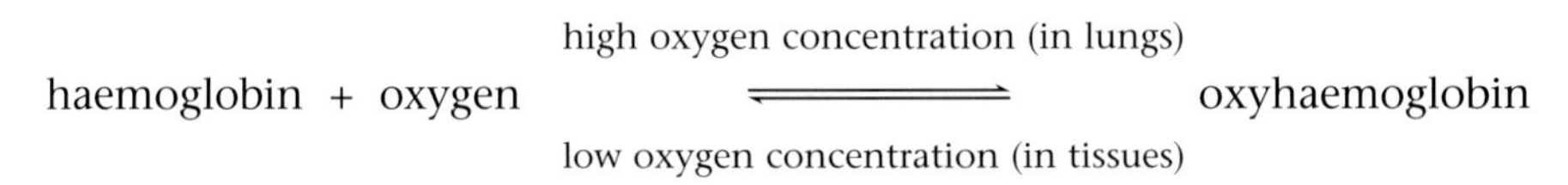

As red blood cells pass through the lungs, they load oxygen. As they pass through active tissues they unload oxygen.

Red blood cells do not contain a nucleus. It is lost during their development in the bone marrow. This means that more haemoglobin can be packed into each red blood cell so more oxygen can be transported. Their biconcave shape allows efficient exchange of oxygen in and out of the cell. Each red blood cell has a high surface area to volume ratio, giving a large area for diffusion. The thinness of the cell gives a short diffusion distance to the centre of the cell. In addition, red blood cells have very thin cell surface membranes which allow oxygen to diffuse through easily. The role of white blood cells in immunity is described in chapter 25.

End of Chapter Checklist

If you haven't got a copy of your specification, read the introduction on page vi.

You will need to be able to do some or all of the following. Check your Awarding Body's specification (syllabus) to find out exactly what you need to know.

- Describe the main components of the human circulatory system.
- Describe the differences between single and double circulatory systems and explain the advantages of a double circulatory system.
- Identify and describe the functions of the main structures in a human heart.
- Describe the events of the cardiac cycle.
- Explain how the structure of the heart adapts it to its function.
- Describe how the heart rate and force of the heart beat can be varied.
- Describe and explain the effects of exercise on heart function.
- Understand the importance of the work of William Harvey.
- Describe the structure of arteries, veins and capillaries and be able to relate the structure of each to its function.
- Describe the structure and function of the different types of blood cells.

Questions

More questions on blood and circulation can be found at the end of Section B on page 119.

1 Some animals have a single circulatory system, some have a double circulatory system and some organisms have no circulatory system at all.

a) Name one type of animal with a single circulatory system and one type of animal with a double circulatory system.

b) Explain:

i) the difference between single and double circulatory systems

ii) why a double circulatory system is more efficient than a single circulatory system.

c) Explain why single-celled organisms do not need a circulatory system.

2 Blood transports oxygen and carbon dioxide around the body. Oxygen is transported by the red blood cells.

a) Give three ways in which a red blood cell is adapted to its function of transporting oxygen.

b) Describe how oxygen:

i) enters a red blood cell from the alveoli in the lungs

ii) passes from a red blood cell to an actively respiring muscle cell.

c) Describe how carbon dioxide is transported around the body.

3 Blood is carried around the body in arteries, veins and capillaries.

a) Describe two ways in which the structure of an artery is adapted to its function.

b) Describe three differences between arteries and veins.

c) Describe two ways in which the structure of a capillary is adapted to its function.

4 The diagram shows a section through a human heart.

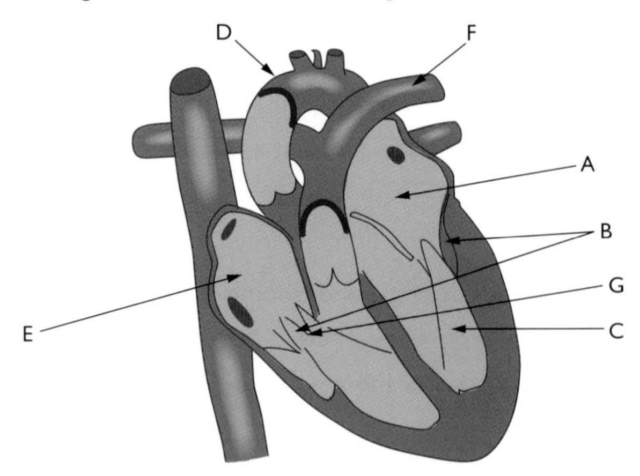

a) Name the structures labelled A, B, C, D and E.

b) What is the importance of the structures labelled B and F?

c) Which letters represent the chambers of the heart to which blood returns:

i) from the lungs

ii) from all the other organs of the body.

5 The diagram shows three types of cells found in human blood.

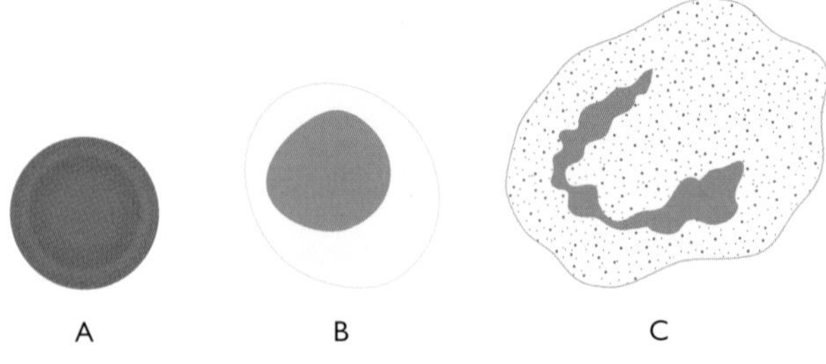

a) Giving a reason for each answer, identify the blood cell which:

i) transports oxygen around the body

ii) produces antibodies to destroy bacteria

iii) engulfs and digests bacteria.

b) Name one other component of blood found in the plasma and state its function.

6 The graph shows changes in a person's heart rate over a period of time.

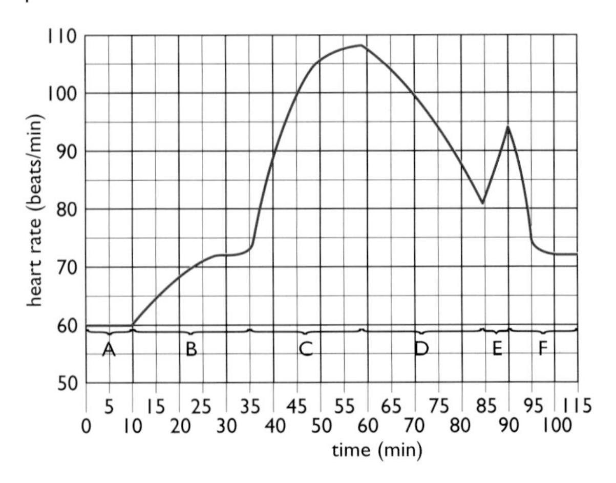

Giving reasons for your answers, give the letter of the time period when the person was

a) running

b) frightened by a sudden loud noise

c) sleeping

d) waking.

7 The graph shows the changes that take place in heart rate before, during and after a period of exercise.

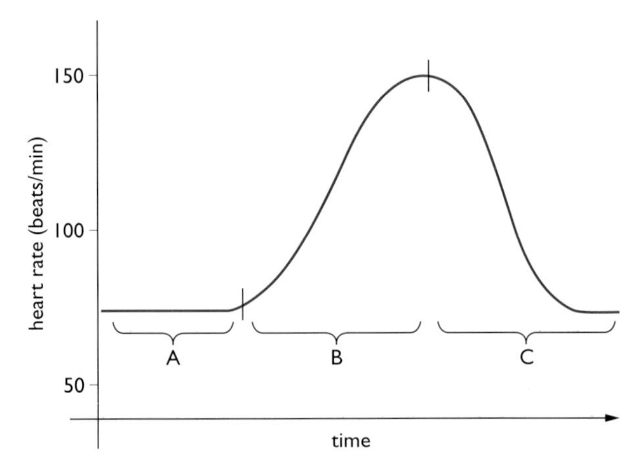

a) Describe and explain the heart rates found.

i) at rest, before exercise (period A)

ii) as the person commences the exercise (period B)

iii) as the person recovers from the exercise (period C).

b) How can the recovery period (period C) be used to assess a person's fitness?

Chapter 6: Coordination

In the body 'coordination' means making things happen at the right time by linking up different body activities. Humans and other animals have two organ systems which do this. The first is the nervous system, which is the subject of this chapter. The second is the hormone or endocrine system, which is dealt with in Chapter 7.

Stimulus and response

Suppose you are walking along when you see a football coming at high speed towards your head. If your nerves are working properly, you will probably move or duck quickly to avoid contact. Imagine another situation where you are very hungry, and you smell food cooking. Your mouth might begin to 'water', in other words secrete saliva.

Each of these situations is an example of a **stimulus** and a **response**. A *stimulus* is a change in an animal's surroundings, and a *response* is a reaction to that change. In the first example, the approaching ball was the stimulus, and your movement to avoid it hitting you was the response. The change in your environment was detected by your eyes, which are an example of a **receptor** organ. The response was brought about by contraction of muscles, which are an **effector** organ (they produce an effect). Linking the two is the nervous system, an example of a coordination system. A summary of the sequence of events is:

stimulus → receptor → coordination → effector → response

In the second example, the receptor for the smell of food was the nose, and the response was secretion of saliva from glands. Glands secrete (release) chemical substances, and they are the second type of effector organ. Again, the link between the stimulus and the response is the nervous system. The information in the nerve cells is transmitted in the form of tiny electrical signals called nerve **impulses**.

Receptors

The role of any receptor is to detect the stimulus by changing its energy into the electrical energy of the nerve impulses. For example, the eye converts light energy into nerve impulses, and the ear converts sound energy into nerve impulses. When energy is changed from one form into another, this is called **transduction**. All receptors are **transducers** of energy (Table 6.1).

Receptor	Type of energy transduced
eye (retina)	light
ear (organ of hearing)	sound
ear (organ of balance)	movement (kinetic)
tongue (taste buds)	chemical
nose (organs of smell)	chemical
skin (touch/pressure/pain receptors)	movement (kinetic)
skin (temperature receptors)	heat
muscle (stretch receptors)	movement (kinetic)

Table 6.1: *Human receptors and the energy they transduce into electrical impulses.*

Notice how a 'sense' like touch is made up of several components. When we touch a warm surface we will be stimulating several types of receptor, including

Some animals can detect changes in their environment that are not sensed by humans. Insects such as bees can see ultraviolet (UV) light. The wavelengths of UV are invisible to humans (Figure 6.1).

Figure 6.1 *This yellow flower (a) looks very different to a bee, which sees patterns on the petals reflecting UV light (b).*

Some organisms can even detect the direction of magnetic fields. Many birds, such as pigeons, have a built-in compass in their brain, which they use for navigation. A species of bacterium can also do this, but as yet no one can explain why this might be an advantage to it!

touch and temperature receptors, as well as stretch receptors in the muscles (see section on skin in Chapter 9). As well as this, each sense detects different aspects of the energy it receives. For example, the ears don't just detect sounds, but different loudness and frequencies of sound, while the eye not only forms an image, but also detects intensity of light and in humans can tell the difference between different light wavelengths (colours). Senses tell us a great deal about changes in our environment.

The central nervous system

The biological name for a nerve cell is a **neurone**. The impulses that travel along a neurone are not an electric current, as in a wire. They are caused by movements of charged particles (ions) in and out of the neurone. Impulses travel at speeds between about 10 and 100 m/s, which is much slower than an electric current, but fast enough to produce a rapid response.

Impulses from receptors pass along nerves containing **sensory neurones**, until they reach the **brain** and **spinal cord**. These two organs are together known as the **central nervous system**, or **CNS** (Figure 6.2).

Figure 6.2 *The brain and spinal cord form the central nervous system. Cranial and spinal nerves lead to and from the CNS. The CNS sorts out information from the senses and sends messages to muscles.*

The CNS is well protected by the skeleton. The brain is encased in the skull or **cranium** (nerves connected to the brain are **cranial** nerves) and the spinal cord runs down the middle of the spinal column, passing through a hole in each vertebra. Nerves connected to the **spinal** cord are called spinal nerves.

Other nerves contain **motor neurones**, transmitting impulses to the muscles and glands. Some nerves contain only sensory or motor cells, while other nerves contain both – they are 'mixed'. A typical nerve contains thousands of individual neurones.

Both sensory and motor neurones can be very long. For instance, a motor neurone leading from the CNS to the muscles in the finger has a fibre about 1 m in length, which is 100 000 times the length of the **cell body** (Figure 6.3).

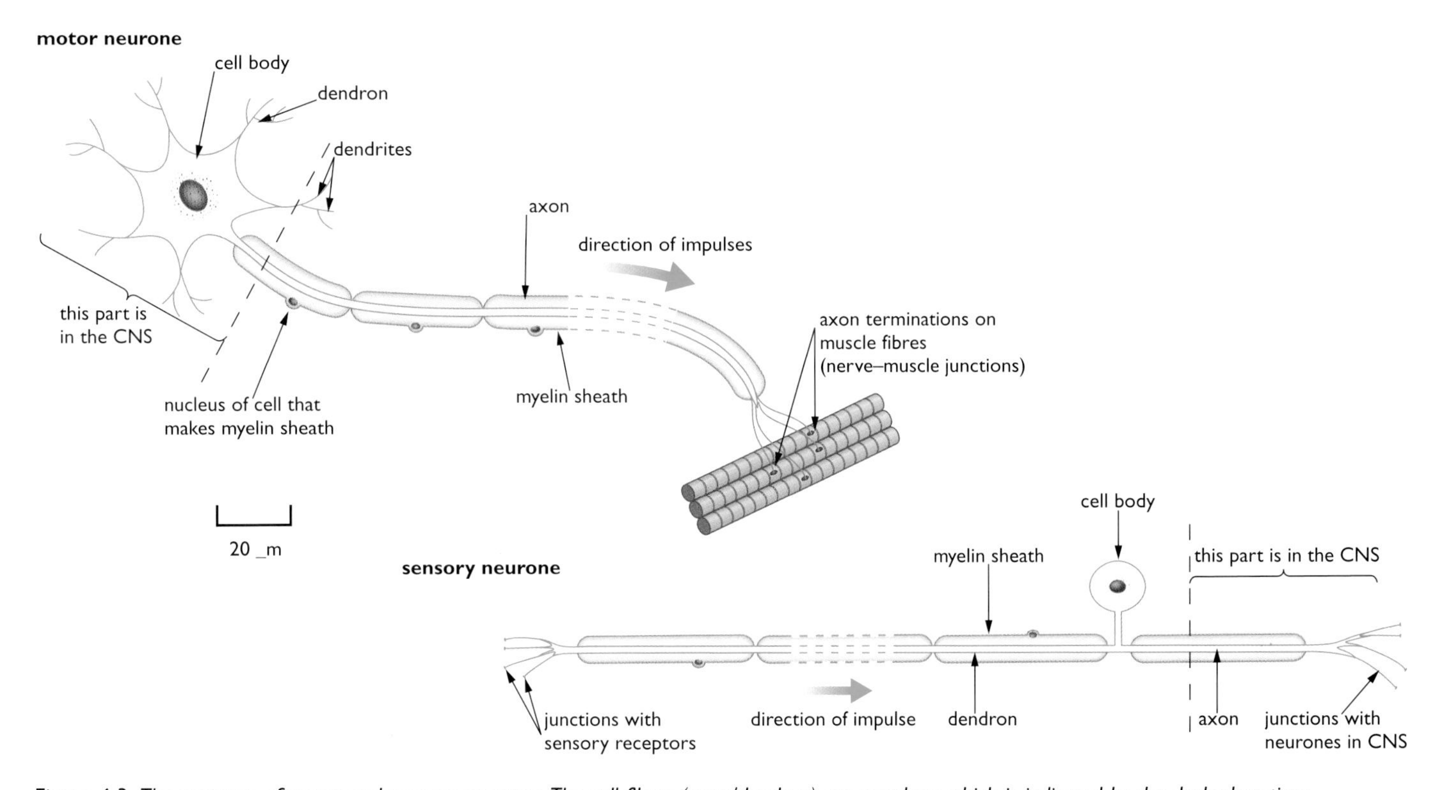

Figure 6.3 *The structure of motor and sensory neurones. The cell fibres (axon/dendron) are very long, which is indicated by the dashed sections.*

The cell body of a motor neurone is at one end of the fibre, in the CNS. The cell body has fine cytoplasmic extensions, called **dendrons**. These in turn form finer extensions, called **dendrites**. There can be junctions with other neurones on any part of the cell body, dendrons or dendrites. These junctions are called **synapses**. Later in this chapter we will deal with the importance of synapses in nerve pathways. One of the extensions from the motor neurone cell body is much longer than the other dendrons. This is the fibre that carries impulses to the effector organ, and is called the **axon**. At the end of the axon furthest from the cell body, it divides into many nerve endings. These fine branches of the axon connect with a muscle at a special sort of synapse called a **nerve-muscle junction**. In this way impulses are carried from the CNS out to the muscle. The signals from nerve impulses are transmitted across the nerve-muscle junction, causing the muscle fibres to contract. The axon is covered by a **sheath** made of a fatty material called **myelin**. The myelin sheath insulates the axon, preventing 'short circuits' with other axons, and also speeds up the conduction of the impulses. The sheath is formed by the membranes of special cells that wrap themselves around the axon as it develops.

A **sensory neurone** has a similar structure to the motor neurone, but the cell body is located on a side branch of the fibre, just outside the CNS. The fibre from the sensory receptor to the cell body is actually a dendron, while the fibre from the cell body to the CNS is a short axon. As with motor neurones, fibres of sensory neurones are often myelinated.

The eye

Many animals have eyes, but few show the complexity of the human eye. Simpler animals such as snails use their eyes to detect light, but cannot form a proper image. Other animals, such as dogs, can form images but cannot distinguish colours. The human eye does all three. Of course it is not really the eye that 'sees' anything at all, but the brain that interprets the impulses from the eye. To find out how light from an object is converted into impulses representing an image, we need to look at the structure of this complex organ (Figure 6.4).

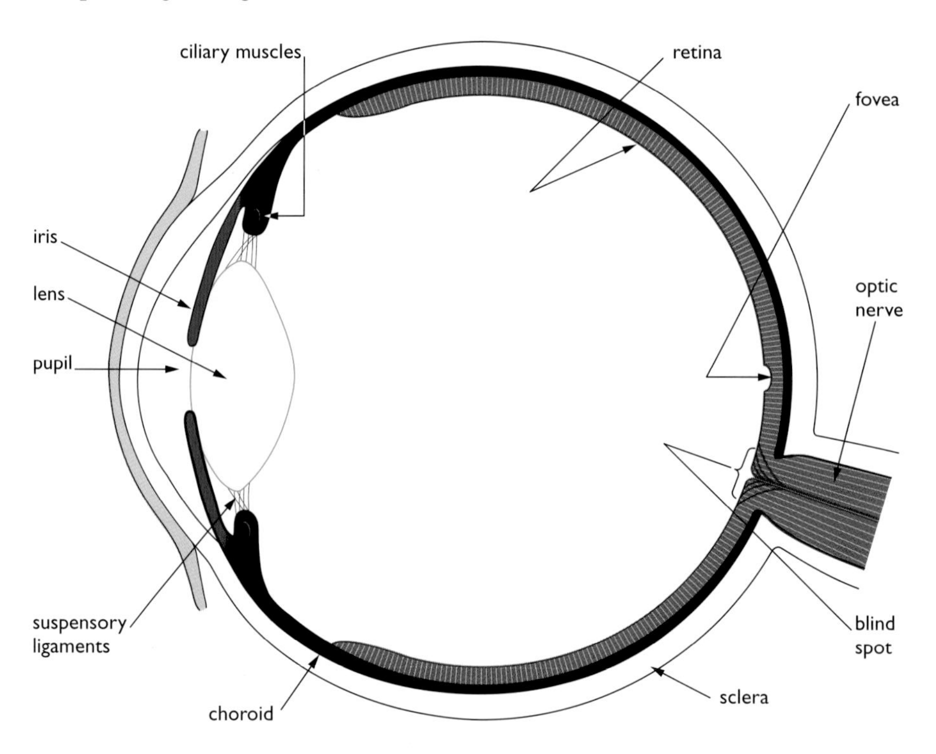

Figure 6.4 *A horizontal section through the human eye.*

The tough outer coat of the eye is called the **sclera**, which is the visible, white part of the eye. At the front of the eye the sclera becomes a transparent 'window' called the **cornea**, which lets light into the eye. Behind the cornea is the coloured ring of tissue called the **iris**. In the middle of the iris is a hole called the **pupil**, which lets the light through. It is black because there is no light escaping from the inside of the eye.

Underneath the sclera is a dark layer called the **choroid**. It is dark because it contains many pigment cells, as well as blood vessels. The pigment stops light being reflected around inside the eye. In the same way, the inside of a camera is painted matt black to stop stray light bouncing around and fogging the image on the film.

The innermost layer of the back of the eye is the **retina**. This is the light-sensitive layer, the place where light energy is transduced into the electrical energy of nerve impulses. The retina contains cells called **rods** and **cones**. These cells react to light, producing impulses in sensory neurones. The sensory neurones then pass the impulses to the brain through the **optic nerve.** Rod cells work well in dim light, but they cannot distinguish between different colours, so the brain 'sees' an image produced by the rods in black and white. This is why we can't see colours very well in dim light: only our rods are working properly. The cones, on the other hand, will only work in bright light, and there are three types which respond to different wavelengths or colours of light – red, green and blue. We can see all the colours of visible light as a result of these three types of cones being stimulated to a different degree. For example, if red, green and blue are stimulated equally, we see white. Both rods and cones are found throughout the retina, but cones are particularly concentrated at the centre of the retina, in an area called the **fovea**. Cones give a sharper image than rods, which is why we can only see objects clearly if we are looking directly at them, so that the image falls on the fovea.

To form an image on the retina, light needs to be bent or **refracted**. Refraction takes place when light passes from one medium to another of a different density. In the eye, this happens first at the air/cornea boundary, and again at the lens (Figure 6.5). In fact the cornea acts as the first lens of the eye.

The fact that the inverted image is seen the right way up by the brain makes the point that it is the brain which 'sees' things, not the eye. An interesting experiment was carried out to test this. Volunteers were made to wear special inverting goggles for long periods. These turned the view of their surroundings upside down. At first this completely disorientated them, and they found it difficult to make even simple coordinated movements. However, after a while their brains adapted, until the view through the goggles looked normal. In fact, when the volunteers removed the goggles, the world then looked upside down!

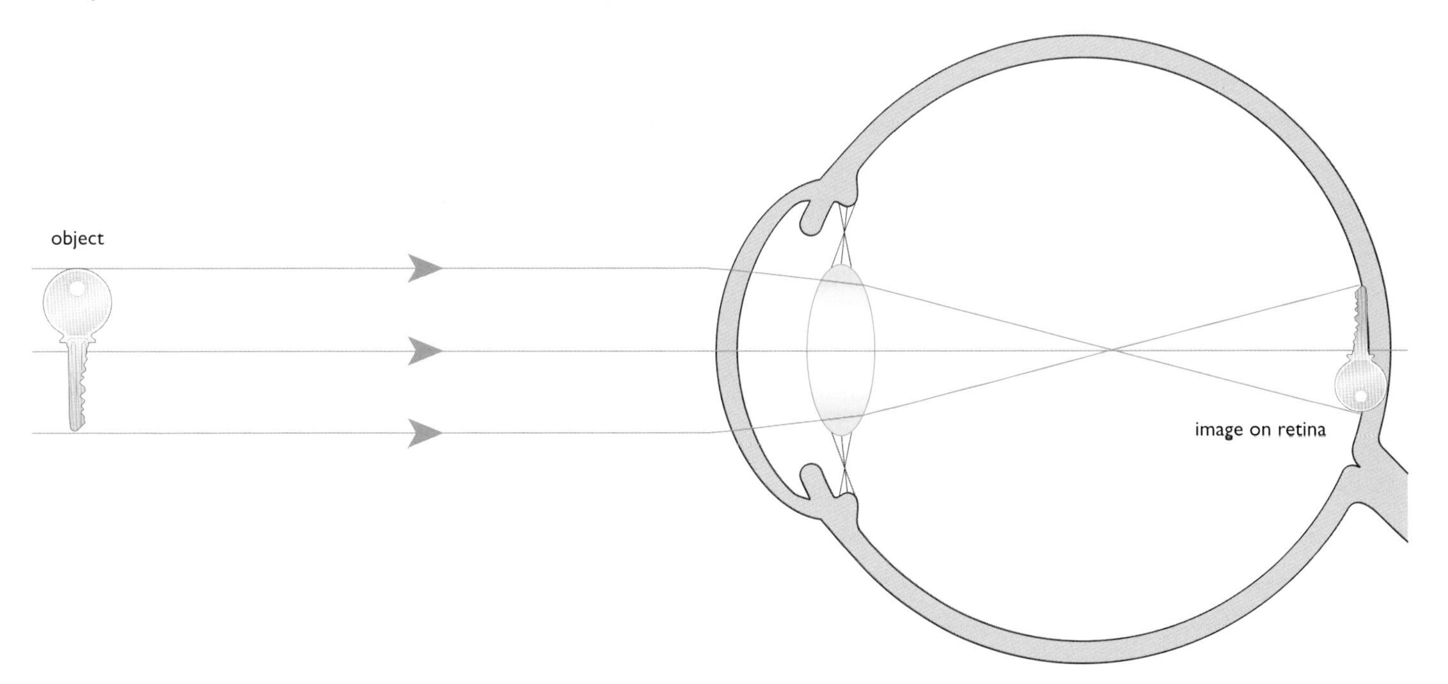

Figure 6.5 *How the eye forms an image. Refraction of light occurs at the cornea and lens, producing an inverted image on the retina.*

As a result of refraction at the cornea and lens, the image on the retina is upside down, or **inverted.** The brain interprets the image the right way up.

The role of the iris is to control the amount of light entering the eye, by changing the size of the pupil. The iris contains two types of muscles. **Circular muscles** form a ring shape in the iris, and **radial muscles** lie like the spokes of a wheel. In bright light, the pupil is made smaller, or **constricted**.

This happens because the circular muscles contract and the radial muscles relax. In dim light, the opposite happens. The radial muscles contract and the circular muscles relax, widening or **dilating** the pupil (Figure 6.6).

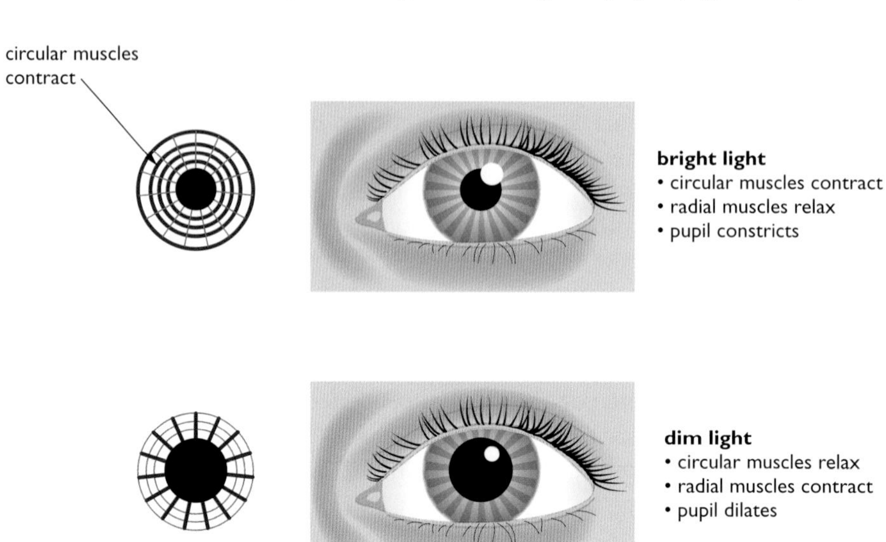

Figure 6.6 *The amount of light entering the eye is controlled by the iris, which alters the diameter of the pupil.*

In the iris reflex, the route from stimulus to response is this:

stimulus (light intensity)
↓
retina (receptor)
↓
sensory neurones in optic nerve
↓
unconscious part of brain
↓
motor neurones in nerve to iris
↓
iris muscles (effector)
↓
response (change in size of pupil)

Whenever our eyes look from a dim light to a bright one, the iris rapidly and automatically adjusts the pupil size. This is an example of a **reflex action**. You will find out more about reflexes later in this chapter. The purpose of the iris reflex is to allow the right intensity of light to fall on the retina. Light that is too bright could damage the rods and cones, and light that is too dim would not form an image. The intensity of light hitting the retina is the stimulus for this reflex. Impulses pass to the brain through the optic nerve, and straight back to the iris muscles, adjusting the diameter of the pupil. It all happens without the need for conscious thought – in fact we are not even aware of it happening.

There is one area of the retina where an image cannot be formed; this is where the optic nerve leaves the eye. At this position there are no rods or cones, so it is called the **blind spot**. The retina of each eye has a blind spot, but they are not a problem, because the brain puts the images from each eye together, cancelling out the blind spots of both eyes. As well as this, the optic nerve leaves the eye towards the edge of the retina, where vision is not very sharp anyway. To 'see' your own blind spot you can do a simple experiment. Cover or close your right eye. Hold this page about 30 cm from your eyes and look at the black dot below. Now, without moving the book or turning your head, read the numbers from left to right by moving your left eye slowly towards the right.

A way to prove to yourself that the eyes form two overlapping images is to try the 'sausage test'. Focus your eyes on a distant object. Place your two index fingers tip to tip, and bring them up in front of your eyes, about 30 cm from your face, while still focusing at a distance. You should see a finger 'sausage' between the two fingers. Now try this with one eye closed. What is the difference?

•   1   2   3   4   5   6   7   8   9   10   11   12   13   14   15

You should find that when the image of the dot falls on the blind spot it disappears. If you try doing this with both eyes open, the image of the dot will not disappear.

Accommodation

The changes that take place in the eye which allow us to see objects at different distances are called **accommodation**.

You have probably seen the results of a camera or projector not being in focus – a blurred picture. In a camera we can focus light from objects that are different distances away by moving the lens backwards or forwards, until the picture is sharp. In the eye a different method is used. Rather than altering its position, the shape of the lens can be changed. A lens that is fatter in the middle (more convex) will refract light rays more than a thinner (less convex) lens. The lens in the eye can change shape because it is made of cells containing an elastic crystalline protein.

Figure 6.4 shows that the lens is held in place by a series of fibres called the **suspensory ligaments**. These are attached like the spokes of a wheel to a ring of muscle, called the **ciliary muscle**. The inside of the eye is filled with a transparent watery fluid which pushes outwards on the eye. In other words there is a slight positive pressure within the eye. The changes to the eye which take place during accommodation are shown in Figure 6.7.

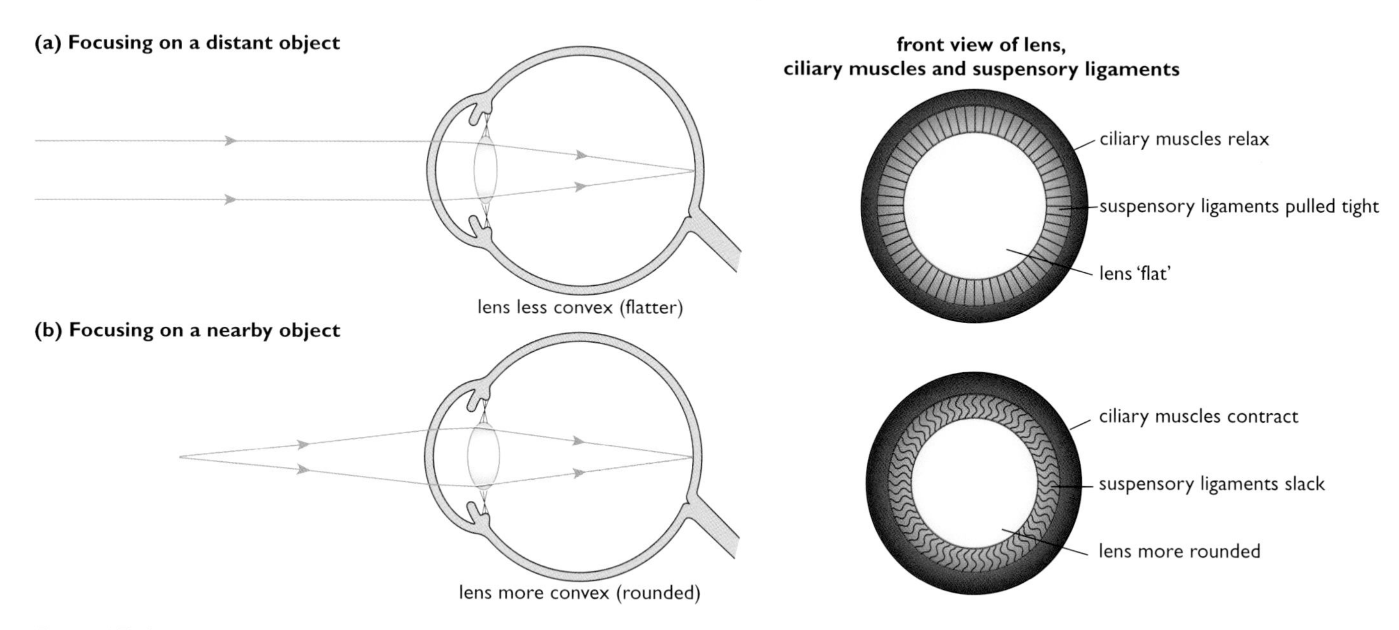

Figure 6.7 *Accommodation: how the eye focuses on objects at different distances.*

When the eye is focused on a distant object, the rays of light from the object are almost parallel when they reach the cornea (Figure 6.7a). The cornea refracts the rays, but the lens does not need to refract them much more to focus the light on the retina, so it does not need to be very convex. The ciliary muscles relax and the pressure in the eye pushes outwards on the lens, flattening it and stretching the suspensory ligaments. This is the condition when the eye is at rest – our eyes are focused for long distances.

When we focus on a nearby object, for example when reading a book, the light rays from the object are spreading out (diverging) when they enter the eye (Figure 6.7b). In this situation, the lens has to be more convex in order to refract the rays enough to focus them on the retina. The ciliary muscles now contract; the suspensory ligaments become slack and the elastic lens bulges outwards into a more convex shape.

A reflex action is a *rapid, automatic* (or *involuntary*) response to a stimulus. The action often (but not always) protects the body. Involuntary means that it is not started by impulses from the brain.

Reflex actions

You saw on page 68 that the dilation and constriction of the pupil by the iris is an example of a reflex action. You now need to understand a little more about the nerves involved in a reflex. The nerve pathway of a reflex is called the **reflex arc**. The 'arc' part means that the pathway goes into the CNS and then straight back out again, in a sort of curve or arc (Figure 6.8).

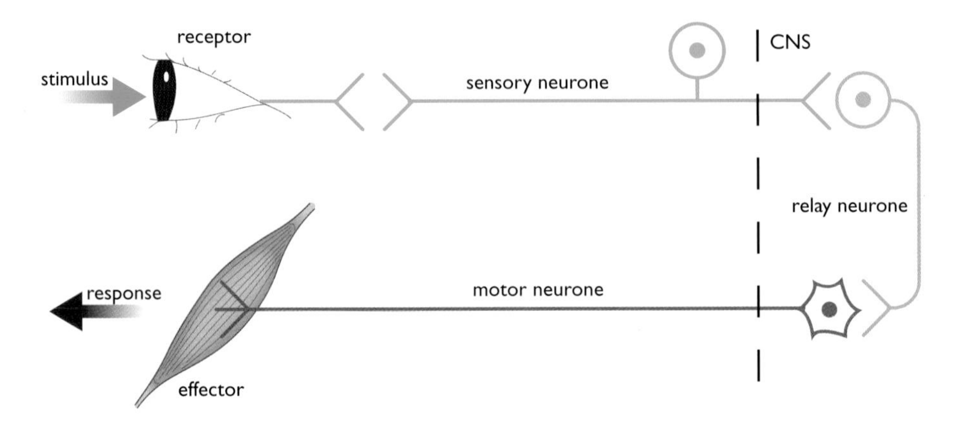

Figure 6.8 *Simplified diagram of a reflex arc.*

The iris–pupil reflex protects the eye against damage by bright light. Other reflexes are protective too, preventing serious harm to the body. Take for example the reflex response to a painful stimulus. This happens when part of your body, such as your hand, touches a sharp or hot object. The reflex results in your hand being quickly withdrawn. Figure 6.9 shows the nerve pathway of this reflex in more detail.

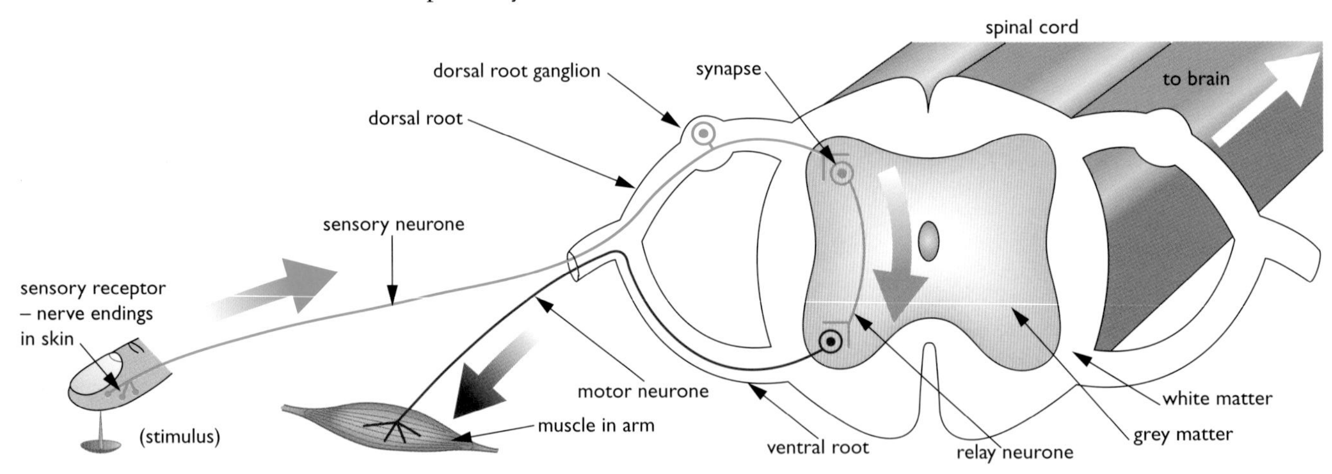

Figure 6.9 *A reflex arc in more detail.*

'Dorsal' and 'ventral' are words describing the back and front of the body. The dorsal roots of spinal nerves emerge from the spinal cord towards the back of the person, while the ventral roots emerge towards the front. Notice that the cell bodies of the sensory neurones are all located in a swelling in the dorsal root, called the **dorsal root ganglion**.

The stimulus is detected by temperature or pain receptors in the skin. These generate impulses in sensory neurones. The impulses enter the CNS through a part of the spinal nerve called the **dorsal root**. In the spinal cord the sensory neurones connect by synapses with short **relay neurones**, which in turn connect with motor neurones. The motor neurones emerge from the spinal cord through the **ventral root**, and send impulses back out to the muscles of the arm. These muscles then contract, pulling the arm away from the harmful stimulus.

The middle part of the spinal cord consists mainly of nerve cell bodies, which gives it a grey colour. This is why it is known as **grey matter**. The outer part of the spinal cord is called **white matter**, and has a whiter appearance because it contains many axons with their fatty myelin sheaths.

Impulses travel through the reflex arc in a fraction of a second, so that the reflex action is very fast, and doesn't need to be started off by impulses from the brain. However, this doesn't mean that the brain is unaware of what is going on. This is because in the spinal cord, the reflex arc neurones also form synapses with nerve cells leading to and from the brain. The brain therefore receives information about the stimulus. This is how we feel the pain.

Movements are sometimes a result of reflex actions, but we can also contract our muscles as a **voluntary action**, using nerve cell pathways from the brain linked to the same motor neurones. A voluntary action is under conscious control.

Synapses

Synapses are critical to the working of the nervous system. The CNS is made of many billions of nerve cells, and these have links with many others, through synapses. In the brain, each neurone may form synapses with thousands of other neurones, so that there are an almost infinite number of possible pathways through the system.

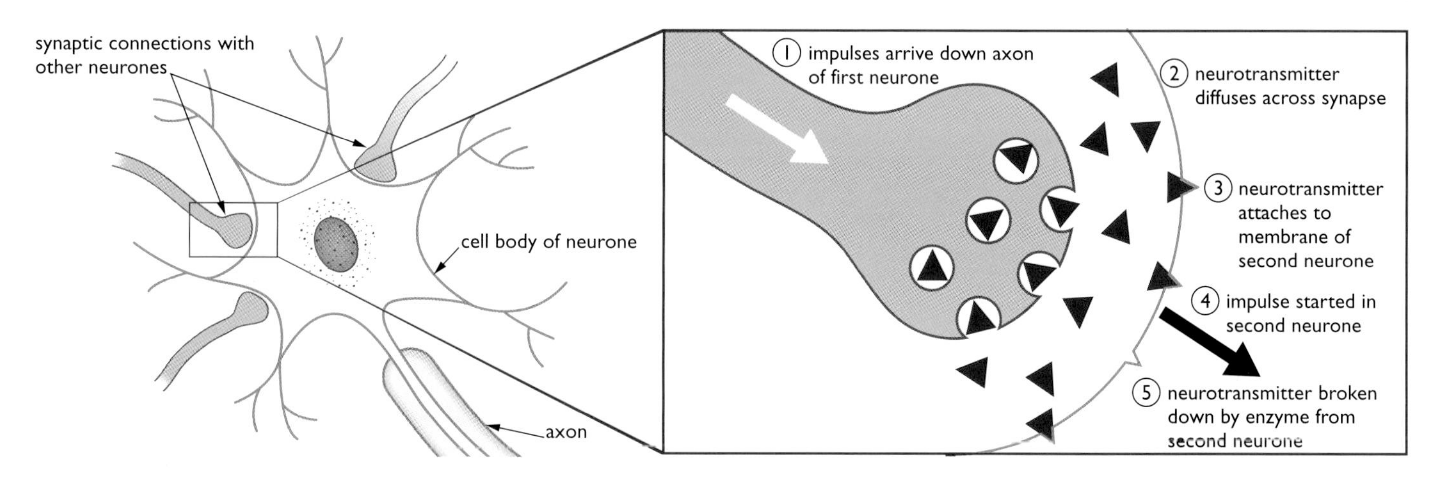

Figure 6.10 *The sequence of events happening at a synapse.*

A synapse is actually a gap between two nerve cells. The gap is not crossed by the electrical impulses passing through the neurones, but by chemicals. Impulses arriving at a synapse cause the ends of the fine branches of the axon to secrete a chemical, called a **neurotransmitter**. This chemical diffuses across the gap and attaches to the membrane of the second neurone. It then starts off impulses in the second cell (Figure 6.10). After the neurotransmitter has 'passed on the message', it is broken down by an enzyme.

Remember that many nerve cells, particularly those in the brain, have thousands of synapses with other neurones. The output of one cell may depend on the inputs from many cells adding together. In this way synapses are important for integrating information in the central nervous system (Figure 6.11).

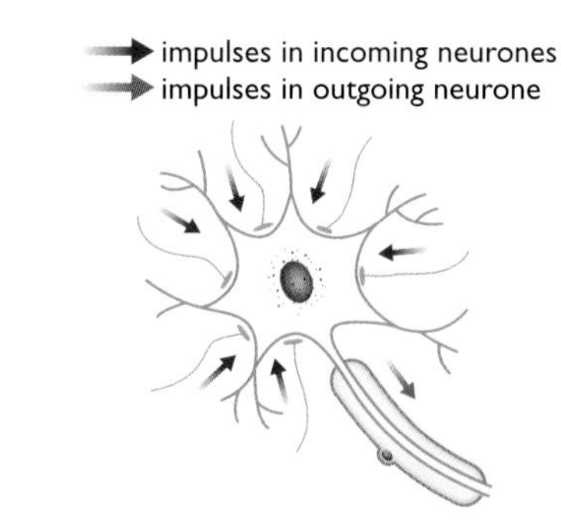

Figure 6.11 *Synapses allow the output of one nerve cell to be a result of integration of information from many other cells.*

Because synapses are crossed by chemicals, it is easy for other chemicals to interfere with the working of the synapse. They may imitate the neurotransmitter, or block its action. This is the way that many well-known drugs, both useful and harmful, work. We will return to this topic later.

The conditioned reflex

Humans and other complex animals show reflexes from birth. This ensures that the body can protect itself from harm from the beginning. Soon after it is born, a doctor or nurse will test some of a baby's reflexes to check that its nervous system is working properly. A baby shows some reflexes that are not protective but help it to survive in other ways. For example, suckling at its mother's breast is a reflex action (as is the release of milk by the mother in response to suckling).

However, there are some reflex actions that are not present at birth, but have to be learned. These are called **conditioned reflexes**, and are one of the simplest forms of learned behaviour. Conditioned reflexes were first investigated by the Russian biologist Ivan Pavlov (1849–1936).

Pavlov presented two stimuli to hungry dogs (Figure 6.12). The first stimulus was the sight of food. The dogs responded to this by salivating (producing saliva), which is a normal reflex action that prepares the animals for feeding. The second stimulus was the sound of a ringing bell. The dogs reacted to the bell by turning towards it and cocking their ears, but did not salivate. However, Pavlov found that he could train the dogs to salivate to the sound of the bell by ringing it whenever they were given food. After a period of time, they would salivate to the sound of the bell alone. The dogs had learned to associate the 'neutral' stimulus (the ringing bell) with food. This type of learning is called **conditioning**, and the response is the conditioned reflex.

The main difference between a conditioned reflex and a simple reflex is that a conditioned reflex is learned, whereas a simple reflex is **innate** (present from birth, or from when the body systems are developed enough to produce the reflex). With Pavlov's dogs, salivating at the sight of food was a simple reflex. Salivating to the sound of a bell had to be learned, so it is a conditioned one.

Conditioned responses like this are used to train animals. A puppy learns to associate words like 'sit' and 'heel' with the correct actions through training it, using a system of rewards. Much human behaviour is thought to be the result of conditioned reflexes. For example, very young children are not afraid of snakes, and will happily stroke a harmless snake. However, they often learn to dislike snakes, associating them with 'bad' as a result of the attitude of adults towards these animals.

Figure 6.12 *Pavlov's experiments on conditioned reflexes.*

The brain

The functions of different parts of the brain were first worked out through studies of people who had suffered brain damage through accident or disease. Nowadays we have very sophisticated electronic equipment that can record the activity in a normal living brain, but we are still relatively ignorant about the workings of this most complex organ of the body.

Your brain is sometimes called your 'grey matter'. This is because the positions of the grey and white matter are reversed in the brain compared with the spinal cord. The grey matter, mainly made of nerve cell bodies, is on the outside of the brain, and the axons that form the white matter are in the middle of the brain. The brain is made up of different parts, each with a particular function (Figure 6.13).

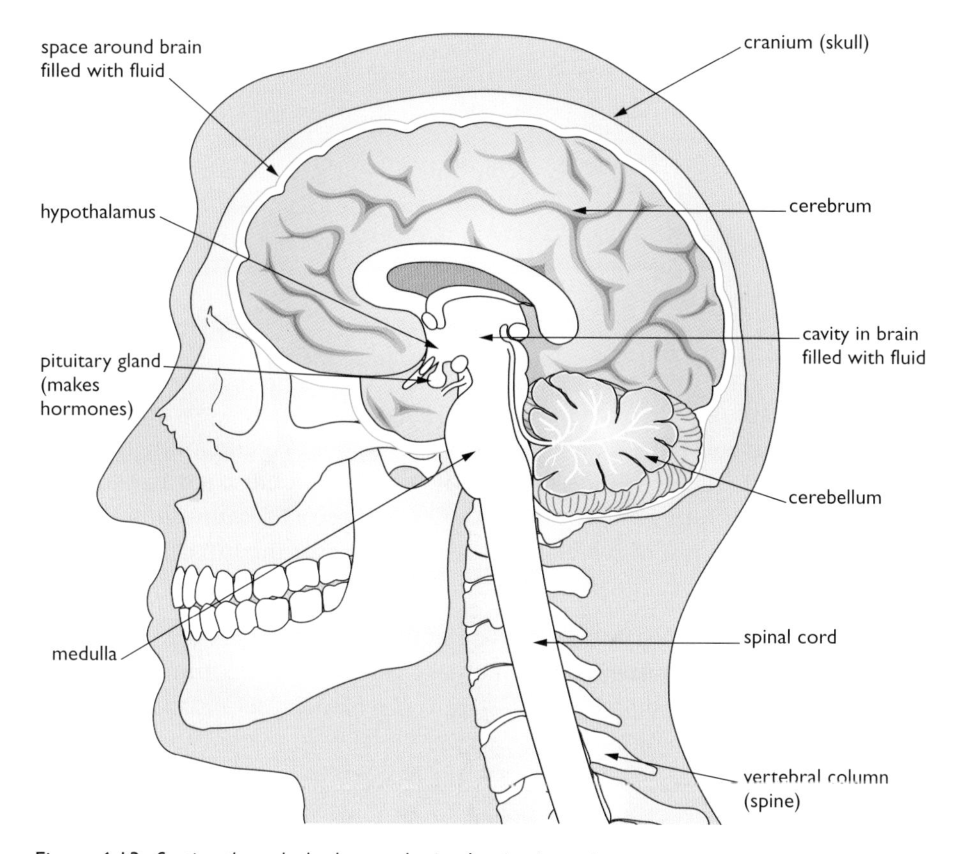

Figure 6.13 *Section through the human brain, showing its main parts.*

The largest part of the brain is the **cerebrum**, made of two **cerebral hemispheres**. The cerebrum is the source of all our conscious thoughts. It has an outer layer called the **cerebral cortex**, with many folds all over its surface (Figure 6.14).

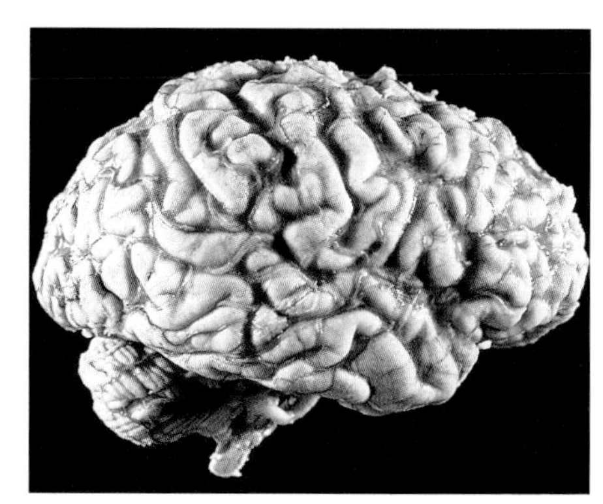

Figure 6.14 *A side view of a human brain. Notice the folded surface of the cerebral cortex.*

The cerebrum has three main functions:

- It contains **sensory areas** that receive and process information from all our sense organs.
- It has **motor areas** which are where all our voluntary actions originate.
- It is the origin of 'higher' activities, such as memory, reasoning, emotions and personality.

Different parts of the cerebrum carry out particular functions. For example, the sensory and motor areas are always situated in the same place in the cortex (Figure 6.15). Some parts of these areas deal with more information than others. Large parts of the sensory area deal with impulses from the fingers and lips, for example. This is illustrated in Figure 6.16.

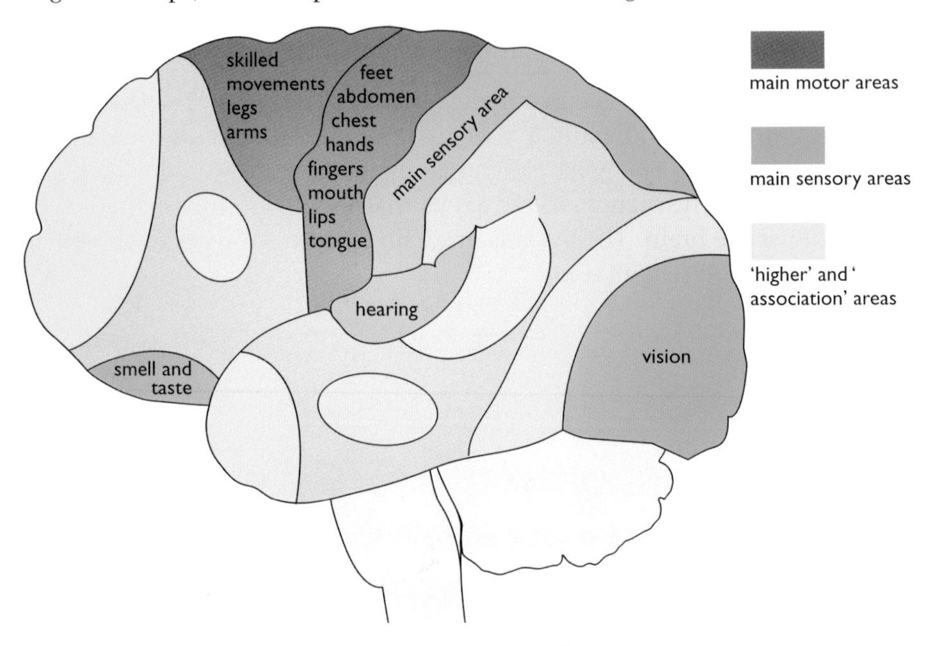

Figure 6.15 *Different parts of the cerebrum carry out specific functions.*

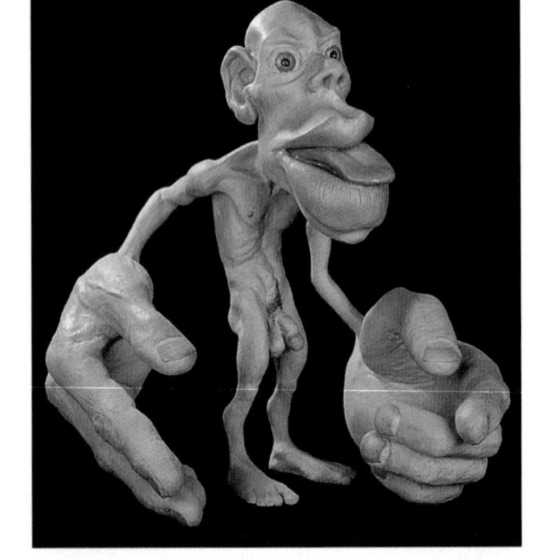

Figure 6.16 *A model of a human with its parts drawn in proportion to the amount of sensory information they send to the cortex of the brain (note that this does not apply to the eyes, which use more cortex than the rest of the body put together).*

Behind the cerebrum is the **cerebellum**. This region is concerned with coordinating the contraction of sets of muscles, as well as maintaining balance. This is important when you are carrying out complicated muscular activities, such as running or riding a bike. Underneath the cerebrum, connecting the spinal cord with the rest of the brain is the brain stem or **medulla**. This controls basic body activities such as heartbeat and breathing rate.

The **pituitary gland** is located at the base of the brain, just below a part of the brain called the **hypothalamus**. The pituitary gland secretes a number of chemical 'messengers' called **hormones**, into the blood. The pituitary and hypothalamus are both discussed in Chapter 7.

Drugs

Drugs are chemicals that affect processes in a person's body. Many drugs are useful. For example, penicillin is an antibiotic that kills many of the bacteria that cause disease, and aspirin is an effective painkiller. However, a number of drugs act by interfering with the nervous system, and some of these can have very harmful side effects. A good example of this is the drug **nicotine**, present in tobacco. When a person smokes a cigarette, especially the first time, they get a 'buzz' from smoking. Their heart beats faster, their blood pressure rises and they feel excited. This is because nicotine is a **stimulant**, meaning that it increases brain activity. It does this by mimicking the action of neurotransmitters at the synapses of nerve cells in the brain. Some other stimulants also affect synapses. **Caffeine**, the drug in tea and coffee, causes more neurotransmitter to be released than normal.

Nicotine is a highly **addictive** drug. A person who smokes develops a craving for the drug. On top of this, their body 'gets used to it' and so more must be taken into the body to satisfy the craving. When the person tries to stop smoking, they suffer **withdrawal symptoms**, such as becoming irritable. Nicotine, along with the other components of tobacco smoke, has a number of harmful effects on the lungs, heart and blood system (see Chapter 2).

Figure 6.17 *Alcohol in the bloodstream increases reaction times and is a common cause of car accidents.*

The opposite of a stimulant is a **depressant** drug. One example is **alcohol** (ethanol). The alcohol in beer, wine and spirits slows down the nervous system, even when drunk in small quantities, and increases the time a person takes to react to a stimulus. That is why driving after drinking alcohol is so dangerous. The driver will not react quickly to sudden danger, such as a person walking into the road (Figure 6.17). Larger amounts of alcohol in the body interfere with the drinker's balance and muscular control, and lead to blurred vision and slurred speech. High concentrations of alcohol in the blood can even cause coma and death.

Many people drink moderate amounts of alcohol to relax. However, to some people alcohol is an addictive drug, and long-term alcohol abuse leads to serious medical problems. Alcohol is quickly absorbed into the blood through the stomach and intestines and taken around the body. The liver breaks the alcohol down, but if a person drinks large amounts regularly, it may not be able to cope. The person can then get a disease called **cirrhosis** of the liver. The liver does not function properly and toxins in the blood build up to high levels. The disease is usually fatal. Alcohol also damages the brain and stomach lining.

Solvents are also depressants. They are liquids used to make industrial products like glue and paint. If a person inhales solvent vapours, they affect the nervous system in the same way as alcohol, producing symptoms like drunkenness. However, 'glue sniffers' risk many unpleasant side effects, including headaches, sickness and vomiting, and may even have fits and stop breathing. Inhaling solvents also damages the heart, lungs, liver and brain. Solvents are addictive, and many young people die every year from solvent abuse.

Some depressant drugs are called **sedatives**. One example is the group of drugs called **barbiturates**. They are used in sleeping pills, but again can be addictive.

Analgesics are painkilling drugs, most of which are not addictive. **Aspirin** is one of the oldest remedies for relief of mild pain, such as headaches. Many drugs were first extracted from plants, and aspirin is one example. It was first obtained from the bark of willow trees. Another mild painkiller is **paracetamol**. This drug is as effective as aspirin, but unlike aspirin, does not irritate the stomach lining. Paracetamol is safe if you take no more than the recommended dose, but an overdose of paracetamol can cause permanent liver damage. Even 'hard' drugs such as heroin have medicinal uses. Heroin and morphine are both very strong analgesics, and are used to give pain relief to terminally ill people, such as cancer patients. However, they are both highly addictive.

There are a number of other types of drug. **Hallucinogens** cause people to see or hear things that are not really there, or have dreams. They work by disrupting synaptic connections in the brain, setting up nerve pathways that shouldn't be there. **Anaesthetics** cause a loss of feeling, or unconsciousness, by affecting sensory nerves. Table 6.2 is a summary of some of the more common types of drug. Notice that some drugs, such as heroin, fit into more than one category.

Type	Examples	Effects
stimulants	caffeine, nicotine, amphetamines, cocaine, ecstasy	increase activity of nervous system
depressants/sedatives	alcohol, barbiturates, valium, morphine, heroin, solvents	decrease activity of nervous system
analgesics	aspirin, paracetamol, morphine, heroin	painkillers
hallucinogens	LSD, marijuana	cause hallucinations
anaesthetics	cocaine, Novocain, nitrous oxide	cause loss of feeling or unconsciousness

Table 6.2: *Types of drug and their uses.*

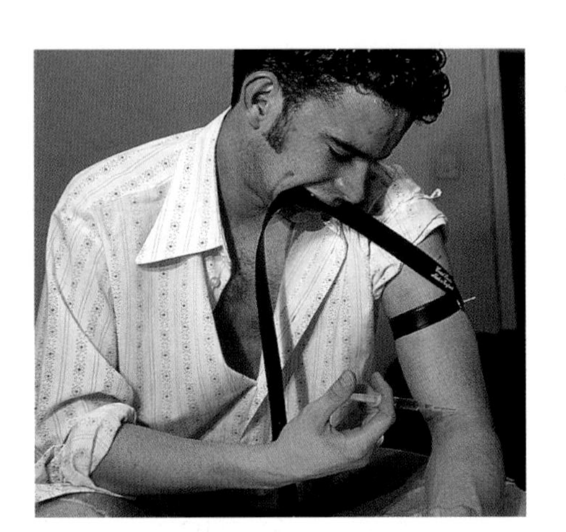

Figure 6.18 *A heroin addict injecting the drug intravenously. Addicts may turn to a life of crime to pay for their habit, and often die from an overdose of the drug.*

Addiction and the harmful effects of drugs on the body are not the only problems faced by drug users. Addicts using drugs such as heroin take it by **intravenous injection** (injecting into a vein – Figure 6.18). If they share needles and syringes with other drug users, they run the risk of contracting infectious diseases such as hepatitis and HIV. The viruses that cause these infections may be passed from one person to another in the blood contaminating a used syringe.

End of Chapter Checklist

If you haven't got a copy of your specification, read the introduction on page vi.

You will need to be able to do some or all of the following. Check your Awarding Body's specification (syllabus) to find out exactly what you need to know.

- Understand the role of the nervous system in reacting to changes in the surroundings (stimuli) and coordination of behaviour to produce a response.
- Understand that receptors contain cells that detect stimuli by transducing energy from the stimulus into nerve impulses in sensory neurones.
- Recall the structure of the eye and the functions of the main parts of the eye.
- Know the role of the iris, pupil, retina and optic nerve in the iris reflex.
- Understand the meaning of accommodation in the eye, and how this is brought about by the action of ciliary muscles, suspensory ligaments and the lens.
- Be able to define 'reflex action' and explain the difference between a reflex and a voluntary action.
- Label a diagram of a simple spinal reflex arc, such as the pain reflex, and describe how it functions. Recall the structure of a neurone.
- Know that synapses are junctions between nerve cells that are crossed by secretion of a chemical, and that certain drugs can interfere with synaptic transmission.
- Be able to describe an example of a conditioned reflex and how it differs from a simple reflex.
- Know the structure and functions of the main regions of the brain, including the cerebrum, cerebellum and medulla.
- Understand that drugs can be useful or harmful, and that they may change the chemical processes in people's bodies so that they become addicted to them and suffer withdrawal symptoms.
- Be able to describe some of the useful and harmful effects of named stimulant, depressant and analgesic drugs.
- Know that there is a danger of contracting HIV and hepatitis from intravenous injection of drugs.

Questions

More questions on coordination can be found at the end of Section B on page 119.

1 A **cataract** is an eye problem suffered by some people, especially the elderly. The lens of the eye becomes opaque (cloudy) which blocks the passage of light. It can lead to blindness. Cataracts can be treated by a simple eye operation, where a surgeon removes the lens. After the operation, the patient is able to see again, but the eye is unable to carry out accommodation, and the patient will probably need to wear glasses.

a) Explain why the eye can still form an image after the lens has been removed.

b) What is meant by 'accommodation'? Why is this not possible after a cataract operation?

c) Will the patient need glasses to see nearby or distant objects clearly? Explain your answer.

2 The diagram shows a section through a human eye.

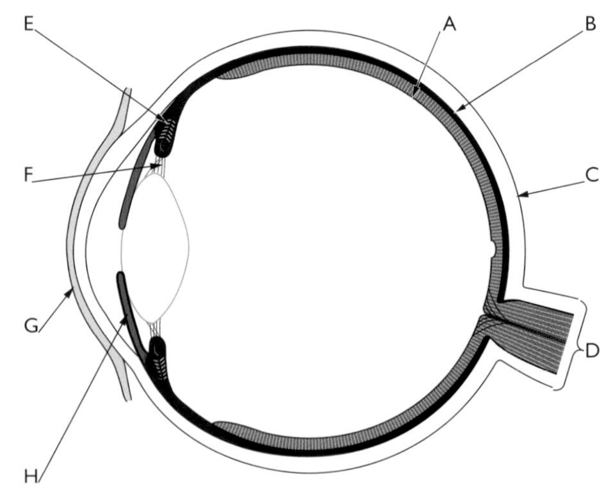

a) The table lists the functions of some of parts A to H. Copy the table and write the letters of the correct parts in the boxes.

Function	Letter
refracts light rays	
converts light into nerve impulses	
contains pigment to stop internal reflection	
contracts to change the shape of the lens	
takes nerve impulses to the brain	

b) *i)* Which label shows the iris?

ii) Explain how the iris controls the amount of light entering the eye.

iii) Why is this important?

3 The diagram shows some parts of the nervous system involved in a simple reflex action that happens when a finger touches a hot object.

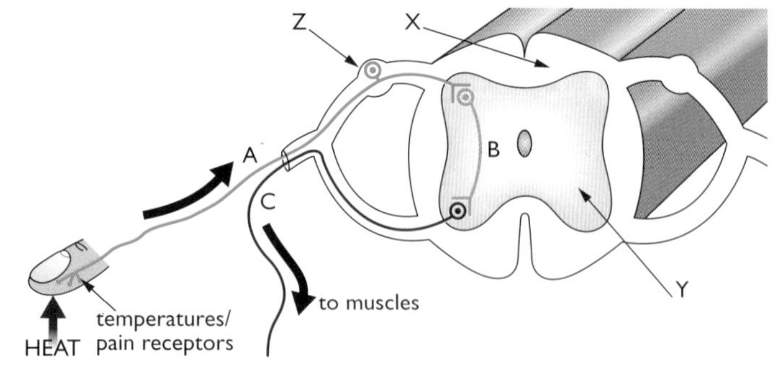

a) What type of neurone is:

i) neurone A *ii)* neurone B *iii)* neurone C?

b) Describe the function of each of these types of neurone.

c) Which parts of the nervous system are shown by the labels X, Y and Z?

d) In what form is information passed along neurones?

e) Explain how information passes from one neurone to another.

f) What is the difference between this reflex and a conditioned reflex?

4 **a)** Which part of the human brain is responsible for controlling each of the following actions:

i) keeping your balance when you walk?

ii) maintaining your breathing when you are asleep?

iii) making your leg muscles contract when you kick a ball?

b) A 'stroke' is caused by a blood clot blocking the blood supply to part of the brain.

i) One patient, after suffering a stroke, was unable to move his left arm. Which part of his brain was affected?

ii) Another patient lost her sense of smell following a stroke. Which part of her brain was affected?

5 **a)** In two sentences explain what a drug is?

b) Explain the meaning of i) addiction ii) withdrawal symptoms.

c) Explain the difference between a stimulant drug and a depressant drug. Give two examples of each and explain the effects that they have on the body.

d) It is recommended that people only drink alcoholic drinks in small amounts. Explain the short-term problems brought about by drinking large volumes of alcohol, and the dangers of regular consumption of large amounts.

6 Find the answers to the following by carrying out an Internet search:

a) Alcoholic drinks are measured in 'units'. What is a unit?

b) How many units are there in *i)* a pint of beer *ii)* a small glass of wine *iii)* a bottle of 'alcopop'?

c) What is the recommended maximum weekly intake of alcohol for an adult man and for an adult woman?

d) In the UK, what is the maximum legal limit of alcohol allowed in the blood of a person driving a car?

7 **a)** List five examples of stimuli that affect the body and state the response produced by each stimulus.

b) For one of your five examples, explain:

i) the nature and role of the receptor

ii) the nature and role of the effector organ.

c) For the same example describe the chain of events from stimulus to response.

Chapter 7: Chemical Coordination

The nervous system (Chapter 6) is a coordination system forming a link between stimulus and response. The body has a second coordination system, which does not involve nerves. This is the **endocrine** system. It consists of organs called endocrine **glands**, which make chemical messenger substances called **hormones**. Hormones are carried in the bloodstream.

The receptors for some hormones are located in the cell membrane of the target cell. Other hormones have receptors in the cytoplasm, and some in the nucleus. Without specific receptors, a cell will not respond to a hormone at all.

Glands and hormones

A gland is an organ that releases or **secretes** a substance. This means that cells in the gland make a chemical which passes out of the cells. The chemical then travels somewhere else in the body, where it carries out its function. There are two types of glands – **exocrine** and **endocrine** glands. Exocrine glands secrete their products through a tube or **duct**. For example, salivary glands in your mouth secrete saliva down salivary ducts, and tear glands secrete tears through ducts that lead to the surface of the eye. Endocrine glands have no duct, and so are called **ductless** glands. Instead, their products, the hormones, are secreted into the blood vessels that pass through the gland (Figure 7.1).

This chapter looks at some of the main endocrine glands and the functions of the hormones they produce. Because hormones are carried in the blood, they can travel to all areas of the body. They usually only affect certain tissues or organs, called 'target organs', which can be a long distance from the gland that made the hormone. Hormones only affect particular tissues or organs if the cells of that tissue or organ have special chemical receptors for the particular hormone. For example, the hormone insulin affects the cells of the liver, which have insulin receptors.

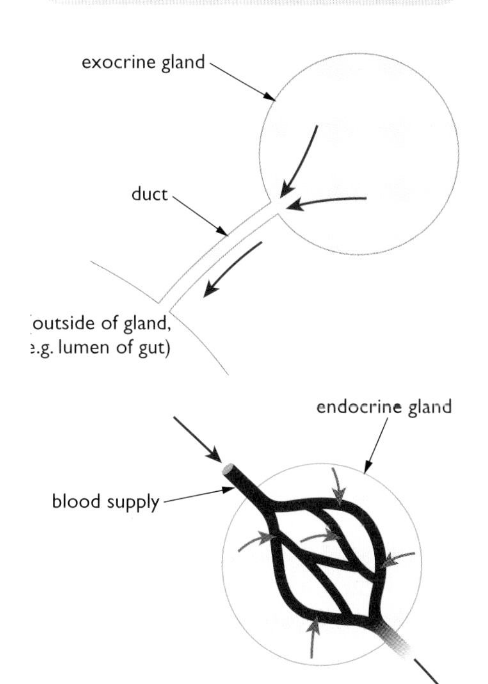

Figure 7.1 *Exocrine glands secrete their products though a duct, while endocrine glands secrete hormones into the blood.*

The differences between nervous and endocrine control

Although the nervous and endocrine systems both act to coordinate body functions, there are differences in the way that they do this. These are summarised in Table 7.1.

Nervous system	Endocrine system
• works by nerve impulses transmitted through nerve cells (although chemicals are used at synapses)	• works by hormones transmitted through the bloodstream
• nerve impulses travel fast and usually have an 'instant' effect	• hormones travel more slowly and generally take longer to act
• response is usually short-lived	• response is usually longer-lasting
• impulses act on individual cells such as muscle fibres, so have a very localised effect	• hormones can have widespread effects on different organs (although they only act on particular tissues or organs if the cells have the correct receptors)

Table 7.1: *The nervous and endocrine systems compared.*

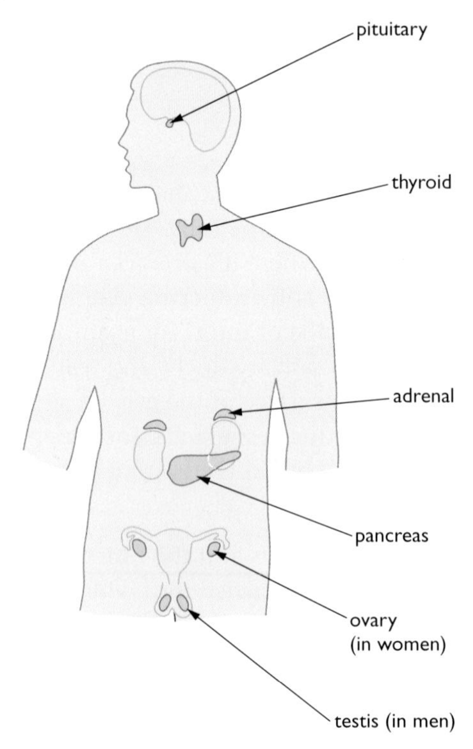

Figure 7.2 *The main endocrine glands of the body.*

The positions of the endocrine glands

The main endocrine glands are shown in Figure 7.2, and a summary of some of the hormones that they make and their functions, is given in Table 7.2.

Gland	Hormone	Some functions of the hormones
pituitary	follicle stimulating hormone (FSH)	stimulates egg development and oestrogen secretion in females and sperm production in males
	luteinising hormone (LH)	stimulates egg release (ovulation) in females and testosterone production in males
	anti-diuretic hormone (ADH)	controls the water content of the blood (see Chapter 9)
	growth hormone (GH)	speeds up the rate of growth and development in children
thyroid	thyroxin	controls the body's metabolic rate (how fast chemical reactions take place in cells)
pancreas	insulin	lowers blood glucose
	glucagon	raises blood glucose
adrenals	adrenaline	prepares body for physical activity
testes	testosterone	controls development of male secondary sexual characteristics
ovaries	oestrogen	controls development of female secondary sexual characteristics; regulates menstrual cycle
	progesterone	regulates menstrual cycle

Table 7.2: *Some of the main endocrine glands, the hormones they produce and their functions.*

The pituitary is a link between the nervous and endocrine coordination systems.

The **pituitary** gland is found at the base of the brain (see Figure 6.13, page 73). It produces a number of hormones, including those that regulate reproduction. The pituitary contains neurones linking it to a part of the brain called the **hypothalamus**, and some of its hormones are produced under the control of the brain.

The **pancreas** is both an endocrine *and* an exocrine gland. It secretes two hormones involved in the regulation of blood glucose, and is also a gland of the digestive system, secreting enzymes through the pancreatic duct into the small intestine (see Chapter 3). The sex organs of males and females, as well as producing sex cells or gametes, are also endocrine organs.

Both the **testes** and **ovaries** make hormones that are involved in controlling reproduction. We will look at the functions of some hormones in more detail.

Adrenaline – the 'fight or flight' hormone

When you are frightened, excited or angry, your **adrenal** glands secrete the hormone **adrenaline**.

'Adrenal' means 'next to the kidneys' which describes where the adrenal glands are located, on top of these organs (see Figure 7.2).

Adrenaline acts at a number of target organs and tissues, preparing the body for action. In animals other than humans this action usually means dealing with an attack by an enemy, where the animal can stay and fight or run away – hence 'fight or flight'. This is not often a problem with humans, but there are plenty of other times when adrenaline is released (Figure 7.3).

Figure 7.3 *Many human activities cause adrenaline to be produced, not just a 'fight or flight' situation!*

Figure 7.4 *Adrenaline affects the body of an animal in many ways.*

If an animal's body is going to be prepared for action, the muscles need a good supply of oxygen and glucose for respiration. Adrenaline produces several changes in the body that cause this to happen (Figure 7.4) as well as other changes to prepare for fight or flight:

- The breathing rate increases and breaths become deeper, taking more oxygen into the body.
- The heart beats faster, sending more blood to the muscles, so that they receive more glucose and oxygen for respiration.
- Blood is diverted away from the intestine and into the muscles.
- In the liver, stored carbohydrate is changed into glucose and released into the blood. The muscle cells absorb more glucose and use it for respiration.
- The pupils dilate, increasing visual sensitivity to movement.
- Body hair stands on end, making the animal look larger to an enemy.
- Mental awareness is increased, so reactions are faster.

In humans, adrenaline is not just released in a 'fight or flight' situation, but in many other stressful activities too, such as preparing for a race, going for a job interview or taking an exam.

Controlling blood glucose

You saw earlier that adrenaline can raise blood glucose from stores in the liver. The liver cells contain carbohydrate in the form of **glycogen**. Glycogen is made from long chains of glucose sub-units joined together (see Chapter 3) producing a large insoluble molecule. Being insoluble makes glycogen a good storage product. When the body is short of glucose, the glycogen can be broken down into glucose, which then passes into the bloodstream.

Most of the cells of the pancreas are concerned with making digestive enzymes. However, in the pancreas tissue, there are small groups of cells called the **Islets of Langerhans**. These contain two types of cell. Larger α (alpha) cells secrete glucagon, and smaller β (beta) cells secrete insulin.

Adrenaline raises blood glucose concentration in an emergency, but two other hormones act all the time to control the level, keeping it fairly constant. Both of these hormones are made by the pancreas. **Insulin** stimulates removal of glucose from the bloodstream into cells and causes the liver cells to convert glucose into glycogen. This lowers the glucose concentration in the blood when it is too high.

The other hormone is **glucagon**. This stimulates the liver cells to break down glycogen into glucose, raising the concentration of glucose in the blood if it is too low. Together, they work to keep the blood glucose approximately constant, at a little less than 1 g of glucose in every dm^3 of blood. Both hormones are released by special cells in the pancreas, in direct response to the level of glucose in the blood passing through this organ. In other words:

$$\text{glucose} \underset{\text{glucagon}}{\overset{\text{insulin}}{\rightleftharpoons}} \text{glycogen}$$

Insulin is a protein, and if it were to be taken by mouth in tablet form, it would be broken down in the gut. Instead it is injected into muscle tissue, where it is slowly absorbed into the bloodstream.

The concentration of glucose in your blood will start to rise after you have had a meal. Sugars from digested carbohydrate pass into the blood and are carried to the liver in the hepatic portal vein (Chapter 5). Here the glucose is converted to glycogen, so the blood leaving the liver in the hepatic vein will have a lower concentration of glucose.

Diabetes

Some people have a disease where their pancreas cannot make enough insulin to keep their blood glucose level constant – it rises to very high concentrations. The disease is called **diabetes**, or sometimes 'sugar diabetes'. One symptom of diabetes can be detected by a chemical test on urine. Normally, people have no glucose at all in their urine. Someone suffering from diabetes may have such a high concentration of glucose in the blood that it is excreted in their urine. This can be shown up by using coloured test strips (Figure 7.5).

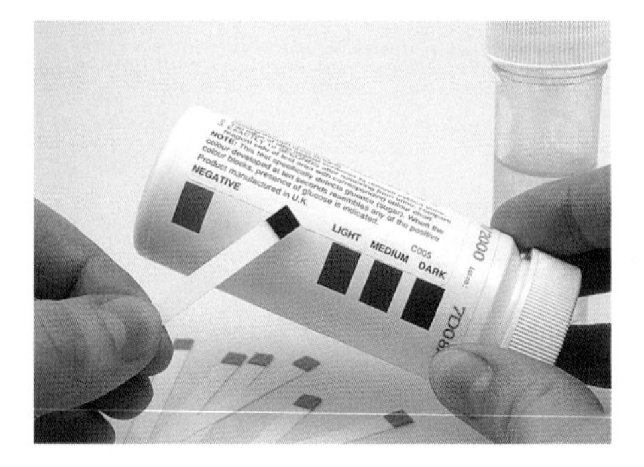

Figure 7.5 *Coloured test strips are used to detect glucose in urine.*

Another symptom of this kind of diabetes is a constant thirst. This is because the high blood glucose concentration stimulates receptors in the hypothalamus of the brain. These 'thirst' centres are stimulated, so that by drinking, the person will dilute their blood.

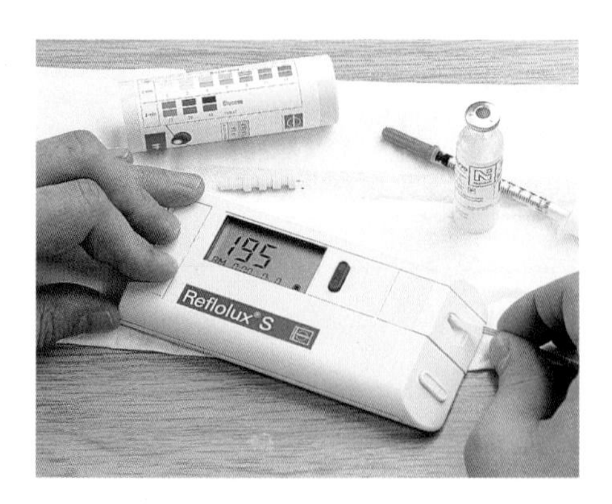

Figure 7.6 *Sensor for measuring blood glucose.*

Severe diabetes is very serious. If it is untreated, the sufferer loses weight and becomes weak and eventually lapses into a coma and dies.

Glucose in the blood is derived from carbohydrates such as starch in the diet, so mild forms of the disease can be treated by controlling the patient's diet, limiting the amount of carbohydrate that they eat. More serious cases of diabetes need daily injections of insulin to keep the glucose level in the blood at the correct level. People with diabetes can check their blood glucose using a special sensor. They prick their finger and place a drop of blood onto a test strip. The strip is then put into the sensor, which gives them an accurate reading of how much glucose is in their blood (Figure 7.6). They can then tell whether or not they need to inject insulin.

Insulin for the treatment of diabetes has been available since 1921, and has kept millions of people alive. It was originally extracted from the pancreases of animals such as pigs and cows, and much insulin is still obtained in this way. However, since the 1970s, human insulin has been produced commercially, from genetically modified (GM) bacteria. The bacteria have their DNA 'engineered' to contain the gene for human insulin (see Chapter 23).

Hormones controlling reproduction

Most animals are unable to reproduce when they are young. We say that they are sexually immature. When a baby is born, it is recognisable as a boy or girl by its sex organs (Figure 7.7).

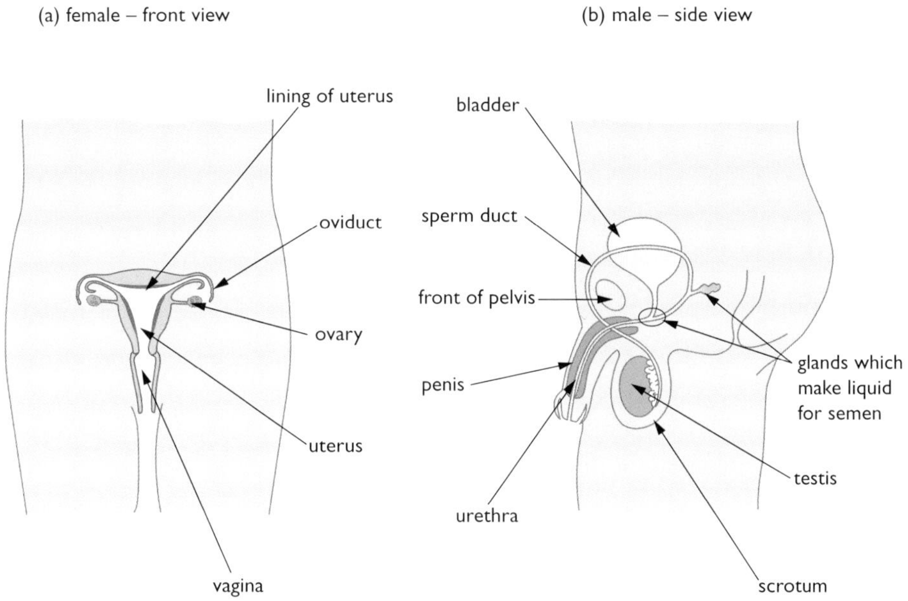

(a) female – front view

lining of uterus
oviduct
ovary
uterus
vagina

(b) male – side view

bladder
sperm duct
front of pelvis
penis
urethra
glands which make liquid for semen
testis
scrotum

Figure 7.7 *The human female and male reproductive systems.*

The presence of male or female sex organs is known as the primary sex characteristics. During their teens, changes happen to boys and girls that lead to sexual maturity. These changes are controlled by hormones, and the time when they happen is called **puberty**. Puberty involves two developments. The first is that the sex cells (eggs and sperm) start to be produced. The second is that the bodies of both sexes adapt to allow reproduction to take place. These events are triggered by hormones released by the pituitary gland (Table 7.2) called **follicle stimulating hormone** (**FSH**) and **luteinising hormone** (**LH**).

In boys, FSH stimulates sperm production, while LH instructs the testes to secrete the male sex hormone, **testosterone**. Testosterone controls the development of the male **secondary sexual characteristics**. These include growth of the penis and testes, growth of facial and body hair, muscle development and breaking of the voice (Table 7.3).

In girls, the pituitary hormones control the release of a female sex hormone called **oestrogen**, from the ovaries. Oestrogen produces the female secondary sex characteristics, such as breast development and the beginning of menstruation ('periods').

Sperm production is most efficient at a temperature of about 34°C, just below the core body temperature (37°C). This is why the testes are outside the body in the scrotum, where the temperature is a little lower.

In boys	In girls
sperm production starts	the menstrual cycle begins, and eggs are released by the ovaries every month
growth and development of male sexual organs	growth and development of female sexual organs
growth of armpit and pubic hair, and chest and facial hair (beard)	growth of armpit and pubic hair
increase in body mass; growth of muscles, e.g. chest	increase in body mass; development of 'rounded' shape to hips
voice breaks	voice deepens without sudden 'breaking'
sexual 'drive' develops	sexual 'drive' develops
	breasts develop

Table 7.3: *Changes at puberty.*

The age when puberty takes place can vary a lot, but it is usually between about 11 and 14 years in girls and 13 and 16 years in boys. It takes several years for puberty to be completed. Some of the most complex changes take place in girls, with the start of menstruation.

Hormones and the menstrual cycle

'Menstrual' means 'monthly', and in most women the cycle takes about a month, although it can vary from as little as two weeks to as long as six weeks (Figure 7.8). In the middle of the cycle is an event called **ovulation**, which is the release of a mature egg cell, or **ovum**.

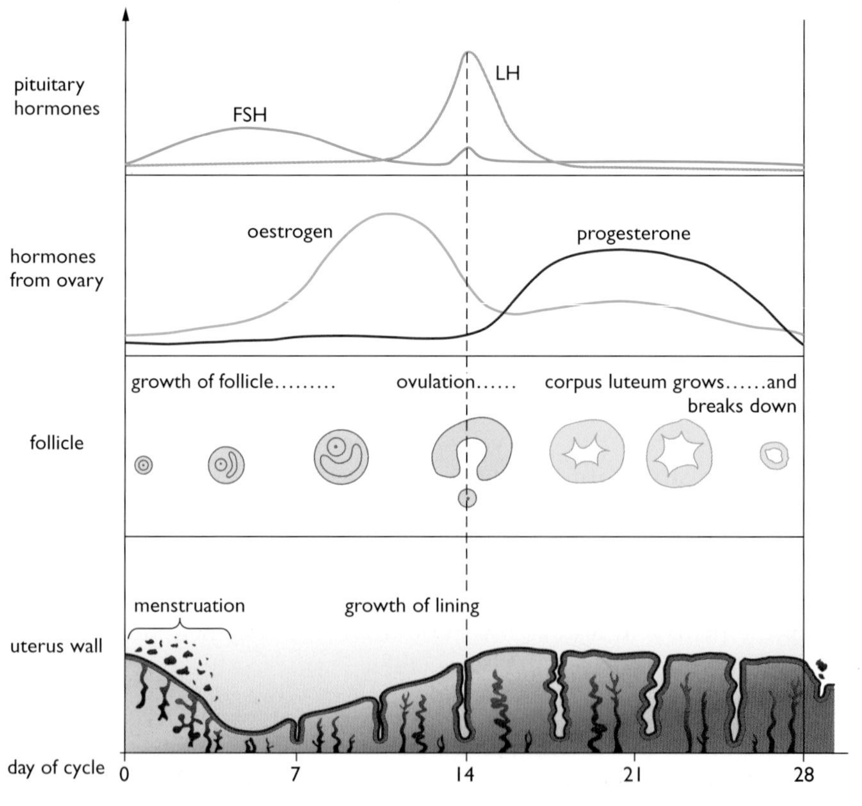

Figure 7.8 *Changes taking place during the menstrual cycle.*

One function of the cycle is to control the development of the lining of the uterus (womb), so that if the ovum is fertilised, the lining will be ready to receive the fertilised egg. If the egg is not fertilised, the lining of the uterus is lost from the woman's body as the flow of menstrual blood and cells of the lining, called a **period**.

A cycle is a continuous process, so it doesn't really have a beginning, but the first day of menstruation is usually called day 1.

Inside a woman's ovaries are hundreds of thousands of cells that could develop into mature eggs. Every month, one of these grows inside a ball of cells called a **follicle**. This is why the pituitary hormone which switches on the growth of the follicle is called 'follicle stimulating hormone'. At the middle of the cycle (about day 14) the follicle moves towards the edge of the ovary and the egg is released as the follicle bursts open. This is the moment of ovulation.

A small percentage of women are able to feel the exact moment that ovulation happens, as the egg bursts out of an ovary.

While this is going on, the lining of the uterus has been repaired after menstruation, and has thickened. This change is brought about by the hormone oestrogen, which is secreted by the ovaries in response to FSH. Oestrogen also has another job. It slows down production of FSH, while stimulating secretion of LH. It is a peak of LH that causes ovulation.

After the egg has been released, it travels down the oviduct to the uterus. It is here in the oviduct that fertilisation may happen, if sexual intercourse has taken place. What's left of the follicle now forms a structure in the ovary called the **corpus luteum**. The corpus luteum makes another hormone called **progesterone**. Progesterone completes the development of the uterus lining, which thickens ready for the fertilised egg to sink into it and develop into an embryo. Progesterone also inhibits (prevents) the release of FSH and LH by the pituitary, stopping ovulation.

'Corpus luteum' is Latin for 'yellow body'. A corpus luteum appears as a large yellow swelling in an ovary after the egg has been released. The growth of the corpus luteum is under the control of luteinising hormone (LH) from the pituitary.

If the egg is not fertilised, the corpus luteum breaks down and stops making progesterone. The lining of the uterus is then shed through the woman's vagina, during menstruation. If, however, the egg is fertilised, the corpus luteum carries on making progesterone, the lining is not shed, and menstruation doesn't happen. The first sign that tells a woman she is pregnant is when her monthly periods stop. Later on in pregnancy, the **placenta** secretes progesterone, taking over the role of the corpus luteum.

The placenta is an organ attached to the lining of the uterus. Blood from the fetus passes through the placenta, as does blood from the mother. A thin membrane separates these two blood supplies. This acts as an exchange surface, allowing nutrients and oxygen to pass from the mother's blood to the blood of the fetus, and waste materials to pass the other way.

Using hormones to control reproduction

Scientists can make artificial human hormones that can be used as contraceptives or to treat women with fertility problems.

The **contraceptive pill** usually contains a mixture of oestrogen and progesterone. These hormones prevent the production of FSH and LH from the pituitary gland. This means that the follicles in the ovary do not develop and ovulation does not take place. Without an egg being released, a woman cannot get pregnant. She takes the pill for 21 days of the menstrual cycle (Figure 7.9) then stops taking it for the last 7 days. During this time she will have a period.

There are many causes of infertility (when a woman is unable to get pregnant). One is that she does not produce enough FSH to start egg development in the ovary. One way that this can be treated is by giving

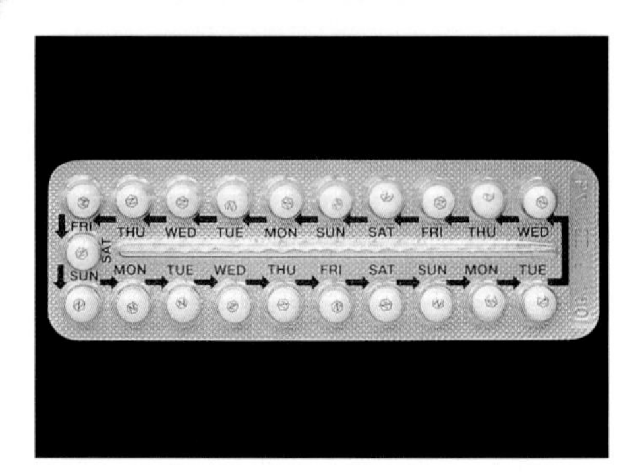

Figure 7.9 *Synthetic hormones can be used to prevent pregnancy.*

'The pill' has been linked to a number of health problems in women, the main one being an increased risk of blood clots (thromboses) forming in blood vessels, which can be fatal. The increased risk is only very slight for most women, but is higher if they are smokers. The 'mini pill' contains only progesterone. It is not quite as reliable as oral contraceptives that contain oestrogen, but it does not increase the chances of blood clots developing.

injections of FSH. Care has to be taken, because too much FSH can produce multiple ovulations (many eggs are released at once). If they are all fertilised, this results in several embryos developing in the uterus, and 'multiple births' (Figure 7.10). Although the children shown in Figure 7.10 were born healthy, multiple births may be premature (early). Even if the fetuses survive the full term of pregnancy (9 months), the mass of each baby is likely to be very low, and they may have health problems.

Figure 7.10 *Hormones can also be used to help a woman get pregnant. These children were born to a mother who had treatment for infertility.*

Hormones and sport

One of the effects of the male sex hormone, testosterone, is that it stimulates the growth of muscles. Because of this, testosterone is known as an **anabolic steroid**. 'Anabolic' means that it causes reactions which build up proteins in the cells (anabolism). A 'steroid' is a type of lipid molecule. Other sex hormones, such as oestrogen, are also steroids.

In the 1950s, some athletes started to inject themselves with testosterone to develop their muscles and gain extra strength. The testosterone also made them more aggressive and determined to win. The use of testosterone and other steroids to improve performance was soon banned by sporting authorities. Not only is it unfair, it is also highly dangerous. The high level of steroid drugs used by some athletes can lead to sterility, heart disease, kidney damage and liver cancer. When used by female athletes, steroids have caused loss of periods and even development of male characteristics.

New kinds of illegal drugs are being developed all the time to give athlete's an 'edge' over their competitors. Many are anabolic steroids; others are based on other types of natural hormone, such as growth hormone (Table 7.2). They are illegal, and people who are caught using any of them are usually banned from taking part in their sport.

End of Chapter Checklist

If you haven't got a copy of your specification, read the introduction on page vi.

You will need to be able to do some or all of the following. Check your Awarding Body's specification (syllabus) to find out exactly what you need to know.

- Understand the nature and role of hormones in coordination within the body.
- Describe the effects of adrenaline on the respiratory and circulatory systems and how it prepares the body for action.
- Describe the control of blood glucose concentration by insulin and glucagon.
- Understand the cause, nature and treatment of diabetes.
- Describe the role of hormones in the development of the secondary sex characteristics in humans.
- Recall the involvement of FSH, LH, oestrogen and progesterone in the maintenance of the menstrual cycle.
- Understand the use of hormones in controlling fertility, including use of FSH as a fertility drug, and hormones in oral contraceptives.
- Describe how hormones can be used illegally to enhance sporting performance.

Questions

More questions on chemical coordination can be found at the end of Section B on page 119.

1 ***a)*** *Hormones* are *secreted* by *endocrine glands*. Explain the meaning of the four words in italics.

b) Identify the hormones A to D in the table.

Hormone	One function of this hormone
A	stimulates the liver to convert glycogen to glucose
B	controls the 'fight or flight' responses
C	in boys, controls the breaking of the voice at puberty
D	completes the development of the uterus lining during the menstrual cycle

2 The number of sperm cells per cm^3 of semen (the fluid containing sperm) is called the 'sperm count'. Some scientists believe that over the last 50 years, the sperm counts of adult male humans have decreased. They think that this is caused by a number of factors, including drinking water polluted with oestrogens and other chemicals. Carry out an Internet search to find out the evidence for this. Download information into a word processor and summarise your findings in no more than two sides of A4, including graphs.

3 The graph shows the changes in blood glucose in a healthy woman over a 12-hour period.

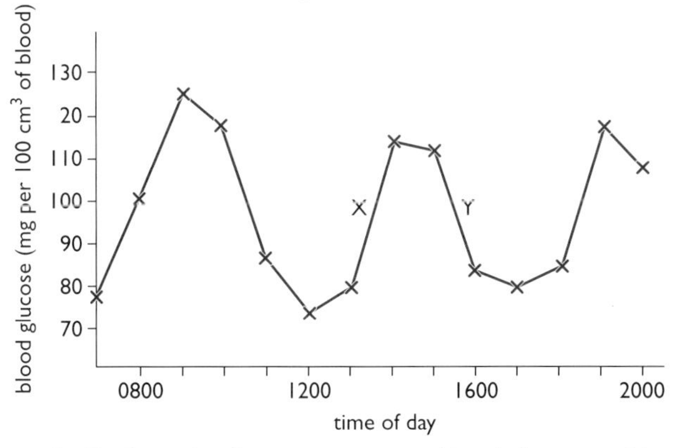

a) Explain why there was a rise in blood glucose at X.

b) How does the body bring about a decrease in blood glucose at Y? Your answer should include the words insulin, liver and pancreas.

c) Diabetes is a disease where the body cannot control the concentration of glucose in the blood.

i) Why is this dangerous?

ii) Describe two ways a person with diabetes can monitor their blood glucose level.

iii) Explain two ways that a person with diabetes can help to control their blood glucose level.

4 The graph shows some of the changes taking place during the menstrual cycle.

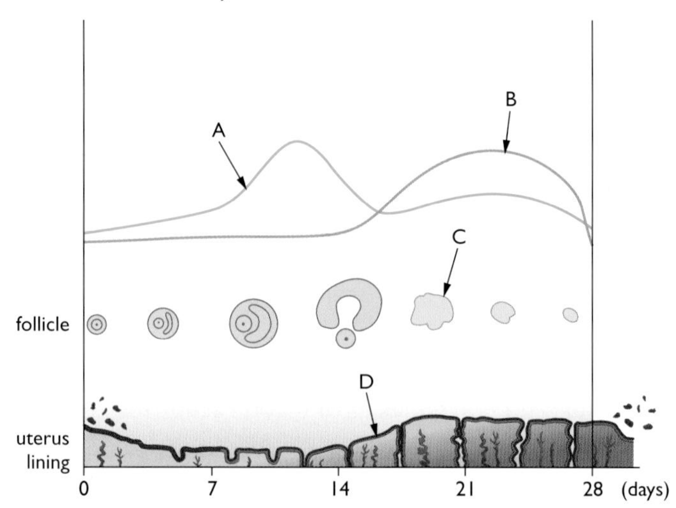

a) Identify the two hormones produced by the ovary, which are shown by the lines A and B on the graph.

b) Name the structure C.

c) What is the purpose of the thickening of the uterus lining at D?

d) When is sexual intercourse most likely to result in pregnancy, at day 6, 10, 13, 20 or 23?

e) Why is it important that the level of progesterone remains high in the blood of a woman during pregnancy? How does her body achieve this:

i) just after she becomes pregnant?

ii) later on in pregnancy?

5 **a)** Complete each of the sentences below with the word 'increases' or 'decreases':

i) The secretion of glucagon from the pancreas after a meal rich in carbohydrates ....

ii) The size of the pupil in response to adrenaline ....

iii) The level of oestrogen in the blood of a woman in the days after menstruation ....

iv) The level of progesterone in the blood of a woman in the days before menstruation ....

b) For each statement, explain what the function of the increase/decrease is.

6 **a)** Explain how artificial hormones can be used for the following:

i) as a method of contraception

ii) to treat women who have difficulty getting pregnant

iii) to illegally improve the performance of athletes.

b) What problems can the treatment described in part ***a)*** *i)* bring?

Chapter 8: Support and Movement

Movement in vertebrates is made possible by the actions of muscles attached to a framework of bones – the skeleton. This chapter deals with the working of human bones and muscles and looks at how movement takes place in other vertebrates.

Vertebrates are animals with a vertebral column or 'backbone'. They include all fish, amphibians, reptiles, birds and mammals. All other animals have no vertebral column or other bones. They are called **invertebrates**.

The vertebrate skeleton is a framework which supports the body and allows movement. Attached to the bones are the body's main effector organs, the muscles. When nerve impulses arrive at the muscles through motor nerves, they cause the muscles to shorten or **contract**, moving the bones.

The skeleton

The vertebrate skeleton (Figure 8.1) has several functions. It protects many vital organs. For example, the cranium protects the brain, the vertebrae protect the spinal cord, and the ribcage protects the heart and lungs. As well as providing protection, bone makes some components of the blood. Red blood cells and platelets are made in the marrow of larger bones, such as the sternum, femur and pelvis. However, the skeleton's more obvious roles are in support and movement.

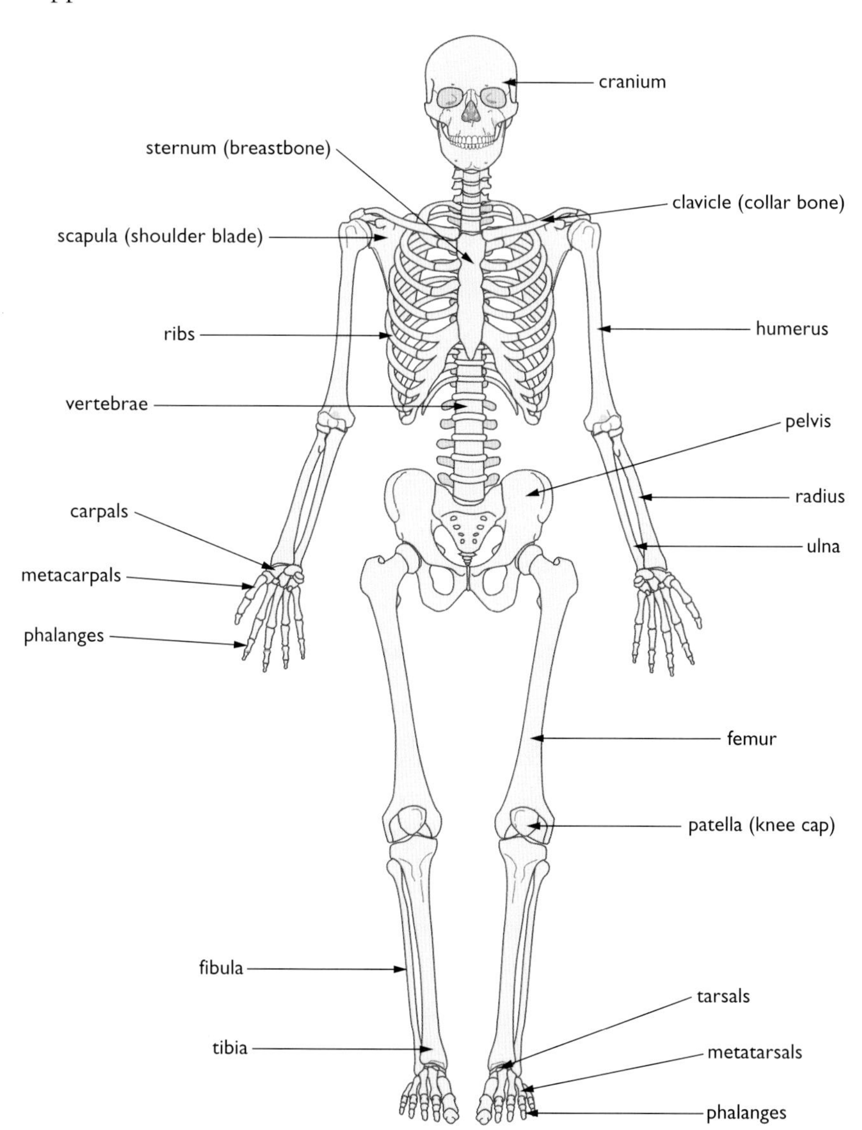

Figure 8.1 *The human skeleton.*

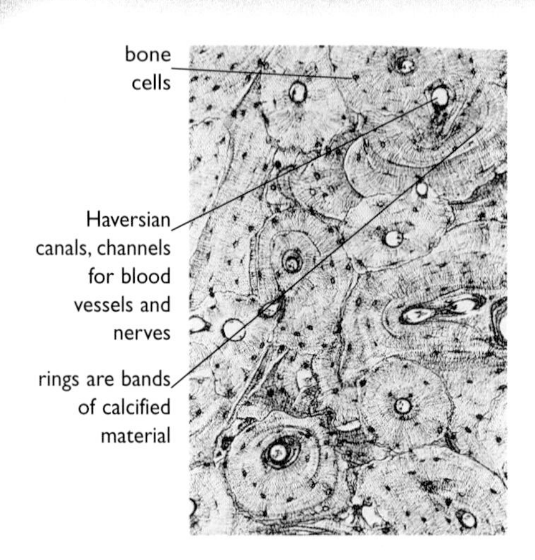

Figure 8.2 *This is a cross section through a normal bone. The rings are calcified material.*

Bone

Bone is a hard substance because it contains calcium salts, mainly calcium phosphate. This results in a rigid material that resists bending and compression (squashing) forces. Although a bone from an animal looks dead, in the body it is a living tissue, made of cells. These cells, along with protein fibres, stop the bone being too brittle. If you dissolve the calcium salts from a bone (Figure 8.2), the fibres that are left are tough but flexible. The presence of living cells also means that bones can repair themselves if they are broken.

Joints

When a vertebrate moves, its bones move relative to each other. The point where two bones meet is called a **joint**. We say that the bones **articulate** at joints. A movable joint such as the hip or elbow needs to have certain features. These include:

- a way to keep the ends of the bones held together, so that they don't separate (i.e. **dislocate**)
- a means of reducing friction between the ends of the moving bones
- a shock-absorbing surface between the two bones.

You can see the structures that do these things in Figure 8.3.

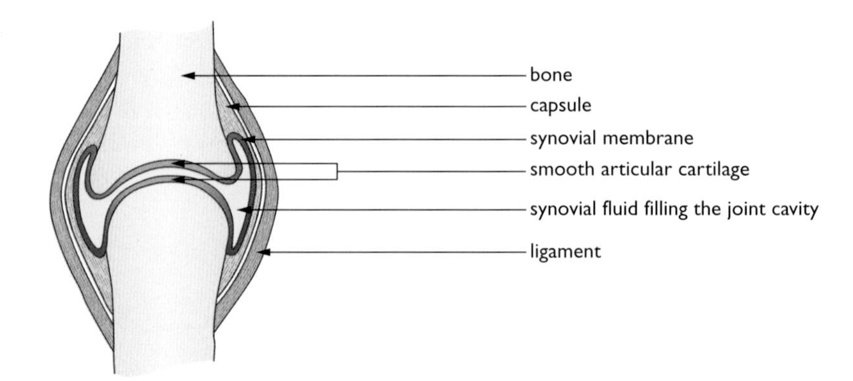

Figure 8.3 *The structure of a synovial joint. The size of the space filled with synovial fluid has been exaggerated in the diagram.*

Movable joints are called **synovial** joints. They contain a liquid called **synovial fluid**, which is secreted by the **synovial membrane**, lining the space in the middle of the joint. Synovial fluid is oily, and acts as a lubricant, reducing the friction between the ends of the bones. The end of each bone has an articulating surface covered with a smooth layer of **cartilage** (gristle). Cartilage is a strong material, but it is not brittle. It acts as a shock absorber between the ends of the bones, rather like a rubber gym mat compresses to absorb the shock when you fall over.

The joint is surrounded by a tough fibrous **capsule**, and held together by **ligaments**, which run from one bone to the other across the joint. Ligaments are composed of fibres that make them very tough. They have great strength to resist stretching, called **tensile** strength. However, ligaments have some elasticity, so that they allow joints to bend without the bones becoming dislocated.

The correct definition of an 'elastic' material is one which, when you bend or stretch it, will return to its original shape. In this chapter we use the word 'elastic' to mean 'easily stretched'.

Damaged joints

Joints are highly stressed parts of the body. Some, like the knee and hip, have to take much of the body's weight, while also allowing movement to continue. A number of injuries can happen at a joint as a result of sudden or unexpected forces being applied to them.

If you are walking over rough ground, and stumble, you might sprain your ankle. A **sprain** is usually a tear in the tissue of a ligament, and the synovial membrane may be damaged. This is painful, but heals after a period of rest. First aid for a sprain normally involves reducing the swelling by applying an ice pack.

A sudden force can also pull a bone out of its joint. This is the **dislocation** discussed above. Dislocations are common in contact sports, such as rugby. During a dislocation, the ligaments and synovial membrane may also be damaged. A doctor can usually put a dislocated bone back into place. Again, the joint needs to be rested afterwards, until any damage heals.

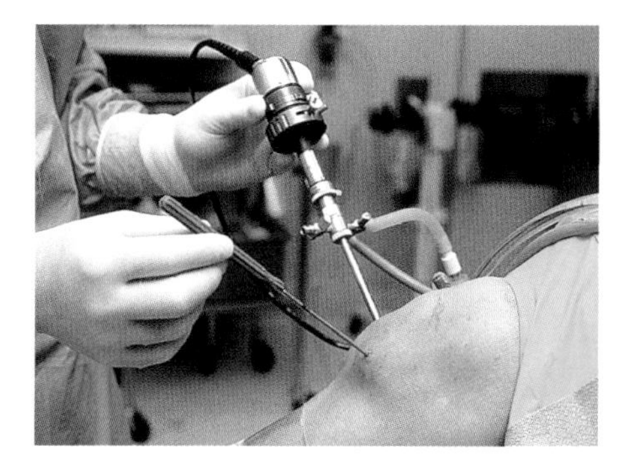

Figure 8.4 *This footballer is having keyhole surgery to remove a piece of torn cartilage from his knee.*

Another injury often seen in contact sports is a **torn cartilage**. Here, the layer of articular cartilage at the end of a bone becomes damaged. The small piece of torn cartilage causes pain and difficulty in moving the joint. It is a common footballer's injury, and is treated by cutting away the damaged tissue. Nowadays this is done by 'keyhole surgery' using a fibre optic camera and fine surgical tools to operate on the joint through a tiny opening (Figure 8.4). This is much less damaging than making a larger cut through the tissues to reach the inside of the joint.

Joints can also become damaged by disease. Elderly people sometimes develop a disease called **osteoarthritis**, where the articular cartilage wears away, leaving the ends of the bones to rub against each other. This causes the joint to be stiff and very painful, eventually preventing movement altogether. If the damage is severe, treatment involves replacing the joint with an artificial one. For example, a replacement hip joint (Figure 8.5) uses a metal ball on a shaft, which is inserted into the femur. The ball fits into a plastic cup placed in the pelvis. The operation makes a tremendous difference to the patient's life. Whereas before the operation they might have been in great pain and unable to walk, afterwards the pain is gone and they can often walk normally.

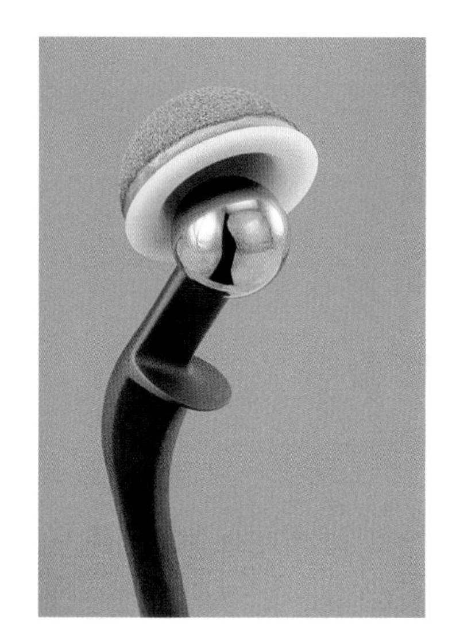

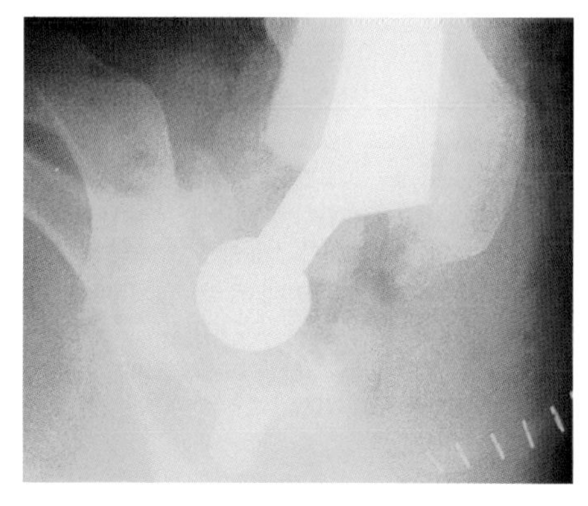

Figure 8.5 *This artificial hip is made of a stainless steel shaft with a ball shape at the end. The ball fits into a plastic socket that will be placed in the patient's pelvis. The X-ray shows a similar artificial hip in place after the operation.*

Muscles

Muscles are organs that are attached to bones and move them by contracting, pulling on the bone. Muscles cannot push, they can only pull – in other words they are not able to expand actively. When muscles get longer, it is because they are stretched by the contraction of another muscle. When a muscle is being stretched it is relaxed (the opposite of contracted). Because of this, muscles usually work in pairs; one contracting while the other relaxes. These are called **antagonistic pairs**. One of the simplest examples of an antagonistic pair of muscles is the arrangement of the **biceps** and **triceps** muscles in the arm (Figure 8.6).

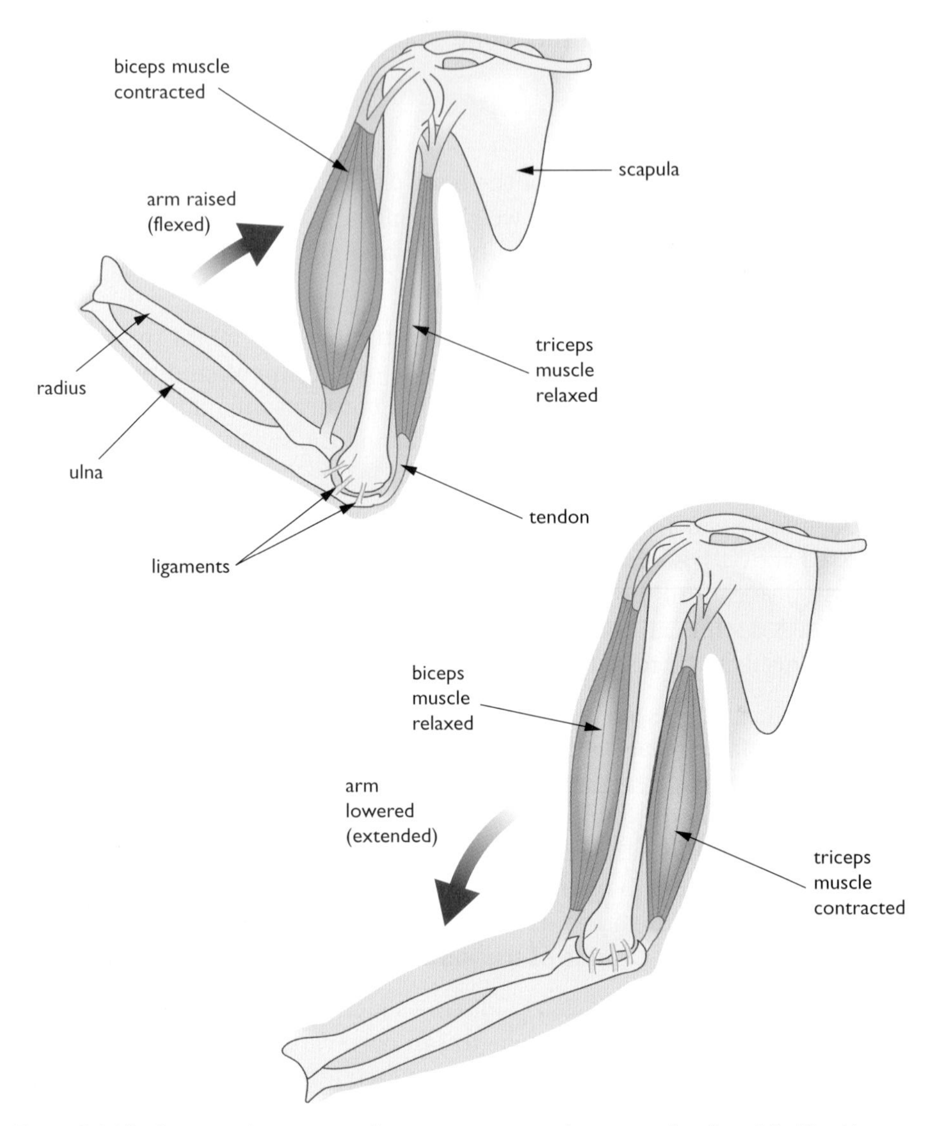

Figure 8.6 *The biceps and triceps muscles contract to move the arm at the elbow joint. The biceps flexes the arm, while its antagonistic partner, the triceps, extends the arm.*

When the biceps muscle contracts it bends, or **flexes** the arm at the elbow joint. Contraction of the triceps straightens, or **extends** the arm. Of course, there are many more muscles in the arm apart from these two. The other muscles produce movement in other directions.

The word we use to mean 'not very elastic' is **inelastic**. Both ligaments and tendons have a high tensile strength, but ligaments are fairly elastic, while tendons are inelastic. Ligaments join bone to bone across a joint, while tendons join muscle to bone.

At the end of a muscle there are **tendons**. A tendon attaches the muscle to the bone. Tendons have very high tensile strength, like ligaments, but unlike ligaments they are not very elastic. This means that they don't stretch when the muscle contracts.

When a muscle contracts, the bone at one end of the muscle moves and the bone at the other end stays still. The place where the muscle is attached to the stationary bone is called the **origin**. The place where it is attached to the moving bone is called the **insertion**. When a muscle contracts, the insertion moves towards the origin.

You can identify other antagonistic pairs of muscles in the body. For example, when we run we use several sets of muscles that cause bending at the hip, knee and ankle joints (Figure 8.7).

Figure 8.7 *Antagonistic muscles are used in running. The main muscle that flexes the knee joint is the biceps femoris, while its antagonistic partner, extending the knee, is the quadriceps. Can you work out which muscles flex and extend at the hip and ankle?*

Muscle contraction and exercise

Skeletal muscle is made up of highly specialised muscle cells or **fibres**, arranged in bundles in a connective tissue sheath. Muscle fibres are adapted for contraction. Under very high magnification, using an electron microscope, we can see that they are composed of fine protein filaments (Figure 8.8).

There are two types of filaments, thick and thin. When a muscle contracts, the thin filaments slide past the thick filaments, making the fibres shorter.

There are three types of muscle in the body – **skeletal**, **cardiac** and **smooth** muscle. The first of these is the muscle attached to the skeleton. It is described as **voluntary** muscle, because it is under the conscious control of the brain (see Chapter 6). Cardiac muscle is only found in the heart. It is involuntary, meaning not under conscious control. Smooth muscle is also **involuntary**. It is found in the wall of the gut, bladder, uterus, sperm ducts, blood vessels and other organs.

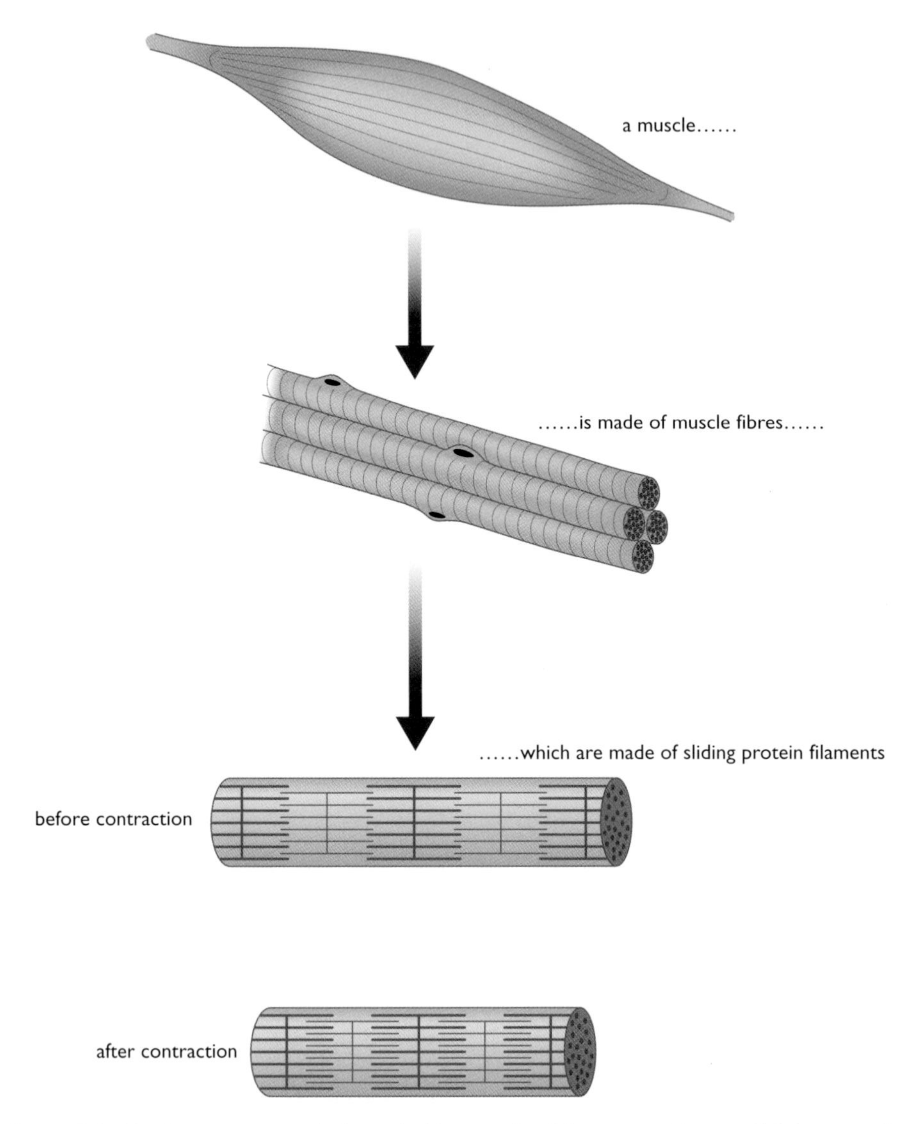

Figure 8.8 *Muscles are composed of muscle fibres, made of fine protein filaments. When a muscle contracts, the filaments slide over each other.*

If the muscles are made to contract very strenuously, too little oxygen reaches the muscle cells for them to respire aerobically. They then begin to respire anaerobically (without using oxygen), producing **lactic acid** (see Chapter 1). Anaerobic respiration provides enough energy for short 'bursts' of activity, such as sprinting. However, it cannot be maintained for long, because lactic acid is toxic, and builds up in the muscle cells and in the blood. Eventually the lactic acid is broken down to carbon dioxide and water. This happens aerobically. The oxygen needed for this is called the **oxygen debt**. The time it takes for this to happen is called the **recovery period**.

Contraction of muscle fibres needs energy; this comes from respiration (Chapter 1). Blood vessels supply glucose and oxygen to the muscle fibres. The fibres respire, converting the glucose and oxygen into carbon dioxide and water. Energy is released for the fibres to contract, but the process is not 100% efficient, and some energy is lost as heat. When we carry out strenuous exercise, our muscles demand a greater supply of glucose and oxygen than usual. In addition, carbon dioxide and heat needs to be removed at a faster rate. To achieve this, various changes take place in the body:

- The breathing rate increases, so more oxygen is taken into the blood by the lungs, and more carbon dioxide is lost. The volume of each breath also increases.
- The heart rate increases, pumping more oxygenated blood to the muscles.
- Blood is diverted away from places like the gut, and towards the muscles.
- The skin carries out processes such as vasodilation and sweating (see Chapter 9) which remove excess heat from the body.

Even when a muscle is relaxed, some of its fibres are contracted. This state of partial contraction is called muscle **tone**. It keeps our muscles taut, but not enough to cause movement. Muscle tone keeps us upright when we are standing or sitting.

Regular exercise keeps muscles toned. This has other benefits. The partial tension in the muscle fibres means that they are ready to contract more quickly. Exercise also develops muscles, because they make more protein filaments, and are able to produce a stronger contraction. If you never exercise, and then have a sudden 'work out', your muscles will feel sore and stiff afterwards. You can avoid this by regular exercise, which also improves the circulation to the muscles, heart and lungs, and keeps the joints working smoothly.

How fish swim

Fish are vertebrates that show outstanding adaptations to movement through water. The shape of a fish is streamlined, which reduces friction so it can move quickly through water. It has a number of fins that have different functions (Figure 8.9).

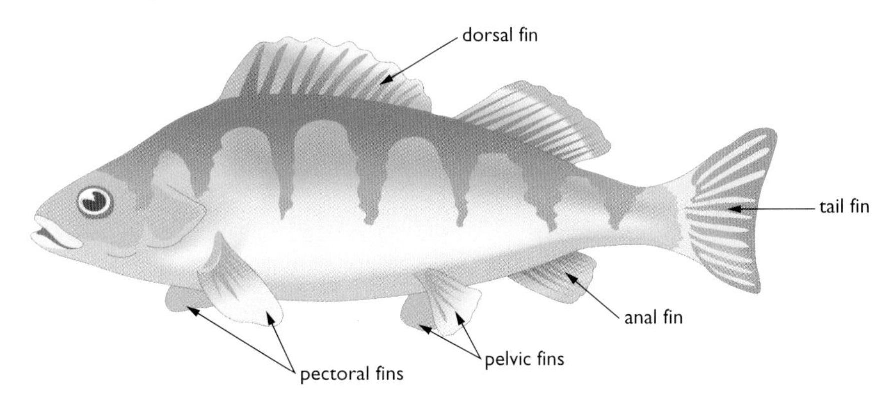

Figure 8.9 *A fish has fins for propulsion and steering. Its shape is streamlined for ease of movement through water.*

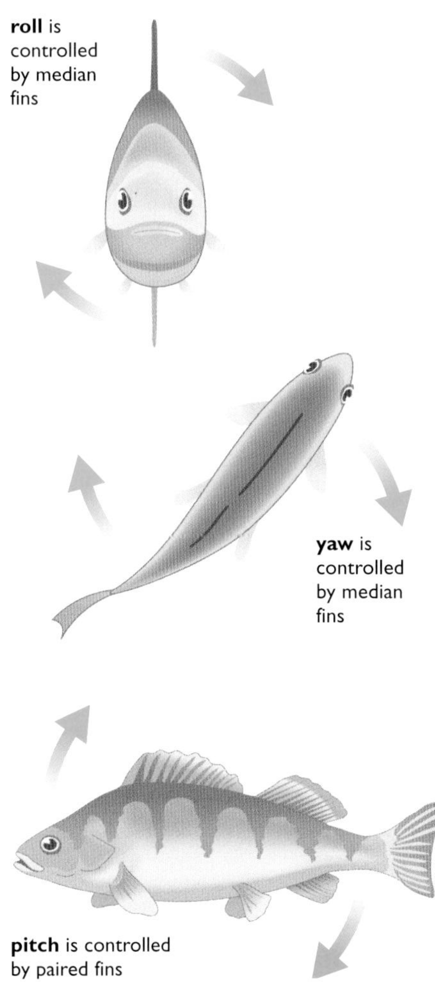

Figure 8.10 *Median and paired fins control movement of the fish in all three planes.*

The **tail fin** has a large surface area, so that when it is moved from side to side by the body muscles, it pushes backwards against the water, providing **thrust**. The other fins allow the fish to change direction. The single **median** fins (**dorsal** and **anal**) produce a large surface area in a vertical plane. This helps the fish to keep upright, controlling side-to-side tilting of the body, called roll, as well as sideways movements, or yaw. The **paired** fins (**pectoral** and **pelvic**) control nose to tail tilting, called pitch (Figure 8.10).

By carefully balanced movements of all its fins, a fish can move up or down or to one side as it passes through the water. Species of fish that are slower moving can even use the paired fins to swim backwards.

The tissues of a fish are slightly denser than water, so that if they are not actively swimming to gain lift, they would sink. In fact this is only true for some types of fish, such as sharks (Figure 8.11). Other species of fish have an ingenious adaptation that allows them to keep their depth in the water without swimming. This is an organ called the **swim bladder**. It is a gas-filled

Figure 8.11 *Fish such as this shark have no swim bladder and sink when they stop swimming.*

Sharks belong to an ancient group of fish that have a skeleton made entirely of cartilage. None of the fish in the group, which includes skates, rays and dogfish, have evolved a swim bladder. The swim bladder is found in 'bony fish' such as cod, herring and tuna. They have a skeleton made of bone.

bag that gives them buoyancy. The fish secretes gas into the bag from its blood to give it extra buoyancy, or reabsorbs the gas back into the blood to allow it to go deeper into the water. In this way it doesn't need to use up extra energy in swimming just to keep its depth.

As in other vertebrates, the force to generate movement in a fish is provided by contraction of muscles. Sideways movements of the tail are brought about by contraction of blocks of muscle either side of the vertebral column. The fibres in each muscle block run from front to back of the block, attached to a thin sheet of connective tissue at each end (Figure 8.12).

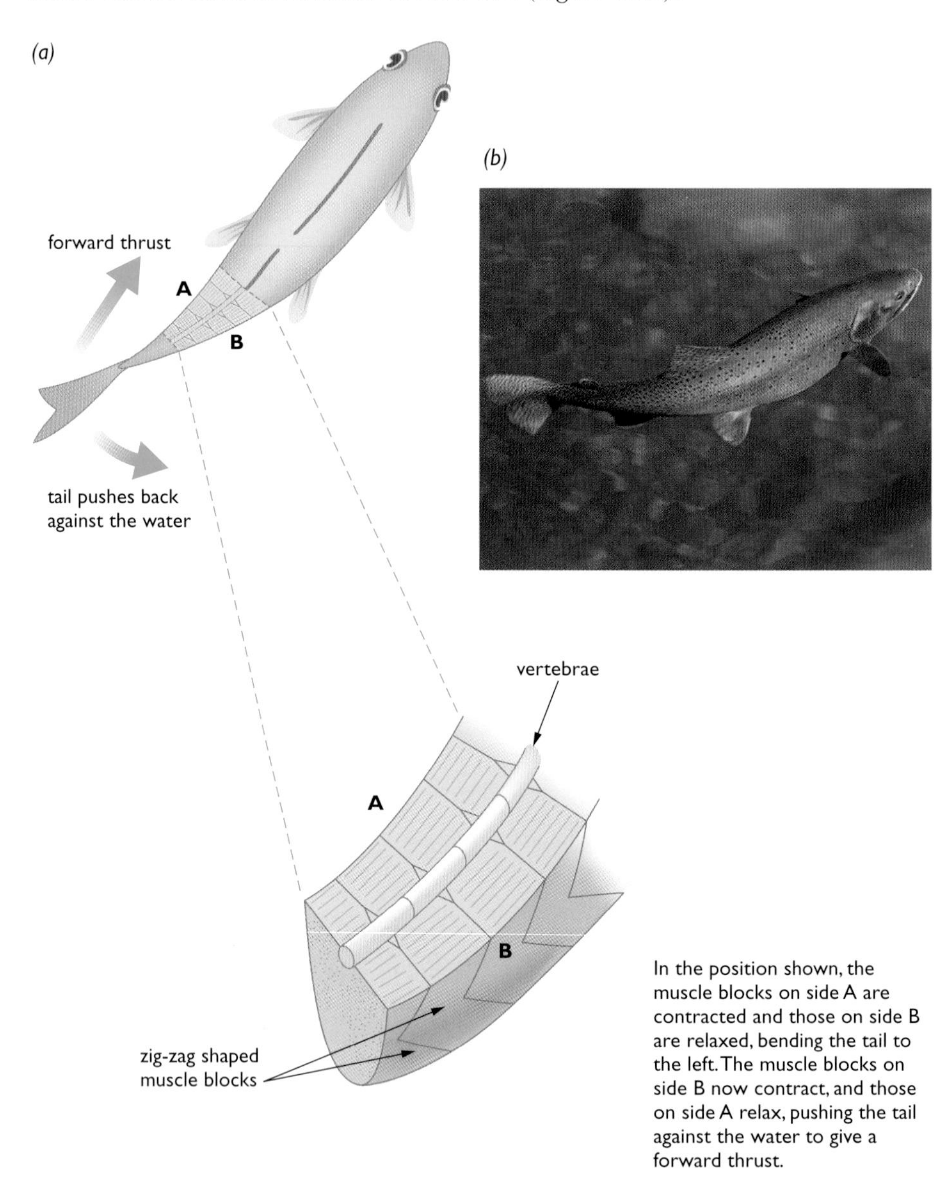

Figure 8.12 *A fish moves its tail against the water by contraction of blocks of muscle on one side of the vertebral column. This produces a forward force, or thrust. At the same time, the muscle blocks on the opposite side of the vertebral column are relaxed.*

If you eat fish, you will have seen these blocks of muscle. They are the large flakes of tissue that you get in a cod fillet.

By alternately contracting and relaxing the muscle blocks down the length of the body, an S-shaped wave passes down the body as the fish swims along. The greatest amplitude (height) of the wave happens at the tail, where most of the thrust is produced.

Flight in birds

If you are not a vegetarian, take a look at the bones next time you eat a chicken wing. The wing is a highly modified front limb, with feathers attached to it (Figure 8.13).

The feathers give the wing a large surface area that can push down on the air to give lift. There are three types of feathers. The **primary feathers** are attached to the second and third fingers, while the **secondary feathers** are attached to two of the arm bones (humerus and ulna). There are also smaller **contour feathers**, which cover the wing, giving it a smooth surface (Figure 8.14).

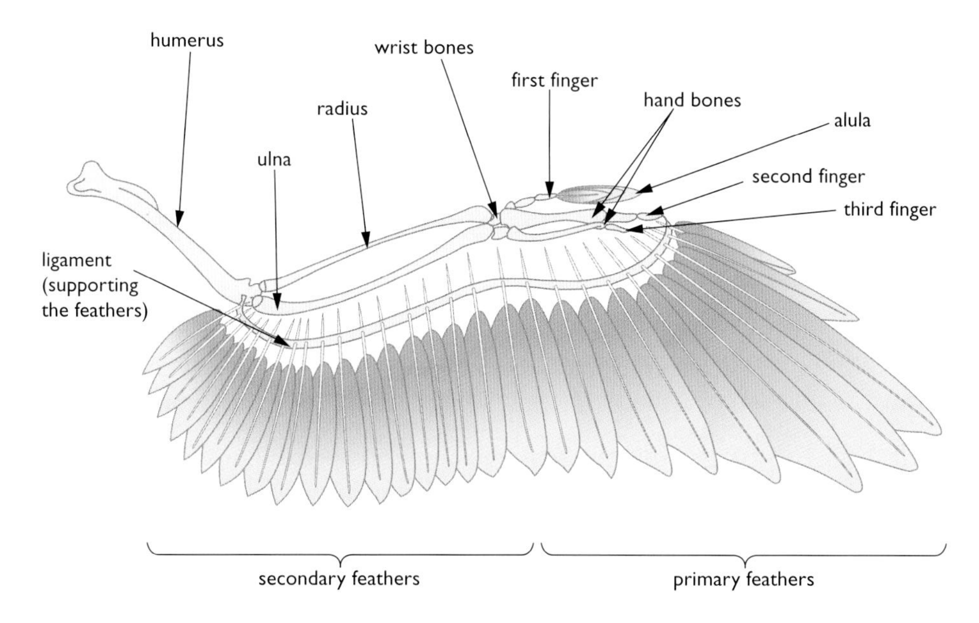

Figure 8.13 *The wing bones and feathers of a bird.*

Figure 8.14 *You can see the three types of feathers in this Arctic tern wing.*

Feathers are a remarkable adaptation for flight. The flight feathers have a very large surface area and are also strong and light, with hollow shafts. The fine branches from the shaft, called **barbs**, are held together by **barbules**. Each barbule has hooks that interlock with grooves on the next barbule (Figure 8.15), keeping the vane of the feather flat. If the barbules become unhooked, a bird can hook them together again by preening with its beak.

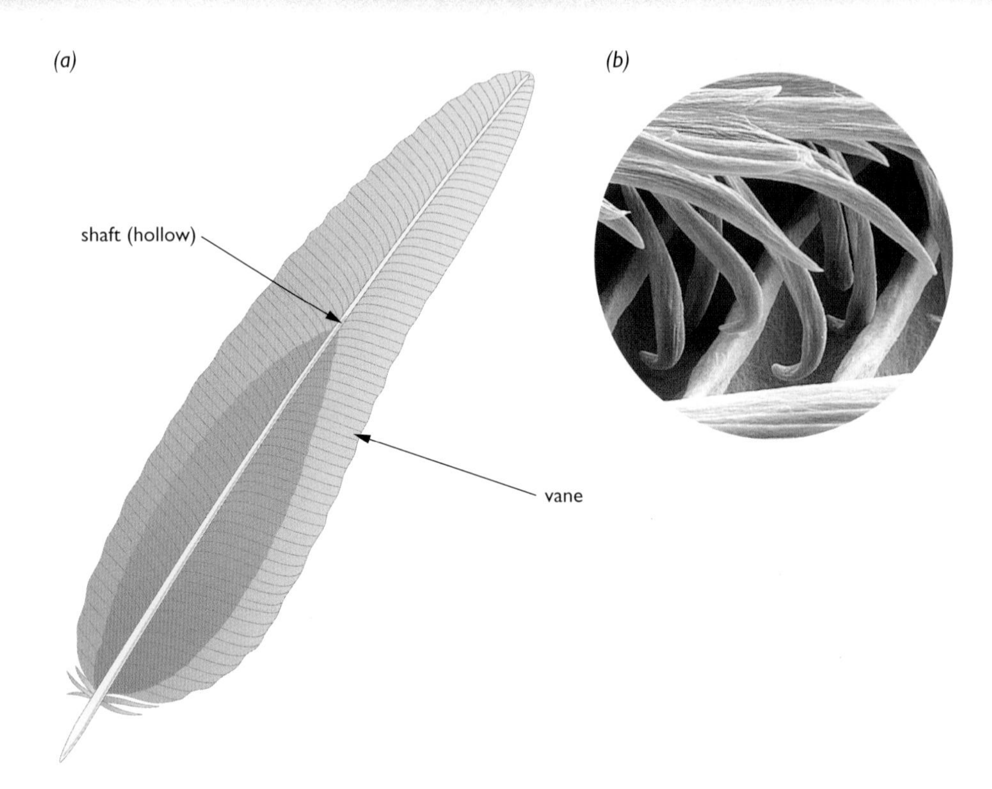

Figure 8.15 *The vane or blade of a flight feather is made up of many fine filaments or barbs. Through a microscope you can see that the barbs are held together by interlocking barbules.*

Birds have two types of flight – powered and unpowered. Powered flight involves flapping the wings to produce a movement of air over them. Unpowered flight means gliding, using natural air currents acting over the wing. Both methods depend on the shape of the bird's wing.

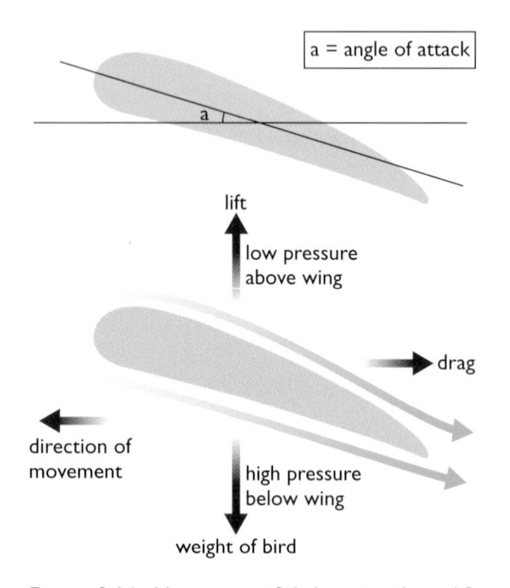

Figure 8.16 *How an aerofoil shape produces lift.*

When a bird is flying, the wing acts as an **aerofoil**. This word describes the shape of the wing in cross section (Figure 8.16). It is the same shape as the wing of an aeroplane, flatter on the lower surface and more rounded on the top.

When air passes over the wing, it travels faster over the top of the wing and more slowly underneath. The air molecules are further apart above the wing, so the air pressure there is lower. The higher pressure under the wing produces an upward force – lift. If the **angle of attack** (Figure 8.16) is increased, the wing produces more lift. However, this cannot go on indefinitely. If the angle of attack is too great, the flow of air over the wing becomes rough or **turbulent** (Figure 8.17). The feathers start to become ruffled and the wing **stalls** – it no longer gives lift. The bird has evolved a way of overcoming this problem. It has feathers attached to the first finger at the front of the wing, called the **alula**, or bastard wing. When the wing starts to stall, the bird raises the alula. The 'slot' between the alula and the leading edge of the wing smoothes out the airflow over the wing and prevents stalling.

However, the wing of a bird is not stiff like an aeroplane wing. To take off and maintain height, a bird has to flap its wings. This is achieved by the contraction of enormous breast muscles pulling down on the humerus. These flight muscles are attached to the breastbone or sternum, which has a deep **keel** that provides a large surface area for muscle attachment (Figure 8.18).

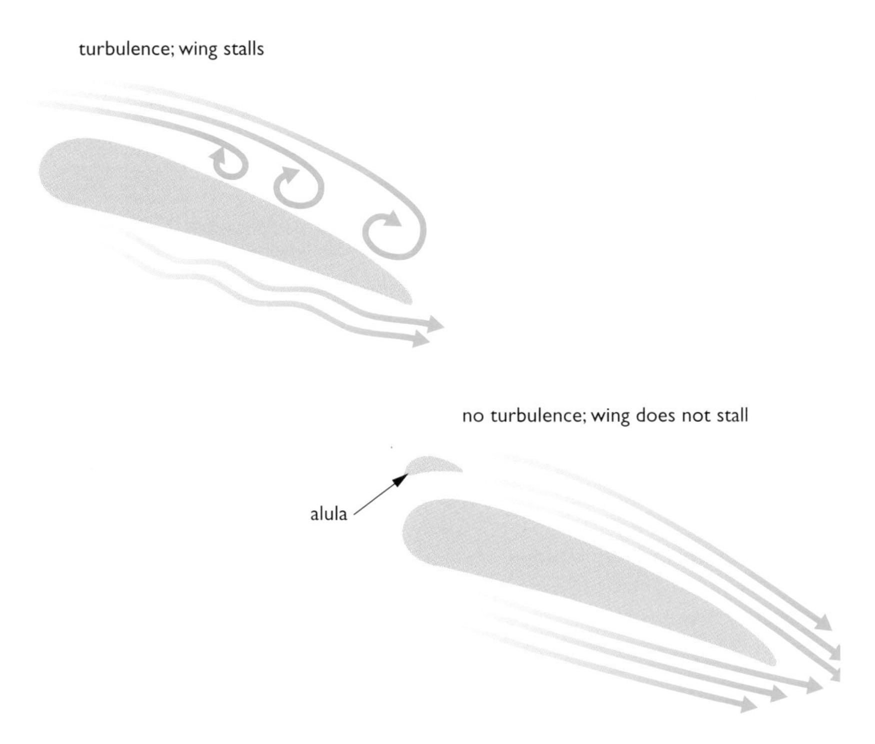

Figure 8.17 *The alula reduces turbulent airflow over the wing, preventing stalling.*

To pull the wings up again, there are smaller breast muscles which have tendons passing through holes between the shoulder bones, so that they attach to the top of the humerus (Figure 8.19). As a result of this arrangement, when they contract they pull the wing up instead of down.

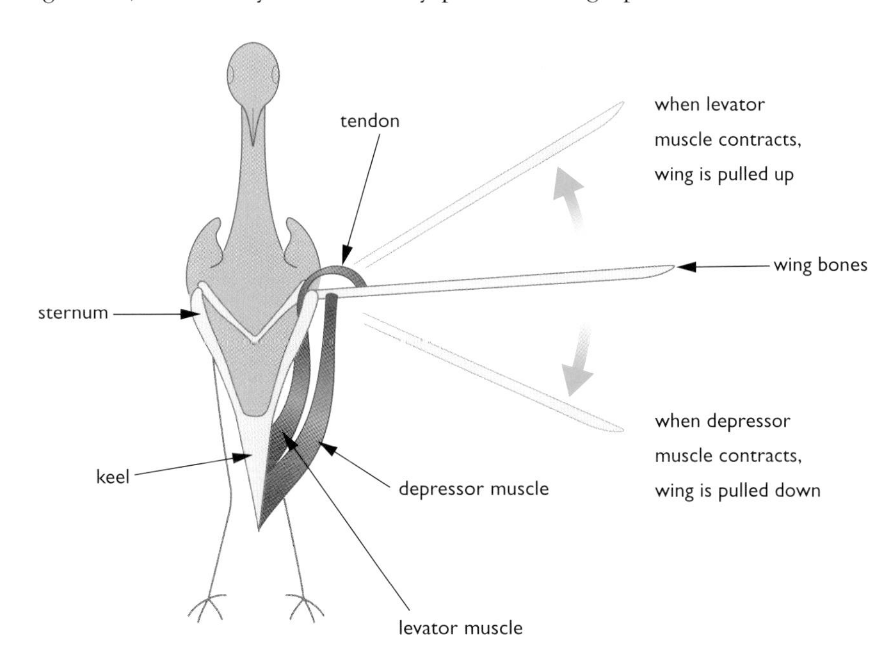

Figure 8.19 *Muscles moving a bird's wing. The depressor muscle pulls the wing down. The levator muscle pulls it up.*

Flapping the wings is not a simple up and down movement. Instead, the edge of the wing moves through a figure-of-eight pattern, first downwards and forwards and then upwards and backwards (Figure 8.20).

Depending on species, the flight muscles make up as much as a third of the mass of a bird.

Figure 8.18 *In this pigeon skeleton you can see the large sternum and keel, which provides a rigid framework and increased area for muscle attachment.*

Figure 8.20 *Freeze-frame photography of a Java dove in flapping flight. Notice that the wing movement is different in the downstroke compared with the upstroke.*

The primary and secondary feathers enable the wing to produce both lift and forward propulsion (thrust) during the downstroke. On the upstroke, the wings have less air resistance and move more quickly. This is because the feathers open up, so that air can flow between them (Figure 8.21). This can only take place because of the structure and arrangement of the feathers. The wider vane of each feather lies underneath the narrower vane of the next feather, so that air can pass between the feathers on the upstroke, but is blocked by the feathers on the downstroke.

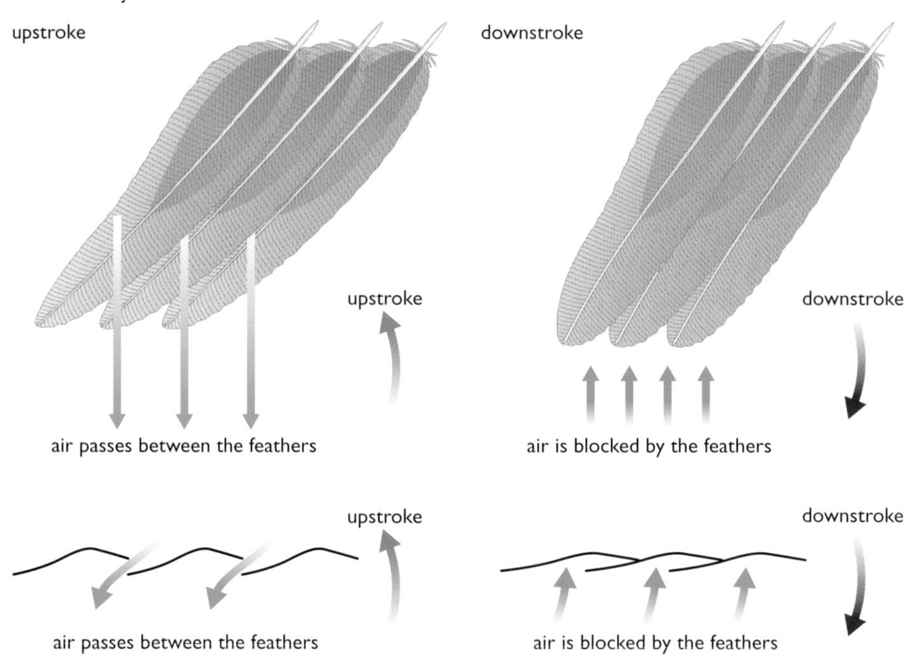

Figure 8.21 *During the downstroke, muscles close the feathers so that they block air movement. On the upstroke, the feathers are opened up like the slats of a Venetian blind, letting air pass between them.*

The 'air sacs' of a bird are not to be confused with the air sacs (alveoli) in a mammal's lungs. The bird's air sacs consist of five pairs of large, thin-walled bags squeezed between the muscles and organs of the body. Contraction of the body muscles pushes air in and out of the sacs and through the bird's lungs. Oxygen is extracted from the air twice: once as it passes in through the lungs to the sacs and again as it passes out through the lungs from the sacs.

Birds have a number of other adaptations for flight. For example, their bones are hollow, with a honeycomb of struts inside. This structure is strong but light. Birds breathe through a system of lungs and **air sacs**, which are very efficient in extracting the oxygen needed to maintain flight. Their tail feathers help to keep their balance during flight.

The bird's wing has developed from the same arrangement of bones as in the human arm and hand. Many other vertebrates share this arrangement, called the **pentadactyl limb**. 'Pentadactyl' is from the Greek for 'five digits', i.e. five fingers or toes. During the course of evolution, humans have kept all five digits, whereas birds have only three fingers remaining on their front limbs. The other fingers have decreased in size and eventually disappeared altogether. The bird's wings are supported partly by the finger bones and partly by the arm bones, while the wrist bones are also reduced in size and number. It is interesting to compare the bones of humans and birds with another group that has evolved flight: the bats. These mammals have wings supported by their finger bones, which are very long (Figure 8.22).

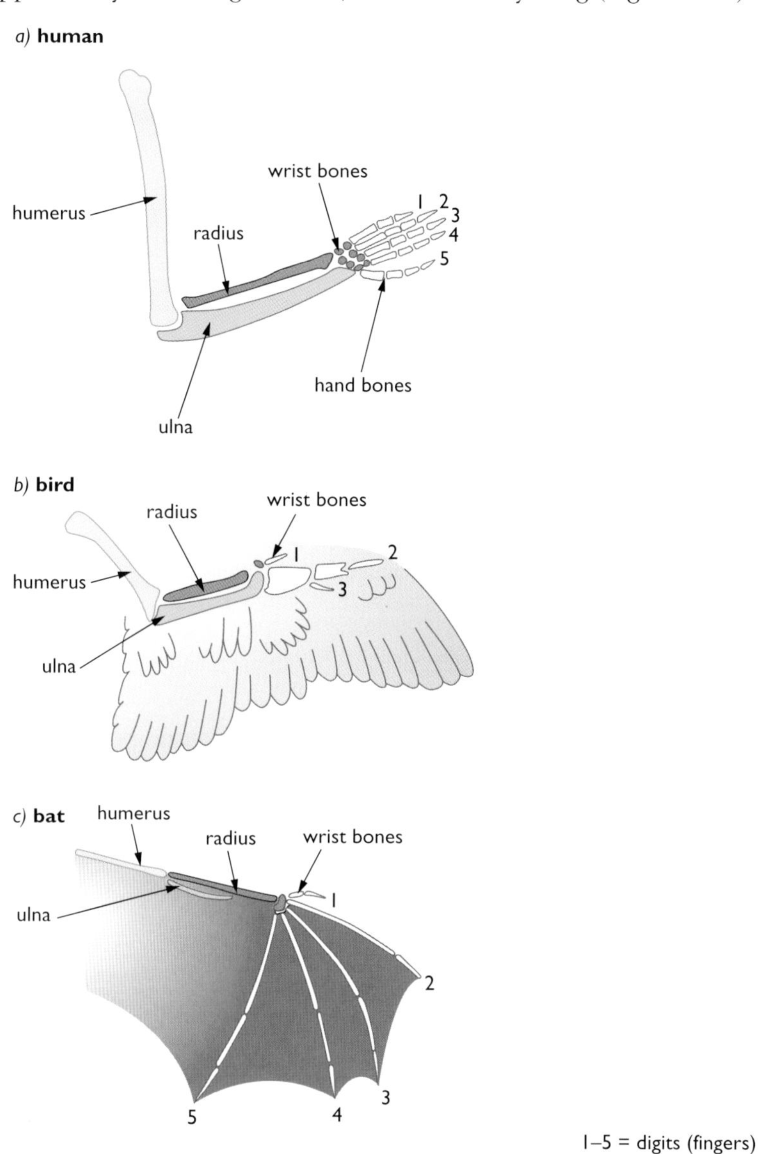

Figure 8.22 *Development of the pentadactyl limb in humans, birds and bats.*

These bones are known as **homologous** structures. They show that all three groups of animals have evolved from a common ancestor with the basic pentadactyl limb structure. You will read more about evolution in Chapter 19.

End of Chapter Checklist

If you haven't got a copy of your specification, read the introduction on page vi.

You will need to be able to do some or all of the following. Check your Awarding Body's specification (syllabus) to find out exactly what you need to know.

- Know that vertebrates have an internal skeleton acting as a framework for movement and support and how the composition of bone makes it rigid without being brittle.
- Describe the structure and function of a synovial joint.
- Understand the meaning of a sprain and a dislocation and be able to outline treatment for these.
- Describe the differences between skeletal, cardiac and smooth muscle.
- Describe the structure and function of skeletal muscles, including muscle fibres and tendons, and the reasons why there is a need for increased respiration and blood flow during exercise.
- Understand the principle of antagonistic muscle pairs (e.g. biceps and triceps) and be able to interpret the action of muscles and bones in other situations.
- Know the effects of regular exercise on muscle tone and strength, joints, blood supply, heart and lungs.
- Describe the adaptations of fish for movement in water, including the roles of streamlining, muscle blocks, fins and swim bladder.
- Describe the adaptations of birds for flight, including wings with large surface area and aerofoil shape, flight feathers, wing muscles and sternum/keel, honeycombed bones and interlocking barbs on feathers.
- Explain how the arrangement of primary and secondary feathers enables the downbeat to provide lift and thrust, while allowing air to flow between the feathers during the upstroke.
- Understand how wing bones have evolved from the pentadactyl limb.

Questions

More questions on support and movement can be found at the end of Section B on page 119.

1 ***a)*** Which parts of the body are protected by

i) the cranium

ii) the vertebral column

iii) the ribs?

b) Between the vertebrae are discs of cartilage. From your knowledge of the properties of cartilage, suggest what their function is.

c) Name three components of bone.

2 ***a)*** The synovial membrane, synovial fluid and ligamants are parts of a synovial joint. Explain the function of each.

b) What is a sprain?

3 Copy and complete the following paragraph, putting the most suitable word or words in the spaces:

Muscles are normally found in .... pairs, such as the triceps and biceps of the arm. When the .... contracts, it flexes the arm, whereas when the .... contracts, it straightens or extends the arm. When muscles are exercised, they need an increased supply of energy. This is supplied by the process of cell .... , which uses .... and .... and makes carbon dioxide. Exercise also produces heat.

4 The diagram shows the bones and some of the muscles of the back leg of a rabbit.

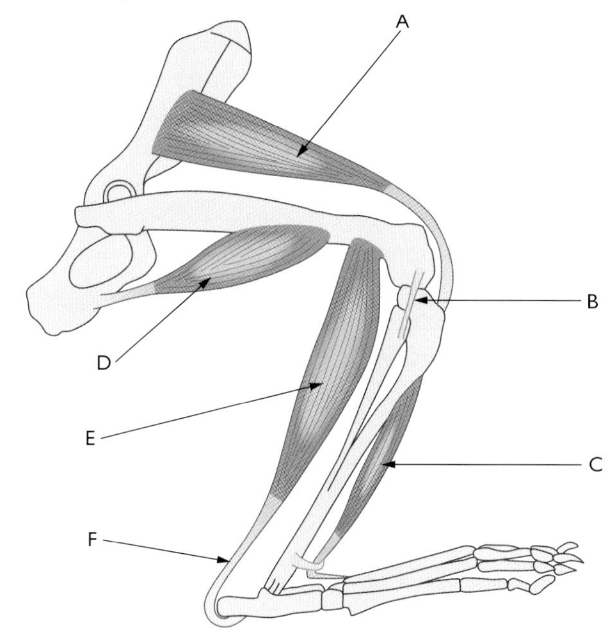

a) Which label (A to F) is a muscle which does the following:

i) straightens the knee joint

ii) flexes the ankle joint?

b) Which label shows a tendon?

c) Is this tendon at the origin or the insertion of the muscle? Explain your answer.

d) Tendons are *inelastic*. Why is this an important property for them to have?

5 Write an essay about the advantages of taking regular exercise. This chapter has outlined some of the benefits, but you can research this important topic more thoroughly and summarise the effects on the muscles, joints, heart and blood system, and the lungs. The length of your essay should be about two sides of writing, plus diagrams. Use other books or the Internet as sources of information.

6 The following are all adaptations that a fish has for swimming: median fins, paired fins, tail fin, contraction of muscle blocks, swim bladder.

Which structure or structures:

a) is used to propel the fish through the water

b) gives neutral buoyancy

c) controls pitch

d) prevents roll?

7 The diagram shows the movements of a bird's wing during flight.

Match each of the statements in column A of the table with the correct ending in column B.

A	B
the downstroke of a wing...	...creates a low pressure above the wing
the arrangement of feathers during the upstroke...	...prevents turbulence that can result in stalling
the aerofoil shape of the wing...	...has a large keel for muscle attachment
the alula...	...allows air flow between the feathers
the sternum...	...creates strength without increasing mass
a hollow wing bone with internal struts...	...provides forward propulsion

Chapter 9: Homeostasis and Excretion

The kidneys play a major part in homeostasis (maintaining a balance of substances in the body) and excretion (removal of waste products from cell metabolism). This chapter is mainly concerned with the activities of the kidneys. It also deals with another important aspect of homeostasis, that of maintaining a steady body temperature.

Inside our bodies, conditions are kept relatively constant. This is called **homeostasis**. The kidneys are organs which have a major role to play in both homeostasis and in the removal of waste products, or **excretion**. They filter the blood, removing substances and controlling the concentration of water and solutes in the blood and other body fluids.

Homeostasis

If you were to drink a litre of water and wait for half an hour, your body would soon respond to this change by producing about the same volume of urine. In other words it would automatically balance your water input and water loss. Drinking is the main way that our bodies gain water, but there are other sources (Figure 9.1). Some water is present in the food that we eat, and a small amount is formed by cell respiration. The body also loses water, mostly in urine, but also smaller volumes in sweat, faeces and exhaled air. Every day, we gain and lose about the same volume of water, so that the total content of our bodies stays more or less the same. This is an example of homeostasis. The word 'homeostasis' means 'steady state', and refers to keeping conditions inside the body relatively constant.

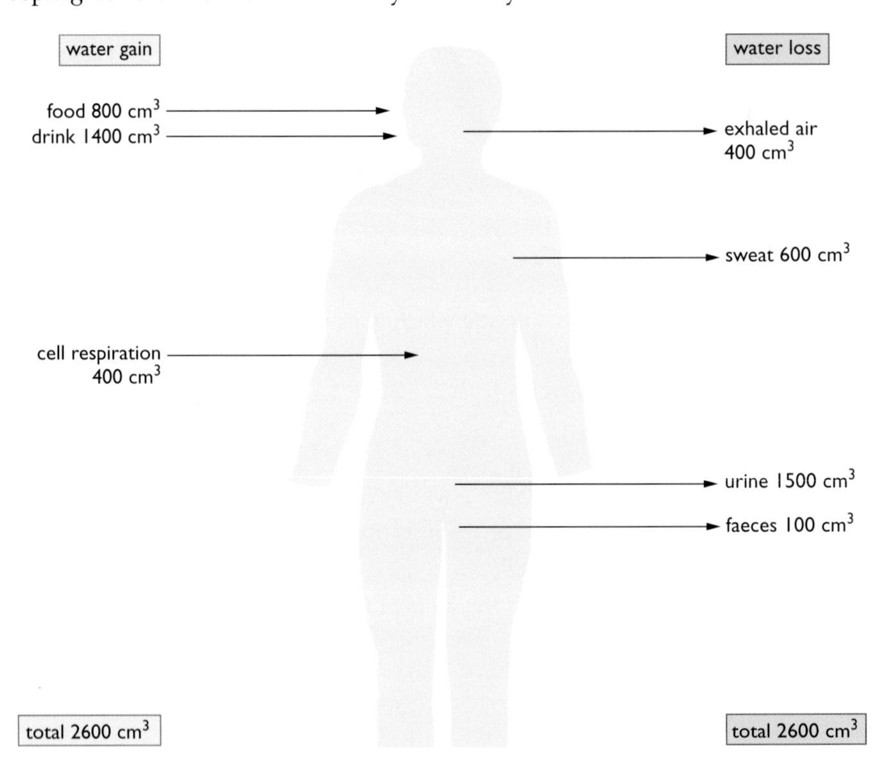

Figure 9.1 *The daily water balance of an adult.*

Homeostasis means 'keeping the conditions in the internal environment of the body relatively constant'.

Inside the body is known as the **internal environment**. You have probably heard of the 'environment', which means the 'surroundings' of an organism. The *internal* environment is the surroundings of the cells inside the body. It particularly means the blood, together with another liquid called **tissue fluid**.

Tissue fluid is a watery solution of salts, glucose and other solutes. It surrounds all the cells of the body, forming a pathway for the transfer of nutrients between the blood and the cells. Tissue fluid is formed by leakage from blood capillaries. It is similar in composition to blood plasma, but lacks the plasma proteins.

It is not just water and salts that are kept constant in the body. Many other components of the internal environment are maintained. For example, the level of carbon dioxide in the blood is regulated, along with the blood pH, the concentration of dissolved glucose (see Chapter 7) and the body temperature.

Homeostasis is important because cells will only function properly if they are bathed in a tissue fluid which provides them with their optimum conditions. For instance, if the tissue fluid contains too many solutes, the cells will lose water by osmosis, and become dehydrated. If the tissue fluid is too dilute, the cells will swell up with water. Both conditions will prevent them working efficiently and might cause permanent damage. If the pH of the tissue fluid is not correct, it will affect the activity of the cell's enzymes, as will a body temperature much different from 37°C. It is also important that excretory products are removed. Substances such as urea must be prevented from building up in the blood and tissue fluid, where they would be toxic to cells.

Urine

An adult human produces about 1.5 dm^3 of urine every day, although this volume depends very much on the amount of water drunk and the volume lost in other forms, such as sweat. Every litre of urine contains about 40 g of waste products and salts (Table 9.1).

Substance	**Amount (g/dm^3)**
urea	23.3
ammonia	0.4
other nitrogenous waste	1.6
sodium chloride (salt)	10.0
potassium	1.3
phosphate	2.3

Table 9.1: *Some of the main dissolved substances in urine.*

'Salts' in urine or in the blood are present as ions. For example, the sodium chloride in Table 9.1 will be in solution as sodium ions (Na^+) and chloride ions (Cl^-). Urine contains many other ions, such as potassium (K^+) phosphate (HPO_4^{2-}) and ammonium (NH_4^+), and removes excess ions from the blood.

Notice the words **nitrogenous waste**. Urea and ammonia are two examples of nitrogenous waste. It means that they contain the element **nitrogen**. All animals have to excrete a nitrogenous waste product.

The reason behind this is quite involved. Carbohydrates and fats only contain the elements carbon, hydrogen and oxygen. Proteins, on the other hand, also contain nitrogen. If the body has too much carbohydrate or fat, these substances can be stored, for example as glycogen in the liver, or as fat under the skin and around other organs. Excess proteins, or their building blocks (called amino acids) *cannot* be stored. The amino acids are first broken down in the liver. They are converted into carbohydrate (which is stored as glycogen) and the main nitrogen-containing waste product, urea.

Excretion is the process by which waste products of metabolism are removed from the body. The main nitrogenous excretory substance is urea. Another excretory product is carbon dioxide, produced by cell respiration and excreted by the lungs (Chapter 2).

The urea passes into the blood, to be filtered out by the kidneys during the formation of urine. Notice that the urea is made by chemical reactions in the cells of the body (the body's metabolism). 'Excretion' means getting rid of waste of this kind. When the body gets rid of solid waste from the digestive system (faeces), this is not excretion, since it contains few products of *metabolism*, just the 'left over remains' of undigested food, along with bacteria and dead cells.

So the kidney is really carrying out two functions. It is a *homeostatic* organ, controlling the water and salt (ion) concentration in the body as well as an *excretory* organ, concentrating nitrogenous waste in a form that can be eliminated.

The urinary system

The human urinary system is shown in Figure 9.2.

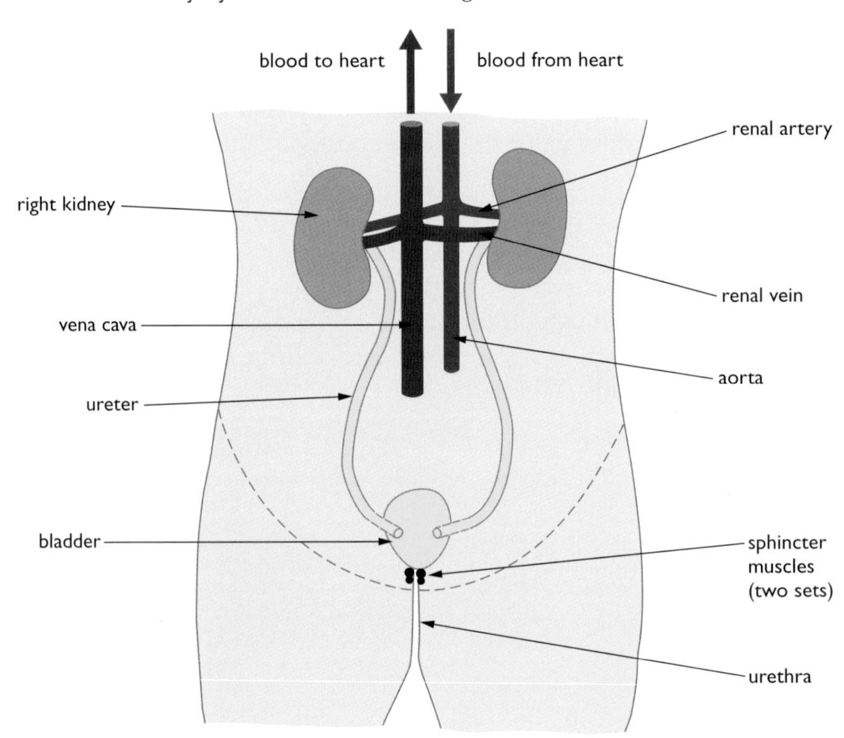

Figure 9.2 *The human urinary system.*

Each kidney is supplied with blood through a short **renal artery**. This leads straight from the body's main artery, the aorta, so the blood entering the kidney is at a high pressure. Inside each kidney the blood is filtered, and the 'cleaned' blood passes out through each **renal vein** to the main vein, or vena cava. The urine passes out of the kidneys through two tubes, the **ureters**, and is stored in a muscular bag called the **bladder**.

A baby cannot control its voluntary sphincter. When the bladder is full, the baby's involuntary sphincter relaxes, releasing the urine. A toddler learns to control this muscle and hold back the urine.

The bladder has a tube leading to the outside, called the **urethra**. The wall of the urethra contains two ring-like muscles, called **sphincters**. They can contract to close the urethra and hold back the urine. The lower sphincter muscle is consciously controlled, or voluntary, while the upper one is involuntary – it automatically relaxes when the bladder is full.

The kidneys

If you cut a kidney lengthwise as in Figure 9.3 you should be able to find the structures shown.

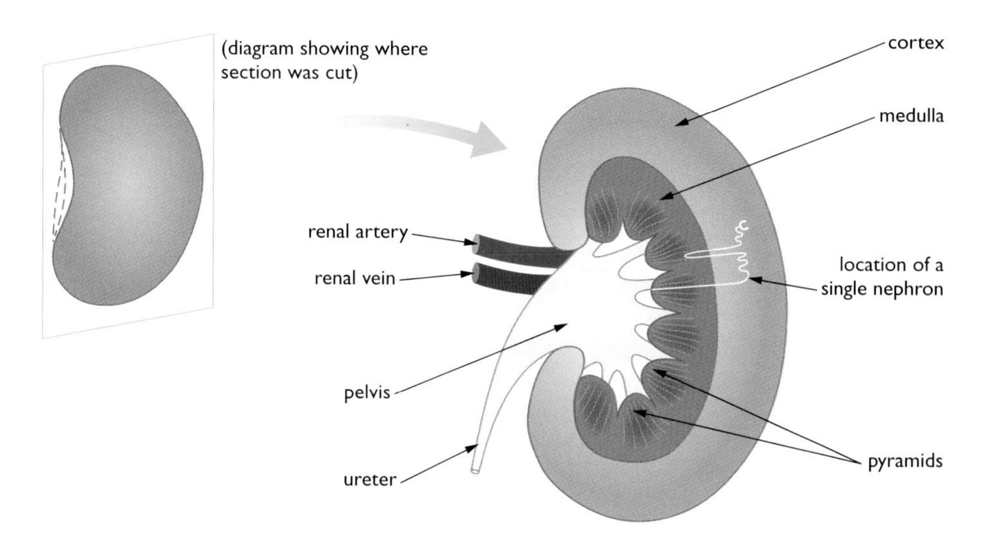

Figure 9.3 *Section through a kidney cut along the plane shown.*

There is not much that you can make out without the help of a microscope. The darker outer region is called the **cortex**. This contains many tiny blood vessels that branch from the renal artery. It also contains microscopic tubes that are not blood vessels. They are the filtering units, called **kidney tubules** or **nephrons** (from the Greek word *nephros*, meaning kidney). The tubules then run down through the middle layer of the kidney, called the **medulla**. The medulla has bulges called **pyramids** pointing inwards towards the concave side of the kidney. The tubules in the medulla eventually join up and lead to the tips of these pyramids, where they empty urine into a space called the **pelvis**. The pelvis connects with the **ureter**, carrying the urine to the **bladder**.

By careful dissection, biologists have been able to find out the structure of a single tubule and its blood supply (Figure 9.4). There are about a million of these in each kidney.

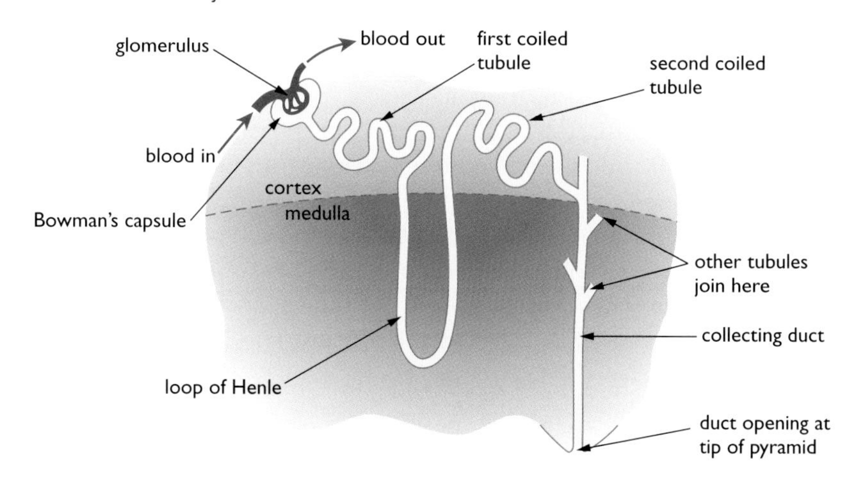

Figure 9.4 *A single nephron, showing its position in the kidney. Each kidney contains about a million of these filtering units.*

At the start of the nephron is a hollow cup of cells called the **Bowman's capsule**. It surrounds a ball of blood capillaries called a **glomerulus** (plural = glomeruli). It is here that the blood is filtered. Blood enters the kidney through the renal artery, which divides into smaller and smaller arteries. The smallest arteries, called **arterioles**, supply the capillaries of the glomerulus (Figure 9.5).

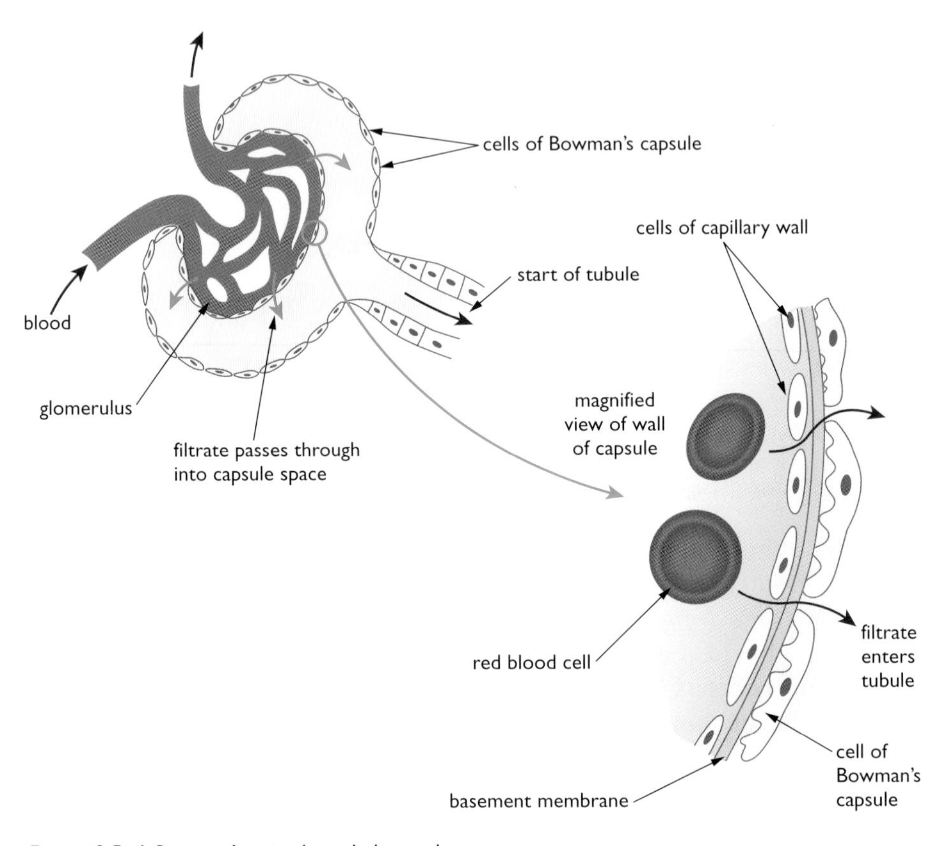

Figure 9.5 *A Bowman's capsule and glomerulus.*

The cells of the glomerulus capillaries do not fit together very tightly, there are spaces between them making the capillary walls much more permeable than others in the body. The cells of the Bowman's capsule also have gaps between them, so only act as a coarse filter. It is the basement membrane which is the fine molecular filter.

A blood vessel with a smaller diameter carries blood away from the glomerulus, leading to capillary networks which surround the other parts of the nephron. Because of the resistance to flow caused by the glomerulus, the pressure of the blood in the arteriole leading to the glomerulus is very high. This pressure forces fluid from the blood through the walls of the capillaries and the Bowman's capsule, into the space in the middle of the capsule. Blood in the glomerulus and the space in the capsule are separated by two layers of cells, the capillary wall and the wall of the capsule. Between the two cell layers is a third layer called the **basement membrane**, which is not made of cells. These layers act like a filter, allowing water, ions and small molecules to pass through, but holding back blood cells and large molecules such as proteins. The fluid that enters the capsule space is called the **glomerular filtrate**. This process, where the filter separates different sized molecules under pressure, is called **ultrafiltration**.

The kidneys produce about 125 cm^3 (0.125 dm^3) of glomerular filtrate per minute. This works out at 180 dm^3 per day. Remember though, only 1.5 dm^3 of urine is lost from the body every day, which is less than 1% of the volume filtered through the capsules. The other 99% of the glomerular filtrate is *reabsorbed* back into the blood.

We know this because scientists have actually analysed samples of fluid from the space in the middle of the nephron. Despite the diameter of the space being only 20 μm (0.02 mm), it is possible to pierce the tubule with microscopic glass pipettes and extract the fluid for analysis. Figure 9.6 shows the structure of the nephron and the surrounding blood vessels in more detail.

There are two **coiled regions** of the tubule in the cortex, separated by a U-shaped loop that runs down into the medulla of the kidney, called the **loop of Henlé**. After the second coiled tubule, several nephrons join up to form a **collecting duct**, where the final urine passes out into the pelvis.

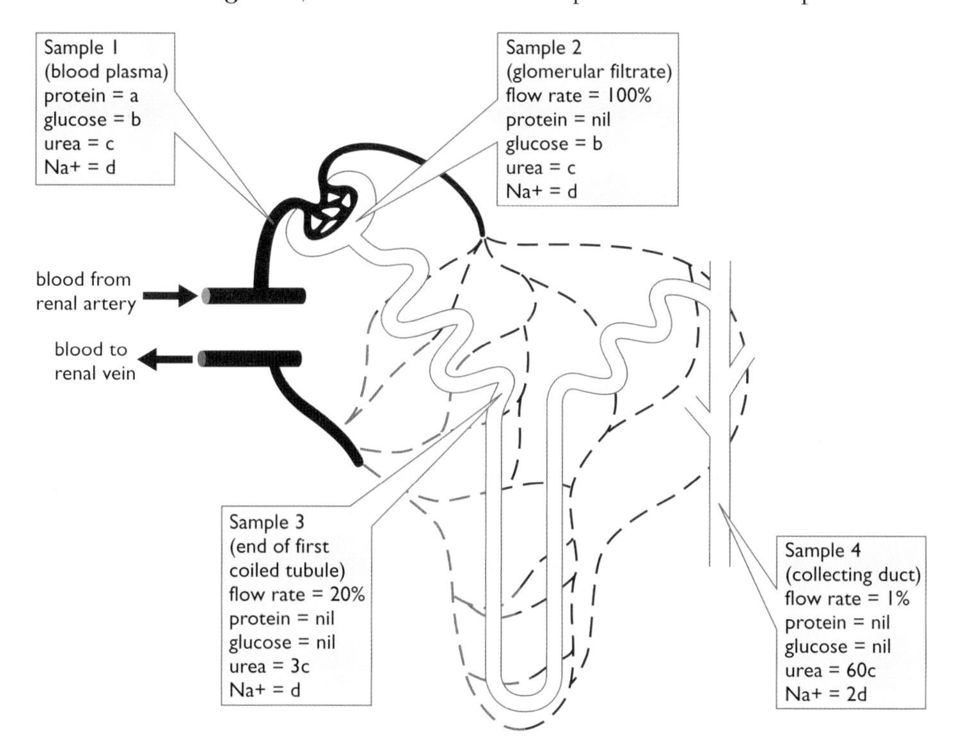

Figure 9.6 *A nephron and its blood supply. Samples 1–4 show what is happening to the fluid as it travels along the nephron.*

Samples 1–4 show the results of analysing the blood before it enters the glomerulus, and the fluid at three points inside the tubule. The flow rate is a measure of how much water is in the tubule. If the flow rate falls from 100% to 50%, this is because 50% of the water in the tubule has gone back into the blood. To make the explanation easier, the concentrations of dissolved protein, glucose, urea and sodium are shown by different letters (a to d). You can tell the relative concentration of one substance at different points along the tubule from this. For example, urea at a concentration '3c' is three times more concentrated than when it is 'c'.

In the blood (sample 1) the plasma contains many dissolved solutes, including protein, glucose, urea and salts (just sodium ions, Na^+, are shown here). As we saw above, protein molecules are too big to pass through into the tubule, so the protein concentration in sample 2 is zero. The other substances are at the same concentration as in the blood.

Now look at sample 3, taken at the end of the first coiled part of the tubule. The flow rate that was 100% is now 20%. This must mean that 80% of the water in the tubule has been reabsorbed back into the blood. If no solutes were reabsorbed along with the water, their concentrations should be *five times* what they were in sample 2. Since the concentration of sodium hasn't changed, 80% of this substance must have been reabsorbed (and some of the urea too). However, the glucose concentration is now zero – *all* of the glucose is taken back into the blood in the first coiled tubule. This is necessary because glucose is a useful substance that is needed by the body.

Finally, look at sample 4. By the time the fluid passes through the collecting duct, its flow rate is only 1%. This is because 99% of the water has been reabsorbed. Protein and glucose are still zero, but most of the urea is still in the fluid. The level of sodium is only 2d, so not all of it has been reabsorbed, but it is still twice as concentrated as in the blood.

This is a summary of what happens in the kidney nephron:

Part of the plasma leaves the blood in the Bowman's capsule and enters the nephron. The filtrate consists of water and small molecules. As the fluid passes along the nephron, all the glucose is absorbed back into the blood in the first coiled part of the tubule, along with most of the sodium and chloride ions. In the rest of the tubule, more water and ions are reabsorbed, and some solutes like ammonium ions are secreted into the tubule. The final urine contains urea at a much higher concentration than in the blood. It also contains controlled quantities of water and ions.

This description has only looked at a few of the more important substances. Other solutes are concentrated in the urine by different amounts. Some, like ammonium ions, are secreted *into* the fluid as it passes along the tubule. The concentration of ammonium ions in the urine is about 150 times what it is in the blood.

You might be wondering what the role of the loop of Henlé is. The full answer to this is too complicated for a GCSE textbook, and a simple explanation will have to be sufficient for now. It is involved with concentrating the fluid in the tubule by causing more water to be reabsorbed into the blood. Mammals with long loops of Henlé can make a more concentrated urine than ones with short loops. Desert animals have many long loops of Henlé, so they are able to produce very concentrated urine, conserving water in their bodies. Animals which have easy access to water, such as otters or beavers, have short loops of Henlé. Humans have a mixture of long and short loops.

Control of the body's water content

Not only can the kidney produce urine that is more concentrated than the blood, it can also *control* the concentration of the urine, and so *regulate* the water content of the blood. This chapter began by asking you to think what would happen if you drank a litre of water. The kidneys respond to this 'upset' to the body's water balance by making a larger volume of more dilute urine. Conversely, if the blood becomes too concentrated, the kidneys produce a smaller volume of urine. These changes are controlled by a hormone produced by the pituitary gland, at the base of the brain. The hormone is called **anti-diuretic hormone**, or **ADH**.

'Diuresis' means the flow of urine from the body, so 'anti-diuresis' means producing less urine. ADH starts to work when your body loses too much water, for example if you are sweating heavily and not replacing lost water by drinking.

As well as causing the pituitary gland to release ADH, the receptor cells in the hypothalamus also stimulate a 'thirst centre' in the brain. This makes the person feel thirsty, so that they will drink water, diluting the blood.

The loss of water means that the concentration of the blood starts to increase. This is detected by special cells in a region of the brain called the **hypothalamus** (see Chapter 6). These cells are sensitive to the solute concentration of the blood, and cause the pituitary gland to release more ADH. The ADH travels in the bloodstream to the kidney. At the kidney tubules, it causes the collecting ducts to become more permeable to water, so that more water is reabsorbed back into the blood. This makes the urine more concentrated, so that the body loses less water and the blood becomes more dilute.

When the water content of the blood returns to normal, this acts as a signal to 'switch off' the release of ADH. The kidney tubules then reabsorb less water. Similarly, if someone drinks a large volume of water, the blood will become too dilute. This leads to lower levels of ADH secretion, the kidney tubules become less permeable to water, and more water passes out of the body in the urine. In this way, through the action of ADH, the level of water in the internal environment is kept constant.

The action of ADH illustrates the principle of **negative feedback.** A change in conditions in the body is detected, and starts a process which works to return conditions to normal. When the conditions are returned to normal, the corrective process is switched off (Figure 9.7). In the situation described, the blood becomes too concentrated. This switches on ADH release, which acts at the kidneys to correct the problem. The word 'negative' means that the process works to eliminate the change. When the blood returns to normal, ADH release is switched off. The feedback pathway forms a 'closed loop'. Many conditions in the body are regulated by negative feedback loops like this.

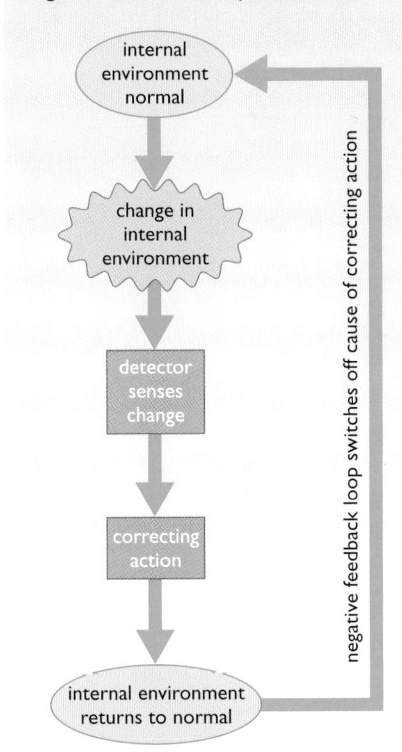

Figure 9.7 *In homeostasis, the extent of a correction is monitored by negative feedback.*

Kidney failure – what happens when things go wrong

A kidney can stop working as a result of disease or an accident. We can live perfectly happily with only one kidney, but if both stop working we will die within a week or so, because poisonous waste builds up in the blood. If a person's kidneys fail, there are two ways they can be kept alive. One is to carry out a kidney **transplant**, and the other is for their blood to be filtered through an artificial kidney machine, in a process called **renal dialysis**.

Kidney transplants

A kidney transplant is an operation where a patient receives another person's kidney. This is often donated by a close relative, or it may be from a person who has had a fatal accident (Figure 9.8).

A kidney transplant is a straightforward operation, and kidneys were one of the first human organs to be successfully transplanted. The main problem comes soon after the operation, when the new kidney can be **rejected** by the patient's immune system (see Chapter 25). The immune system recognises the new kidney as being made of 'foreign' tissues and tries to destroy it. There are a number of ways doctors can try to stop the kidney being rejected:

- The donated organ is taken from a person with tissues that are as genetically similar to the patient's as possible. This is done by **tissue typing**, which classifies the patient's tissues and tries to find a close 'match' with possible donors' tissues. This is why close relatives are often used as donors – they are more likely to have similar tissues. The ideal donor is an identical twin!
- The patient can be given **immuno-suppressant** drugs, until the kidney is 'accepted' by the body. These are drugs which stop the immune system working properly, so that the kidney is not rejected. However, it leaves the patient open to catching infectious diseases. Transplant patients are kept in hospital under very sterile conditions for the first few weeks after the operation, while they are being given these drugs.
- The bone marrow of the patient can be treated with radiation. This reduces the production of white blood cells, which are the cells that are responsible for the immune response (see Chapter 25). Again, this increases the likelihood of infections.

Kidney transplants have a very high success rate, with over 80% of transplants surviving for longer than three years. They are the most simple and effective way of treating a patient whose kidneys have failed permanently. Unfortunately there are not enough donor organs available,

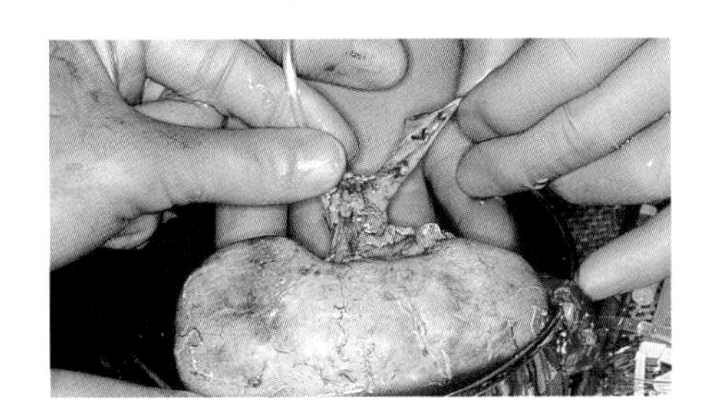

Figure 9.8 *'A kidney being got ready for a transplant operation'.*

Figure 9.9 *If a person is killed in an accident and they carry a donor card, it shows the hospital that they have agreed to allow their organs to be used for transplants. It is important that a transplant is carried out as soon as possible after a person has died.*

The word 'dialysis' means 'splitting into two', and refers to the way the patient's blood is purified by separating off the unwanted waste products such as urea, by the dialysis membrane.

despite the number of people who carry organ donor cards (Figure 9.9). Many people are on a waiting list for a kidney transplant. While on the list, they are treated by renal dialysis.

Kidney dialysis machines

The artificial kidney, or renal dialysis machine, filters the patient's blood, removing urea and other waste, as well as excess water and salts. The filter is a special **dialysis membrane** called **Visking tubing**, which looks rather like cellophane. This thin material has millions of tiny holes in it. The holes will let small molecules like water, ions and urea pass through, but not larger molecules such as proteins, or blood cells.

Blood from the patient flows on one side of the membrane, and a watery liquid called **dialysis solution** flows past the other side, in the opposite direction (Figure 9.10). This is a solution of salts and glucose in exactly the concentrations that the body needs.

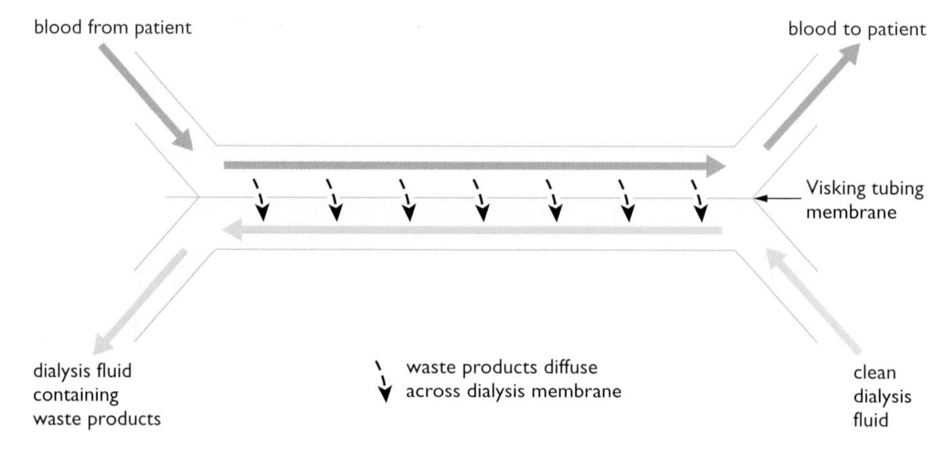

Figure 9.10 *The principle of renal dialysis. The dialysis membrane filters the blood, removing toxic waste.*

As the blood flows past the membrane, urea and unwanted water and salts diffuse through the holes in the membrane into the dialysis fluid. Cells and large molecules such as proteins are kept back in the blood. The dialysis fluid is replaced with fresh solution all the time, so that after several hours, the patient's blood has been 'cleaned' of toxic waste and the correct balance of water and salts established.

The surface area of the dialysis membrane separating the blood and dialysis fluid must be large to filter the blood enough. To achieve this, the membrane can be arranged in different ways. In some dialysis machines it is in the form of many long narrow tubes, while in others it is arranged as a stack of flat sheets.

In order to carry out dialysis, it is easier to take blood from a vein than an artery, because veins are closer to the skin, and have a wider diameter than arteries (see Chapter 5). However, the blood pressure in veins is too low, so an operation is first carried out to join an artery to a vein, which raises the blood pressure. A tube is then permanently connected to the vein, so that the patient can be linked to the machine without having to use a needle each time. The purified blood returns to the patient through a second tube joined to the vein, and the 'used' dialysis fluid is discarded.

The kidney machine is a complex and expensive piece of apparatus (Figure 9.11). It has pumps to keep the blood and dialysis fluid flowing, traps to prevent air bubbles getting into the blood, and the oxygenation and temperature of the blood is controlled. Although dialysis will keep a person whose kidneys have failed alive, it is a time consuming and unpleasant process. The patient has to have their blood 'cleaned' for many hours, two or three times a week. A transplant, if one becomes available, is a much better option.

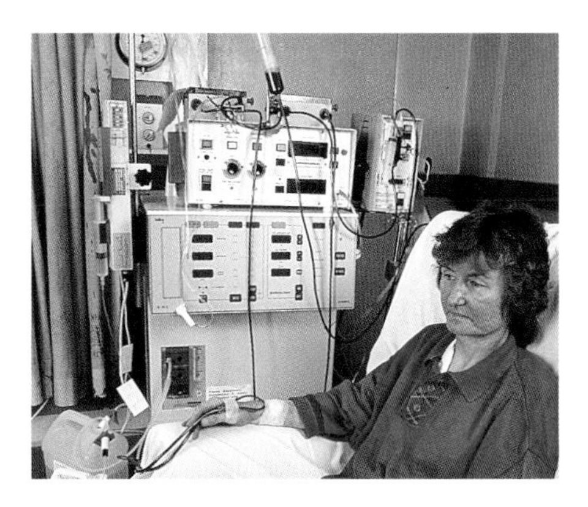

Figure 9.11 *A patient connected to a kidney dialysis machine.*

Control of body temperature

You may have heard mammals and birds described as 'warm blooded'. A better word for this is **homeothermic**. It means that they keep their body temperature constant, despite changes in the temperature of their surroundings. For example, the body temperature of humans is kept steady at about 37°C, give or take a few tenths of a degree. This is another example of homeostasis. All other animals are 'cold blooded'. For example, if a lizard is kept in an aquarium at 20°C, its body temperature will be 20°C too. If the temperature of the aquarium is raised to 25°C, the lizard's body temperature will rise to 25°C as well. We can show this difference between homeotherms and other animals as a graph (Figure 9.12).

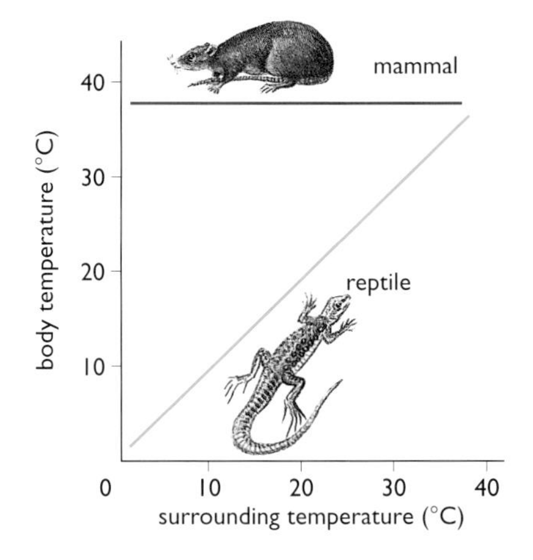

Figure 9.12 *The temperature of a homeotherm such as a mammal is kept constant at different external temperatures, whereas the lizard's body temperature changes.*

In the wild, lizards keep their temperature more constant than in Figure 9.13, by adapting their behaviour. For example, in the morning they may bask in the sun to warm their bodies, or at midday, if the sun is too hot, retreat to holes in the ground to cool down.

The real difference between homeotherms and all other animals is that homeotherms can keep their temperatures constant by using **physiological** changes for generating or losing heat. For this reason, mammals and birds are also called **endotherms**, meaning 'heat from inside'.

Physiology is a branch of biology that deals with how the bodies of animals or plants work, for example how muscles contract, how nerves send impulses, or how xylem carries water through plants. In this chapter you have read about kidney physiology.

An endotherm uses heat from the chemical reactions in its cells to warm its body. It then controls its heat loss by regulating processes like sweating and blood flow through the skin. Endotherms use behavioural ways to control their temperature too. For example, penguins 'huddle' to keep warm, and humans put on extra clothes in winter.

What is the advantage of a human maintaining a body temperature of 37°C? It means that all the chemical reactions taking place in the cells of the body can go on at a steady, predictable rate. The metabolism doesn't slow down in cold environments. If you watch goldfish in a garden pond, you will notice that in summer, when the pond water is warm, they are very active, swimming about quickly. In winter, when the temperature drops, the fish slow down and become very sluggish in their actions. This would happen to mammals too, if their body temperature was not kept steady.

It is also important that the body does not become *too* hot. The cells' enzymes work best at 37°C. At higher temperatures enzymes, like all proteins, are destroyed by **denaturing** (see Chapter 1). Endotherms have all evolved a body temperature around 40°C (Table 9.2) and enzymes that work best at this temperature.

Species	Average and normal range of body temperature (°C)
brown bear	38.0±1.0
camel	37.5±0.5
elephant	36.2±0.5
fox	38.8±1.3
human	36.9±0.7
mouse	39.3±1.3
polar bear	37.5±0.4
shrew	35.7±1.2
whale	35.7±0.1
duck	43.1±0.3
ostrich	39.2±0.7
penguin	39.0±0.2
thrush	40.0±1.7
wren	41.0±1.0

Table 9.2: *The body temperatures of a range of mammals and birds.*

Monitoring body temperature

In humans and other mammals the core body temperature is monitored by a part of the brain called the **thermoregulatory centre**. This is located in the hypothalamus of the brain. It acts as the body's thermostat.

A **thermostat** is a switch that is turned on or off by a change in temperature. It is used in electrical appliances to keep their temperature steady. For example, a thermostat in an iron can be set to 'hot' or 'cool' to keep the temperature of the iron set for ironing different materials.

If a person goes into a warm or cold environment, the first thing that happens is that temperature receptors in the skin send electrical impulses to the hypothalamus, which stimulates the brain to alter our behaviour. We start to feel hot or cold, and usually do something about it, such as finding shade or having a cold drink.

If changes to our behaviour are not enough to keep our body temperature constant, the thermoregulatory centre in the hypothalamus detects a change in the temperature of the blood flowing through it. It then sends signals via nerves to other organs of the body, which regulate the temperature by physiological means.

The skin and temperature control

The human skin has a number of functions related to the fact that it forms the outer surface of the body. These include:

- forming a tough outer layer able to resist mechanical damage
- acting as a barrier to the entry of disease-causing microorganisms
- forming an impermeable surface, preventing loss of water
- acting as a sense organ for touch and temperature changes
- controlling the loss of heat through the body surface.

Figure 9.13 shows the structure of human skin. It is made up of three layers, the epidermis, dermis and hypodermis.

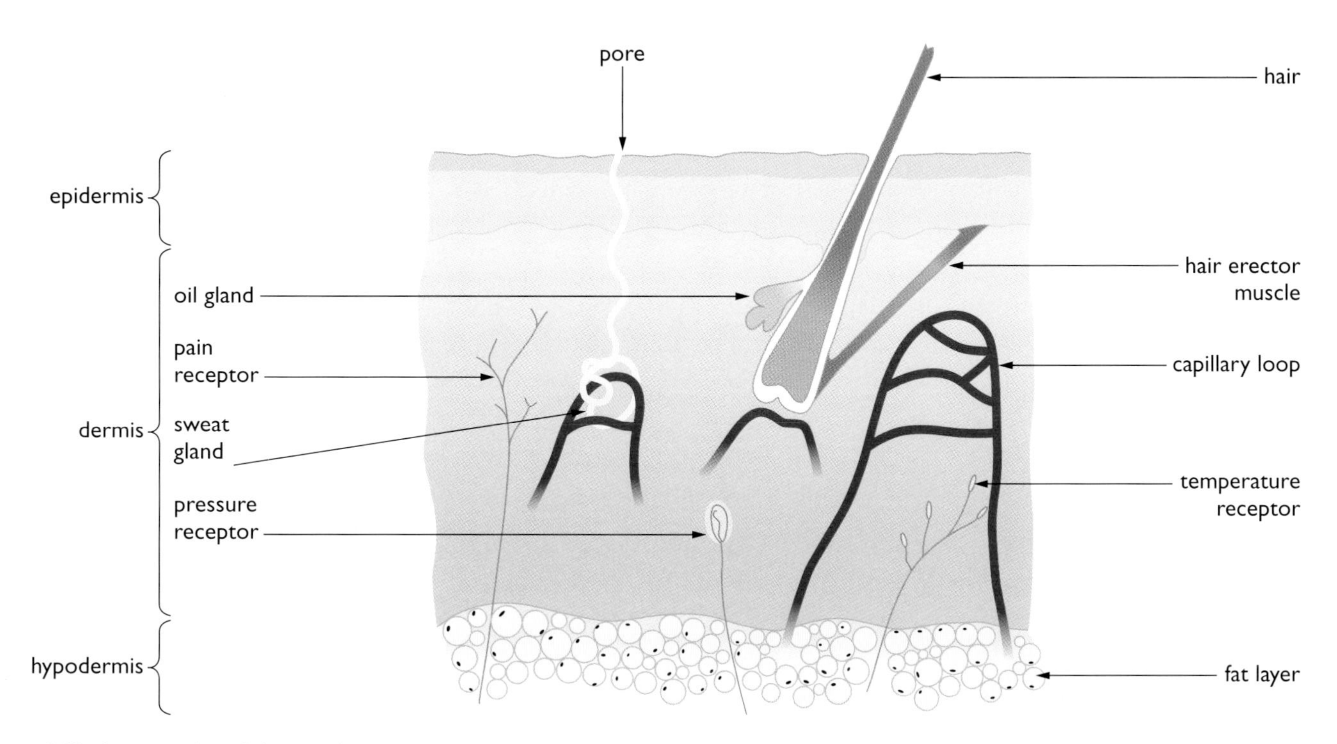

Figure 9.13 *A section through human skin.*

The outer **epidermis** consists of dead cells that stop water loss and protect the body against invasion by microorganisms such as bacteria. The **hypodermis** contains fatty tissue, which insulates the body against heat loss and is a store of energy. The middle layer, the **dermis**, contains many sensory receptors. It is also the location of sweat glands and many small blood vessels, as well as hair follicles. These last three structures are involved in temperature control.

Imagine that the hypothalamus detects a rise in the central (core) body temperature. Immediately it sends nerve impulses to the skin. These bring about changes to correct the rise in temperature.

First of all, the **sweat glands** produce greater amounts of sweat. This liquid is secreted onto the surface of the skin. When a liquid evaporates, it turns into a gas. This change needs energy, called the **latent heat of vaporisation**. When sweat evaporates, the energy is supplied by the body's heat, cooling the body down. It is not that the sweat is cool – it is secreted at body temperature. It only has a cooling action when it evaporates. In very humid atmospheres (e.g. a tropical rainforest) the sweat stays on the skin and doesn't evaporate. It then has very little cooling effect.

Secondly, hairs on the surface of the skin lie flat against the skin's surface. This happens because of the relaxation of tiny muscles called **hair erector muscles** attached to the base of each hair. In cold conditions these contract and the hairs are pulled upright. The hairs trap a layer of air next to the skin, and since air is a poor conductor of heat, this acts as insulation. In warm conditions the thinner layer of trapped air means that more heat will be lost. This is not very effective in humans, because the hairs over most of our body do not grow very large. It is very effective in hairy mammals like cats or dogs. The same principle is used by birds, which 'fluff out' their feathers in cold weather.

Students often describe vasodilation incorrectly. They talk about the blood vessels 'moving nearer the surface of the skin'. They don't *move* at all, it's just that more blood flows through the surface vessels.

Lastly, there are tiny blood vessels called capillary loops in the dermis. Blood flows through these loops, radiating heat to the outside, and cooling the body down. If the body is too hot, small arteries (arterioles) leading to the capillary loops **dilate** (widen). This increases the blood flow to the skin's surface. At the same time blood flow through deeper capillaries is reduced by arterioles narrowing (Figure 9.14). This is called **vasodilation**.

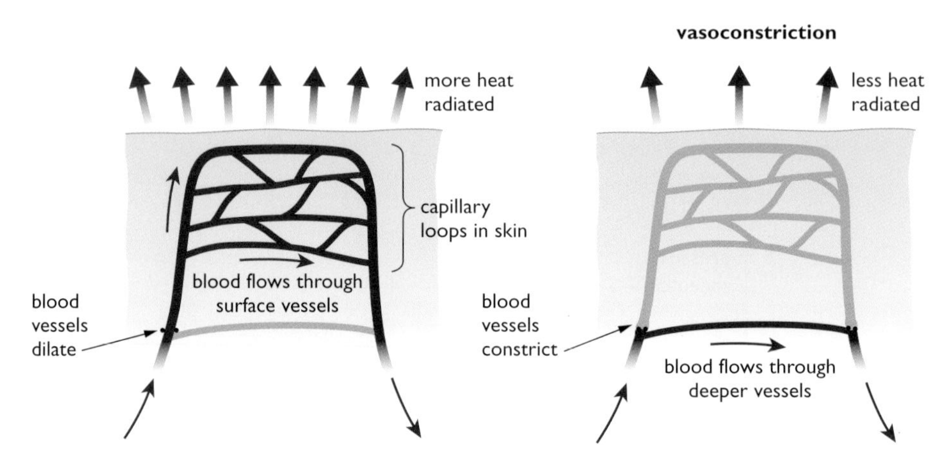

Figure 9.14 *Blood flow through the surface of the skin is controlled by vasodilation or vasoconstriction.*

In cold conditions the opposite happens. The surface blood vessels undergo **vasoconstriction**, so that less heat is lost. Vasoconstriction and vasodilation are brought about by tiny rings of muscles in the walls of the arterioles, called sphincter muscles, like the sphincters you met earlier in this chapter, at the outlet of the bladder.

There are other ways that the body can control heat loss and heat gain. In cold conditions the body's **metabolism** speeds up, generating more heat. The liver, a large organ, can produce a lot of metabolic heat in this way. The hormone **adrenaline** stimulates the increase in metabolism (see Chapter 7). **Shivering** also takes place, where the muscles contract and relax rapidly. This also generates a large amount of heat.

Sweating, vasodilation and vasoconstriction, hair erection, shivering and changes to the metabolism, along with behavioural actions, work together to keep the body temperature to within a few tenths of a degree of the 'normal' 37°C. If the difference is any bigger than this it shows that something is wrong. For instance, a temperature of 39°C might be due to an illness.

End of Chapter Checklist

If you haven't got a copy of your specification, read the introduction on page vi.

You will need to be able to do some or all of the following. Check your Awarding Body's specification (syllabus) to find out exactly what you need to know.

- Define homeostasis and recognise the importance of maintaining a constant internal environment, including control of the water and ion (salt) content of the body, and body temperature.
- Understand the term excretion, including carbon dioxide produced from respiration, and urea formed in the liver from the breakdown of excess amino acids.
- Describe the structure and function of the kidneys, including the structure and role of an individual nephron. The functions to include ultrafiltration in the Bowman's capsule and glomerulus and reabsorption of glucose, water and salts in the coiled tubules and collecting duct.
- Understand how people who suffer from kidney failure may be treated by dialysis or a kidney transplant, including the precautions that can be taken to prevent rejection of a transplanted kidney.
- Be able to explain the role of ADH from the pituitary gland in regulating the water content of the blood.
- Understand the advantages of maintaining a constant body temperature. Know that the body temperature is monitored by the thermoregulatory centre in the brain.
- Be able to explain how humans control body temperature by sweating, vasodilation and vasoconstriction, alteration in metabolic rate and shivering.

Questions

More questions on homeostasis and excretion can be found at the end of Section B on page 119.

1 Explain the meaning of the following terms:

a) homeostasis

b) excretion

c) ultrafiltration

d) renal dialysis

e) endotherm.

2 The diagram below shows a simple diagram of a nephron (kidney tubule).

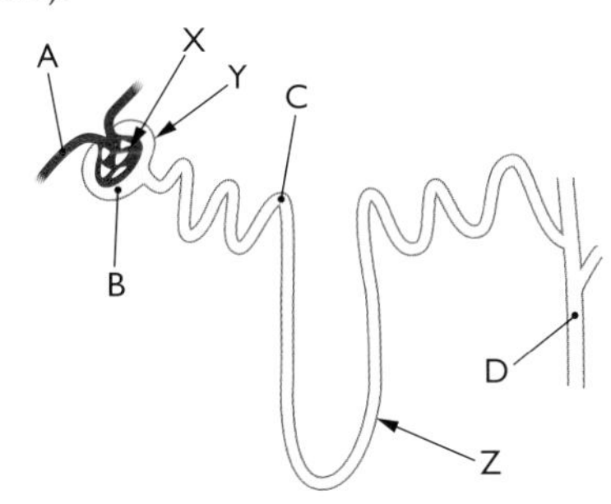

a) What are the names of the parts labelled X, Y and Z?

b) Four places in the nephron and its blood supply are labelled A, B and C and D. Which of the following substances are found at each of these four places?

water urea protein glucose salt

3 The hormone ADH controls the amount of water removed from the blood by the kidneys. Write a short description of the action of ADH in a person who has lost a lot of water by sweating, but has been unable to replace this water by drinking. Explain how this is an example of negative feedback. (You need to write about half a page to answer this question fully.)

4 The bar chart shows the volume of urine collected from a person before and after drinking 1000 cm^3 (1 dm^3) of distilled water. The person's urine was collected immediately before the water was drunk and then at 30 minute intervals for four hours.

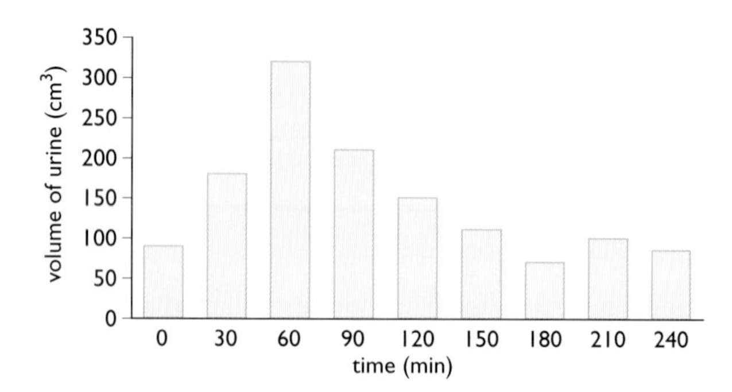

a) Describe how the output of urine changed during the course of the experiment.

b) Explain the difference in urine produced at 60 minutes and at 90 minutes.

c) The same experiment was repeated with the person sitting in a very hot room. How would you expect the volume of urine collected to differ from the first experiment? Explain your answer.

d) Between 90 and 120 minutes, the person produced 150 cm^3 of urine. If the rate of filtration at the glomeruli during this time was 125 cm^3 per minute, calculate the percentage of filtrate reabsorbed by the kidney tubules.

5 Use a word processor to construct a table showing the changes that take place when a person is put in a hot or cold environment. Your table should have three columns:

Changes taking place	Hot environment	Cold environment
sweating		
blood flow through capillary loops		Vasoconstriction decreases blood flow through surface capillaries so that less heat is radiated from the skin.
hairs in skin		
shivering		
metabolism		

6 Look at the body temperatures of mammals and birds shown in Table 9.2 on page 114. Use the information in the table to answer these questions:

a) How does the average temperature of birds differ from the average temperature of mammals? Can you suggest why this is an advantage for birds?

b) Is there a relationship between the body temperature of a mammal and the temperature of its habitat? Give an example to support your answer.

c) Polar bears have thick white fur covering their bodies. Explain two ways in which this is an adaptation to their habitat.

7 Carry out an Internet search to find out about kidney transplants.

a) When was the first successful kidney transplant carried out?

b) Where is a transplanted kidney placed in the body? (It is not put in its normal location in the body.)

c) How many kidney transplants are performed in the UK every year?

End of Section Questions

1 The table shows the concentration of gases in inhaled and exhaled air:

Gas	Inhaled air	Exhaled air
nitrogen	78	79
oxygen		
carbon dioxide		
other gases (mainly argon)	1	1

a) Copy the table and fill in the gaps by choosing from the following numbers:

21 4 0.04 16 *(2 marks)*

b) Explain why the concentration of carbon dioxide is so different. *(2 marks)*

c) Explain why exhaling is a form of excretion. *(2 marks)*

d) The following features can be seen in the lungs:

i) thin membranes between the alveoli and the blood supply

ii) a good blood supply

iii) a large surface area.

In each case explain how the feature helps gas exchange to happen quickly. *(6 marks)*

Total 12 marks

2 Digestion is brought about by enzymes converting large insoluble molecules into smaller soluble molecules that can be more easily absorbed.

a) The activity of enzymes is influenced by pH and temperature. The graph shows the activity of two human enzymes from different regions of the gut at different pHs.

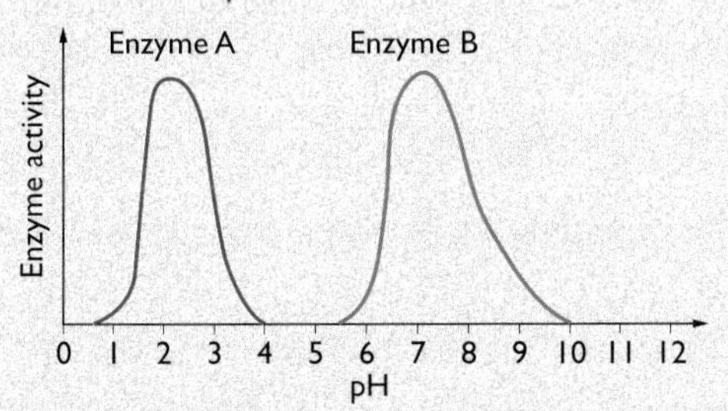

i) Suggest which regions of the gut the two enzymes come from. Explain your answer. *(4 marks)*

ii) Which nutrient does enzyme A digest? *(1 mark)*

b) Ruminants are a kind of herbivore. They have a large four-chambered stomach. The first is the rumen and it contains billions of microorganisms. Periodically, some of the microorganisms die and pass into the other chambers of the stomach and then the small intestine. This pathway is shown in the diagram.

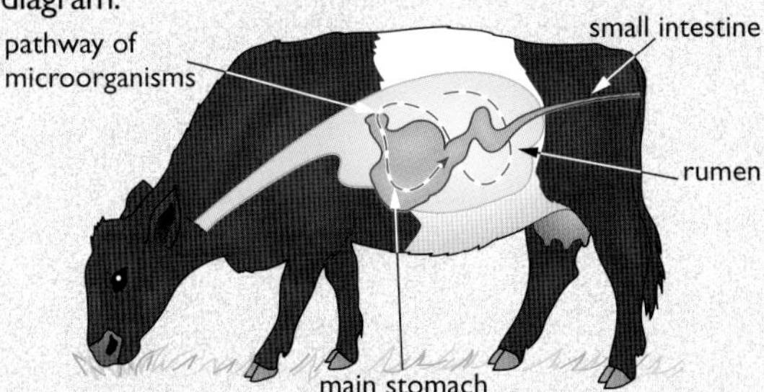

i) What is the benefit to the cow of having billions of microorganisms in the rumen? *(2 marks)*

ii) What is the benefit to the microorganisms of living in the rumen? *(2 marks)*

iii) Name the type of nutritional association between the cow and the microorganisms. Explain your answer. *(2 marks)*

c) Farmers sometimes include urea in cattle food. The microorganisms in the rumen can use urea to make protein.

i) In mammals, where in the body is urea made? *(1 mark)*

ii) What is urea made from? *(1 mark)*

iii) Suggest how feeding urea to cattle can result in an increased growth rate. *(1 mark)*

iv) The Bowman's capsule and the Loop of Henlé are both parts of a nephron. Explain how each of them help to remove urea from the bloodstream. *(4 marks)*

Total 18 marks

3 Coronary heart disease is one of the major causes of death in the western world. One of the risk factors associated with coronary heart disease is a diet containing too much saturated fat. However, some fat is essential in the diet.

a) The graph shows the percentage of deaths due to coronary heart disease in different age groups in Britain in 1983.

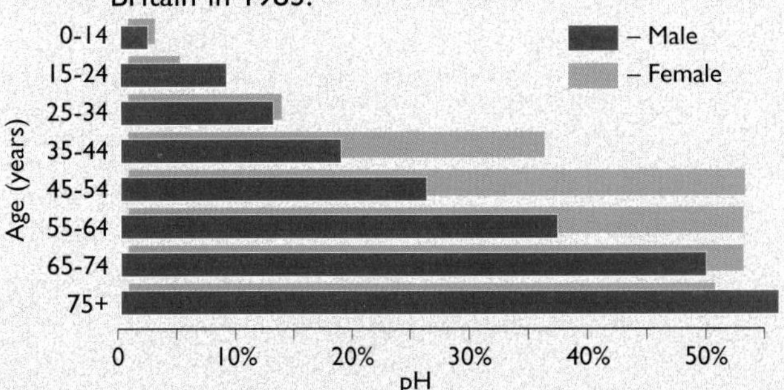

i) Compare the death rate from coronary heart disease in males and females below the age of 35 and above the age of 35. *(6 marks)*

ii) Other than eating a healthy diet, suggest two ways that we can avoid coronary heart disease. *(2 marks)*

b) Give two reasons why we need to eat some fat in our diet. *(2 marks)*

c) Fat is digested by the enzyme lipase.

i) Name the organ which produces lipase. *(1 mark)*

ii) Name the products of digestion of fats. *(2 marks)*

iii) Bile salts are important in the digestion of fats. Explain why. *(2 marks)*

Total 15 marks

4 The circulation system carries nutrients, oxygen and carbon dioxide around the body.

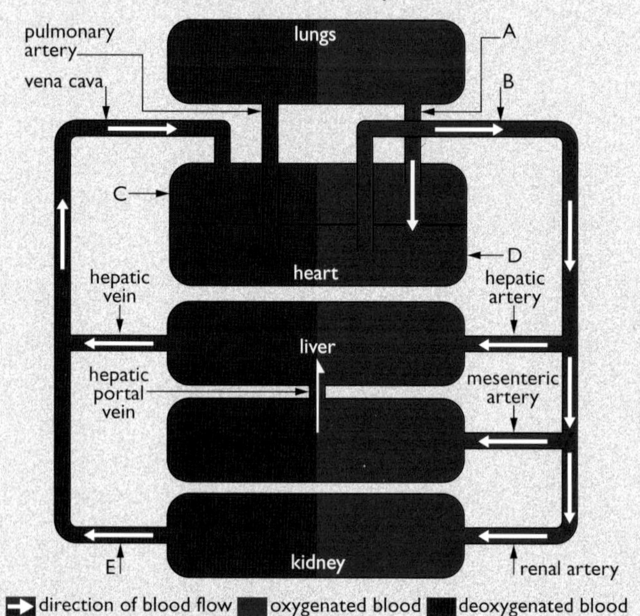

a) Write out the correct labels for A to E. *(5 marks)*

b) Give two differences between the blood vessels at point X and point Y. *(2 marks)*

c) During exercise, the adrenal gland releases the hormone adrenaline. Reflexes involving the medulla of the brain influence the heartbeat and breathing.

i) Describe two effects of adrenaline on the heartbeat. *(2 marks)*

ii) Describe two other effects of adrenaline on the body. *(2 marks)*

d) How is a reflex action different from a voluntary action? *(2 marks)*

e) Following exercise there is a recovery period in which breathing rate and heart rate gradually return to pre-exercise levels. Explain why they do not return immediately to these levels. *(3 marks)*

Total 16 marks

5 Humans and other mammals are able to maintain a constant body temperature which is usually higher than that of their surroundings.

a) Explain the advantage in maintaining a constant, high body temperature. *(1 mark)*

b) The temperature of the blood is constantly monitored by the brain. If it detects a drop in blood temperature, the following things happen: the arterioles leading to the skin capillaries constrict, less sweat is formed and shivering begins.

i) Explain how each response helps the body to keep warm. (2 marks)

ii) Explain how the structure of arterioles allows them to constrict. (2 marks)

c) When the weather is hot we produce less urine.

i) What is the name of the hormone that controls the amount of urine produced by the body? *(1 mark)*

ii) Explain why the body produces less urine in hot weather. *(1 mark)*

iii) Explain how the hormone in *i)* works in the kidney to produce less urine. *(3 marks)*

Total 10 marks

6 The brain controls most of the actions in our bodies.

a) The diagram shows a human brain in section.

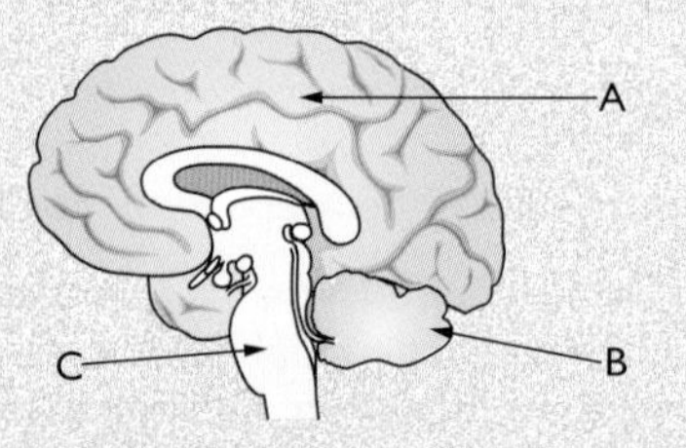

Name, and describe the functions of, the parts of the brain labelled A, B and C. *(6 marks)*

b) The brain contains billions of cells, each of which forms synapses with thousands of other cells.

i) Make a labelled drawing of a synapse. *(3 marks)*

ii) Explain how information is transmitted along a nerve cell and then across a synapse. *(3 marks)*

c) Many drugs can influence activity in the brain, usually by influencing transmission at synapses.

i) Explain what an analgesic is. Give two examples. *(3 marks)*

ii) Explain what a stimulant is. Give two examples. *(3 marks)*

Total 18 marks

Chapter 10: Plants and Food

This chapter looks at photosynthesis, the process by which plants make starch, and the structure of leaves in relation to photosynthesis. It also describes the nature and method of obtaining other nutrient materials and their uses in the plant.

Plants make starch

All the foods shown in Figure 10.1 are products of plants. Some, such as potatoes, rice and bread, form the staple diet of humans. They all contain *starch*, which is the main storage carbohydrate made by plants. Starch is a good way of storing carbohydrate because it is not soluble, is compact and can be broken down easily.

Figure 10.1 *All these foods are made by plants and contain starch.*

Testing leaves for starch

You can test for starch in food by adding a few drops of red-brown iodine solution (see Chapter 3). If the food contains starch, a blue-black colour is produced.

Leaves that have been in sunlight also contain starch, but you can't test for it by adding iodine solution to a fresh leaf. The outer waxy surface of the leaf will not absorb the solution, and besides, the green colour of the leaf would hide the colour change. To test for starch in a leaf, the outer waxy layer needs to be removed and the leaf decolourised. You can do this by placing the leaf in boiling ethanol. The steps in the method are as follows (see Figure 10.2):

- Set up a beaker of water on a tripod and gauze and heat the water until it boils.
- Remove a leaf from the plant, and holding it with forceps, kill it by placing it in the boiling water for 30 seconds (this stops all chemical reactions in the leaf).

- Turn off the Bunsen burner, place the leaf in a boiling tube containing ethanol and stand the boiling tube in a hot water bath. The tube containing ethanol *must not be heated directly, since ethanol is highly flammable*. The boiling point of ethanol (about 78 °C) is lower than that of water (100 °C) so the ethanol will boil for a few minutes until the water bath cools down. This is long enough to remove most of the **chlorophyll** from the leaf.
- When the leaf has turned colourless or pale yellow, remove it and wash it with cold water to soften it.
- Spread the leaf out on a white tile or Petri dish. Cover the leaf with a few drops of iodine solution and leave it for a few minutes, noting any colour change.

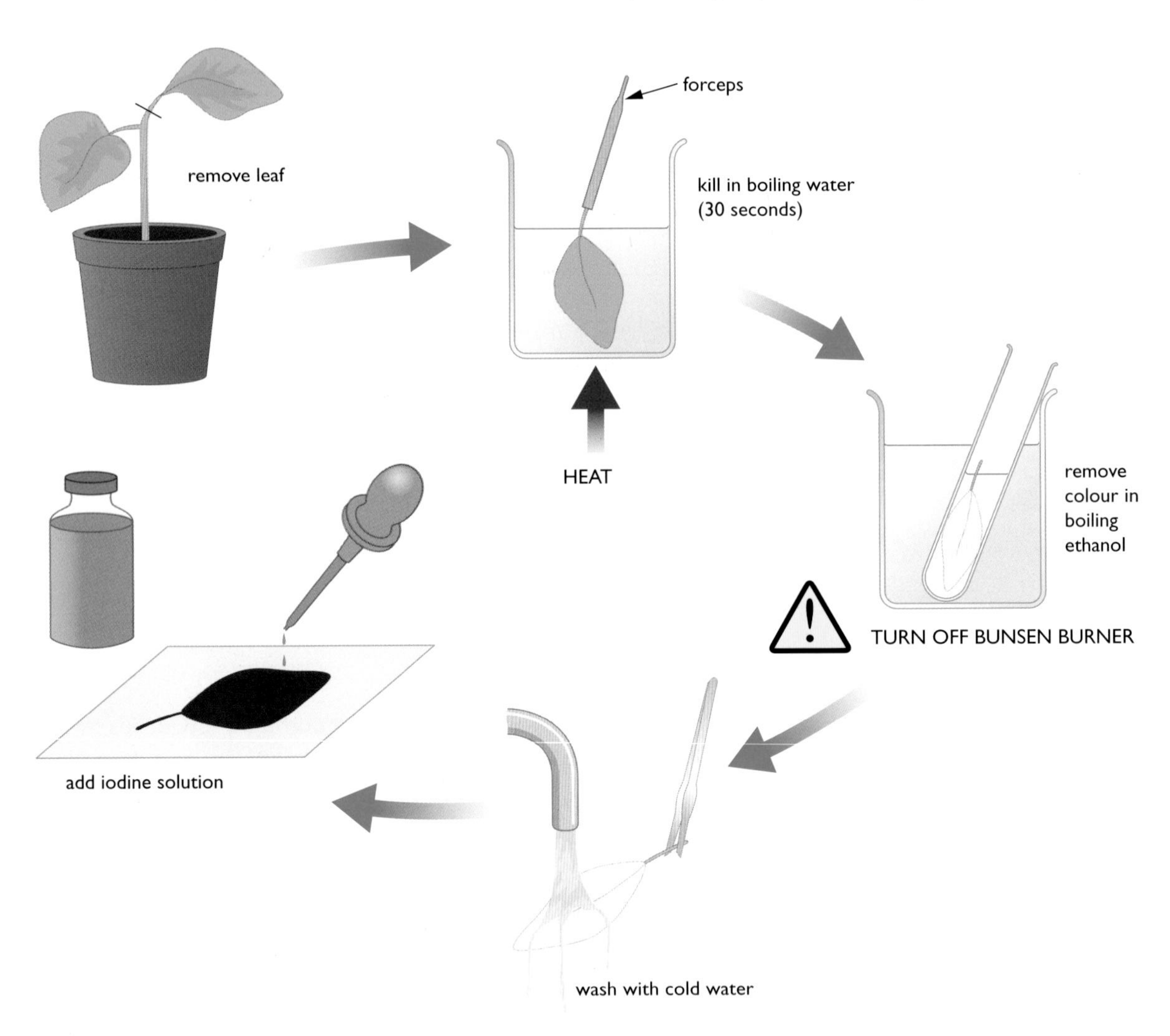

Figure 10.2 *How to test a leaf for starch.*

You can 'destarch' a plant by placing it in the dark for 2 or 3 days. The plant uses up the starch stores in its leaves.

When you try this method, you will see that the parts of the leaf that contain starch turn a very dark 'blue-black' colour as the iodine reacts with the starch. This will only work if the plant has had plenty of light for some time before the test.

Starch is only made in the parts of leaves that are green. You can show this by testing a **variegated** leaf which has green and white areas. The white regions, which lack the green pigment called chlorophyll, give a negative starch test. The results of the starch tests on three leaves are shown in Figure 10.3.

Figure 10.3 *The leaf on the left was taken from a plant that was left under a bright light for 48 hours. The middle leaf is from a plant that was put in a dark cupboard for the same length of time. The third leaf is variegated, and only contains starch in the parts which were green.*

You might think that the results of the test on the variegated leaf prove that chlorophyll is needed for photosynthesis. However, this is not really a 'fair test'. The leaf could have photosynthesised in the white areas and transported the sugars elsewhere in the plant. Similarly, the green areas may not be photosynthesising at all, but simply laying down starch from glucose made somewhere else. All it really shows is that starch is made in the green areas and not in the white areas of the leaf. We *assume* this is because chlorophyll is needed for photosynthesis.

Depriving a plant of light is not the only way you can prevent it making starch in its leaves. You can also place the plant in a closed container containing a chemical called soda lime. This substance absorbs carbon dioxide from the air around the plant. If the plant is kept under a bright light but with no carbon dioxide, it will again be unable to make starch.

Where does the starch come from?

You have now found out three important facts about starch production by leaves:

- it uses carbon dioxide from the air
- it needs light
- it needs chlorophyll in the leaves.

As well as starch, there is another product of this process which is essential to the existence of most living things on the Earth – oxygen. When a plant is in the light, it makes oxygen gas. You can show this using an aquatic plant such as *Elodea* (Canadian pondweed). When a piece of this plant is placed in a test tube of water under a bright light, it produces a stream of small bubbles. If the bubbles are collected and their contents analysed, they are found to contain a high concentration of oxygen (Figure 10.4).

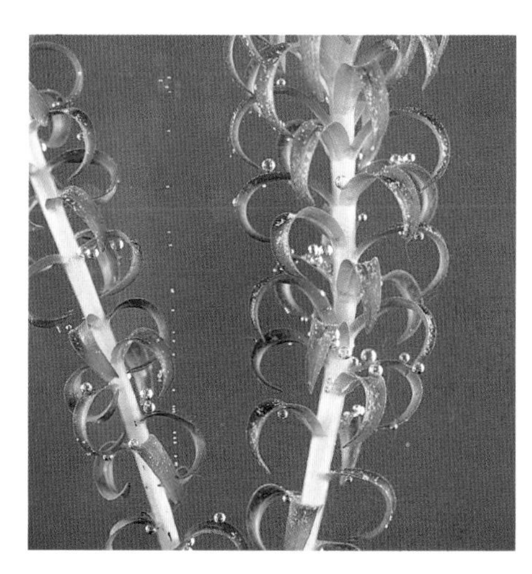

Figure 10.4 *The bubbles of gas released from this pondweed contain a higher concentration of oxygen than in atmospheric air.*

Starch is composed of long chains of glucose (see Chapter 3). A plant does not make starch directly, but first produces glucose, which is then joined together in chains to form starch molecules. A carbohydrate made of many sugar sub-units is called a **polysaccharide**. Glucose has the formula $C_6H_{12}O_6$. The carbon and oxygen atoms of the glucose come from the carbon dioxide gas in the air around the plant. The hydrogen atoms come from another molecule essential to the living plant – water.

It would be very difficult in a school laboratory to show that a plant uses water to make starch. If you deprived a plant of water in the same way as you deprived it of carbon dioxide, it would soon wilt and die. However, scientists have proved that water is used in photosynthesis. They have done this by supplying the plant with water with 'labelled' atoms, for example using the 'heavy' isotope of oxygen (^{18}O). This isotope ends up in the oxygen gas produced by the plant. A summary of the sources of the atoms in the glucose and oxygen looks like this:

Isotopes are forms of the same element with the same atomic number but different mass numbers (due to extra neutrons in the nucleus). Isotopes of some elements are radioactive and can be used as 'labels' to trace chemical pathways, others like ^{18}O are identified by their mass.

$$CO_2 \rightarrow C_6H_{12}O_6 + O_2^* \leftarrow H_2O^*$$

(*oxygen labelled with ^{18}O)

Photosynthesis

The 'photo' in photosynthesis comes from the Greek word *photos*, meaning light, and a 'synthesis' reaction is one where small molecules are built up into larger ones.

Plants use the simple inorganic molecules carbon dioxide and water, in the presence of chlorophyll and light, to make glucose and oxygen. This process is called **photosynthesis**.

It is summarised by the equation:

$$\text{carbon dioxide} + \text{water} \xrightarrow[\text{chlorophyll}]{\text{light}} \text{glucose} + \text{oxygen}$$

$$\text{or:} \quad 6CO_2 + 6H_2O \longrightarrow C_6H_{12}O_6 + 6O_2$$

The role of the green pigment, chlorophyll, is to absorb the light energy needed for the reaction to take place. The products of the reaction (glucose and oxygen) contain more energy than the carbon dioxide and water. You will probably have noticed that the equation for photosynthesis is the reverse of the one for aerobic respiration (see Chapter 1):

$$C_6H_{12}O_6 + 6O_2 \longrightarrow 6CO_2 + 6H_2O \text{ (plus energy)}$$

Respiration, which is carried out by both animals and plants, *releases* energy (but not as light) from the breakdown of glucose. The chemical energy in the glucose came originally from light 'trapped' by the process of photosynthesis.

The structure of leaves

Most green parts of a plant can photosynthesise, but the leaves are the plant organs which are best adapted for this function. To be able to photosynthesise efficiently, leaves need to have a large surface area to absorb light, many chloroplasts containing the chlorophyll, a supply of water and carbon dioxide, and a system for carrying away the products of photosynthesis to other parts of the plant. They also need to release oxygen (and water vapour) from the leaf cells. Most leaves are thin, flat structures supported by a leaf stalk which can grow to allow the blade of the leaf to be angled to receive the maximum amount of sunlight (Figure 10.5).

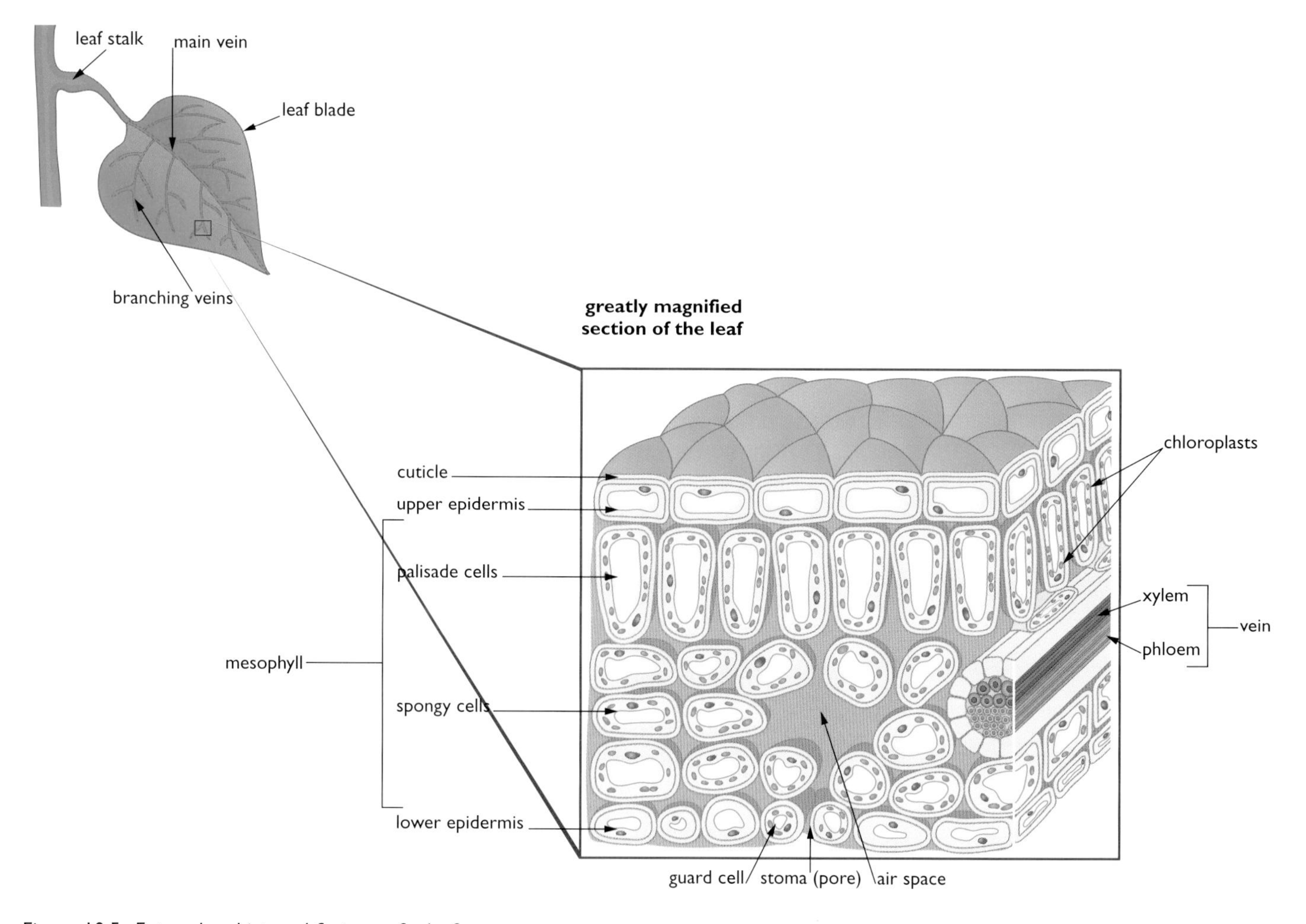

Figure 10.5 *External and internal features of a leaf.*

Inside the leaf are layers of cells with different functions:

- The two outer layers of cells (the upper and lower **epidermis**) have few chloroplasts and are covered by a thin layer of a waxy material called the **cuticle**. This reduces water loss by evaporation, and acts as a barrier to the entry of disease-causing microorganisms such as bacteria and fungi.
- The lower epidermis has many pores called **stomata** (a single pore is a **stoma**). Usually the upper epidermis contains fewer or no stomata. The stomata allow carbon dioxide to diffuse into the leaf, to reach the photosynthetic tissues. They also allow oxygen and water vapour to diffuse out. Each stoma is formed as a gap between two highly specialised cells called **guard cells**, which can alter their shape to open or close the stoma (see Chapter 11).
- In the middle of the leaf are two layers of photosynthetic cells called the **mesophyll** ('mesophyll' just means 'middle of the leaf'). Just below the upper epidermis is the **palisade** layer. This is a tissue made of elongated cells, each containing hundreds of chloroplasts, and is the main site of photosynthesis. The palisade cells are close to the source of light, and the upper epidermis is relatively transparent, allowing light to pass through to the enormous numbers of chloroplasts which lie below.

A 'gas exchange' surface is a tissue that allows gases (usually oxygen, carbon dioxide and water vapour) to pass across it between the plant or animal and the outer environment. Gas exchange surfaces all have a large surface area in proportion to their volume, which allows large amounts of gases to diffuse across. Examples include the alveoli of the lungs, the gills of a fish and the spongy mesophyll of a leaf.

Starch is insoluble and so cannot be transported around the plant. The phloem carries only soluble substances such as sugars. These are converted into other compounds when they reach their destination.

- Below the palisade layer is a tissue made of more rounded, loosely packed cells, with air spaces between them, called the **spongy** layer. These cells also photosynthesise, but have fewer chloroplasts than the palisade cells. They form the main **gas exchange surface** of the leaf, absorbing carbon dioxide and releasing oxygen and water vapour. The air spaces allow these gases to diffuse in and out of the mesophyll.
- Water and mineral ions are supplied to the leaf by vessels in a tissue called the **xylem**. This forms a continuous transport system throughout the plant. Water is absorbed by the roots and passes up through the stem and through veins in the leaves in the **transpiration stream**. In the leaves, the water leaves the xylem and supplies the mesophyll cells.
- The products of photosynthesis, such as sugars, are carried away from the mesophyll cells by another transport system, the **phloem**. The phloem supplies all other parts of the plant, so that tissues and organs that can't make their own food receive products of photosynthesis. The veins in the leaf contain both xylem and phloem tissue, and branch again and again to supply all parts of the leaf.

You can find out more about both plant transport systems in Chapter 11.

Photosynthesis and respiration

Through photosynthesis, plants supply animals with two of their essential needs – food and oxygen, as well as removing carbon dioxide from the air. But remember that living cells, including plant cells, respire *all the time*, and they need oxygen for this. When the light intensity is high, a plant carries out photosynthesis at a much higher rate than it respires. So in bright light, there is an overall uptake of carbon dioxide from the air around a plant's leaves, and a surplus production of oxygen that animals can use. A plant only produces more carbon dioxide than it uses up in dim light. We can show this as a graph of carbon dioxide exchanged at different light intensities (Figure 10.6).

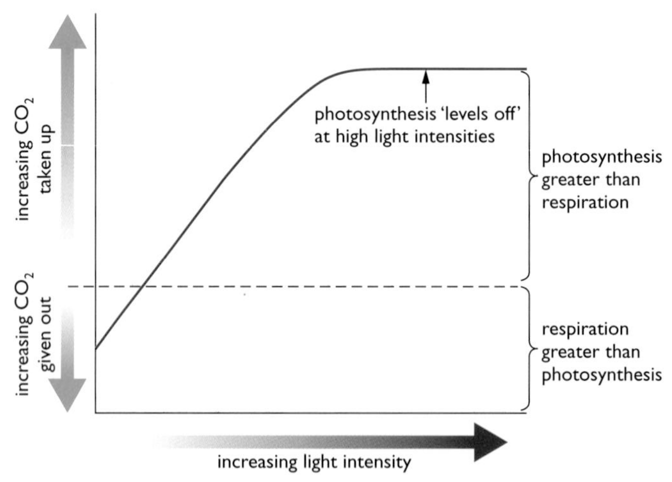

Figure 10.6 *As the light intensity gets higher, photosynthesis speeds up, but eventually levels off in very bright light.*

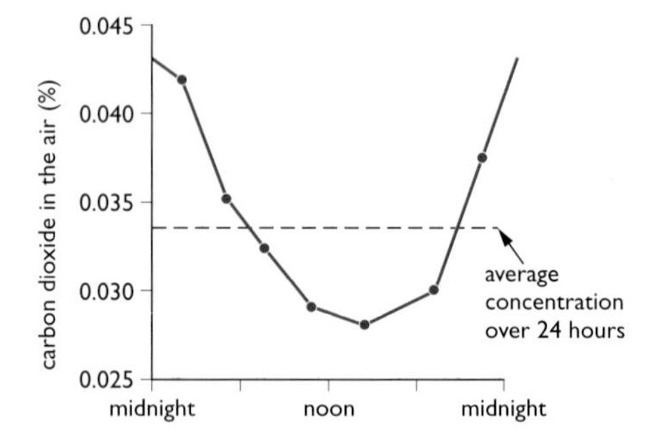

Figure 10.7 *Photosynthesis affects the concentration of carbon dioxide in the air around plants. Over a 24-hour period, the concentration rises and falls, as a result of the relative levels of photosynthesis and respiration.*

The concentration of carbon dioxide in the air around plants actually changes throughout the day. Scientists have measured the level of carbon dioxide in the air in the middle of a field of long grass in summer. They found that the air contained least carbon dioxide in the afternoon, when photosynthesis was happening at its highest rate (Figure 10.7). At night when there was no photosynthesis, the level of carbon dioxide rose. This rise is due to less carbon dioxide being absorbed by the plants, while carbon dioxide was added to the air from the respiration of all organisms in the field.

Factors affecting the rate of photosynthesis

ideas
evidence

In Figure 10.6, you can see that when the light intensity rises, the rate of photosynthesis starts off rising too, but eventually it reaches a maximum rate. What makes the rate 'level off' like this? It is because some other factor needed for photosynthesis is in short supply, so that increasing the light intensity does not affect the rate any more. Normally, the factor which 'holds up' photosynthesis is the concentration of carbon dioxide in the air. This is only about 0.03 to 0.04%, and the plant can only take up the carbon dioxide and fix it into carbohydrate at a certain rate. If the plant is put in a closed container with a higher than normal concentration of carbon dioxide, it will photosynthesise at a faster rate. If there is both a high light intensity and a high level of carbon dioxide, the temperature may limit the rate of photosynthesis, by limiting the rate of the chemical reactions in the leaf. A rise in temperature will then increase the rate. With normal levels of carbon dioxide, very low temperatures (close to 0 °C) slow the reactions, but high temperatures (above about 35 °C) also reduce photosynthesis by denaturing enzymes in the plant cells (see Chapter 1).

Light intensity, carbon dioxide concentration and temperature can all act as what are called **limiting factors** in this way. This is easier to see as a graph (Figure 10.8).

A limiting factor is the component of a reaction that is in 'shortest supply' so that it prevents the rate of the reaction increasing, in other words sets a 'limit' to it.

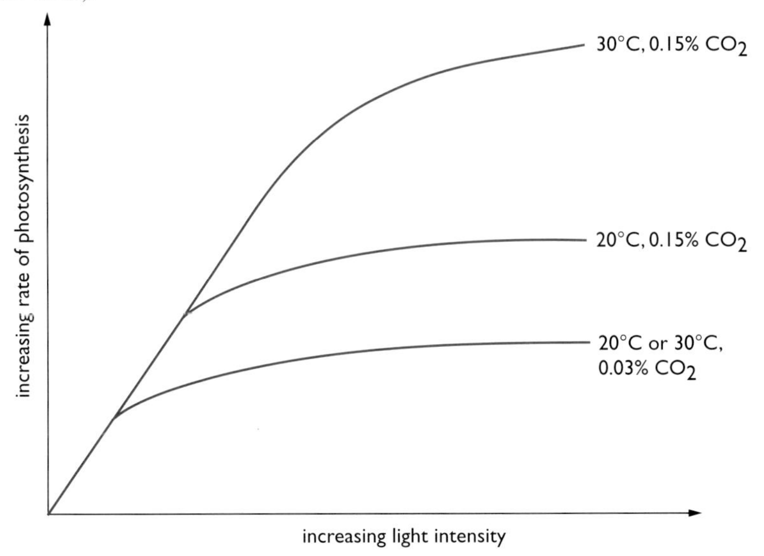

Figure 10.8 *Light intensity, carbon dioxide concentration and temperature can all act as limiting factors on the rate of photosynthesis*

Knowledge of limiting factors is used in some glasshouses (greenhouses) to speed up the growth of crop plants such as tomatoes and lettuces. Extra carbon dioxide is added to the air around the plants, by using gas burners. The higher concentration of carbon dioxide, along with the high temperature in the glasshouse, increases the rate of photosynthesis and the growth of the leaves and fruits.

The plant's uses for glucose

As you have seen, some glucose that the plant makes is used in respiration to provide the plant's cells with energy. Some glucose is quickly converted into starch for storage. However, a plant is not made up of just glucose and starch, and must make all of its organic molecules, starting from glucose.

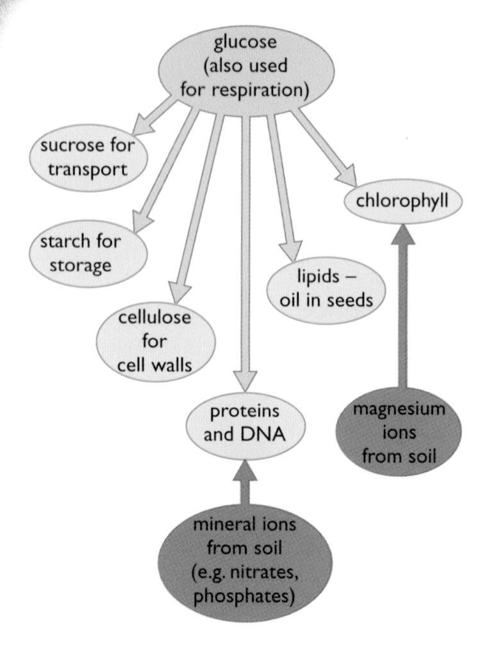

Figure 10.9 *Compounds that plant cells can make from glucose.*

Glucose is a single sugar unit (a **monosaccharide**). Plant cells can convert it into other sugars, such as a monosaccharide called **fructose** (found in fruits) and the **disaccharide** (double sugar unit) **sucrose**, which is the main sugar carried in the phloem. It can also be changed into another polymer, the polysaccharide called **cellulose**, which forms plant cell walls.

All these compounds are carbohydrates. Plant cells can also convert glucose into lipids (fats and oils). Lipids are needed for the membranes of all cells, and are also an energy store in many seeds and fruits, such as peanuts, sunflower seeds and olives.

Carbohydrates and lipids both contain only three elements – carbon, hydrogen and oxygen – and so they can be inter-converted without the need for a supply of other elements. Proteins contain these elements too, but all amino acids (the building blocks of proteins) also contain nitrogen. This is obtained as nitrate ions from the soil, through the plant's roots see Chapter 11. Other compounds in plants contain other elements. For example, chlorophyll contains magnesium ions, which are also absorbed from water in the soil. Some of the products that a plant makes from glucose are summarised in Figure 10.9.

Mineral nutrition

Nitrate ions are absorbed from the soil water, along with other minerals such as phosphate, potassium and magnesium ions. The element phosphorus is needed for the plant cells to make many important compounds, including DNA. Potassium ions are required for enzymes in respiration and photosynthesis to work, and magnesium forms a part of the chlorophyll molecule.

Glucose from photosynthesis is not just used as the *raw material* for the production of molecules such as starch, cellulose, lipids and proteins. Reactions like these, which synthesise large molecules from smaller ones, also need a source of energy. This energy is provided by the plant's *respiration* of glucose.

Water culture experiments

A plant takes only water and mineral ions from the soil for growth. Plants can be grown in soil-free cultures (water cultures) if the correct balance of minerals is added to the water. In the nineteenth century the German biologist Wilhelm Knop invented one example of a culture solution. Knop's solution contains the following chemicals (per dm^3 of water):

0.8 g	calcium nitrate
0.2 g	magnesium sulphate
0.2 g	potassium nitrate
0.2 g	potassium dihydrogen phosphate
(trace)	iron(III) phosphate

Notice that these chemicals provide all of the main elements that the plant needs to make proteins, DNA and chlorophyll, as well as other compounds, from glucose. It is called a *complete* culture solution. If you were to make up a similar solution, but to replace, for example, magnesium sulphate with more calcium sulphate, this would produce a culture solution which was *deficient* (lacking) in magnesium. You could then grow plants in the complete and deficient solutions, and compare the results. There are several ways to grow the plants, such as using the apparatus shown in Figure 10.10, which is useful for plant cuttings. Seedlings can be grown by packing cotton wool around the seed, instead of using a rubber bung.

In fact, in addition to the ions listed in Knop's solution, plants need very small amounts of other mineral ions for healthy growth. Knop's culture solution only worked because the chemicals he used to make his solutions weren't very pure, and supplied enough of these additional ions by mistake!

The plant is kept in bright light, so that it can photosynthesise. The covering around the flask prevents algae from growing in the culture solution, and the aeration tube is used for short periods to supply the roots with oxygen for respiration of the root cells. Using methods like this, it soon becomes clear that mineral deficiencies result in poor plant growth. A shortage of a particular mineral results in particular symptoms in the plant, called a **mineral deficiency disease**. For example, lack of magnesium means that the plant won't be able to make chlorophyll, and the leaves will turn yellow. Some of the mineral ions that a plant needs, their uses, and the deficiency symptoms are shown in Table 10.1. Compare the photographs of the mineral deficient plants to those of the control plants in Figure 10.12a.

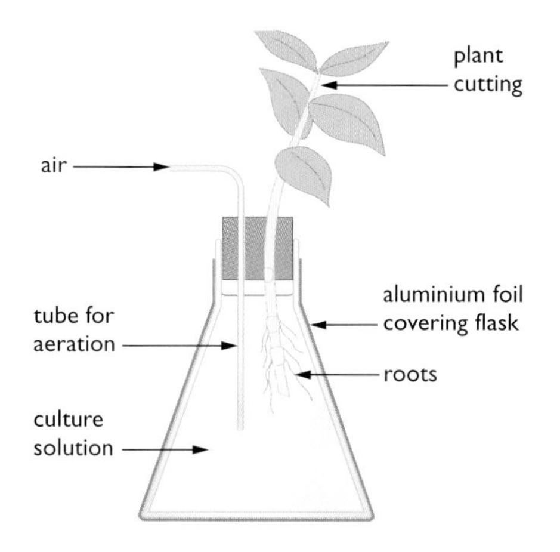

Figure 10.10 *A simple water culture method.*

Mineral ion	Use	Deficiency symptoms
nitrate	making amino acids, proteins, chlorophyll, DNA and many other compounds	stunted growth of plant; older leaves turn yellow Figure 10.12 *(b) A plant showing symptoms of nitrate deficiency.*
phosphate	making DNA and many other compounds; part of cell membranes	poor root growth; younger leaves turn purple Figure 10.12 *(c) A plant showing symptoms of phosphate deficiency.*
potassium	needed for enzymes of respiration and photosynthesis to work	leaves turn yellow withdead spots Figure 10.12 *(d) A plant showing symptoms of potassium deficiency.*
magnesium	part of chlorophyll molecule	leaves turn yellow Figure 10.12 *(e) A plant showing symptoms of magnesium deficiency.*

Table 10.1: *Mineral ions needed by plants.*

Figure 10.12 *(a) A healthy bean plant*

Some commercial crops such as lettuces can be grown without soil, in culture solutions. This is called **hydroponics**. The plants' roots grow in a long plastic tube which has culture solution passing through it (Figure 10.11). The composition of the solution can be carefully adjusted to ensure the plants grow well. Pests, which might live in soil, are also less of a problem.

Figure 10.11 *Lettuce plants grown by hydroponics.*

End of Chapter Checklist

If you haven't got a copy of your specification, read the introduction on page vi.

You will need to be able to do some or all of the following. Check your Awarding Body's specification (syllabus) to find out exactly what you need to know.

- Write a word equation and balanced symbol equation for photosynthesis, and understand the role of light and chlorophyll in the process.
- Describe the structure of a leaf, including cuticle, epidermis, palisade and spongy layers, stomata, veins, xylem and phloem.
- Explain how the structure of the leaf is adapted for photosynthesis.
- Recognise how light intensity, temperature and carbon dioxide concentration can affect the rate of photosynthesis, and understand the idea of limiting factors.
- Interpret data from experiments relating to photosynthesis.
- Explain the pattern of gas exchange between the plant and the atmosphere resulting from photosynthesis and respiration over a 24 hour period.
- Know that glucose made by photosynthesis can be used for respiration, or converted into starch for storage, cellulose for plant cell walls, protein for growth, lipids for storage in seeds, and used to make chlorophyll.
- Recognise that the production of many important compounds by the plant needs elements that are obtained from soil minerals, including nitrogen for amino acid and protein synthesis, phosphorus to make DNA and cell membranes, and magnesium for chlorophyll. Know that potassium is needed to help enzymes in photosynthesis and respiration work.
- Know that mineral deficiencies result in poor plant growth. Recognise the symptoms of nitrate, phosphate and potassium deficiency. Interpret data from water culture experiments.

Questions

More questions on plants and food can be found at the end of Section C on page 149.

1 A plant with variegated leaves had a piece of black paper attached to one leaf as shown in the diagram opposite.

The plant was kept under a bright light for 24 hours. The leaf was then removed, the paper taken off and the leaf was tested for starch.

a) Name the chemical used to test for starch, and describe the colour change if the test is positive.

b) Copy the leaf outline and shade in the areas which would contain starch.

c) Explain how you arrived at your answer to (b).

d) What is starch used for in a plant? How do the properties of starch make it suitable for this function?

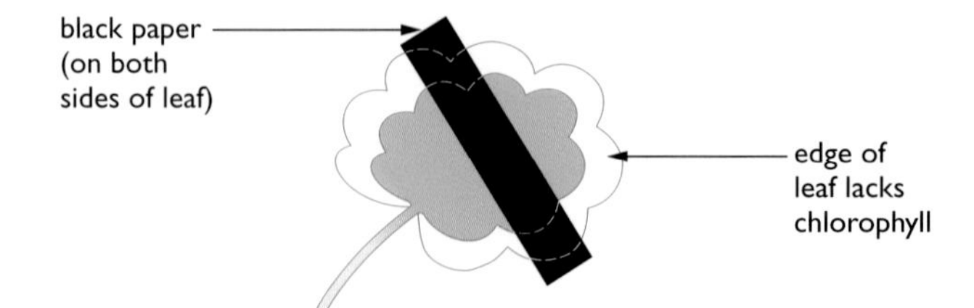

2 Copy and complete the following table to show the functions of different parts of a leaf. One has been done for you.

Part of leaf	Function
palisade mesophyll layer	main site of photosynthesis
spongy mesophyll layer	
stomata	
xylem	
phloem	

3 The graph shows the changes in the concentration of carbon dioxide in a field of long grass throughout a 24-hour period in summer.

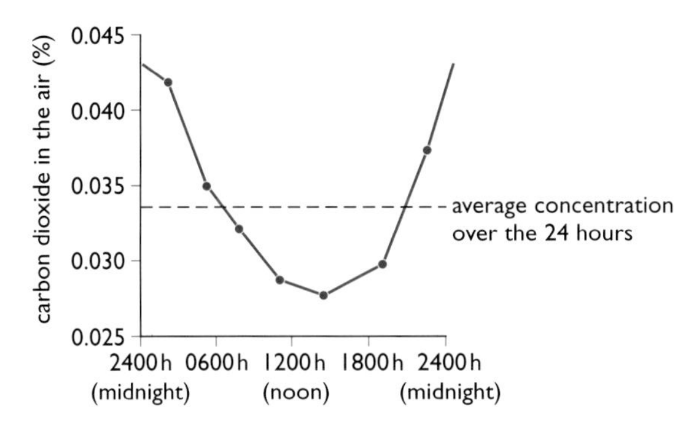

a) Explain why the levels of carbon dioxide are high at 0200 hours and low at 1200 hours.

b) What factor will limit the rate of photosynthesis at 0400 hours and at 1400 hours?

4 The table below shows some of the substances that can be made by plants. Give one use in the plant for each. The first has been done for you.

Substance	Use
glucose	oxidised in respiration to give energy
sucrose	
starch	
cellulose	
protein	
lipid	

5 The apparatus shown in the diagram was used to grow a pea seedling in a water culture experiment.

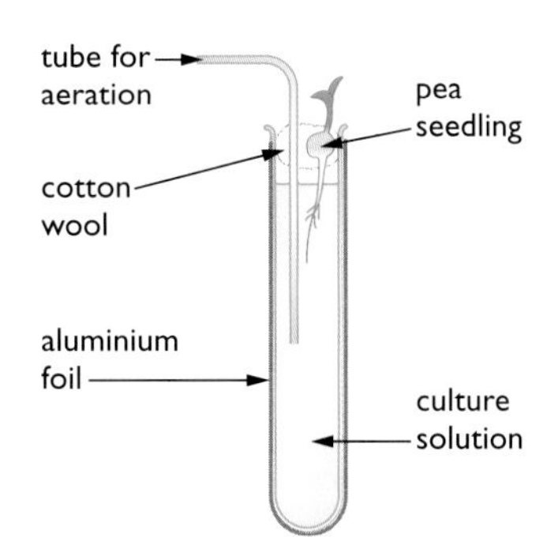

a) Explain the purpose of the aeration tube and the aluminium foil around the test tube.

b) After two weeks, the roots of the pea seedling had grown less than normal, although the leaves were well developed. What mineral ion is likely to be deficient in the culture solution?

6 A piece of Canadian pondweed was placed upside down in a test tube of water, as shown in the diagram. Light from a bench lamp was shone onto the weed, and bubbles of gas appeared at the cut end of the stem. The distance of the lamp from the weed was changed, and the number of bubbles produced per minute was recorded. The results are shown in the table below:

Distance of lamp (cm) (D)	Number of bubbles per minute
5	126
10	89
15	64
20	42
25	31
30	17
35	14
40	10

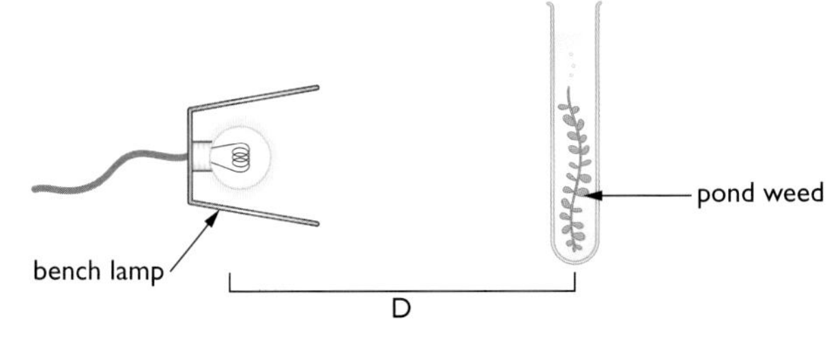

a) Plot a graph of the number of bubbles per minute against the distance of the weed from the lamp. You could do this on graph paper, or use a spreadsheet.

b) Using your graph, predict the number of bubbles per minute that would be produced if the lamp was placed 17 cm from the weed.

c) The student who carried out this experiment arrived at the following conclusion:

'The gas made by the weed is oxygen from photosynthesis, so the faster production of bubbles shows that the rate of photosynthesis is greater at higher light intensities.'

Write down three reasons why his conclusion could be criticised (Hint: think about the bubbles, and whether the experiment was a 'fair test').

7 Write a summary account of photosynthesis. You should include a description of the process, a summary equation, an account of how a leaf is adapted for photosynthesis and a note of how photosynthesis is important to other organisms, such as animals. You must keep your summary to less than two sides. It will be easier to organise your account if you use a word processor.

Chapter 11: Transport in Plants

Chapter 10 described how plants make food by photosynthesis. Photosynthesis needs a supply of water from the roots, and a way of carrying away the sugars and other products. This chapter explains how these materials are moved through the plant.

Osmosis

Osmosis is the name of a process by which water moves into and out of cells. To be able to understand how water moves through a plant, you need to understand the mechanism of osmosis. Osmosis happens when a material called a **partially permeable membrane** separates two solutions. One artificial partially permeable membrane is called Visking tubing. This is used in kidney dialysis machines (see Chapter 9). Visking tubing has microscopic holes in it, which let small molecules like water pass through (it is *permeable* to them) but is not permeable to some larger molecules, such as the sugar sucrose. This is why it is called 'partially' permeable. You can show the effects of osmosis by filling a Visking tubing 'sausage' with concentrated sucrose solution, attaching it to a capillary tube and placing the Visking tubing in a beaker of water (Figure 11.1).

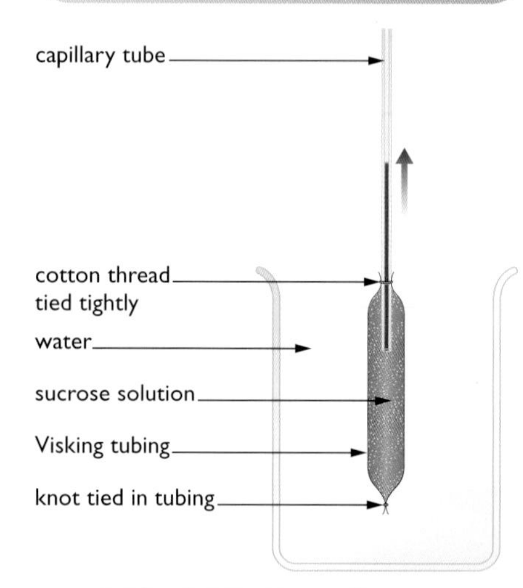

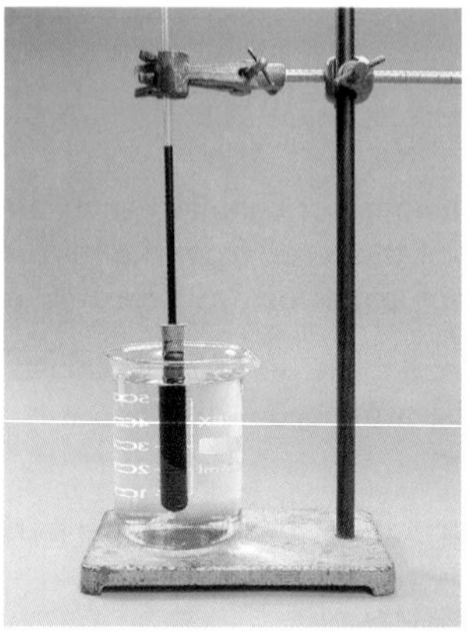

Figure 11.1 *Water enters the Visking tubing 'sausage' by osmosis. This causes the level of liquid in the capillary tube to rise. In the photograph, the contents of the Visking tubing have had a red dye added to make it easier to see the movement of the liquid.*

The level in the capillary tube rises as water moves from the beaker to the inside of the Visking tubing. This movement is due to osmosis. You can understand what's happening if you imagine a highly magnified view of the Visking tubing separating the two liquids (Figure 11.2).

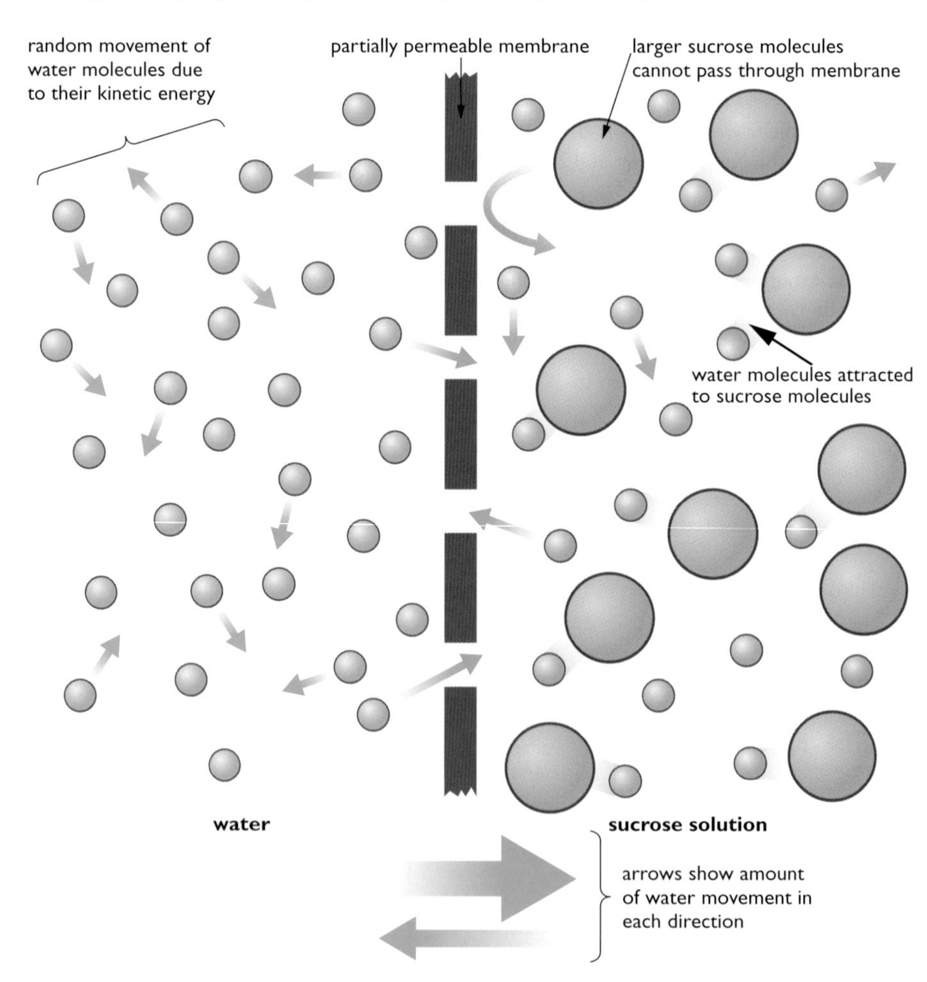

Figure 11.2 *In this model of osmosis, more water molecules diffuse from left to right than from right to left.*

Diffusion is the net movement of particles (molecules or ions) due to their kinetic energy, down a concentration gradient (see Chapter 1).

The sucrose molecules are too big to pass through the holes in the partially permeable membrane. The water molecules can pass through the membrane in either direction, but those on the right are attracted to the sugar molecules. This slows them down and means that they are less free to move – they have less kinetic energy. As a result of this, more water molecules diffuse from left to right than from right to left. In other words, there is a greater diffusion of water molecules from the more dilute solution (in this case pure water) to the more concentrated solution.

How 'free' the water molecules are to move is called the **water potential**. The molecules in pure water can move most freely, so pure water has the highest water potential. The more concentrated a solution is, the lower is its water potential. In the model in Figure 11.2, water moves from a high to a low water potential. This is a law which applies whenever water moves by osmosis. We can bring these ideas together in a definition of osmosis:

Osmosis is the net diffusion of water across a partially permeable membrane, from a solution with a high water potential to one with a lower water potential.

It is important to realise that neither of the two solutions has to be pure water. As long as there is a difference in their concentrations (and their water potentials), and they are separated by a partially permeable membrane, osmosis can still take place.

Osmosis in plant cells

So far we have only been dealing with osmosis through Visking tubing. However, there are partially permeable membranes in cells too. The cell surface membranes of both animal and plant cells are partially permeable, and so is the inner membrane around the plant cell's sap vacuole (Figure 11.3).

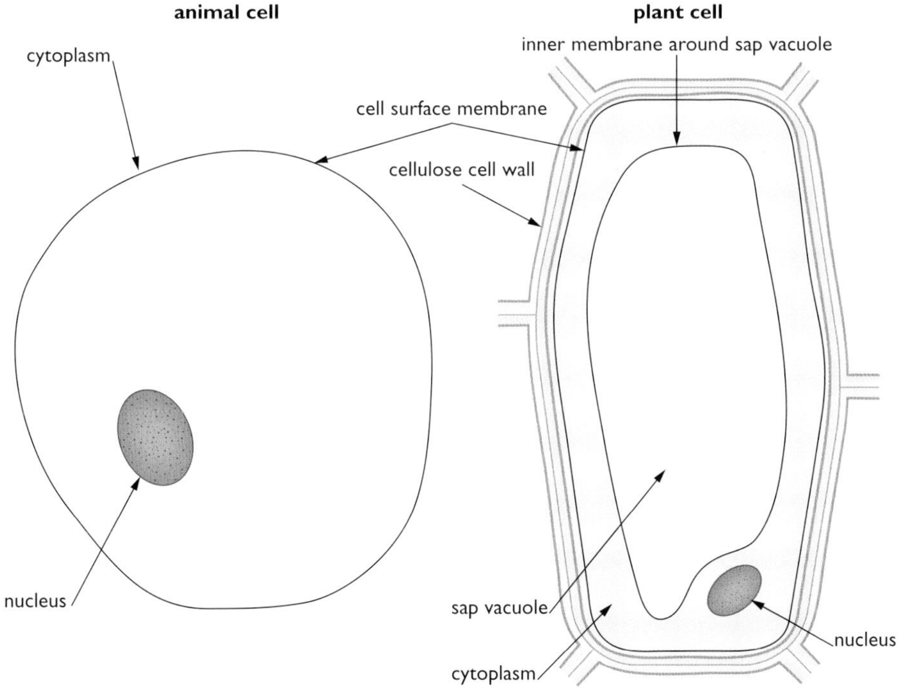

Figure 11.3 *Membranes in animal and plant cells.*

Around the plant cell is the tough cellulose cell wall. This outer structure keeps the shape of the cell, and can resist changes in pressure inside the cell. This is very important, and critical in explaining the way that plants are supported. The cell contents, including the sap vacuole, contain many dissolved solutes, such as sugars and ions.

The cell *wall* has large holes in it, making it fully permeable to water and solutes. Only the cell *membranes* are partially permeable barriers that allow osmosis to take place.

If a plant cell is put into pure water or a dilute solution, the contents of the cell have a lower water potential than the external solution, so the cell will absorb water by osmosis (Figure 11.4). The cell then swells up and the cytoplasm pushes against the cell wall. A plant cell that has developed an internal pressure like this is called **turgid**.

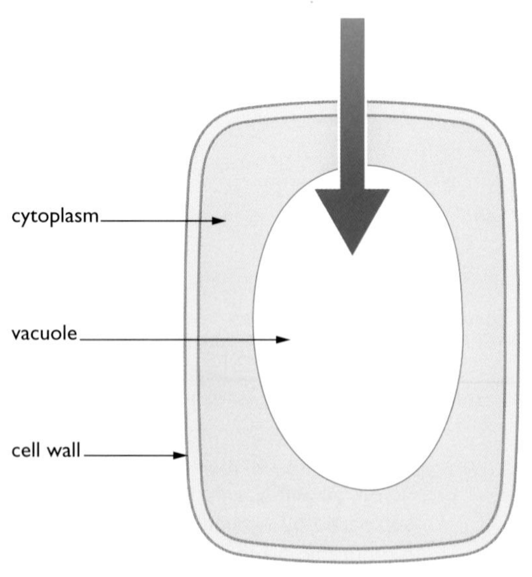

cell placed in dilute solution, or water, absorbs water by osmosis and becomes turgid

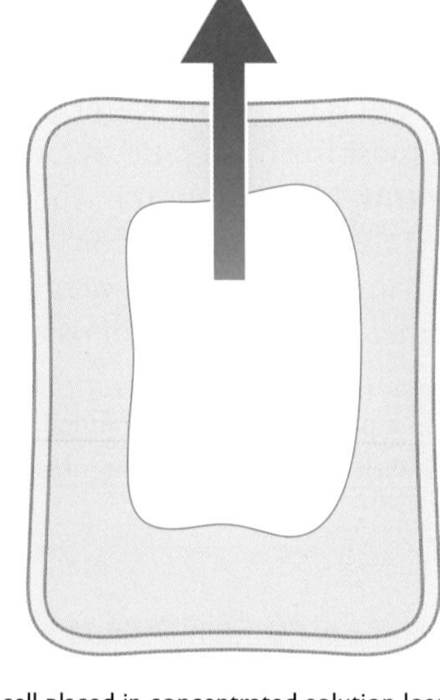

cell placed in concentrated solution loses water by osmosis and becomes flaccid

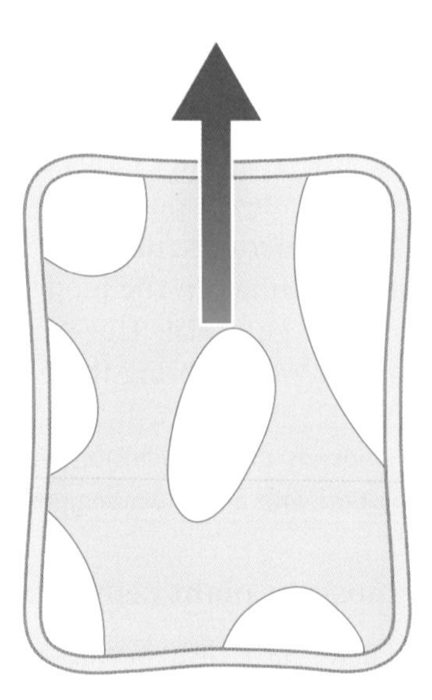

excessive loss of water by osmosis causes the cell to become plasmolysed

Figure 11.4 *The effects of osmosis on plant cells.*

On the other hand, if the cell is placed in a concentrated sucrose solution that has a lower water potential than the cell contents, it will *lose* water by osmosis. The cell decreases in volume and the cytoplasm no longer pushes against the cell wall. In this state, the cell is called **flaccid**. Eventually the cell contents shrink so much that the membrane and cytoplasm split away from the cell wall and gaps appear between the wall and the membrane. A cell like this is called **plasmolysed**. You can see plasmolysis happening in the plant cells shown in Figure 11.5. The space between the cell wall and the cell surface membrane will now be filled with the sucrose solution.

Turgor (the state a plant is in when its cells are turgid) is very important to plants. The pressure inside the cells pushes neighbouring cells against each other, like a box full of inflated balloons. This supports the non-woody parts of the plant, such as young stems and leaves, and holds stems upright, so the leaves can carry out photosynthesis properly. Turgor is also important in the functioning of stomata. If a plant loses too much water from its cells so that they become flaccid, this makes the plant **wilt**. You can see this in a pot plant which has been left for too long without water. The leaves droop and collapse. In fact this is a protective action. It cuts down water loss by reducing the exposed surface area of the leaves and closing the stomata.

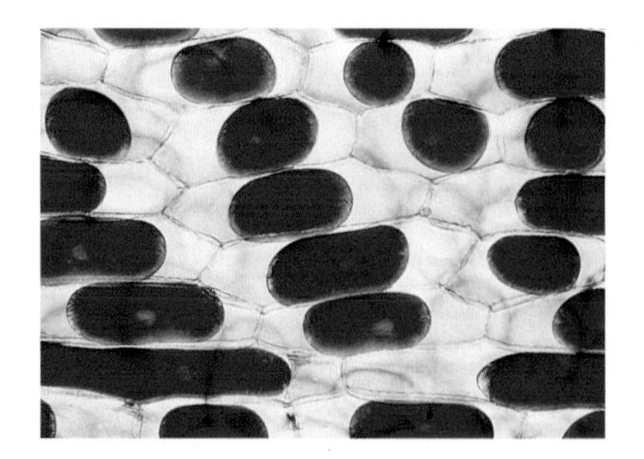

Figure 11.5 *Cells of rhubarb epidermis, showing plasmolysis. The cell membranes can be seen pulling away from the cell walls.*

Inside the plant, water moves from cell to cell by osmosis. If a cell has a higher water potential than the cell next to it, water will move from the first cell to the second. In turn, this will dilute the contents of the second cell, so that it has a higher water potential than the next cell. In this way, water can move across a plant tissue, down a gradient of water potential (Figure 11.6).

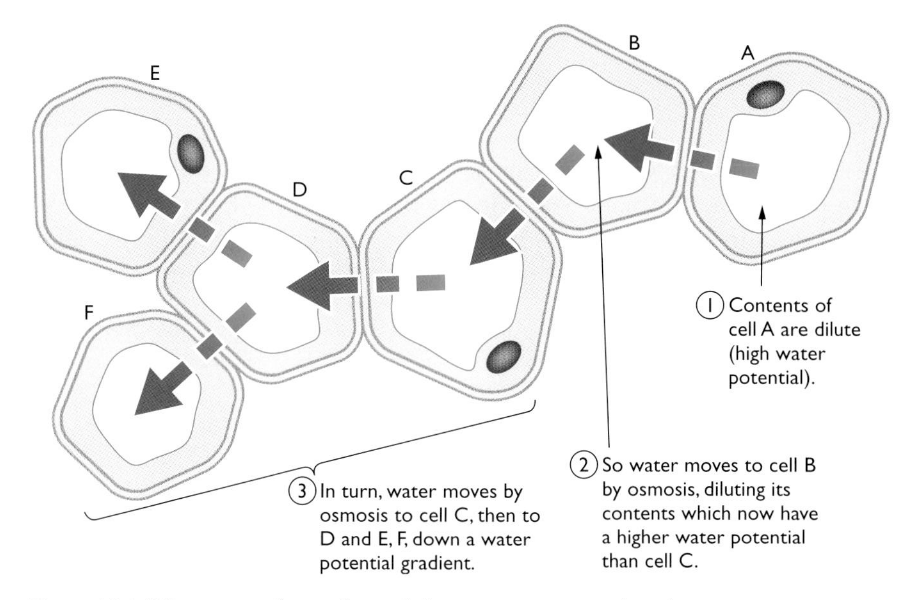

Figure 11.6 *Water moves from cell to cell down a water potential gradient.*

This chapter only deals with plant cells. Osmosis also happens in animal cells, but there is much less water movement. This is because animal cells do not have a strong cell wall around them, and can't resist the changes in internal pressure resulting from large movements of water. For example, if red blood cells are put into water, they will swell up and burst. If the same cells are put into a concentrated salt solution, they lose water by osmosis and shrink, producing cells with crinkly edges (Figure 11.7).

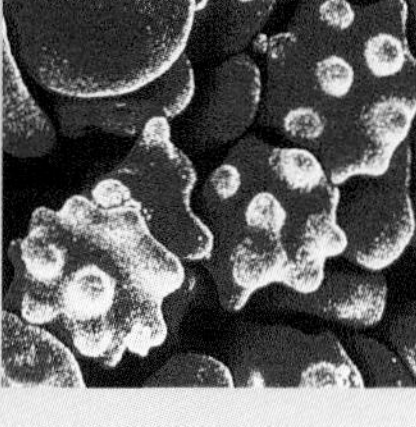

Figure 11.7 *Compare the red blood cells on the right, which were placed in a 3% salt solution, with the normal cells on the left. Blood plasma has a concentration equal to a 0.85% salt solution*

Uptake of water by roots

The regions just behind the growing tips of the roots of a plant are covered in thousands of tiny root hairs (Figure 11.8). These areas are the main sites of water absorption by the roots, where the hairs greatly increase the surface area of the root epidermis.

Each hair is actually a single, specialised cell of the root epidermis. The long, thin outer projection of the root hair cell penetrates between the soil particles, reaching the soil water. The water in the soil has some solutes dissolved in it, such as mineral ions, but their concentrations are much lower than the concentrations of solutes inside the root hair cell. The soil water therefore has a higher water potential than the inside of the cell. This allows water to enter the root hair cell by osmosis. In turn, this water movement dilutes the contents of the cell, increasing its water potential. Water then moves out of the root hair cell into the outer tissue of the root (the root cortex). Continuing in this way, a gradient of water potential is set up across the root cortex, kept going by water being taken up by the xylem in the middle of the root (Figure 11.9).

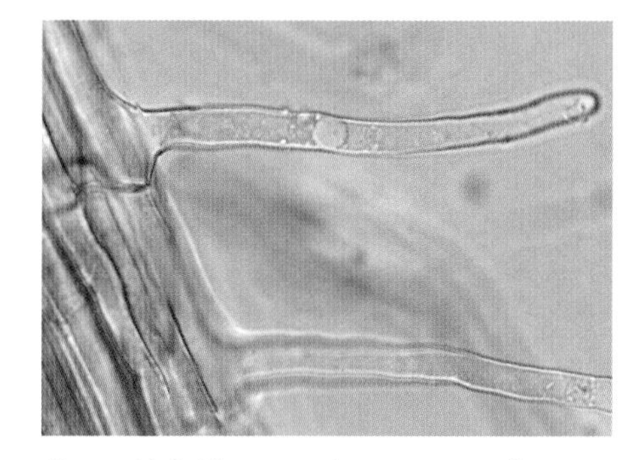

Figure 11.8 *These root hairs increase the surface area for water absorption.*

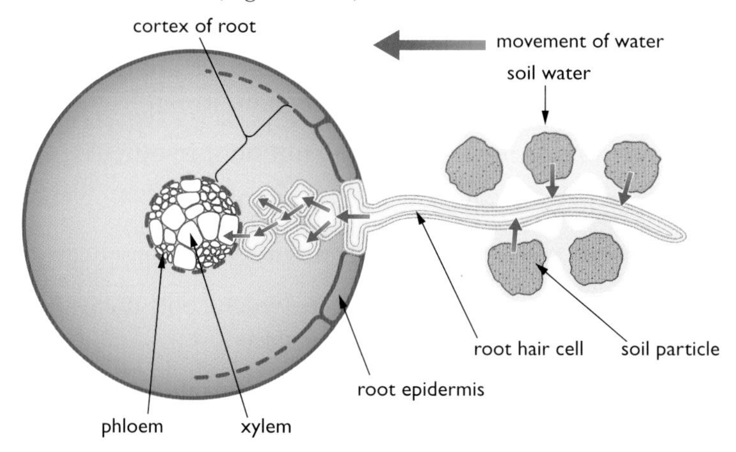

Figure 11.9 *Water is taken up by root hairs of the plant epidermis and carried across the root cortex by a water potential gradient. It then enters the xylem and is transported to all parts of the plant.*

Plants that live in dry habitats, such as cacti and succulents, usually have a very thick layer of waxy cuticle on their leaves, which reduces water loss as much as possible. Cacti have leaves reduced to spines, which have a low surface area and cut down water loss.

Loss of water by the leaves – transpiration

Osmosis is also involved in the movement of water through leaves. The epidermis of leaves is covered by a waxy cuticle (see Chapter 10), which is impermeable to water. Most water passes out of the leaves as water vapour through pores called stomata. Water leaves the cells of the leaf mesophyll and evaporates into the air spaces between the spongy mesophyll cells. The water vapour then diffuses out through the stomatal pores (Figure 11.10).

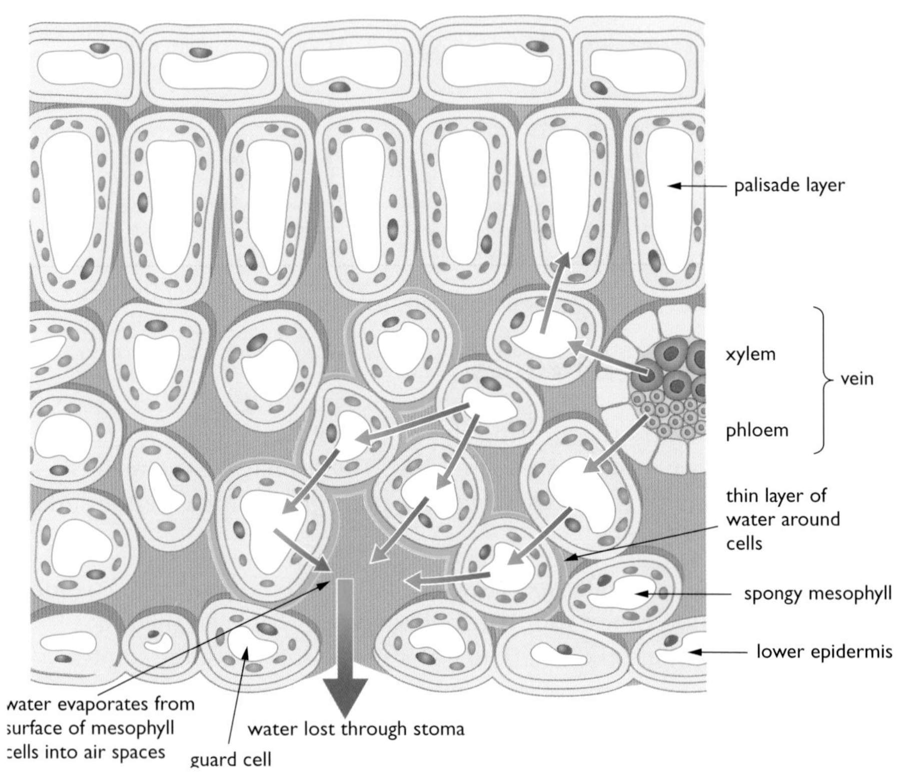

Figure 11.10 *Passage of water from the xylem to the stomatal pores of a leaf.*

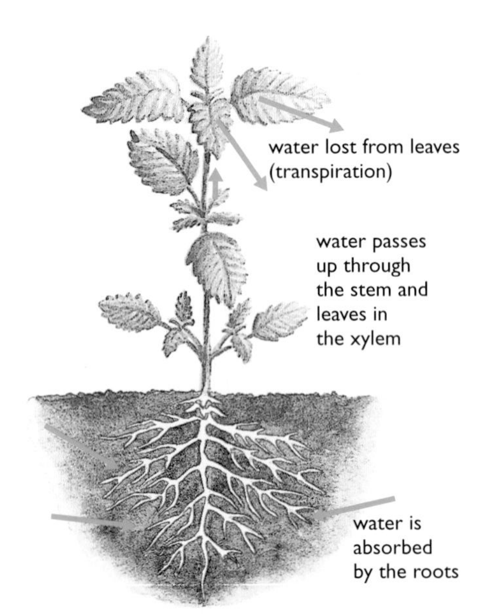

Figure 11.11 *The transpiration stream.*

Loss of water from the mesophyll cells sets up a water potential gradient which 'draws' water by osmosis from surrounding mesophyll cells. In turn, the xylem vessels supply the leaf mesophyll tissues with water.

This loss of water vapour from the leaves is called **transpiration**. Transpiration causes water to be 'pulled up' the xylem in the stem and roots in a continuous flow known as the **transpiration stream** (Figure 11.11). The transpiration stream has more than one function. It:

- supplies water for the leaf cells to carry out photosynthesis
- carries mineral ions dissolved in the water
- provides water to keep the plant cells turgid
- allows evaporation from the leaf surface, which cools the leaf, in a similar way to sweat cooling the human skin.

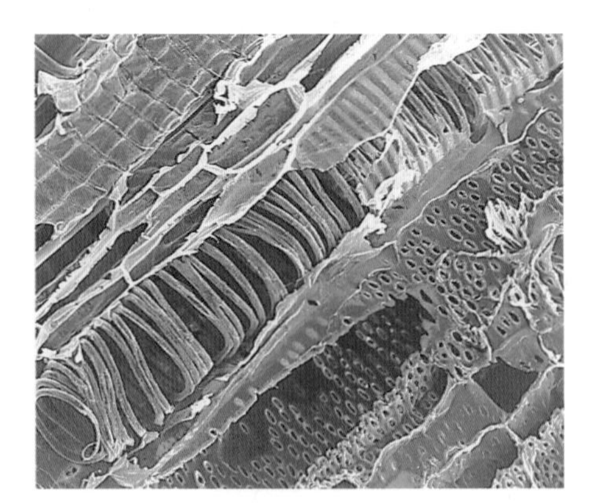

Figure 11.12 *Xylem vessels in a stem.*

Xylem contains dead cells arranged end-to-end, forming continuous vessels. When they are mature, the vessels contain no cytoplasm. Instead, they have a hollow central space or **lumen** through which the water passes. The walls of the xylem vessels contain a woody material called **lignin** (Figure 11.12).

The xylem vessels begin life as living cells with normal cytoplasm and cellulose cell walls. As they develop they become elongated, and gradually their original cellulose cell walls become impregnated with lignin, made by the cytoplasm. As this happens, the cells die, forming hollow tubes. Lignification makes them very strong, and enables them to carry water up tall plants without collapsing. Lignin is also impermeable to water.

The other plant transport tissue, the phloem, consists of living cells at all stages in its development. Tubes in the phloem are also formed by cells arranged end-to-end, but they have cell walls made of cellulose, and retain their cytoplasm. The end of each cell is formed by a cross-wall of cellulose with holes, called a **sieve plate**. The living cytoplasm extends through the holes in the sieve plates, linking each cell with the next, forming a long **sieve tube** (Figure 11.13). The sieve tubes transport the products of photosynthesis from the leaves to other parts of the plant. Sugars for energy, or amino acids for building proteins, are carried to young leaves and other growing points. Sugar may also be taken to the roots and converted into starch for storage. Despite being living cells, the phloem sieve tubes have no nucleus. They seem to be controlled by other cells that lie alongside the sieve tubes, called **companion cells** (Figure 11.13).

In a young stem, xylem and phloem are grouped together in areas called **vascular bundles**. Unlike in the root, where the vascular tissue is in the central core, the vascular bundles are arranged in a circle around the outer part of the stem (Figure 11.14).

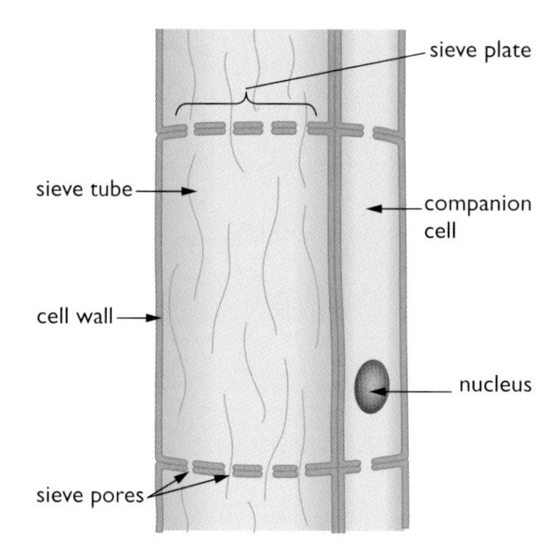

Figure 11.13 *Phloem is living tissue, responsible for carrying the products of photosynthesis around the plant.*

Remember: Xylem carries water and minerals up from the roots. Phloem carries products of photosynthesis away from the leaves. The contents of the phloem can travel up or down the plant.

'Vascular' means 'made of vessels'. A vascular bundle is a group of vessels or tubes (xylem and phloem).

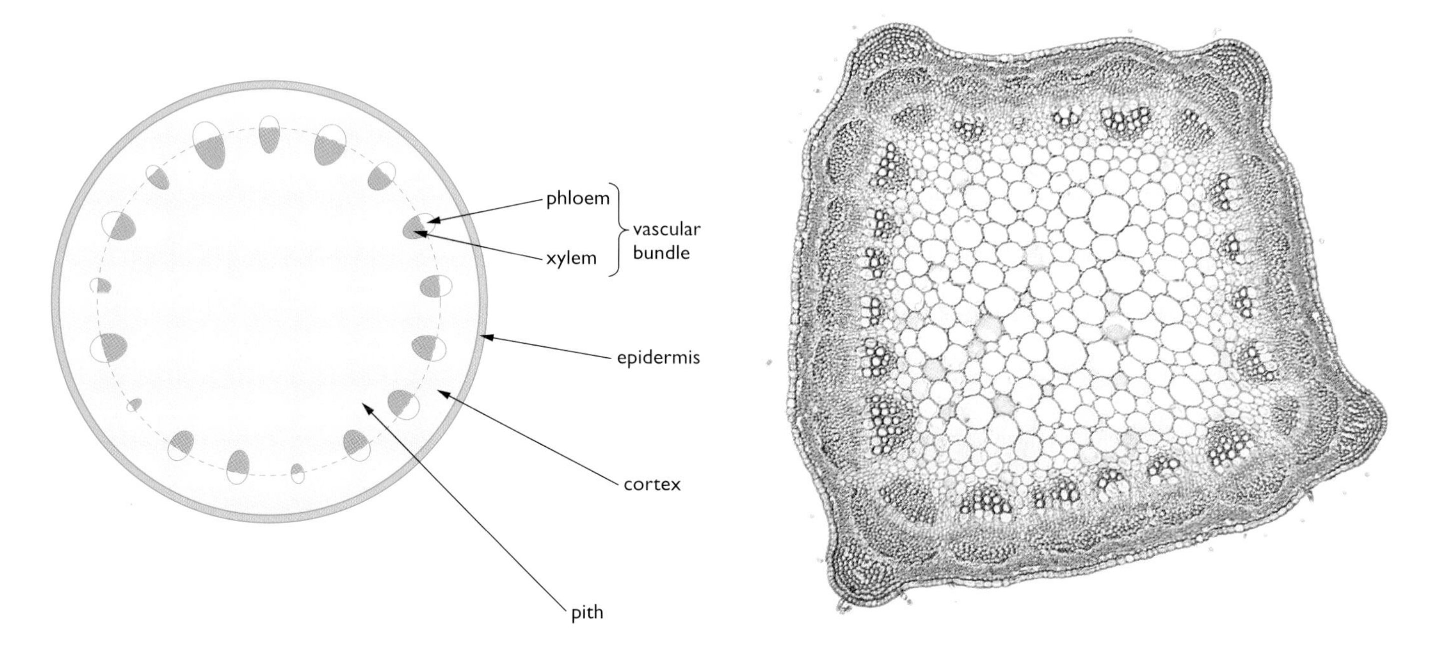

Figure 11.14 *This cross-section of a stem shows the arrangement of xylem and phloem tissue in vascular bundles.*

In older stems, the vascular tissue grows to form complete rings around the stem. The inner xylem forms the woody central core of a stem, with the living layer of phloem outside this.

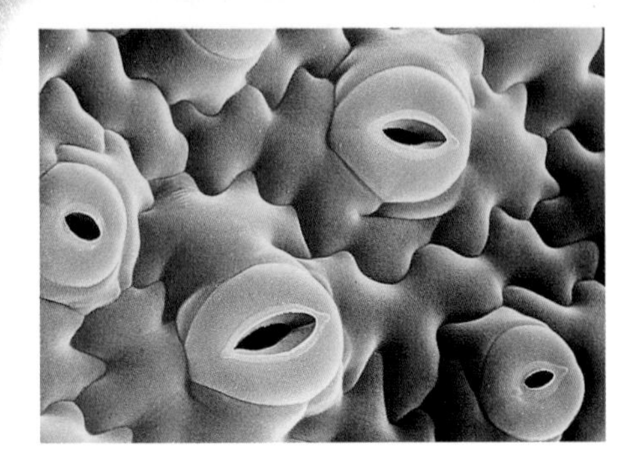

Figure 11.15 *Stomata in the surface of the lower epidermis of a leaf.*

Control of transpiration by stomata

As you saw in Chapter 10, there are usually more stomata on the lower surface of the leaves than the upper surface in most plant species (Figure 11.15). If they were mainly on the upper leaf surface, the leaf would lose too much water. This is because the stomata would be exposed to direct sunlight, which would produce a high rate of evaporation from the exposed stomata. There is also less air movement on the underside of leaves. The evolution of this arrangement of stomata is an adaptation that reduces water loss.

The stomata can open and close. The guard cells that surround each stoma have an unusual 'banana' shape, and the part of their cell wall nearest the stoma is particularly thick. In the light, water enters the guard cells by osmosis from the surrounding epidermis cells. This causes the guard cells to become turgid, and, as they swell up, their shape changes. They bend outwards, opening up the stoma. In the dark, the guard cells lose water again, they become flaccid and the stoma closes. No one knows for sure how this change is brought about, but it seems to be linked to the fact that the guard cells are the only cells in the lower epidermis that contain chloroplasts. In the light, the guard cells use energy to accumulate solutes in their vacuoles, causing water to be drawn in by osmosis (Figure 11.16).

You can make a 'mould' of a leaf surface using colourless nail varnish. Paint an area of nail varnish about 1 cm^2 on the lower surface of a leaf (privet leaves work well). Leave this to dry for half an hour and then use forceps to peel off the nail varnish. Place this on a slide and look at it through a microscope (you don't need to use a cover slip). Under medium or high power you should be able to see impressions of the stomata and guard cells.

You can model the action of a guard cell using a long balloon and some sticky tape. Stick a few pieces of tape down one side of the balloon and then blow it up. As it inflates, the balloon will curve outwards like a turgid guard cell. The tape represents the thick inner cell wall.

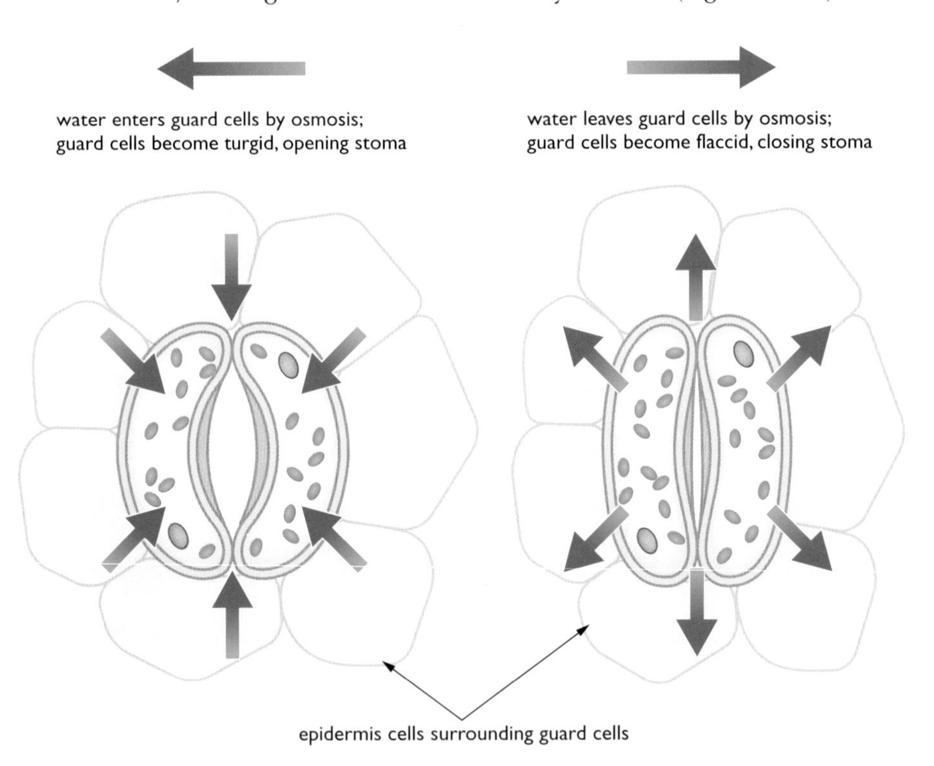

Figure 11.16 *When the guard cells become turgid, the stoma opens. When they become flaccid, it closes.*

Closure of stomata in the dark is a useful adaptation. Without the sun there is no need for loss of water vapour from the stomata to cool the leaves. In addition, leaves cannot photosynthesise in the dark, so they don't need water for this purpose. Therefore, it doesn't matter if the transpiration stream is shut down by closure of the stomata. As you will see on page 137, other physical factors can also affect the rate of transpiration.

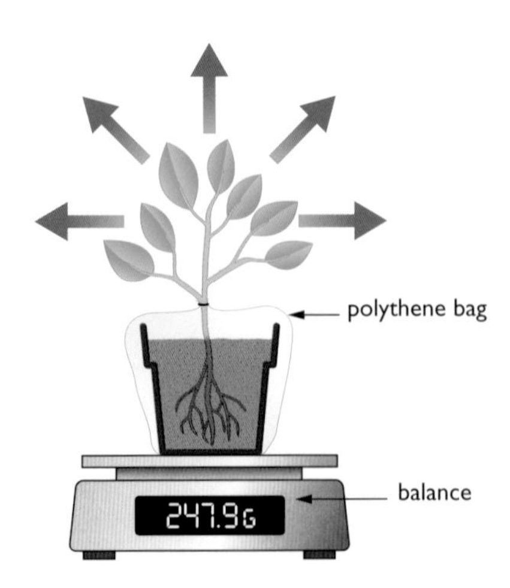

Figure 11.17 *A 'weight' potometer.*

Measuring the rate of transpiration: potometers

A potometer is a simple piece of apparatus which measures the rate of transpiration or the rate of uptake of water by a plant. (These are not the same thing – some of the water taken up by the plant may stay in the plant cells, or be used for photosynthesis.) There are two types, 'weight' and volume potometers.

A 'weight' potometer measures the rate of loss of mass from a potted plant or leafy shoot over a longish period of time, usually several hours (Figure 11.17).

The polythene bag around the pot prevents loss of moisture by evaporation from the soil. Most of the mass lost by the plant will be due to water evaporating from the leaves during transpiration (although there will be small changes in mass due to respiration and photosynthesis, since both of these processes exchange gases with the air).

A volume potometer is used to find the rate of uptake of water by a leafy shoot, by 'magnifying' this uptake in a capillary tube. These potometers come in various shapes and sizes. The simplest is a straight vertical tube joined to the shoot by a piece of rubber tubing. More 'deluxe' versions have a horizontal capillary tube and a way of refilling the capillary to re-set the water at its starting position (Figure 11.18).

To set up a volume potometer, the whole apparatus is placed in a sink of water and any air in the tubing removed. A shoot is taken from a plant and the end of the stem cut at an angle. This makes it easier to push the stem into the rubber tubing, which is done under water to stop air entering. The apparatus is removed from the sink, and vaseline used to seal any joins. The movement of the water column in the capillary tube can be timed. If the water moves more quickly, this shows a faster rate of transpiration. The plant can then be exposed to different conditions to see how they affect the rate.

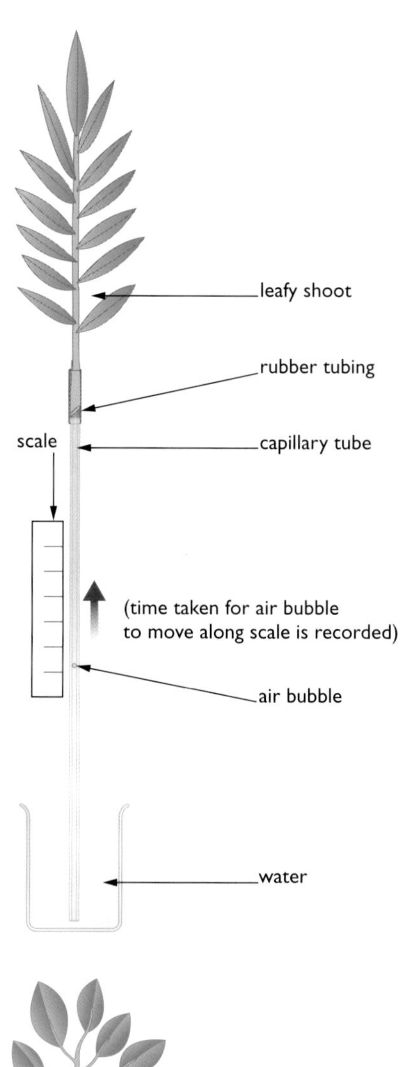

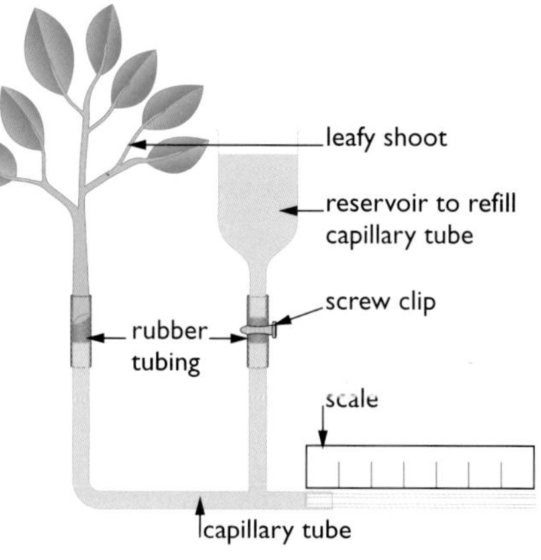

Figure 11.18 *Two types of volume potometer.*

Factors affecting the rate of transpiration

There are four main factors which affect the rate of transpiration:

- **Light intensity** The rate of transpiration increases in the light, because of the opening of the stomata in the leaves.
- **Temperature** High temperatures increase the rate of transpiration, by increasing the rate of evaporation of water from the mesophyll cells.
- **Humidity** When the air around the plant is humid, this reduces the diffusion gradient between the air spaces in the leaf and the external air. The rate of transpiration therefore decreases in humid air and speeds up in dry air.
- **Wind speed** The rate of transpiration increases with faster air movements across the surface of the leaf. The moving air removes any water vapour which might remain near the stomata. This moist air would otherwise reduce the diffusion gradient and slow down diffusion.

It is easy to use a potometer to demonstrate these effects. For example, you can use a fan or hair drier on a plant to show the effects of moving air, or put the plant under a bright light or in the dark to find the effect of light intensity.

The concentration of water vapour inside the leaf is greater than the concentration in the air outside. This is a diffusion (or concentration) gradient. A big difference in concentration means there is a steep gradient. The molecules of water vapour will diffuse out of the leaf more quickly if the gradient is steep.

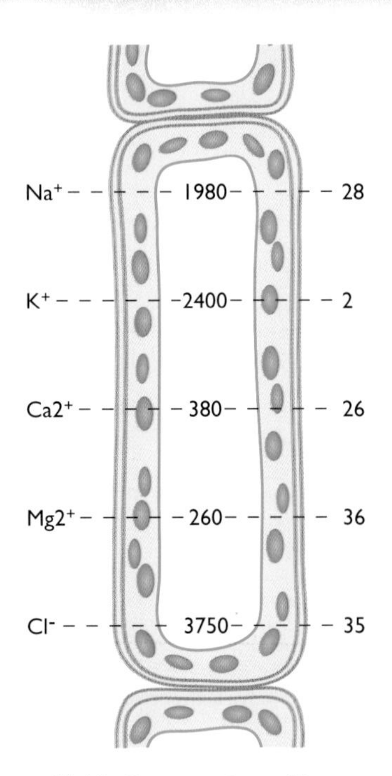

Figure 11.19 *Concentrations of ions are given in mg/dm^3. The algal cell accumulates ions against a concentration gradient.*

A substance moves by diffusion from a place where it is at a high concentration to a place where it is at a lower concentration, i.e. down a concentration gradient. This is because molecules and ions use their own kinetic energy to diffuse. If they are moved *against* a concentration gradient, this needs another, *external* source of energy.

Active transport is the uptake of molecules or ions from a region of low concentration to a region of high concentration across a cell membrane. It uses energy from cell respiration.

You could use a spreadsheet to plot a bar chart of the concentrations of the ions in Table 11.1. You should be able to plot both the concentrations in the cell sap and in the pond water on the same graph. You could then use the data in the spreadsheet to calculate how much more concentrated each ion is inside the cells compared with outside.

Uptake of mineral ions

Mineral ions are needed for plants to make a range of organic compounds from the products of photosynthesis (see Chapter 10). These mineral ions are taken up from the soil, along with the water that enters by osmosis. However, *only* water can move by osmosis. Mineral ions have to enter the root cells by other means.

One way an ion might enter the root cells would be by simple diffusion. For this to happen there needs to be a concentration gradient, with the soil water containing a higher concentration of the ion than the root hair cells. An experiment with an alga (a small plant-like organism) showed that plant cells (or strictly algal cells) could concentrate ions in the cell sap to a much higher level than in the water surrounding the alga (Table 11.1 and Figure 11.19).

	Concentration of ion (mg/dm^3)				
	sodium (Na^+)	**potassium (K^+)**	**calcium (Ca^{2+})**	**magnesium (Mg^{2+})**	**chloride (Cl^-)**
in cell sap	1980	2400	380	260	3750
in pond water	28	2	26	36	35

Table 11.1: *Concentration of ions in the cell sap of an alga and in the surrounding pond water.*

None of these ions could have entered the cells of the alga by diffusion, because the concentration gradient for each ion is in the wrong direction.

To accumulate the ions inside its cells, the alga must have 'pumped' the ions in. Scientists now know that the cell membranes of algae and plant cells, including the root hair cell, contain molecular pumps which push the ions into the cells, against a concentration gradient. This process is called **active transport**. Unlike diffusion, which just uses the kinetic energy of the diffusing particles, active transport needs another source of energy. It uses chemical energy from respiration.

Active transport is responsible for the uptake of several mineral ions into the root hair cells. However, it depends upon their concentration in the soil water. If an ion is present in the soil water in large amounts, there may be a downward concentration gradient allowing it to enter by diffusion. For example, if nitrate fertiliser is added to the soil, active transport may not be needed for the roots to absorb the nitrate ions.

Table 11.1 shows another interesting fact about transport across the plant cell membrane. If you look at potassium ions, they are 1200 times more concentrated inside the cells than outside, whereas magnesium ions are only about 7 times more concentrated. The cell membrane is able to concentrate different ions by different amounts. Because the membrane is being 'selective' in how much of each ion it allows in, some scientists prefer to call it **selectively permeable**, rather than partially permeable.

Once inside the root, the mineral ions pass across the root cortex mainly by diffusion, and enter the xylem vessels. They are then carried all around the plant in the transpiration stream.

End of Chapter Checklist

If you haven't got a copy of your specification, read the introduction on page vi.

You will need to be able to do some or all of the following. Check your Awarding Body's specification (syllabus) to find out exactly what you need to know.

- Remember the definition of osmosis and understand the diffusion model of osmosis.
- Understand the role of water in maintaining cell turgidity to support plant tissues.
- Explain how roots take up water and the role of root hair cells.
- Know that water and mineral ions are transported in xylem vessels, which are dead cells with no cytoplasm.
- Know that phloem is living tissue, which conducts sugars and amino acids produced by photosynthesis up and down stems to growing and storage areas of the plant.
- Remember the position of xylem and phloem in vascular bundles of the stem, and know that xylem and phloem are transport systems that are continuous throughout roots, stems and leaves.
- Explain how leaves lose water by transpiration, and how the leaf structure is adapted for this, including the role of the waxy cuticle, spongy mesophyll, guard cells and stomata.
- Understand how certain conditions (light, temperature, air movement and humidity) affect the rate of transpiration. Interpret data from experiments on transpiration, including the use of potometers.
- Describe the uptake of mineral ions by active transport and know that this process needs energy from respiration.

Questions

More questions on transport in plants can be found at the end of Section C on page 149.

1 Three 'chips' of about the same size and shape were cut from the same potato. Each was blotted, weighed and placed in a different sucrose solution (A, B or C). The chips were left in the solutions for an hour, then removed, blotted and re-weighed. Here are the results:

	Starting mass (g)	**Final mass (g)**	**Change in mass (%)**
solution A	7.4	6.5	−12.2
solution B	8.2	8.0	
solution C	7.7	8.5	+10.4

a) Calculate the percentage change in mass for the chip in solution B.

b) Name the process that caused the chips to lose or gain mass.

c) Which solution was likely to have been the most concentrated?

d) Which solution had the highest water potential?

e) Which solution had a water potential most similar to the water potential of the potato cells?

f) The cell membrane is described as 'partially permeable'. Explain the meaning of this.

2 Explain how each of the following cells is adapted for its function:

a) a root hair cell

b) a xylem vessel

c) a guard cell.

3 Suggest reasons for each of the following observations:

a) When transplanting a small plant from a pot to the garden, it is important to dig it up carefully, leaving the roots in a ball of soil. If this is not done, the plant may wilt after it has been transplanted.

b) A plant cutting is more likely to grow successfully if you remove some of its leaves before planting it in compost.

c) Plants that live in very dry habitats often have stomata located in sunken pits in their leaves.

d) Greenflies feed by sticking their hollow tube-like mouthparts into the phloem of a plant stem.

4 The diagram shows a cross-section through the stem of a plant.

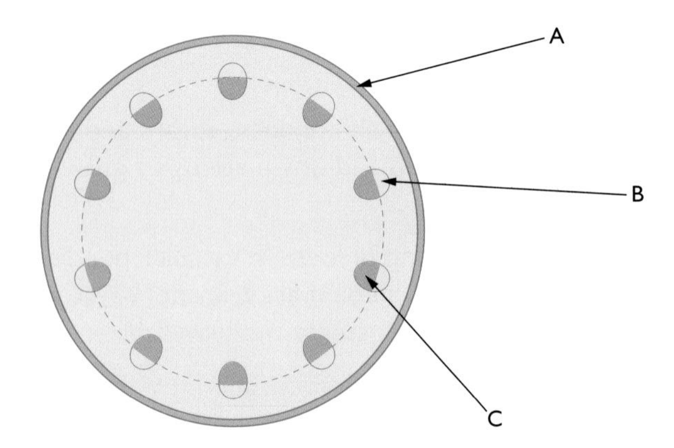

a) Identify the tissues labelled A, B and C.

b) A young stem was placed in a solution of a red dye for an hour. Which tissue in the diagram would be most likely to contain the dye? Explain your answer.

5 A simple volume potometer was used to measure the uptake of water by a leafy shoot under four different conditions. During the experiment, the temperature, humidity and light intensity were kept constant. The conditions were:

1. Leaves in still air with no vaseline applied to them.

2. Leaves in moving air with no vaseline applied to them.

3. Leaves in still air with the lower leaf surface covered in vaseline.

4. Leaves in moving air with the lower surface covered in vaseline.

The results are shown in the graph.

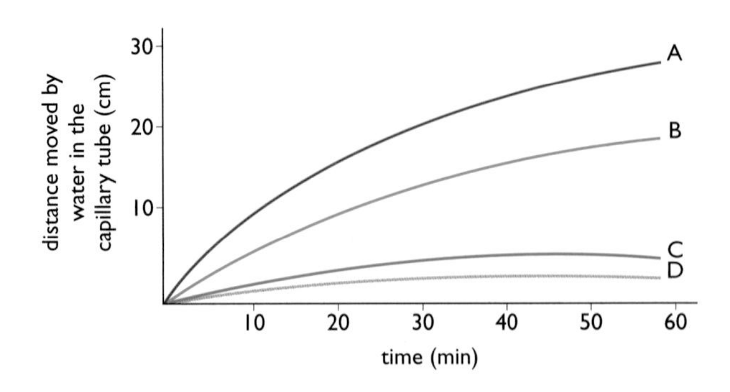

a) Complete the table below to show which condition (1, 2, 3 or 4) is most likely to produce curve A, B, C and D. One has been done for you:

Condition	Curve
1	B
2	
3	
4	

b) Explain why moving air affects the rate of uptake of water by the shoot.

6 During transpiration, water moves through the cells of a leaf and is finally lost through the stomata by evaporation.

a) Explain how water travels from one cell to another in the leaf.

b) How would the rate of transpiration be affected if the air temperature increased from 20 °C to 30 °C? Explain your answer.

c) Describe an adaptation found in plants living in dry habitats that decreases water loss from the leaves.

7 A scientist measured the uptake of sulphate ions by seedlings grown in a liquid culture. Two groups of seedlings were used. The roots of the first group were given oxygen bubbled through the culture solution. The second group had nitrogen gas bubbled through the solution, to produce anaerobic conditions around the roots. The amount of sulphate taken up by the seedlings is shown in the table below:

Time (h)	Sulphate absorbed with roots in aerobic conditions (arbitrary units)	Sulphate absorbed with roots in anaerobic conditions (arbitrary units)
0	0	0
1	108	65
2	143	93
3	176	106
4	195	112
5	216	115
6	237	125
7	249	131
8	267	137

a) Plot a graph of both sets of results on the same axes.

b) What process which uses oxygen takes place in most cells?

c) Explain why the roots in aerobic conditions absorbed more sulphate than when oxygen was not present.

8 Write a short description of how water moves from the soil through the plant. Your description must include these words: xylem, evaporation, root hair cells, water potential, stomata and osmosis.

Underline each of these words in your description.

Chapter 12: Chemical Coordination in Plants

Like animals, plants sense and respond to their environment, but the responses are usually much slower than those of animals because their movements are due to changes in the plant's growth. This chapter is about these growth responses, and the chemicals that coordinate them.

Chapter 6 explains how animals sense and respond to changes in their environment. Animals usually respond very quickly – for example the reflex action resulting from a painful stimulus (page 70) is over in a fraction of a second.

As in animals, some species of plant can respond rapidly to a stimulus, for example the Venus' flytrap (Figure 12.1). This plant has modified leaves, which close quickly around their 'prey', trapping it. The plant then secretes enzymes to digest the insect. The movement is brought about by rapid changes in turgor of specialised cells at the base of the leaves.

Figure 12.1 *The Venus' flytrap catches and digests insects to gain extra nutrients. The plant responds very quickly to a fly landing on one of its leaves.*

Tropisms

Most plants do not respond to stimuli as quickly as this, because their response normally involves changing their rate of growth. Different parts of plants may grow at different rates, and a plant may respond to a stimulus by increasing growth near the tip of its shoot or roots. Imagine a plant growing normally in a pot. Usually, most light will be falling on the plant from above. If you turn the plant on its side and leave it for a day or so, you will see that its shoot starts to grow upwards (Figure 12.2).

Figure 12.2 *This bean has responded to being placed horizontally. The growing shoot has started to bend upwards.*

There are two stimuli acting on the plant in Figure 12.2. One is the direction of the light that falls on the plant. The other stimulus is gravity. Both light and gravity are **directional stimuli** (they act in a particular direction). The growth response of a plant to a directional stimulus is called a **tropism**. If the growth response is *towards* the direction of the stimulus, it is a *positive* tropism, and if it is *away* from the direction of the stimulus, it is a *negative* tropism. The stem of the plant in Figure 12.2 is showing a positive **phototropism** and a negative **geotropism**, which both make the stem grow upwards.

Phototropisms are growth responses to light from one direction. Geotropisms are growth responses to the direction of gravity ('geo' refers to the Earth).

The aerial part of a plant (the 'shoot') needs light to carry out photosynthesis. This means that in most species, a positive phototropism is the strongest tropic response of the shoot. If a shoot grows towards the light, it ensures that

Light from one direction is called 'unidirectional'. If light shines on the plant evenly from all directions, this is called 'uniform' light.

Figure 12.3 *The shoots of these cress seedlings are showing a positive phototropism.*

the leaves, held out at an angle to the stem, will receive the maximum amount of sunlight. This response is easily seen in any plant placed near a window, or another source of 'one-way' or *unidirectional* light (Figure 12.3).

In darkness or uniform light, the shoot shows a negative geotropism. As you might expect, the roots of plants are strongly positively geotropic. This response makes sure that the roots grow down into the soil, where they can reach water and mineral ions, and obtain anchorage.

The roots of some species that have been studied are also negatively phototropic, but most roots don't respond to directional light at all. In the same way some experiments have shown that roots of a few species show positive **hydrotropism** (attraction to water). The common tropisms are summarised in Table 12.1.

Stimulus	Name of response	Response of shoots	Response of roots
light	phototropism	grow towards light source (positive phototropism)	most species show no response; some grow away from light (negative phototropism)
gravity	geotropism	grow away from direction of gravity (negative geotropism)	grow towards direction of gravity (positive geotropism)
water	hydrotropism	none	some species may grow towards water (positive hydrotropism)

Table 12.1: *Common responses of plants to directional stimuli (tropisms).*

Detecting the light stimulus – plant hormones

Plants do not have the obvious sense organs and nervous system of animals, but since they respond to stimuli such as light and gravity, they must have some way of detecting them and coordinating the response. The detection system of phototropism was first investigated by the great English biologist Charles Darwin (see Chapter 19) in the late nineteenth century. Instead of using stems, Darwin (and later scientists) used cereal **coleoptiles**, which are easier to grow and use in experiments.

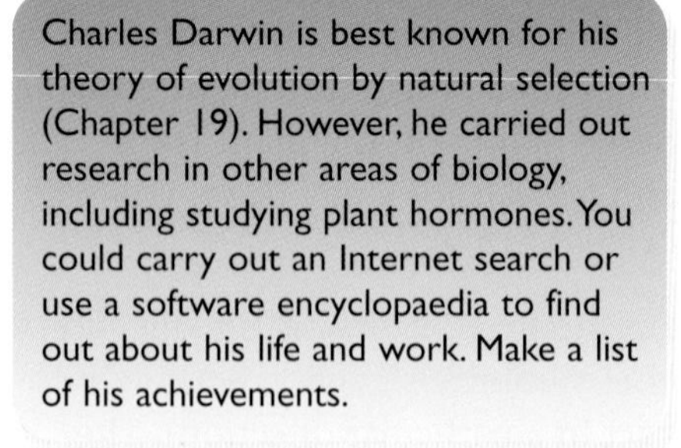

Charles Darwin is best known for his theory of evolution by natural selection (Chapter 19). However, he carried out research in other areas of biology, including studying plant hormones. You could carry out an Internet search or use a software encyclopaedia to find out about his life and work. Make a list of his achievements.

Darwin showed that the stimulus of unidirectional light was detected by the tip of the coleoptile, and transmitted to a growth zone, just behind the tip (Figure 12.5).

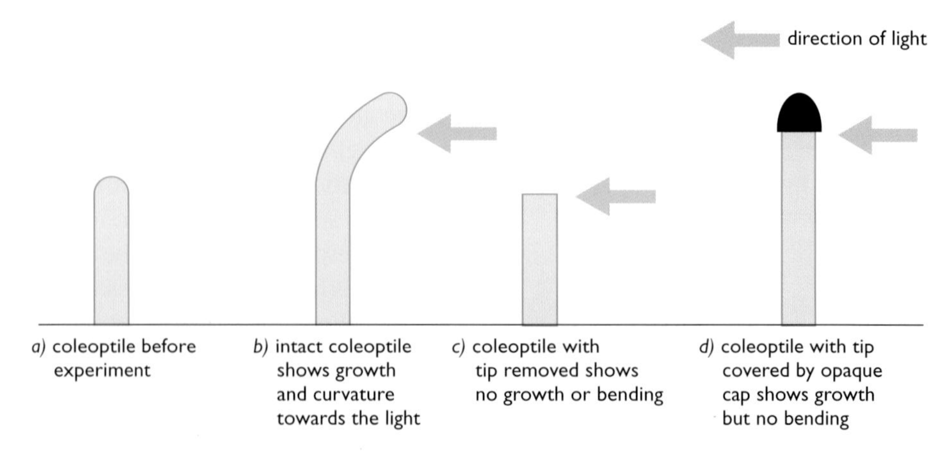

Figure 12.5 *Darwin's experiments with phototropism (1880).*

Since plants don't have a nervous system, biologists began to look for a chemical messenger, or **hormone** that might be the cause of phototropism in coleoptiles. Between 1910 and 1926 several scientists investigated this problem. Some of their results are summarised in Figure 12.6.

A coleoptile is a protective sheath that covers the first leaves of a cereal seedling. It protects the delicate leaves as the shoot emerges through the soil (Figure 12.4).

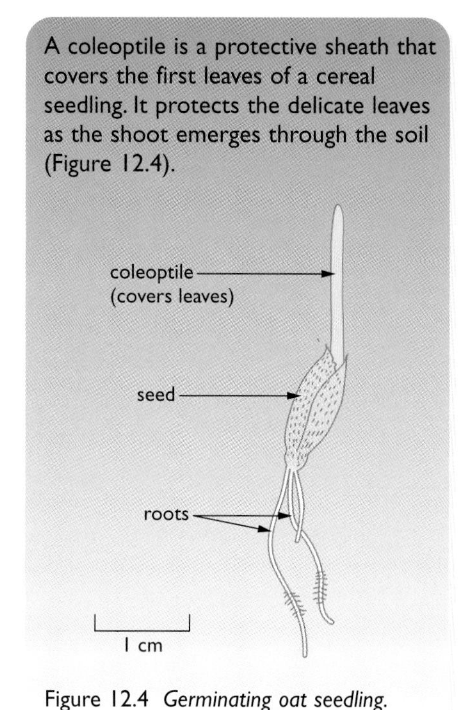

Figure 12.4 *Germinating oat seedling.*

Experiment 1

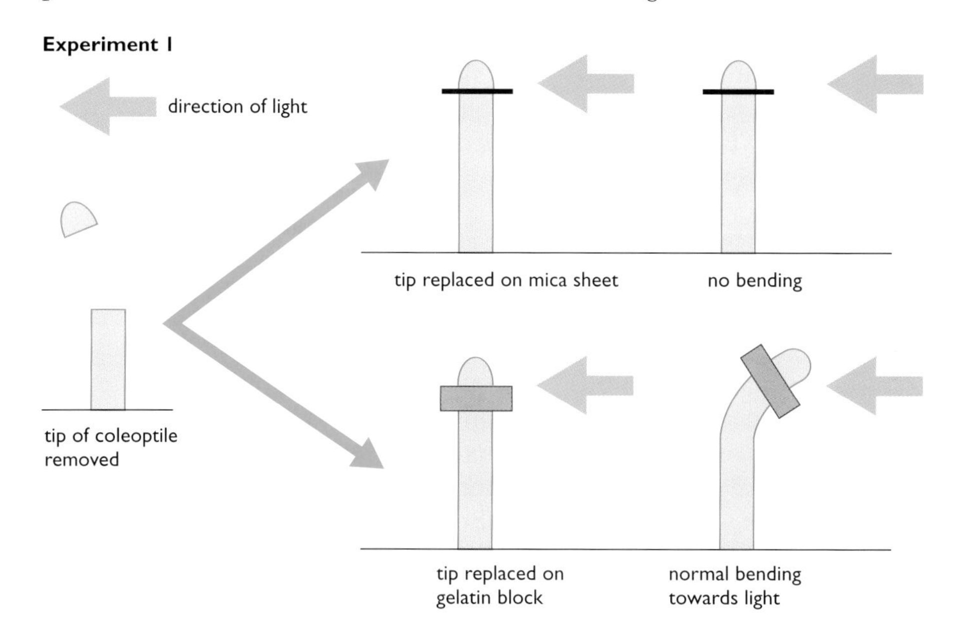

Experiment 2

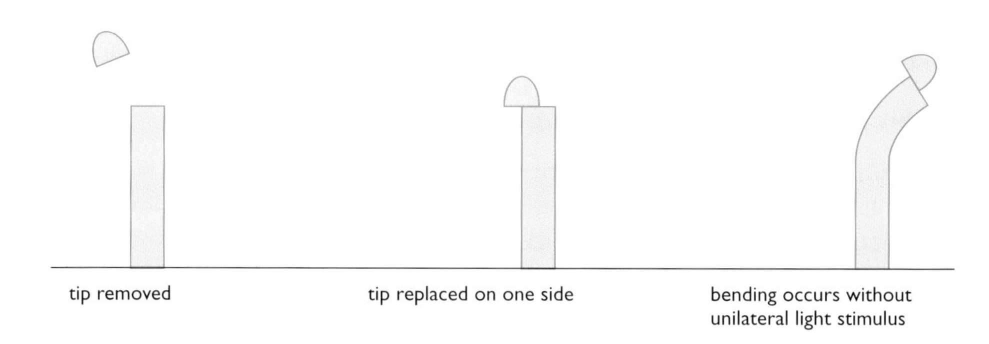

Experiment 3

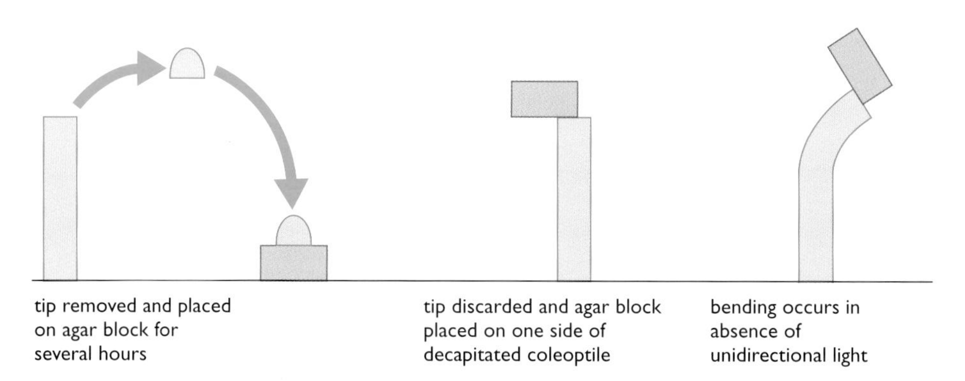

Figure 12.6 *Experiments on coleoptiles that helped to explain the mechanism of phototropism.*

- In experiment 1, the stimulus for growth was found to pass through materials such as gelatin, which absorbs water-soluble chemicals, but not through materials such as mica (a mineral) which is impermeable to water. This made biologists think that the stimulus was a chemical that was soluble in water.
- In experiment 2, it was shown that the phototropic response could be brought about, even *without* unidirectional light, by removing a coleoptile tip ('decapitating' the coleoptile) and placing the tip on one side of the decapitated stalk.
- In experiment 3, it was found that the hormone could be collected in another water-absorbing material (a block of agar jelly). Placing the agar block on one side of the decapitated coleoptile stalk caused it to bend.

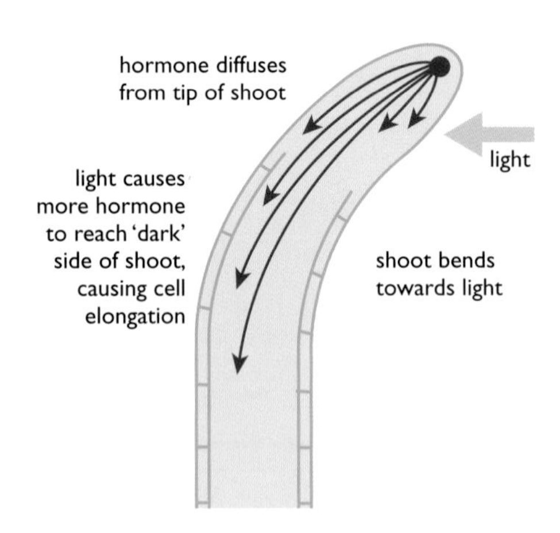

Figure 12.7 *How movement of a plant hormone causes phototropism.*

Experiments like 2 and 3 led scientists to believe that the hormone caused bending by stimulating growth on the side of the coleoptile furthest from the light. The theory is that the hormone is produced in the tip of the shoot, and diffuses back down the shoot. If the shoot is in the dark, or if light is all around the shoot, the hormone diffuses at equal rates on each side of the shoot, so it stimulates the shoot equally on all sides. However, if the shoot is receiving light from one direction, the hormone moves away from the light as it diffuses downwards. The higher concentration of hormone on the 'dark' side of the shoot stimulates cells there to grow, making the shoot bend towards the light (Figure 12.7).

'Auxin' should really be 'auxins', since there are a number of chemicals with very similar structures making up a group of closely related plant hormones.

Since these experiments were carried out, scientists have identified the hormone responsible. It is called **auxin**. Several other types of plant hormone have been found. Like auxin, they all influence growth and development of plants in one way or another, so that many scientists prefer to call them **plant growth substances** rather than plant hormones.

Plant hormones and geotropism

Bending of the root and shoot during geotropism is also thought to be due to plant hormones. If a broad bean seedling is placed in the dark in a horizontal position, its shoot will bend upwards and its root downwards (Figure 12.8).

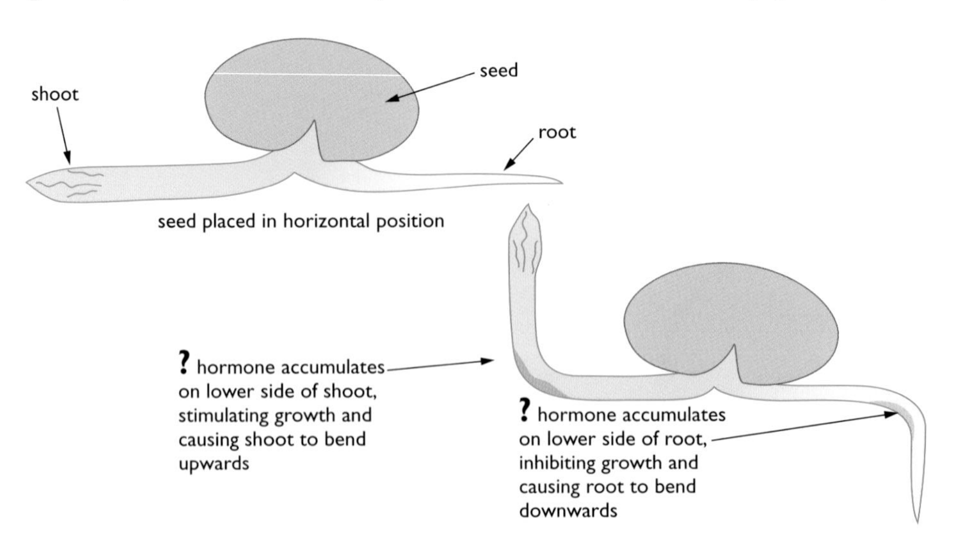

Figure 12.8 *Geotropism in a broad bean seedling. It was once thought that movement of auxin caused this response. We now know that the true explanation is not as simple as this.*

It was once thought that these geotropic responses of the shoot and root were due to auxin. The auxin was supposed to be produced at the tip of the shoot, and to sink under the influence of gravity as it diffused back from the tip. A high concentration of auxin would increase growth on the lower side of the shoot, causing it to bend upwards. The downward growth of the root was supposed to be caused in a similar way, except that auxin *inhibited* growth on the lower side. We know now that these explanations are not the whole story. Although some movement of auxin happens due to the effect of gravity, it is not enough to explain geotropisms, where other hormones seem to be involved.

As well as auxins, there are four other main groups of plant hormones, called gibberellins, cytokinins, abscisic acid and ethene. They control many aspects of plant growth and development, apart from tropisms. These include growth of buds, leaves and fruit, fruit ripening, seed germination, leaf fall and opening of stomata, to name just a few. In addition to auxins, abscisic acid, ethene and gibberellins have all been shown to be involved in geotropisms.

Commercial uses of plant hormones

'Auxin' is actually the name of a group of similar chemical compounds found in plant tissues. Chemists have made synthetic versions of these compounds, with structures like that of natural auxin. Along with other plant hormones, they have a number of commercial uses, some of which are listed below.

- Synthetic auxins are used in high concentrations in some weedkillers. They work in an unexpected way. They cause the cells of the weed to elongate, so that the plant grows very quickly. Its physical structure becomes weakened and its physiological processes stop working properly. It soon dies (Figure 12.9). These compounds are called **selective weedkillers**, since they have a greater effect on broad-leaved plants than grasses and cereals. They are used to kill weeds like dandelions in lawns, without harming the grass, or to kill broad-leaved weeds in a cereal crop.

Figure 12.9 *Selective weedkiller has been sprayed on these plants. The grass is not affected, but the auxin is killing the broad-leaved weeds.*

- Synthetic auxins are also used in **rooting powder**, for taking plant cuttings. The cut end of the stem is dipped into the rooting powder, which stimulates the growth of roots, helping the cutting to become established.
- Plant hormones are used in **plant tissue culture media**, to encourage cell division and growth of tissue.
- Certain synthetic auxins and other hormones (gibberellins) are sprayed on to fruit crops. They stimulate the development of the fruit (called fruit 'set'), so that the fruit sets on all plants at the same time, making it easier to harvest. They can even cause a fruit to form when the flower has not been pollinated. This is used in the growing of **seedless fruits**, such as some varieties of seedless grapes. Other hormones can delay fruit ripening, allowing time for the fruit to be transported to the consumers.
- Another group of plant hormones (cytokinins) affects cell division. These hormones are sprayed on to green vegetables such as cabbage and lettuce, where they delay aging and discolouration of the leaves. They are also used to keep cut flowers fresh.

End of Chapter Checklist

If you haven't got a copy of your specification, read the introduction on page vi.

You will need to be able to do some or all of the following. Check your Awarding Body's specification (syllabus) to find out exactly what you need to know.

- Understand that plants respond to directional stimuli, such as light, gravity and moisture, and know that these responses are called tropisms.
- Understand how a response such as phototropism in a shoot is brought about by auxin.
- Be able to interpret experimental data on how plant hormones affect the growth of roots and shoots.
- Know the commercial uses of plant hormones.

Questions

More questions on chemical coordination in plants can be found at the end of Section C on page 149.

1 ***a)*** What are the main stimuli affecting the growth of

i) the shoot,

ii) the root?

b) How does a plant benefit from a positive phototropism in its stem?

2 Draw a labelled diagram to show how auxin brings about phototropism in a coleoptile that has light shining on it from one direction.

3 An experiment was carried out to investigate phototropism in a coleoptile. The diagram shows what was done. Predict the results you would expect to get in each of the experiments ***a)*** to ***c)***. Explain your answers.

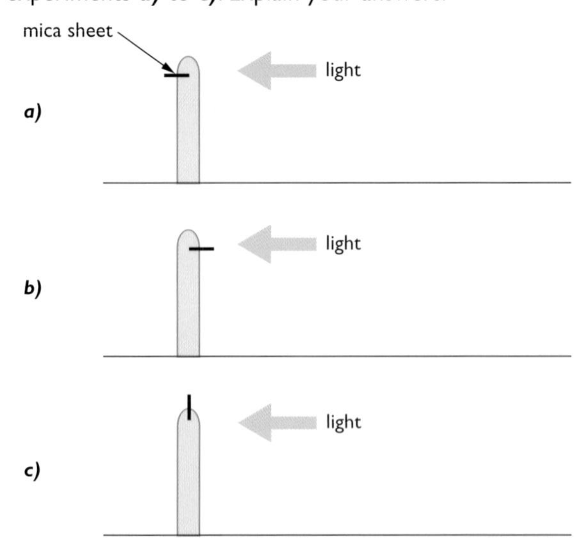

4 Plant hormones have a number of commercial uses. Name three of these. In each case explain briefly how the hormone works.

End of Section Questions

1 Light intensity and the concentration of carbon dioxide in the atmosphere influence the rate of photosynthesis.

a) The graph shows the effect of changing light intensity on the rate of photosynthesis at two different carbon dioxide concentrations.

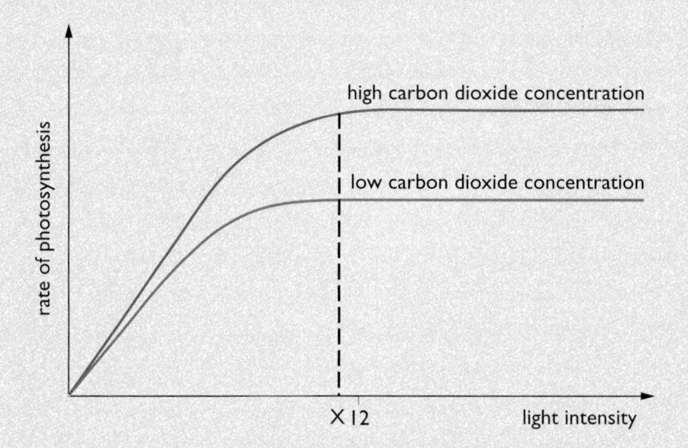

i) Describe the effect of light intensity on the rate of photosynthesis at each concentration of carbon dioxide up to light intensity X and beyond light intensity X. *(8 marks)*

ii) Which factor limits the rate of photosynthesis up to light intensity X and beyond light intensity X? *(2 mark)*

Explain your answer in each case. *(3 marks)*

b) *i)* Describe two other factors which influence the rate of photosynthesis. *(2 marks)*

ii) Explain why each is a limiting factor. *(4 marks)*

c) 'Photosynthesis is a means of transducing light energy into chemical energy.'

Explain what this statement means. *(2 marks)*

Total 21 marks

2 In an investigation to determine the water potential of potato cells, the following procedure was adopted.

- Cylinders of potato tissue were obtained using a cork borer and each was cut to a length of 5 cm.
- Each was dried and then weighed.
- Three potato cylinders were placed in each of seven different concentrations of sucrose solution and left for 2 hours.
- The cylinders were then removed from the solutions, dried and reweighed. The percentage change in mass for each was calculated, and then an average percentage change in mass calculated for each solution.

The graph summarises the results.

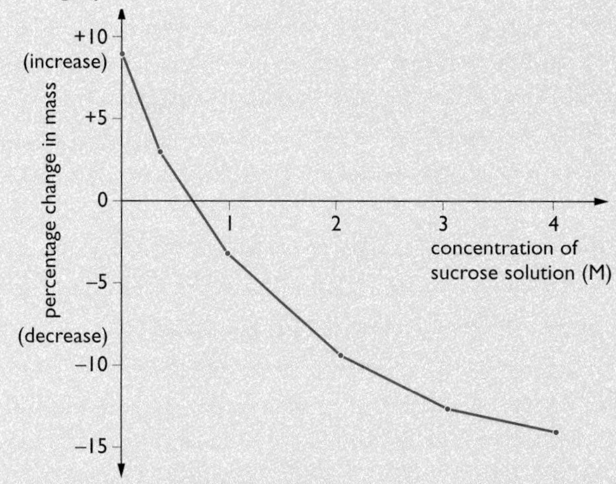

a) Explain why:

i) the cylinders were dried before and after being placed in the sucrose solutions *(1 mark)*

ii) three cylinders were used for each solution, allowing an average to be calculated *(1 mark)*

iii) all the cylinders were obtained with the same sized cork borer and were cut to the same length. *(1 mark)*

b) *i)* In terms of water potential, explain the result obtained with a 3M sucrose solution. *(3 marks)*

ii) What concentration of sucrose has a water potential equivalent to that of the potato cells? Explain your answer. *(3 marks)*

iii) Look at the graph and suggest which result is anomalous. *(1 mark)*

iv) Suggest how this anomaly might have been caused. *(2 marks)*

c) How could the water potential of the potato tissue be determined more accurately? *(2 marks)*

Total 14 marks

3 Transpiration is the process by which water moves through plants from roots to leaves. Eventually, it is lost through the stomata.

a) The diagram shows the main stages in the movement of water through a leaf.

i) Name the tissues labelled A and B. Explain your answers. *(4 marks)*

ii) Describe how water is being moved at each of the stages 1, 2, 3 and 4. *(4 marks)*

b) Describe two ways, visible in the diagram, in which this leaf is adapted to photosynthesise efficiently. *(4 marks)*

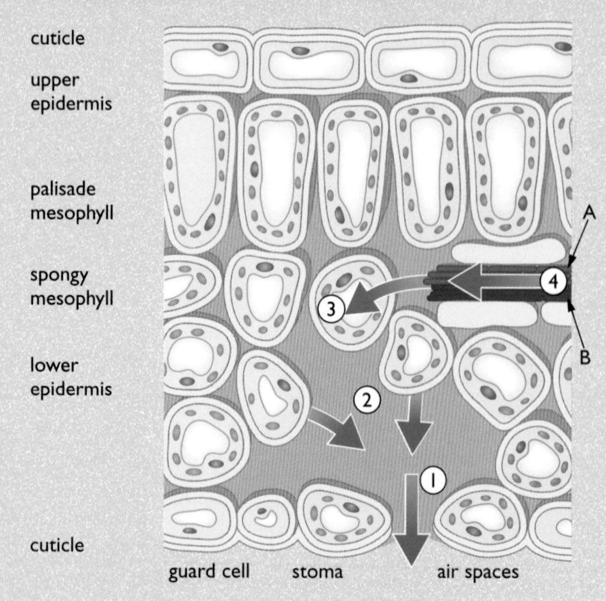

c) For plants living in dry areas, explain a possible conflict between the need to obtain carbon dioxide for photosynthesis and the need to conserve water. *(2 marks)*

Total 14 marks

4 Plants can respond to a range of stimuli.

a) Plant shoots detect and grow towards light.

i) What is this process called? *(1 mark)*

ii) Explain how a plant bends towards the light. *(3 marks)*

iii) Explain the advantage to the plant of this response. *(1 mark)*

b) In an investigation, young plant shoots were exposed to light from one side. The wavelength of the light was varied. The graphs summarises the results of the investigation.

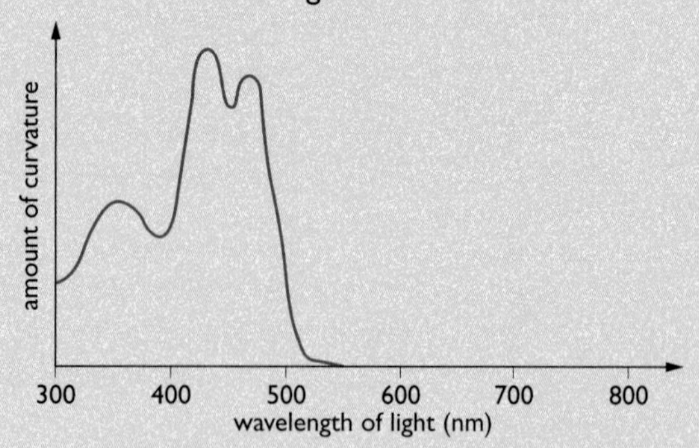

i) Describe the results shown in the graph. *(2 marks)*

ii) Suggest why the results show this pattern. *(2 marks)*

c) *i)* Name two other stimuli that produce growth responses in plants. *(2 marks)*

ii) For each stimulus you name, describe the way that both roots and shoots respond. *(4 marks)*

iii) What is the benefit to the plant of these responses. *(2 marks)*

Total 17 marks

5 Plants produce many hormones. One of these is called gibberellin.

a) In an investigation into the effects of gibberellin, two groups of cabbage plants were used. One was treated with gibberellin once a week for 8 weeks. The other group received no gibberellin treatment. Some results are shown in the table.

Height of cabbage plants at the end of 8 weeks (cm)	
Treated plants	**Untreated plants**
420	25
385	30
415	24
415	22
370	26
365	25
380	32
375	30
345	32
380	25

i) Calculate the average height of cabbage plants in the treated and untreated groups. *(4 marks)*

ii) What effects do gibberellins have on plants? *(1 mark)*

b) Gibberellin is also used to produce seedless grapes. The diagram shows a bunch of untreated seedless grapes and one that had a small injection of gibberellin.

no gibberellin treatment

treated with gibberellin

i) Describe two differences between the two bunches of grapes. *(2 marks)*

ii) Grape crops can be affected by diseases caused by fungi. The grapes treated with gibberellin do not get as many fungal infections as untreated grapes, even when they are kept in identical conditions. Suggest why. You should bear in mind that fungal infections spread more rapidly in moist conditions. *(3 marks)*

Total 10 marks

Section D: Ecology and Adaptation

Chapter 13: Adaptation to Environment

Earth has an infinite variety of environments, from frozen Arctic ice fields, to scorching desert sand dunes, lush tropical rainforest and deep ocean trenches. We find living organisms in even the most inhospitable of these habitats. This chapter looks at the way organisms have adapted to survive in their particular environment.

Adaptations are features that give an organism a good chance of surviving in the environment in which it lives.

- Some adaptations are shared by nearly all species of a phylum (a large group of organisms – see Chapter 19). For example, all fish have gills to enable them to live and breathe in water, and nearly all birds have feathered wings to enable them to fly through the air.
- Some adaptations are shared by a number of species that inhabit similar environments. For example, plants living in a desert environment often have reduced numbers of stomata and thick waxy cuticles to help them to conserve water. These features reduce transpiration (see Chapter 11) and enable the plants to survive in a hot, dry environment.
- Some adaptations are shared by members of just one species or by just one type of one species (Figure 13.1 shows two forms of the peppered moth *Bistan betularia*.).

Figure 13.1 *These moths are different forms of the same species –* Biston betularia, *the peppered moth. Both forms are food for several species of birds. Both are found in areas with clean air and in smoke-polluted areas. Which form is best adapted to which area?*

The dark form has a much better chance of surviving in a sooty area because it is better camouflaged on sooty bark. The same is true for the light form in unpolluted areas. Most peppered moths in a sooty area are dark and most in an unpolluted area are light. This is a result of **natural selection**, (see Chapter 19). The camouflaging is an adaptation to the environment.

Different environments demand different adaptations.

There are many different aquatic environments. Some organisms live on the surface of pond water; others live suspended just under the surface. Some live scavenging on the bottom of ponds. Carry out an Internet search to discover some of the adaptations shown by organisms inhabiting different aquatic environments

Adaptations to aquatic environments

Living in water is very different from living on land. Key differences are:

- Water is a *fluid* – it flows.
- Water is more dense than air.
- Water contains much less oxygen gas than air.

These differences affect movement, support and gas exchange.

Movement in water

Movement on land is more efficient than movement through water for two main reasons. Firstly, when you push against the ground with your foot, the ground stays still and you move forwards. In water, when you push against the water – it moves! Secondly, because water is more dense than air, it offers more resistance to movement.

To help make movement through water as efficient as possible, many aquatic animals have evolved a streamlined shape and a tail or feet with a large surface area. A streamlined shape allows water to flow smoothly over the body and minimises the resistance offered by the water (Figure 13.2). The large area of tail or feet displaces a large volume of water and creates a large forward force.

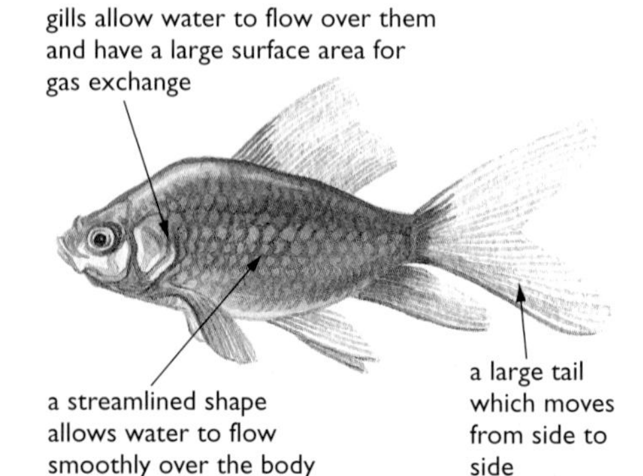

Figure 13.2 *Fish are adapted to their environment.*

Support

Because water is more dense than air, the bodies of animals and plants living in water are partly supported by the surrounding water. Therefore, they need less skeletal tissue to support them than organisms that live on land.

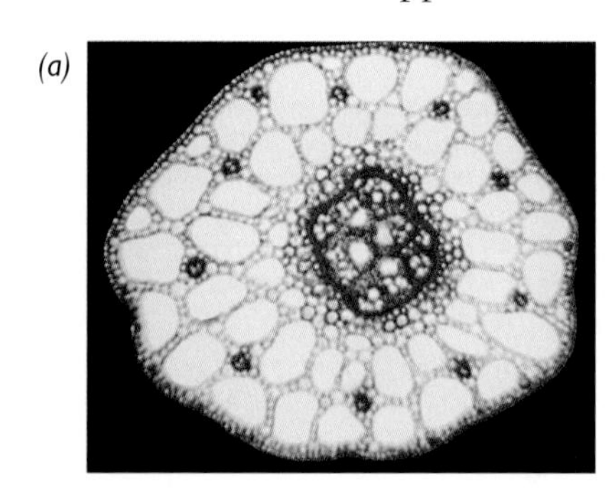

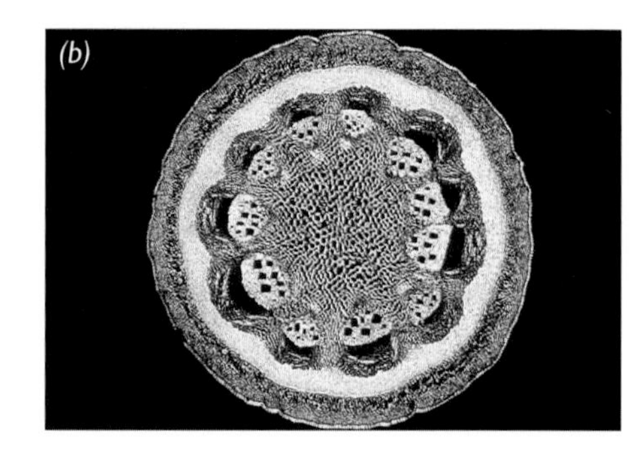

Figure 13.3 *(a) Water provides pondweed with support so less supporting tissues are needed. (b) Air provides less support so more internal support is needed in the form of xylem.*

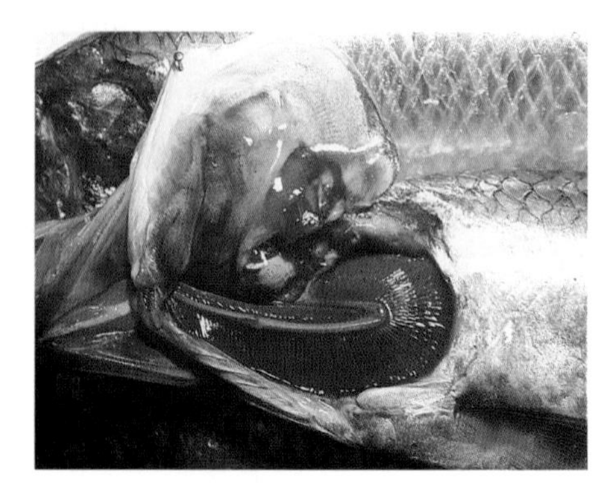

Figure 13.4 *The gill filaments give the gill a large surface area and a rich supply of blood for gas exchange (see Chapter 2).*

Gas exchange

We breathe by moving air in and out of our lungs. It would take too much energy for a fish to move water into and out of a sac like this. Fish move water in a continuous stream over their gills. Gills are specially adapted to obtain oxygen efficiently. They are divided into gill filaments, which give a large surface area for gas exchange, and the filaments have a very thin surface membrane with many capillaries close to the surface (Figure 13.4).

> A salmon begins life as a freshwater fish in a river. When it is approximately 1-year old, it migrates to sea. It spends a considerable time in an estuary before finally leaving for the open sea. Some years later it returns to the same river to mate. Again, it spends a considerable time in the estuary. What problems does the salmon face in its changing environments? Carry out an Internet search to try to find out how the salmon solves these problems.

Different aquatic environments

Living in the sea poses different problems to living in a pond or river. The problems of movement, support and gas exchange are the same in both, but the fact that seawater is salty and river water isn't creates some different problems related to control of water content in the body. Water moves from a liquid with a high water potential to one with a lower water potential across a partially permeable membrane by osmosis (see Chapter 1). In practice, this means it moves from a more dilute solution to a more concentrated one. Seawater is more concentrated than body fluids, while fresh water is less concentrated. As a result, marine organisms tend to lose water by osmosis and therefore have adaptations to conserve it, whereas freshwater organisms tend to gain water by osmosis and therefore have adaptations to remove the excess.

Adaptations to terrestrial environments

The bodies of organisms living on land need more support than those of aquatic organisms of a similar size. The air which surrounds them is less dense than water and so provides little external support. They must move by walking or flying and they must be able to breathe air and conserve water.

Snakes do not walk or fly – they wriggle! They have lost their legs as they evolved from reptiles that had legs. Try to find out a possible advantage of not having legs on land.

The ways in which these problems are solved are described in Chapter 2 (breathing), Chapter 8 (support and movement) and Chapter 9 (homeostasis and excretion).

There are many different terrestrial environments, each with very specific features that affect the organisms living there. For example, living in a hot desert poses very different problems to living in the Arctic. Animals and plants must obtain and conserve water either in a frozen wasteland or from scorching sand. Mammals and birds also need to maintain a constant body temperature in these very different extreme conditions.

Adaptations to Arctic conditions

For about three months of the year, the lands North of the Arctic Circle receive hardly any sunshine. During this period, and for much of the rest of the year it is very cold, water is frozen and there is very little light. The main plants of the Arctic are mosses, lichens and grasses. They grow and flower only during summer when there is sufficient light and heat to 'drive' photosynthesis. During the long Arctic winter they are buried beneath several metres of snow. These plants often grow flat against the soil, have thick cuticles and are covered in hairs. All these features protect them from the drying Arctic winds, which could otherwise dehydrate them in a very short time.

Figure 13.5 *The Arctic is an inhospitable place in which to live. The main plants of the Arctic, mosses, lichens and grasses grow only during the summer.*

As the plants grow they provide food for the land-dwelling animals of the Arctic. Many herbivores depend on these summer plants for food and they, in turn, provide food for the carnivores. The polar bear is the top carnivore of the Arctic ecosystem and is well adapted to life in the extreme cold (Figure 13.6).

Figure 13.6 *The polar bear is the top carnivore of the Arctic food web. The adaptations shown increase its chance of survival in the Arctic.*

In the Arctic summers, temperatures can reach 25°C or more in some areas. The polar bear loses much of the thick winter coat and fat underneath the skin. It also eats less fatty animal tissue and will eat considerable amounts of plant matter. As winter approaches, the polar bear begins to grow another winter coat and reverts to a more carnivorous diet to replace the fat under the skin.

The female polar bear spends the entire winter in a 'den', where her cubs are born (Figure 13.7). During this time she fasts and produces milk for the cubs from accumulated fat. The temperatures in the den are above zero, even though they may be –10°C or lower outside.

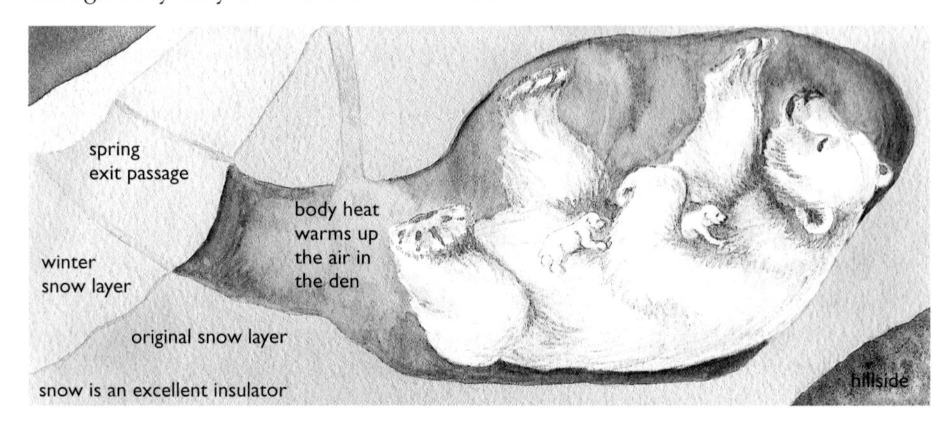

Figure 13.7 *Birth and early life in an igloo! The thick snow layer provides excellent insulation. Body heat keeps the 'den' warm.*

Some mammals of the Arctic hibernate throughout the long winter. In this state, their body temperature falls and their metabolic rate also falls. This considerably reduces the rate at which food is used up and the animals can survive for long periods on stored food.

Adaptations to desert conditions

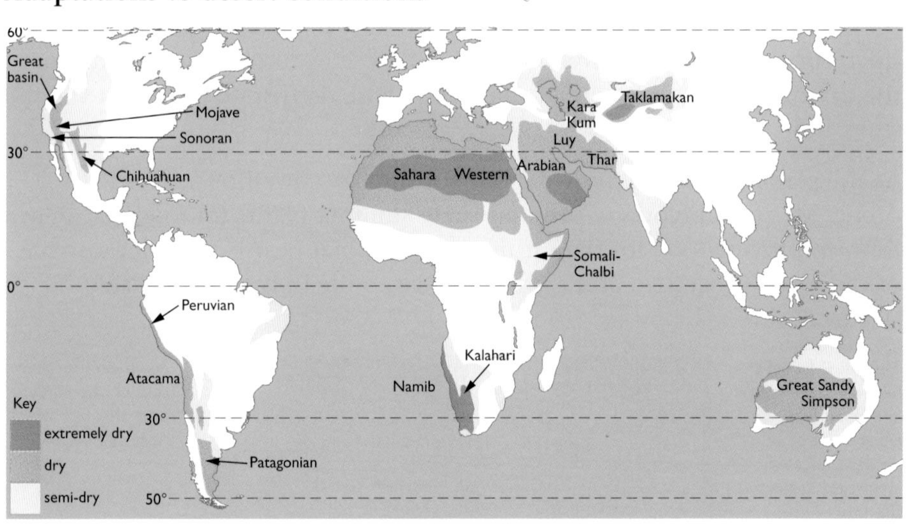

Figure 13.8 *The main hot deserts of the world.*

> Deserts heat up quickly during the day because cloudless skies allow intense sunlight to strike the sands. At night, the same cloudless skies allow heat to escape quickly and the desert cools down rapidly.

The two main threats to survival in a desert environment are:

- the huge variation in temperature during each 24 hour period
- the shortage of water.

Desert plants must solve the problem of being able to open their stomata to obtain carbon dioxide for photosynthesis without losing too much water through transpiration (see Chapter 10 and Chapter 11). They must also obtain enough water to keep all the cells hydrated. Cacti have several adaptations that help them to survive. Some of these are shown in Figure 13.9.

They have a specialised method of photosynthesis which allows them to obtain carbon dioxide during the cooler night. They will then not lose too much water through the open stomata. They convert the carbon dioxide to a storage product and use it in photosynthesis the following day.

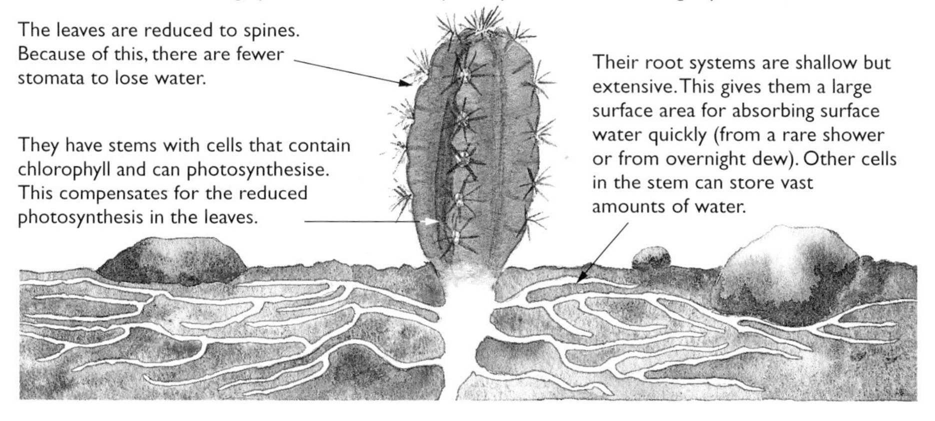

Figure 13.9 *Cacti have several adaptations which enable them to survive in hot deserts.*

Other desert plants have different adaptations. Acacia trees have roots up to 35 m in length that find water deep underground. Some plants have an 'ephemeral' life cycle. They lie dormant as seeds for most of the year. When the short rainy season arrives, the seeds germinate and the plants grow, reproduce, make more seeds, and then die – all within three or four weeks.

Some plants have their leaves rolled into a tube with most stomata facing the inside of the tube. Here the air is moist and little water is lost. Others have stomata sunk below the leaf surface, protected from the drying effects of the sun.

Animals living in the desert must conserve water and also keep their body temperatures constant. This is a particular problem for mammals as they use water (in sweat) to cool down. Camels are adapted to living in the desert in several ways as shown in Figure 13.10.

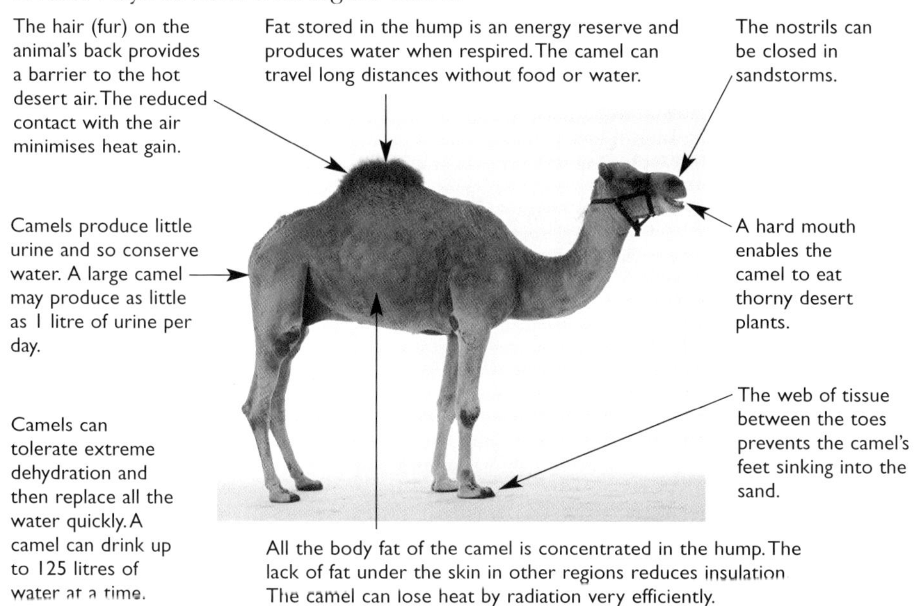

Figure 13.10 *Camels are well adapted to survive in a hot desert.*

The fat in the hump of a camel is an important food store. Fat gives more energy per gram than any other nutrient when it is respired. So the camel can survive longer than if the hump contained a store of carbohydrate. Fats also produce more water per gram in respiration than carbohydrate. This 'metabolic water' is also important to a desert mammal.

Camels can tolerate a much larger rise in body temperatures than humans. A body temperature of 39°C in humans would be extremely serious. A camel does not even begin to sweat until its body temperature reaches 40°C. This 'temperature tolerance' allows the camel to conserve water that other mammals would lose in sweat (Figure 13.11).

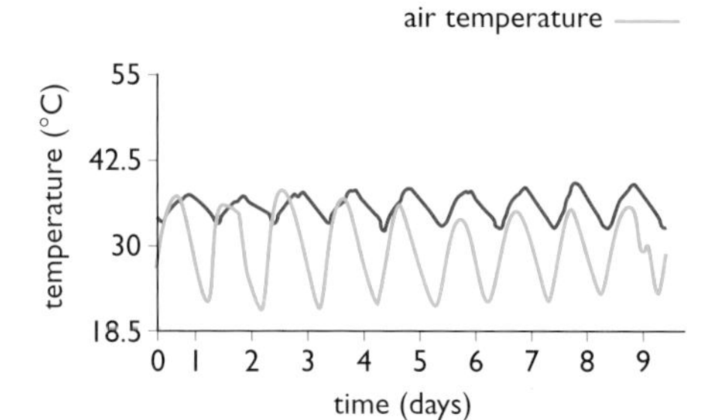

Figure 13.11 *Temperature tolerance in a camel. The large fluctuations in the body temperature of the camel could be fatal in other mammals.*

The gerbil (a small, desert rodent) has some different adaptations that enable it to survive desert conditions. It produces only tiny amounts of concentrated urine each day, has very few sweat glands and so loses little in sweat, and spends most of the hot day underground, where the conditions are cooler and more humid than above ground. It produces enough water from respiring fats and carbohydrates in dry seeds to compensate for the tiny losses (Figure 13.12).

Figure 13.12 *A gerbil in its natural habitat.*

(a)

(b)

Figure 13.13 *The relative sizes of (a) a polar bear and (b) a sun bear.*

Mammals in the desert and Arctic compared

Although many mammals have specific, individual adaptations to a particular environment, there are some general features that are common to most individuals. Table 13.1 illustrates this.

Feature	Arctic mammal	Desert mammal
body fat	thick layer to reduce heat loss by radiation	thin layer to increase heat loss by radiation
body size	large compared to similar mammals in warmer climates	small compared to similar mammals in warmer climates
body surface area to volume ratio	small ratio reduces rate of heat loss through surface	large ratio increases rate of heat loss through surface
thickness of fur	thick fur increases insulation	thin fur decreases insulation
camouflage	white fur	sandy coloured fur

Table 13.1: *Adaptations common to Arctic and desert animals.*

Animals living in cold climates are often larger than similar animals living in warmer climates. The large size reduces the ratio of surface area to volume and so reduces the rate of heat loss. Figure 13.13 shows the relative sizes of a polar bear (from the Arctic) and a sun bear (from south-east Asia).

Adaptations can be limitations

If an animal or plant is very well adapted to one environment, it will almost certainly be less well adapted to other environments. For example, the polar bear is superbly adapted to life in the Arctic, but would die very quickly in a desert. Similarly, a camel could not live for long in the Arctic. Adaptations, therefore, restrict the organism to the environment to which it is adapted.

Fish cannot live on land. Their gills adapt them to breathe in water but are ineffective in air. However, adult amphibians, like frogs and newts, can live in water as well as on the land. They are not particularly well adapted to either environment but can exist in both. The reasons for this are shown in Table 13.2.

Feature	Adaptation / limitation of feature to... ...life on land	...life in water
moist, non-scaly skin	allows gas exchange in air but loses water easily; adults must stay close to water or they will dehydrate and die	allows efficient movement through water; some gas exchange in water is possible
simple lungs	allow gas exchange in air	no role in gas exchange in water
long hind legs with webbed feet; short front legs	allow jumping and slow crawling, but are less effective than an animal without webbed feet	hind limbs give good thrust but are less effective than a fish's tail
sexual reproduction requires external fertilisation	adults cannot breed on land – they must return to water to breed	adults can breed successfully in water
young (tadpoles) have gills and a tail	life on land is impossible – they cannot move or breathe on land	well adapted for life in water as gills allow gas exchange and tail gives efficient movement

Table 13.2: *The adaptations of amphibians and their limitations.*

Adult amphibians must remain close to water, even though they can move about on land. Their adaptations limit them to the areas near to ponds and streams for much of the time.

The amphibian life cycle is an example of how **metamorphosis** allows animals to exploit more than one environment. However, as Table 13.2 shows, amphibians are only able to exploit the land as adults, while the larva (tadpole) can only live in water. The larva **metamorphoses** (changes body form) into an adult, which has features that allow it to move and breathe on land, whilst still being able to move in water.

Other animals also metamorphose and exploit more than one environment. Butterflies begin life as caterpillars and blowflies begin life as maggots. Figures 13.14 and 13.15 show the life cycles of these two insects.

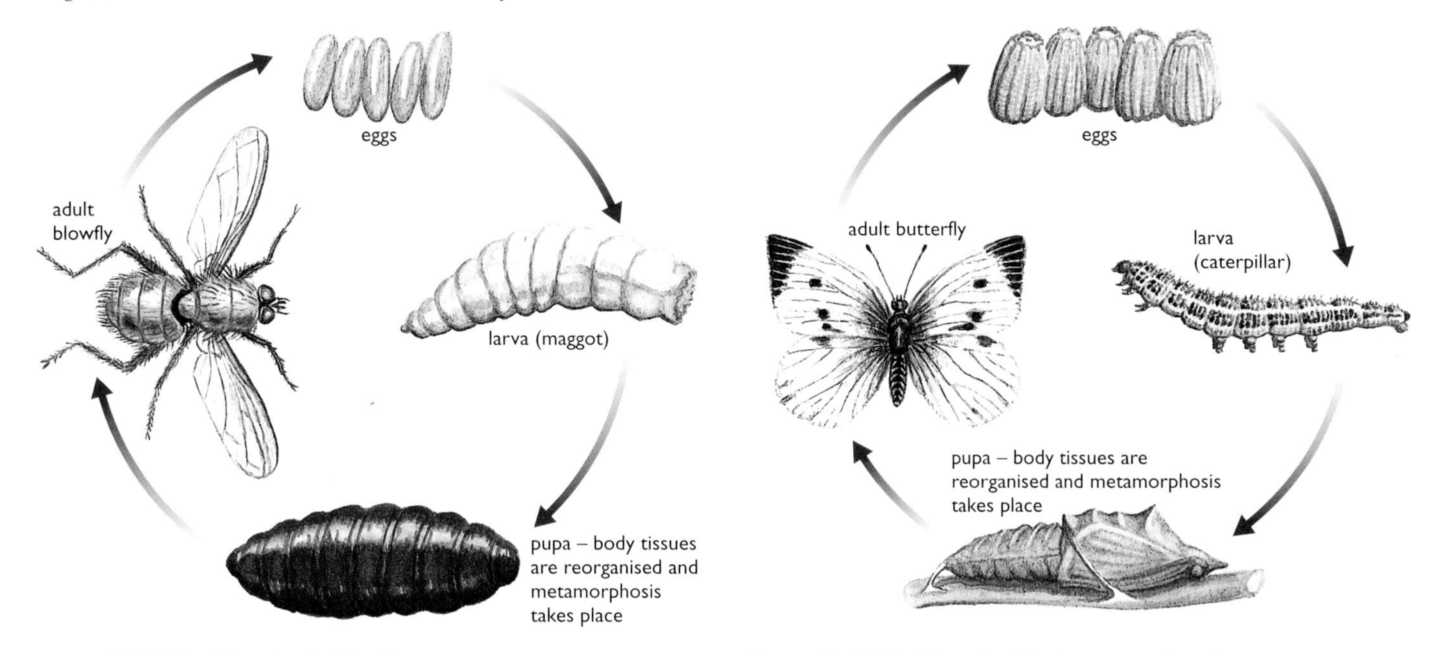

Figure 13.14 *The life cycle of a blowfly.*

Figure 13.15 *The life cycle of the large white butterfly.*

Both life cycles effectively separate the growth phase from the reproductive phase of the animal. The larva just eats and grows; only the adult can reproduce. The life cycle of a butterfly is a more extreme example of separation of growth and reproductive phases. The adult blowfly can dissolve many different food materials using enzymes it secretes, whereas an adult large white butterfly can only digest sucrose (the sugar found in nectar). It cannot digest proteins and so cannot repair or replace damaged cells. It must find a mate within one or two days, before it dies.

End of Chapter Checklist

If you haven't got a copy of your specification, read the introduction on page vi.

You will need to be able to do some or all of the following. Check your Awarding Body's specification (syllabus) to find out exactly what you need to know.

- Explain what is meant by an adaptation.
- Describe and explain the properties of water that affect movement, gas exchange and support.
- Explain why fish living in seawater face different problems to those living in freshwater.
- Describe the main threats to survival faced by organisms living in the Arctic.
- Explain how many plants manage to survive Arctic conditions.
- Describe the adaptations shown by the polar bear and explain how each aids survival in the Arctic.
- Describe the main threats to survival faced by organisms living in a desert.
- Describe and explain how cacti are adapted to survive in a desert.
- Describe and explain some other adaptations shown by desert plants.
- Describe and explain the adaptations shown by a camel.
- Describe and explain some other adaptations shown by desert mammals.
- Explain why adaptations can limit organisms to specific environments.
- Explain how metamorphosis allows organisms to exploit more than one environment.

Questions

More questions on adaptation to environment can be found at the end of section D on page 195.

1 Explain how the following adaptations increase the chances of survival of the organism:

   a) the thick winter fur of a polar bear

   b) the extensive root system of a cactus

   c) the short life cycle of some desert plants.

2 The diagram shows the ear sizes of three similar animals which live in different environments. The Arctic hare lives in the Arctic snow, the Varying hare lives in temperate conditions and the Jack rabbit lives in the desert.

Jack rabbit   Varying hare   Arctic hare

   a) Describe how ear size is related to the mean environmental temperature of each animal.

   b) Explain the benefits to the Arctic hare and the Jack rabbit of their different ear sizes.

   c) Suggest how the coat colour of the Arctic hare and Jack rabbit may be adapted to their environment.

3 The graphs show the changes in body water content of a camel and a human when they suffer dehydration.

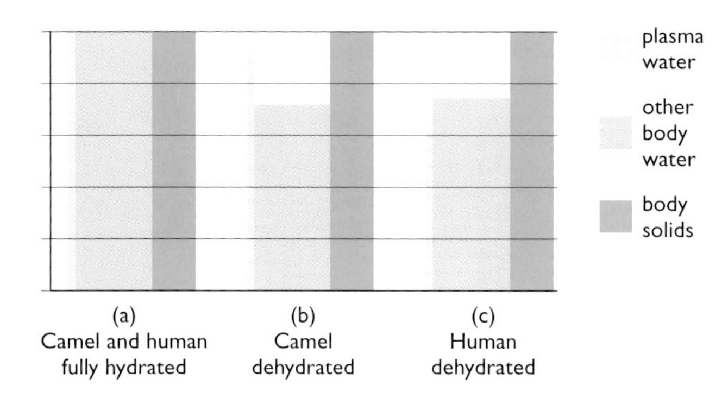

a) Suggest what the 'other body water' might be.

b) From the graphs, describe differences in the ways in which humans and camels lose water.

c) A result of their pattern of water loss is that humans rapidly become unable to cool their bodies. Suggest why.

4 The graphs show the various gains and losses of water of a gerbil kept under conditions of 50% humidity at 25°C. Gerbils spend much of the day in underground burrows with similar conditions to these.

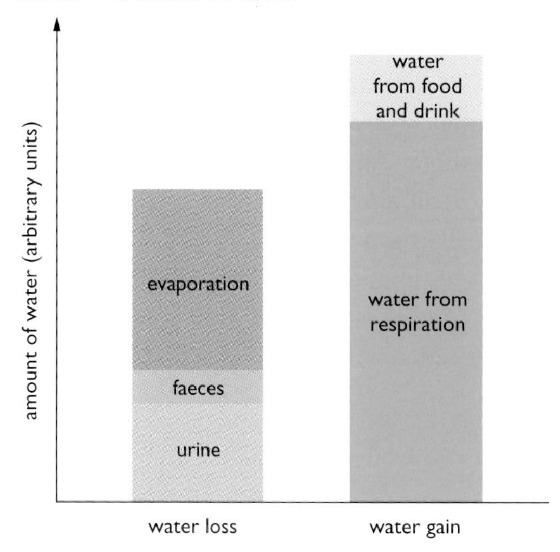

Use the information to explain how a gerbil can survive for long periods in the desert without drinking any water.

5 Write a short essay on the problems of life in the desert. Use the Internet and other books to find your information. It is a good idea to word process your essay. You could include pictures as well.

Chapter 14: Ecosystems

An **ecosystem** is a distinct, self-supporting system of organisms interacting with each other and with a physical environment. An ecosystem can be small, such as a garden pond, or large, such as a large forest. This chapter looks at a variety of ecosystems and the interactions that happen within them.

Large areas of the Earth dominated by a specific type of vegetation are called **biomes**. For example, temperate woodland and tropical rain forest are biomes. You could do an Internet search to find out about these and other biomes.

The components of ecosystems

Figure 14.1 *A garden pond is a small ecosystem.*

Figure 14.2 *A wood is a larger ecosystem.*

Whatever their size, ecosystems usually have the same components:

- **producers** – plants which photosynthesise to produce food
- **consumers** – animals that eat plants or other animals
- **decomposers** – decay dead material and help to recycle nutrients
- a physical **environment** – the sum total of the non-biological components of the ecosystem; for example, the water and soil in a pond or the soil and air in a forest.

Within each ecosystem there is a range of **habitats** – these are the places where specific organisms live. Some are provided by the physical environment. For example, in a pond ecosystem, the habitat of many of the plants is provided partly by the soil at the bottom of the pond (where the roots penetrate) and partly by the water itself (where the stem, leaves and flowers grow). Tadpoles spend most of their time swimming in the surface waters of a pond and that is their habitat.

You may be asked to explain the terms in **bold type** in an examination.

Other habitats are provided by organisms themselves. For example, dead vegetation provides a habitat for many decomposers.

All the organisms of a particular species found in an ecosystem at any one time form the **population** of that species in that ecosystem. For example, in a pond, all the tadpoles swimming in it form a population of tadpoles; all the *Elodea* plants growing in it make up a population of *Elodea*.

The populations of *all* species (all the animals, plants and decomposers) found in a particular ecosystem at any one time form the **community** in that ecosystem. Figure 14.3 illustrates the main components of a pond ecosystem.

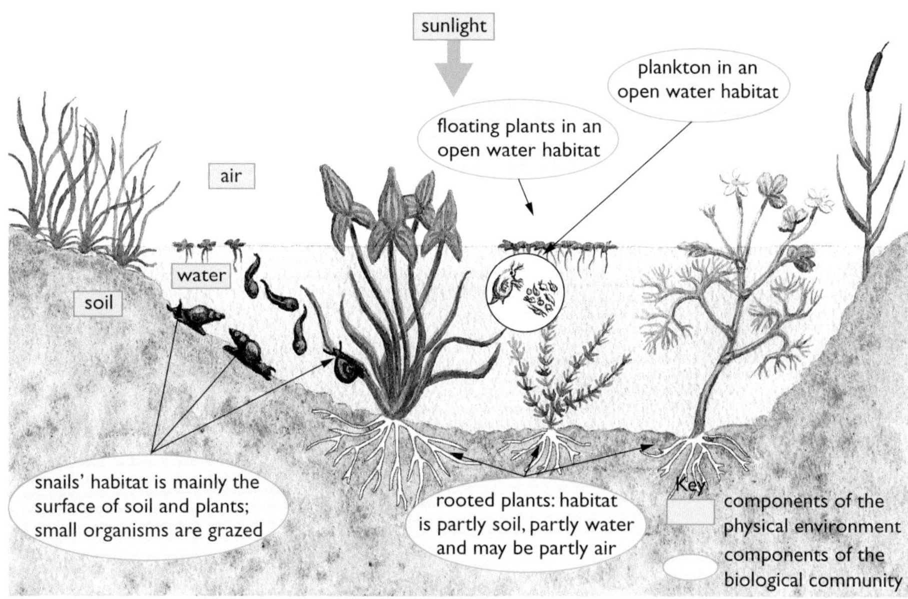

Figure 14.3 *A pond ecosystem.*

The community in an ecosystem can be described by listing the species present. However, this does not give any measure of the size or success of the community. Estimating the size of the population of each species gives that information, but it is difficult to compare different communities using lots of numbers. To get around these problems, a measure called the **species diversity index** has been developed. In this, a single figure is calculated from the numbers of all the organisms in the community. The formula for a species diversity index (d) is given below.

$$d = \frac{N(N-1)}{\Sigma n(n-1)}$$

Where N = the total number of all the organisms in the community
n = the number of organisms of a particular species
Σ = the sum of

A high diversity index suggests a well-established, stable community with many successful species. A low diversity index suggests a less well established, less stable community often dominated by just a few species.

worked example

Calculate a species diversity index for two small areas, A and B, each with the same four species present. The total number of organisms is the same in the two areas.

Populations of species (n)	Area A	Area B
A	86	25
B	10	25
C	3	25
D	1	25

Area A

$N = 86 + 10 + 3 + 1 = 100$
$(N - 1) = 99$
$N(N - 1) = 100 \times 99 = 9900$
For species A, $n(n - 1) = 86 \times 85 = 7310$
For species B, $n(n - 1) = 10 \times 9 = 90$
For species C, $n(n - 1) = 3 \times 2 = 6$
For species D, $n(n - 1) = 1 \times 0 = 0$
$\Sigma n(n - 1) = 7310 + 90 + 6 + 0 = 7406$

$$d = \frac{9900}{7406} = 1.3$$

Area B

$N = 25 + 25 + 25 + 25 = 100$
$(N - 1) = 99$
$N(N - 1) = 100 \times 99 = 9900$
For species A, $n(n - 1) = 25 \times 24 = 600$
For species B, $n(n - 1) = 25 \times 24 = 600$
For species C, $n(n - 1) = 25 \times 24 = 600$
For species D, $n(n - 1) = 25 \times 24 = 600$
$\Sigma n(n - 1) = 600 + 600 + 600 + 600 = 2400$

$$d = \frac{9900}{2400} = 4.1$$

The community in B has a higher diversity index than that of A. In area A, the community is dominated by species A. In area B, the four species are all successful.

Interactions in ecosystems

The organisms in an ecosystem are continually interacting with each other and with their physical environment. Some interactions include:

- Feeding among the organisms – the plants, animals and decomposers are continually recycling the same nutrients through the ecosystem.
- Competition among the organisms – animals compete for food, shelter, mates, nesting sites; plants compete for carbon dioxide, mineral ions, light and water.
- Interactions between organisms and the environment – plants absorb mineral ions, carbon dioxide and water from the environment; plants also give off water vapour and oxygen into the environment; animals use materials from the environment to build shelters; the temperature of the environment can affect processes occurring in the organisms; processes occurring in organisms can affect the temperature of the environment (all organisms give off some heat).

You can find out more about competition in the section on populations, on page 168.

Don't forget that plants take in carbon dioxide and give out oxygen only when there is sufficient light for photosynthesis to occur efficiently. When there is little light, plants take in oxygen and give out carbon dioxide. You should be able to explain why – if not see Chapter 10.

Feeding relationships

The simplest way of showing feeding relationships within an ecosystem is a **food chain** (Figure 14.4).

Figure 14.4 *A simple food chain.*

In any food chain, the arrow (→) means 'is eaten by'. In the food chain illustrated, the grass is the **producer**. It is a plant so it can photosynthesise and produce food materials. The rabbit is the **primary consumer**. It is an animal which eats the producer and is also a **herbivore**. The fox is the **secondary consumer**. It eats the primary consumer and is also a **carnivore**. The different stages in a food chain (producer, primary consumer and secondary consumer) are called **trophic levels**.

You can find out more about herbivores and carnivores in Chapter 4.

Many food chains have more than three links in them. Here are some examples of longer food chains:

filamentous algae → mayfly nymph → caddis fly larvae → salmon

In this freshwater food chain, the extra link in the chain makes the salmon a **tertiary consumer**.

plankton → crustacean → fish → ringed seal → polar bear

In this marine food chain, the fifth link makes the polar bear a **quaternary consumer**. Because nothing eats the polar bear, it is also called the **top carnivore**.

Food chains are a convenient way of showing the feeding relationships between a few organisms in an ecosystem, but they oversimplify the situation. The marine food chain above implies that only crustaceans feed on plankton, which is not true. Some whales and other mammals also feed on plankton. For a fuller understanding, you need to consider how the different food chains in an ecosystem relate to each other. Figure 14.5 gives a clearer picture of the feeding relationships involved in a freshwater ecosystem in which salmon are the top carnivores. This is the **food web** of the salmon.

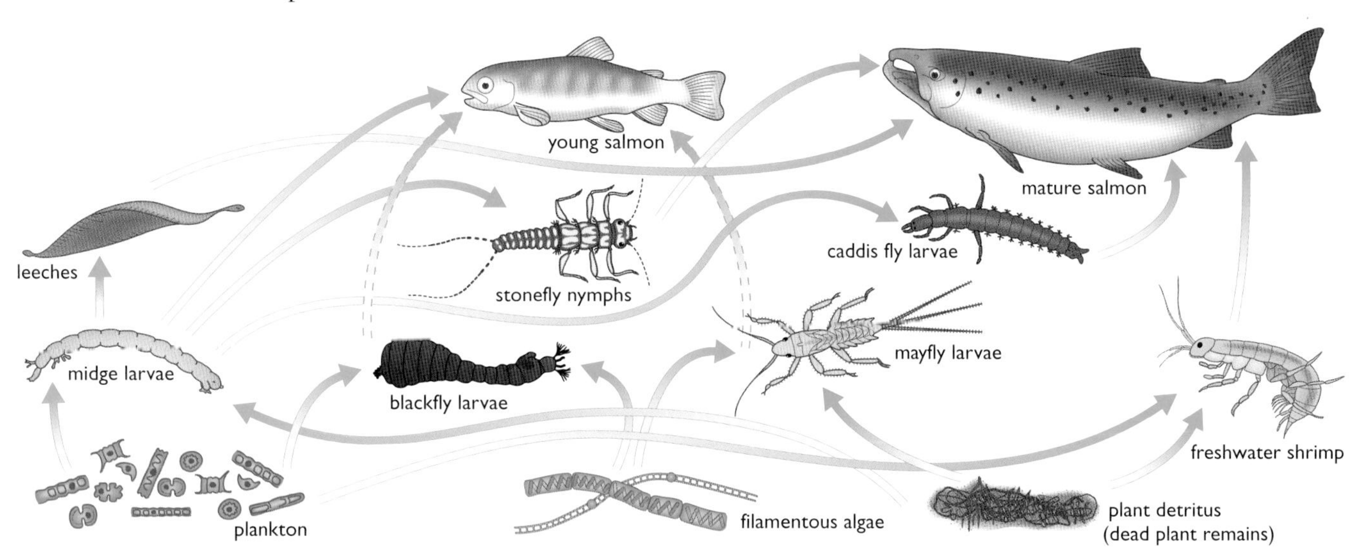

Figure 14.5 *The food web of the salmon. As you can see, young salmon have a slightly different diet to mature salmon.*

This is still a simplification of the true situation, as some feeding relationships are still not shown. It does, however, give some indication of the interrelationships that exist between food chains in an ecosystem. With a little thought, you can predict how changes in the numbers of an organism in one food chain in the food web might affect those in another food chain. For example, if the leech population were to decline through disease, there could be several possible consequences:

- the stonefly nymph population could increase as there would be more midge larvae to feed on
- the stonefly nymph population could decrease as the mature salmon might eat more of them as there would be fewer leeches
- the numbers could stay static due to a combination of the above.

Although food webs give us more information than food chains, they don't give any information about how many, or what mass of organisms is involved. Neither do they show the role of the decomposers. To see this, we must look at other ways of presenting information about feeding relationships in an ecosystem.

Ecological pyramids

Ecological pyramids are diagrams that represent the relative amounts of organisms at each trophic level in a food chain. There are two main types:

- **pyramids of numbers**, which represent the numbers of organisms in each trophic level in a food chain, irrespective of their mass
- **pyramids of biomass**, which show the total mass of the organisms in each trophic level, irrespective of their numbers.

Biomass is the total amount of living material in an organism.

Consider these two food chains:

grass → grasshopper → frog → bird   oak tree → aphid → ladybird → bird

The diagrams below show the pyramids of numbers and biomass for these two food chains.

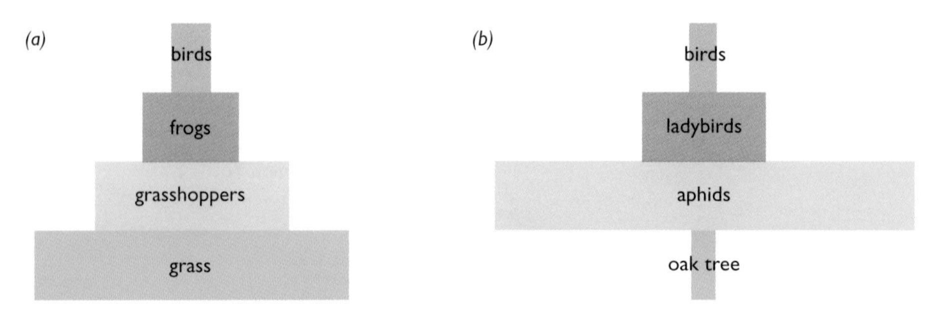

Figure 14.6 *Pyramids of numbers for two food chains.*

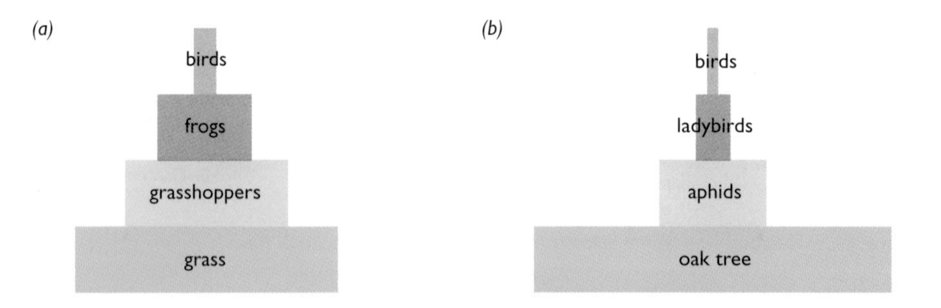

Figure 14.7 *Pyramids of biomass for the two food chains.*

The two pyramids for the 'grass' food chain look the same –the numbers at each trophic level decrease. The *total* biomass also decreases along the food chain – the mass of *all* the grass plants in a large field would be more than that of *all* the grasshoppers which would be more than that of *all* the frogs, and so on.

The two pyramids for the 'oak tree' food chain look different because of the size of the oak trees. Each oak tree can support many thousands of aphids, so the numbers *increase* from first to second trophic levels. But each ladybird will need to eat many aphids and each bird will need to eat many ladybirds, so the numbers *decrease* at the third and fourth trophic levels. However, the total biomass *decreases* at each trophic level – the biomass of one oak tree is much greater than that of the thousands of aphids it supports. The total biomass of all these aphids is greater than that of the ladybirds, which is greater than that of the birds.

Suppose the birds in the second food chain are parasitised by nematode worms. The food chain now becomes:

oak tree → aphid → ladybird → bird → nematode worm

The pyramid of numbers now takes on a very strange appearance (Figure 14.8) because of the large numbers of parasites on each bird. The pyramid of biomass, however, has a true pyramid shape because the total biomass of the nematode worms must be less than that of the birds they parasitise.

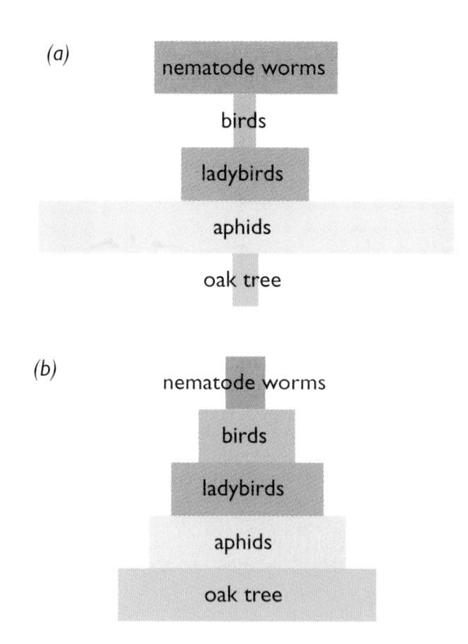

Figure 14.8 *(a) A pyramid of numbers and (b) a pyramid of biomass for the parasitised food chain.*

Why do diagrams of feeding relationships give a pyramid shape?

The explanation is relatively straightforward (Figure 14.9). When a rabbit eats grass, not all of the materials in the grass plant end up as rabbit! There are losses:

- some parts of the grass are not eaten (the roots for example)
- some parts are not digested and so are not absorbed – even though rabbits have a very efficient digestive system
- some of the materials absorbed form excretory products
- many of the materials are respired to release energy, with the loss of carbon dioxide and water.

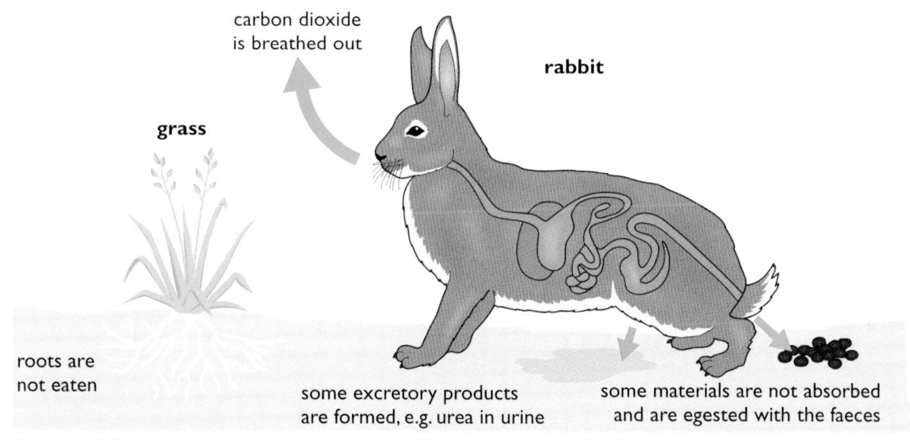

Figure 14.9 *Not all the grass eaten by a rabbit ends up as rabbit tissue.*

In fact, only a small fraction of the materials in the grass ends up in new cells in the rabbit. Similar losses are repeated at each stage in the food chain, so smaller and smaller amounts of biomass are available for growth at successive trophic levels. The shape of pyramids of biomass reflects this.

Feeding is a way of transferring energy between organisms. Another way of modelling ecosystems looks at the energy flow between the various trophic levels.

The flow of energy through ecosystems

This approach focuses less on individual organisms and food chains and rather more on energy transfer between trophic levels (producers, consumers and decomposers) in the whole ecosystem. There are a number of key ideas that you should understand at the outset.

- Photosynthesis 'fixes' sunlight energy into chemicals such as glucose and starch.
- Respiration releases energy from organic compounds such as glucose.
- Almost all other biological processes (e.g. muscle contraction, growth, reproduction, excretion, active transport) use the energy released in respiration.
- If the energy released in respiration is used to produce new cells (general body cells in growth and sex cells in reproduction) then the energy remains 'fixed' in molecules in that organism. It can be passed on to the next trophic level through feeding.
- If the energy released in respiration is used for other processes then it will, once used, eventually escape as heat from the organism. Energy is therefore lost from food chains and webs at each trophic level.

This can be shown in an **energy flow diagram**. Figure 14.10 shows the main ways in which energy is transferred in an ecosystem. It also gives the amounts of energy transferred between the trophic levels of a grassland ecosystem.

All figures given are kilojoules ($\times 10^5$)/m^2/year.

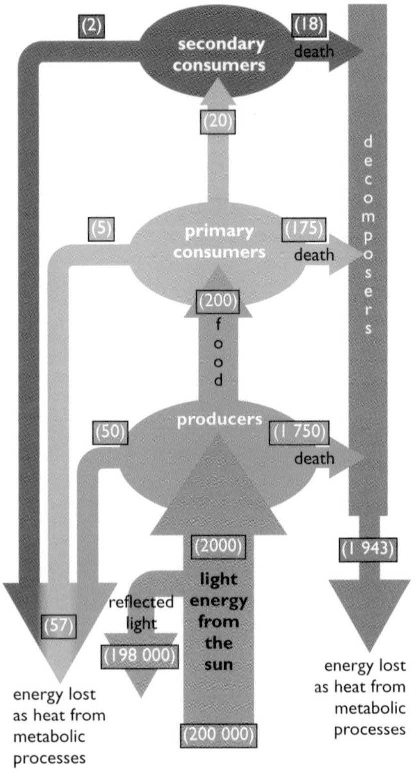

Figure 14.10 *The main ways in which energy is transferred in an ecosystem. The amounts of energy transferred through 1 m^2 of a grassland ecosystem per year are shown in brackets.*

As you can see, only about 10% of the energy entering a trophic level is passed on to the next trophic level. This explains why not many food chains have more than five trophic levels. Think of the food chain:

$$A \rightarrow B \rightarrow C \rightarrow D \rightarrow E$$

If we use the idea that only about 10% of the energy entering a trophic level is passed on to the next level, then, of the original 100% reaching A (a producer), 10% passes to B, 1% (10% of 10%) passes to C, 0.1% passes to D and only 0.001% passes to E. There just isn't enough energy left for another trophic level. In certain parts of the world some marine food chains have six trophic levels because of the huge amount of light energy reaching the surface waters .

Cycling nutrients through ecosystems

The chemicals that make up our bodies have all been around before – probably many times! You may have in your body some carbon atoms that were part of the carbon dioxide molecules breathed out by Winston Churchill making one of his famous speeches in the second World War, or by Geoff Hurst as he scored the winning goal for England in the 1966 World Cup final. This constant recycling of substances is all part of the cycle of life, death and decay.

Microorganisms play a key role in recycling. They break down complex organic molecules in the bodies of dead animals and plants into simpler substances, which they release into the environment.

Organic compounds all contain carbon and hydrogen. Starch and glucose are organic molecules, but carbon dioxide (CO_2) isn't.

The carbon cycle

Carbon is a component of all major biological molecules. Carbohydrates, lipids, proteins, nucleic acids, vitamins and many other molecules all contain carbon. The following processes are important in cycling carbon through ecosystems:

- Photosynthesis 'fixes' carbon atoms from carbon dioxide into organic compounds.
- Feeding and assimilation pass carbon atoms already in organic compounds along food chains.
- Respiration produces inorganic carbon dioxide from organic compounds (mainly carbohydrates) as they are broken down to release energy.
- Fossilisation – sometimes living things do not decay fully when they die due to the conditions in the soil (decay is prevented if it is too acidic) and fossil fuels (coal, oil, natural gas and peat) are formed.
- Combustion releases carbon dioxide into the atmosphere when fossil fuels are burned.

Figures 14.11 and 14.12 show the role of these processes in the carbon cycle in different ways.

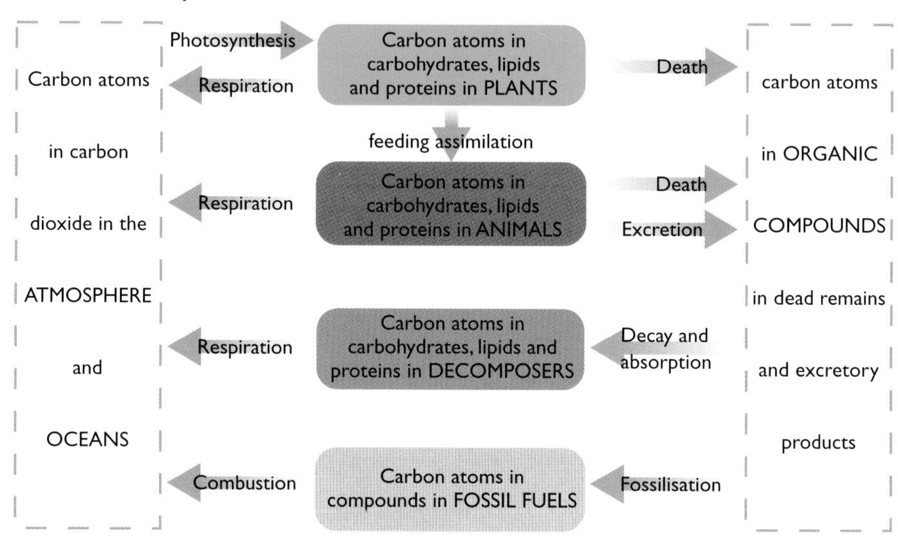

Figure 14.11 *The main stages in the carbon cycle.*

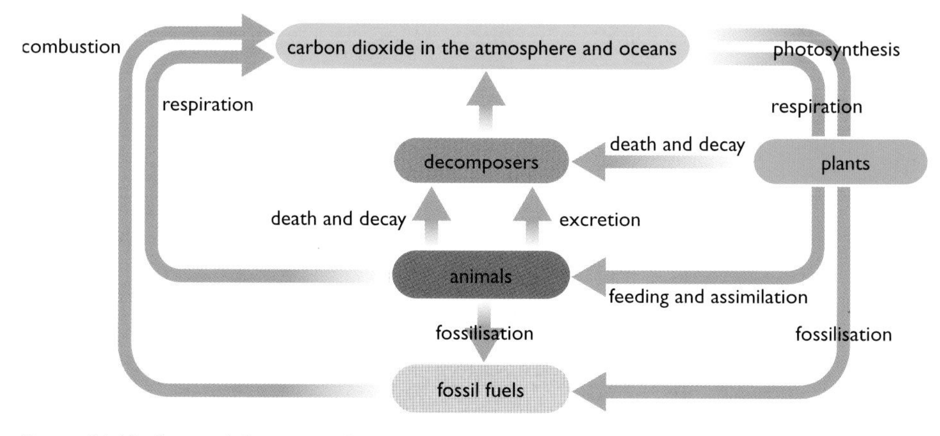

Figure 14.12 *A typical illustration of the carbon cycle.*

Some nitrogen-fixing bacteria are free-living in the soil. Others form associations with the roots of legumes (legumes are plants that produce seeds in a pod, like peas and beans). They form little bumps or 'root nodules' (Figure 14.13).

Figure 14.13 *Root nodules on a clover plant.*

This is an example of **mutualism** (see Chapter 4). The bacteria receive a supply of organic nutrients from the plants and the plants receive a supply of ammonia from the bacteria. They can use the ammonia to form amino acids in the same way as they use nitrates.

To remember if an organic compound contains nitrogen, check to see if the letter **N** (symbol for nitrogen) is present.

Protei**N**s, ami**N**o acids and **DNA** all contain nitrogen. Carbohydrates and fats don't – they have no **N**.

The nitrogen cycle

Nitrogen is a key element in many biological compounds. It is present in proteins, amino acids, most vitamins, DNA, RNA and adenosine triphosphate (ATP). Like the carbon cycle, the nitrogen cycle involves feeding, assimilation, death and decay. Photosynthesis and respiration are not directly involved in the nitrogen cycle as these processes fix and release carbon, not nitrogen. The following processes are important in cycling nitrogen through ecosystems:

- Feeding and assimilation pass nitrogen atoms already in organic compounds along food chains.
- Decomposition (putrefaction) by decomposers produces ammonia from the nitrogen in compounds like proteins, DNA and vitamins.
- The ammonia is oxidised first to nitrite and then to nitrate by **nitrifying bacteria.** This overall process is called **nitrification.**
- Plant roots can absorb the nitrates. They are combined with carbohydrates (from photosynthesis) to form amino acids and, then proteins as well as other nitrogen-containing compounds.

This represents the basic nitrogen cycle, but other bacteria carry out processes that affect the amount of nitrate in the soil that is available to plants:

- **Denitrifying bacteria** use nitrates as an energy source and convert them into nitrogen gas. **Denitrification** *reduces* the amount of nitrate in the soil.
- **Nitrogen-fixing bacteria** convert nitrogen gas in the soil into ammonia. This can then be oxidised by nitrifying bacteria to nitrate. **Nitrogen fixation** *increases* the amount of nitrate available to plants.

In addition to all the processes described so far, lightning converts nitrogen gas in the air into various oxides of nitrogen. These dissolve in rainwater and enter the soil to be converted into nitrates by nitrifying bacteria. Figure 14.14 shows the role of these processes in the nitrogen cycle.

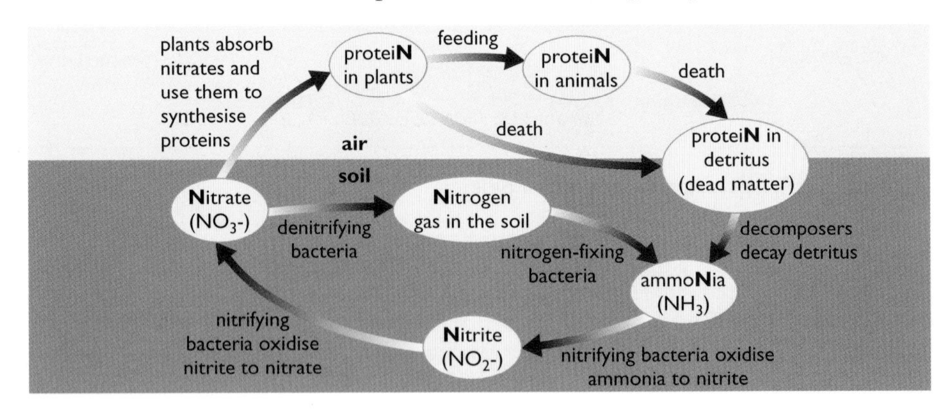

Figure 14.14 *The main stages in the nitrogen cycle.*

Populations

As we saw earlier, a population is a group of organisms of the same species that occupy the same habitat at any one time. However, the numbers of organisms in a population don't remain static, and any population must first establish itself. A species must **colonise** an area and multiply. If it is sufficiently well adapted it will achieve this. If not, it will die out.

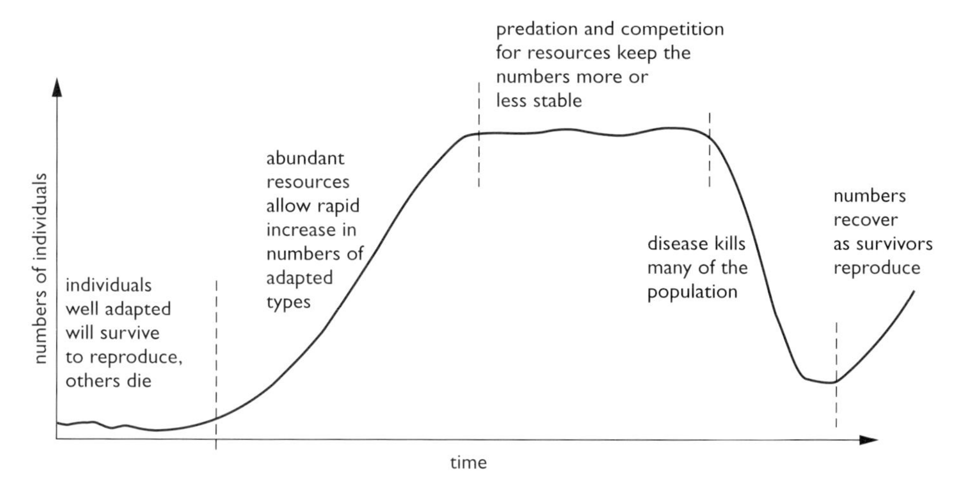

Figure 14.15 *A population growth curve.*

A population cannot increase in numbers indefinitely. Eventually, something will limit its growth and numbers will not increase further unless conditions change. These changes can be shown in a population growth curve, Figure 14.15.

Factors that can limit the growth of a population include predation, disease, limited food supply and limited space for breeding.

Predation is the continued feeding of one animal species on another. For example, lions are predators of antelope, killer whales are predators of seals.

As a population grows, the available food must be shared between more and more organisms. They must *compete* for the food. Eventually, there will not be enough food for any more organisms. Those best able to obtain food will be more likely to survive and others will die. The numbers in the population cannot increase beyond this level.

Plants compete for light. Overcrowding of plants means that some cannot get enough light to photosynthesise effectively. Plants that can grow quickly and can get their leaves into a position to obtain sufficient light will survive at the expense of others.

Disease can drastically reduce a population if few are immune. Numbers recover as those with immunity that survived begin to reproduce.

Predation can also limit the numbers of a population. The relationship between the numbers of a predator and its prey is a very close one. One of the best researched examples of predator–prey numbers involves the snowshoe hare (the prey) and the lynx (the predator) in Canada (Figure 14.16).

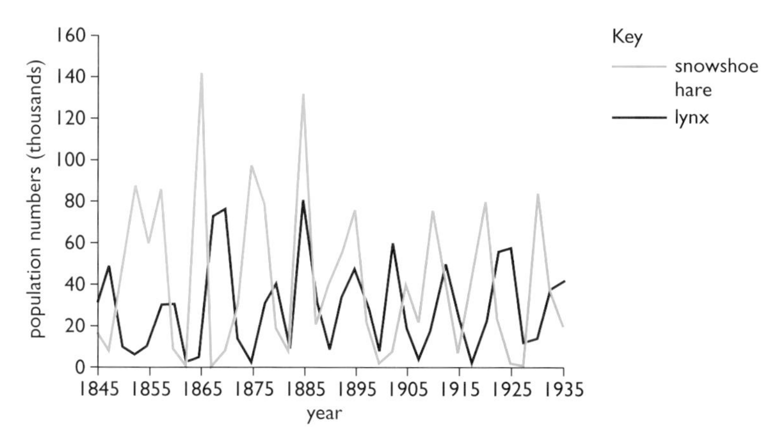

Figure 14.16 *The relationship between the numbers of lynx (b) and its prey, the snowshoe hare (c).*

You may be asked to identify which is the predator and which is the prey from a graph like the one in Figure 14.16. Remember:

- the numbers of prey will usually be higher than those of the predator as they are lower in the food chain
- changes in the numbers of prey occur first.

The patterns in the graph are typical of predator–prey relationships where the prey is the only food or the main food of the predator. They can be explained as follows:

1 As the numbers of prey increase, there is more food available for the predators.

2 The predators kill and eat more prey and so their numbers increase.

3 As more prey are now being eaten, their numbers start to decrease.

4 There is now *less* food for an *increased* number of predators.

5 The numbers of predators decrease as they compete for scarce food.

6 Fewer predators means that there will be fewer prey eaten and so the prey population recovers.

7 There are more prey and so there is more food available for the predators, and the cycle begins again.

Predators keep populations in check. They prevent dramatic surges in numbers. They also help the prey population by killing and eating mainly the weaker members. This means that only the healthier, fitter prey survive.

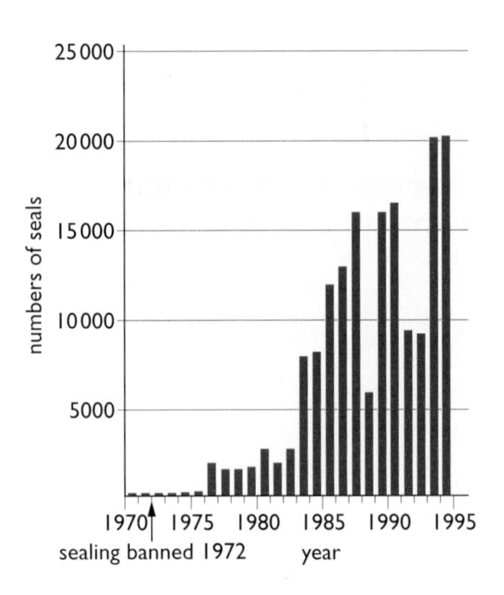

Figure 14.17 *Fur seal numbers at Signy Island in the Antarctic from 1970 to 1995.*

In the case of fur seals in the Antarctic, the effects of predation and food supply coincided. Sealing was banned and so predation of the seals by humans ceased. At the same time, the numbers of whales in the Antarctic decreased. Less plankton was eaten by the whales, so there was more for the fish. The numbers of fish increased, so there was more food for the seals. The numbers of seals surged. Later, as their numbers continued to increase, they had to compete for available food (fish) and their numbers stabilised, even decreasing in some years. These changes are shown in Figure 14.17.

In an established natural ecosystem, nutrients are constantly being recycled and made available for a new generation of organisms. Populations remain more or less stable, showing natural fluctuations typical of predator–prey relationships. Complex food webs mean that fluctuations in numbers of one species do not necessarily have a dramatic effect on others. Sometimes, human influence upsets the balance in an ecosystem. The next chapter deals with some of the ways in which humans influence the environment.

End of Chapter Checklist

If you haven't got a copy of your specification, read the introduction on page vi.

You will need to be able to do some or all of the following. Check your Awarding Body's specification (syllabus) to find out exactly what you need to know.

- Explain what is meant by the term ecosystem.
- Explain the terms producer, consumer, decomposer.
- Explain the terms habitat, population, community and environment.
- Identify the producer, primary consumer and secondary consumer in a food chain.
- Identify specific food chains in food webs.
- Predict the effect of a change in numbers of one of the organisms in a food web on numbers of other organisms in the same food web.
- Draw pyramids of numbers and pyramids of biomass of given food chains.
- Explain why the pyramid of numbers and the pyramid of biomass for the same food chain can have different shapes.
- Explain the way that energy is passed through an ecosystem and identify ways in which energy is gained and lost in an energy flow diagram.
- Calculate energy gains and losses in an energy flow diagram using data supplied (see End of Section Questions page 195).
- Explain the main stages of the carbon cycle and interpret unfamiliar representations of the carbon cycle.
- Explain the main stages of the nitrogen cycle and interpret unfamiliar representations of the nitrogen cycle.
- Identify the main phases in a population growth curve and explain the reasons for the shape of the growth curve.
- Identify, with reasons, predator and prey from a predator–prey graph.

Questions

More questions on ecosystems can be found at the end of Section D on page 195.

1 ***a)*** Explain what is meant by the terms habitat, community, environment and population.

b) What are the roles of plants, animals and decomposers in an ecosystem?

2 A marine food chain is shown below.

plankton → small crustacean → krill → seal → killer whale

a) Which organism is *i)* the producer, *ii)* the secondary consumer?

b) What term best describes the killer whale?

c) Suggest why five trophic levels are possible in this case, when many food chains only have three or four.

3 Part *(a)* of the diagram shows a woodland food web. Part *(b)* shows a pyramid of numbers and a pyramid of biomass for a small part of this wood.

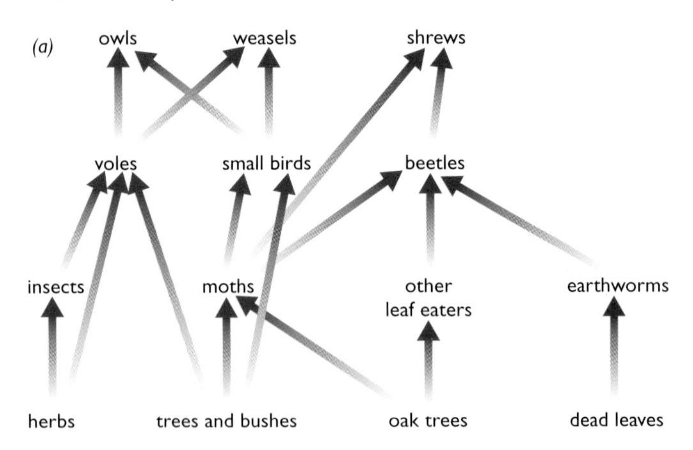

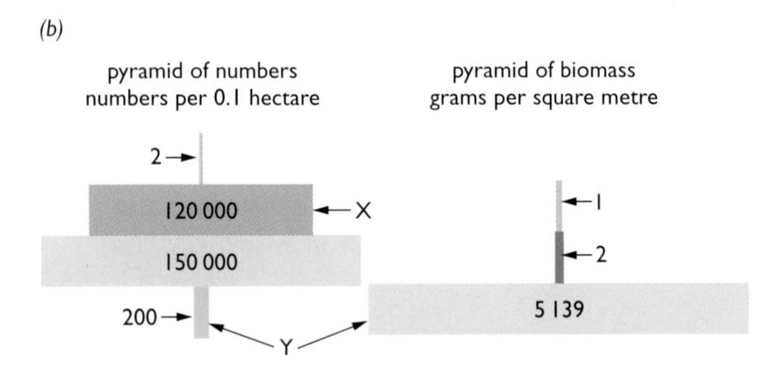

a) Write out **two** food chains (from the food web) containing four organisms, both involving moths.

b) Name **one** organism in the food web which is both a primary consumer and a secondary consumer.

c) Suggest how a reduction in the dead leaves may lead to a reduction in the numbers of voles.

d) In part *(b)* of the diagram, explain why level Y is such a different width in the two pyramids.

4 The diagram shows part of the nitrogen cycle.

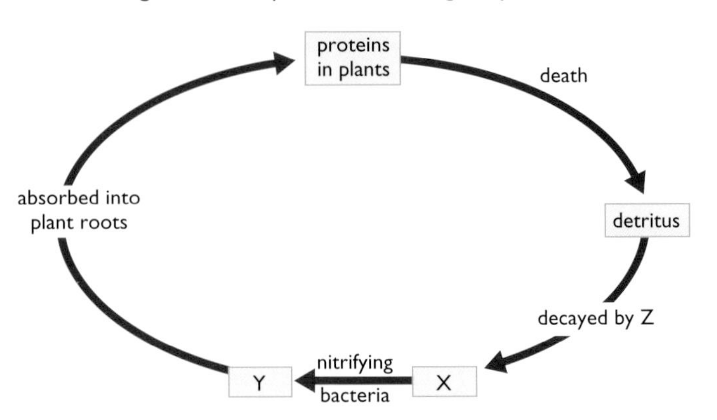

a) What do X, Y and Z represent?

b) Name the process by which plant roots absorb nitrates.

c) What are nitrogen-fixing bacteria?

d) Give **two** ways, not shown in the diagram, in which animals can return nitrogen to the soil.

5 The diagram shows a population growth curve for some deer, which were allowed to colonise a small island where there were no natural predators.

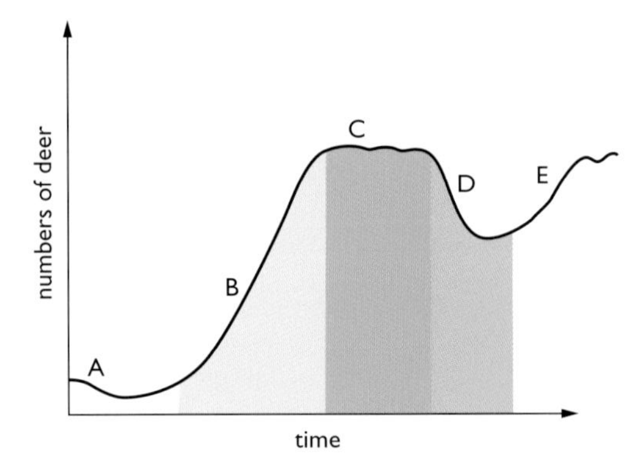

Suggest explanations for:

a) the initial decrease in population (A)

b) the rapid increase in population (B)

c) the steady population (C)

d) the rapid decrease in population (D)

e) the recovery in numbers (E).

Chapter 15: Human Influences on the Environment

Humans have intelligence far beyond that of any other animal on Earth. This chapter looks at the ways in which we have used our intelligence to influence natural environments during the course of our evolution.

ideas
evidence

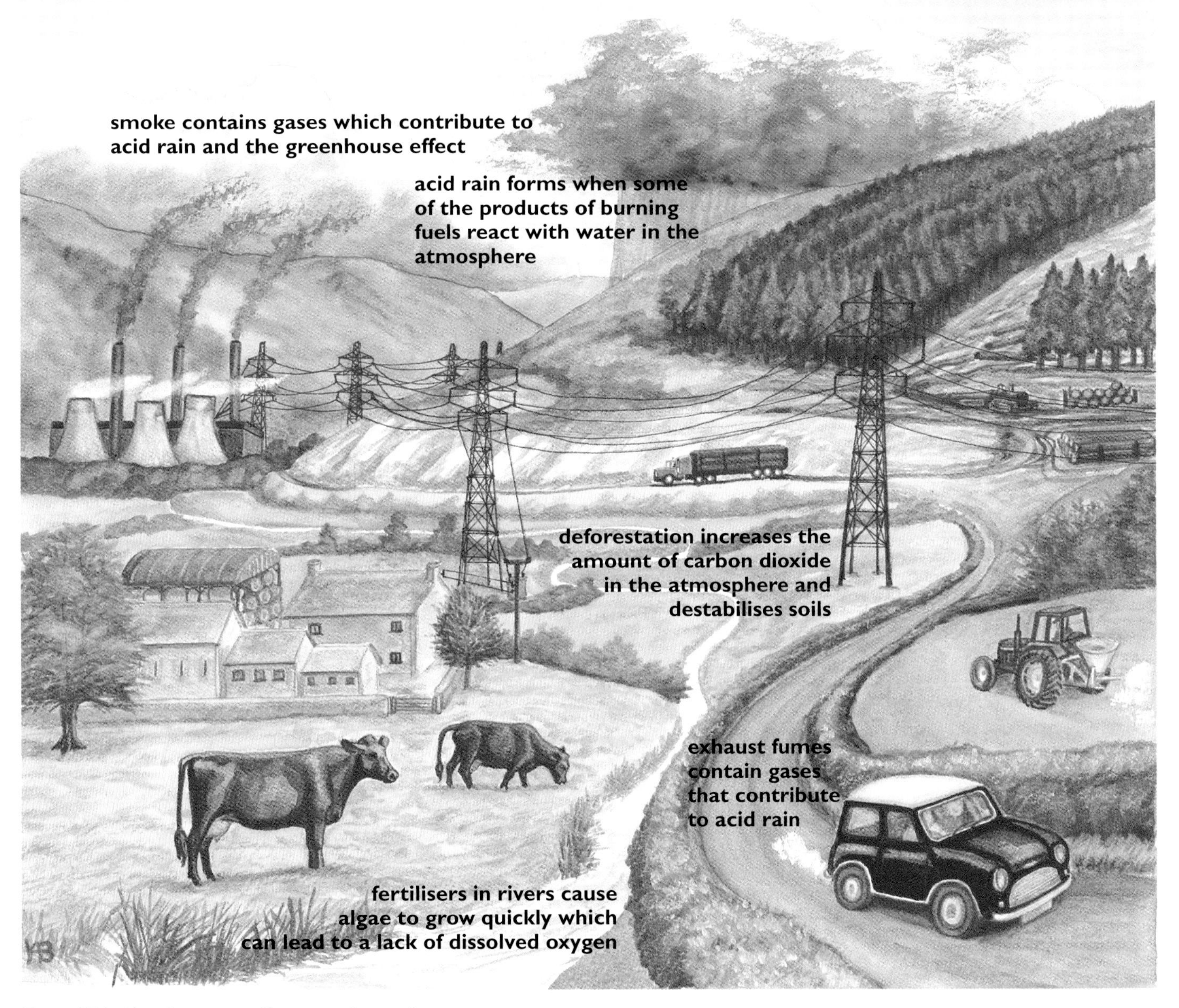

Figure 15.1 *How the actions of humans influence the environment.*

You can find out more about selection, variation and the evolution of modern humans (*Homo sapiens*) in Chapter 19. Chapter 13 gives more information on adaptation.

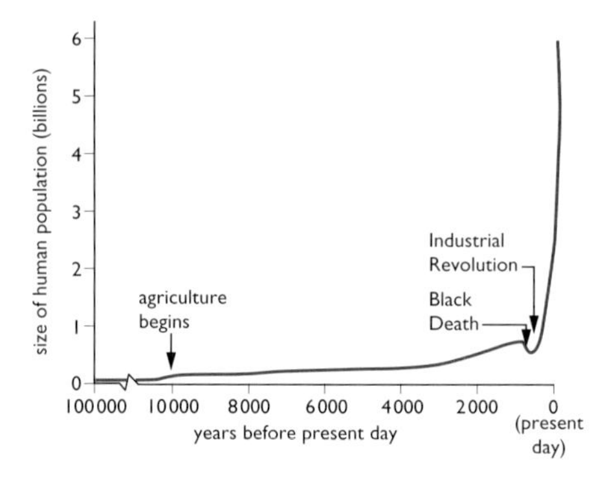

Figure 15.2 *Human population growth.*

Since humans first appeared on Earth, our numbers have grown dramatically (Figure 15.2). The secret of our success has been our intelligence. Unlike other species, we have not adapted to one specific environment, we have changed many environments to suit us.

As our numbers have grown, so has the sophistication of our technology. Early humans made tools from materials readily to hand. Today's technology involves much more complex processes. As a result, we produce ever-increasing amounts of materials that pollute our air, soil and waterways.

Early humans influenced their environment, but the sheer size of the population today and the extent of our industries mean that we affect the environment much more significantly. We make increasing demands on the environment for:

- food to sustain an ever-increasing population
- materials to build homes, schools, industries, etc.
- fuel to heat homes and power vehicles
- space in which to build homes, schools and factories, as well as for our leisure facilities
- space in which to dump our waste materials.

Modern agriculture – producing the food we need

A modern farm can be thought of as a managed ecosystem. Many of the interactions are the same as in natural ecosystems. Crop plants depend on light and mineral ions from the soil as well as other factors in the environment. Stock animals (sheep, cattle and pigs) depend on crop plants or plant products for food (see Figure 15.3).

Farmers must make a profit from their farms. To do this, they try to control the environment in such a way as to maximise the yield from crop plants and livestock.

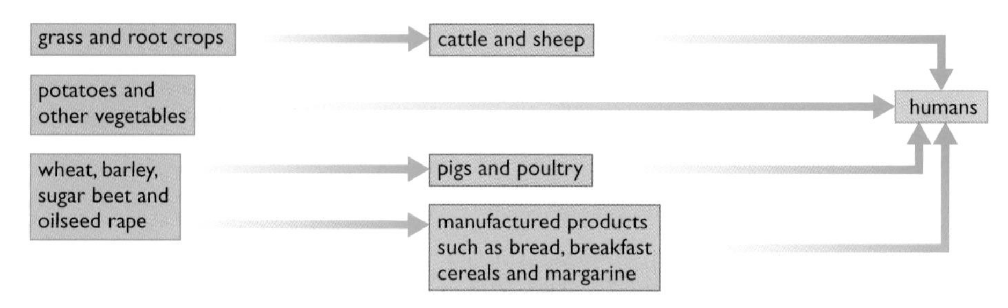

Figure 15.3 *A food web on a farm.*

Controlling the environment to maximise yield

Farmers control the environment of crops and stock animals in a number of ways.

Improving yields from crop plants

Table 15.1 summaries various agricultural features that can be controlled by the farmer in order to maximise yield from crops.

Soil pH can vary between 3.0 and 8.0. Soil pH can be tested using indicator kits and soil can be made more alkaline by adding lime or more acidic by adding peat.

Feature controlled	How it is controlled	Reason for controlling the feature
soil ions (e.g. nitrates)	adding fertilisers (organic or inorganic) to the soil or growing in a hydroponic culture (Figure 15.4a)	extra mineral ions can be taken up and used to make proteins and other compounds for growth
soil structure	ploughing fields to break up compacted soil; adding manure to improve drainage and aeration of heavy, clay soils	good aeration and drainage allow better uptake of mineral ions and water
soil pH	adding lime (calcium salts) to acidic soils; few soils are too alkaline to need treatment	an unsuitable soil pH can affect crop growth as it reduces uptake of mineral ions
carbon dioxide, light and heat	these cannot be controlled for field crops but in a greenhouse (glasshouse), all can be influenced to maximise yield of crops (Figure 15.4b); burning fuels produces both carbon dioxide and heat	all influence the rate of photosynthesis and so influence the production of the organic substances needed for growth

Table 15.1: *How yields from crop plants can be controlled.*

Figure 15.4 *Any kind of greenhouse maintains a favourable environment for plants. (a) Crops grown by hydroponics in a greenhouse. (b) Many crops are produced in huge greenhouses or in tunnels made from transparent polythene.*

Greenhouses provide the right conditions for plants to grow for several reasons. The transparent material allows sufficient natural light in for photosynthesis during the summer months, while additional lighting gives a 'longer day' during the winter. The 'greenhouse effect' also happens in greenhouses! Short wave radiation which enters the greenhouse becomes longer wave radiation as it reflects off surfaces. This longer wave radiation cannot leave as easily and so the greenhouse heats up. Burning fuels to raise the temperature when the external temperature is too low also produces carbon dioxide and water vapour. The water vapour maintains a moist atmosphere and so reduces water loss by transpiration (see Chapter 11). The carbon dioxide is a raw material of photosynthesis.

> By heating greenhouses to the **optimum** temperature for photosynthesis for a crop, a farmer can maximise his yield. Heating above this temperature is a waste of money as there is no further increase in yield.

In addition, growing plants in a hydroponic culture provides *exactly* the right balance of mineral ions for the specific crop being grown.

ideas
evidence

Improving yields from stock animals

Table 15.2 summaries various livestock features that can be controlled by the farmer in order to maximise yield from stock animals.

Feature controlled	How it is controlled	Reason for controlling the feature
breeding cycle	using artificial insemination	allows insemination by sperm from a 'superior' male from another herd
	using reproductive hormones to ensure that all females are 'in season' together	allows insemination of many females in a short period of time and production of offspring at more or less the same time. The farmer can prepare for an intensive period of lambing/calving
temperature	heating the sheds where stock animals are kept	less energy is used in maintaining body temperature, more used in growth
diet	adding pelleted supplements to fodder (food from crops)	ensures adequate vitamins and minerals in diet

Table 15.2 *How livestock features can be controlled.*

Cycling nutrients on a farm

Chapter 14 describes how the elements nitrogen and carbon are cycled in nature. On a farm, the situation is quite different, particularly with regard to the circulation of nitrogen.

Nitrates from the soil become part of proteins in the plants. Some of these are crops to be sold, others are used as fodder for the stock animals. When the crops are sold, the nitrogen in the proteins goes with them and is lost from the farm ecosystem. Similarly, when livestock is sold, the nitrogen in their proteins (gained from the fodder) goes with them and is lost from the farm ecosystem. To replace the lost nitrogen, a farmer usually adds some kind of fertiliser. Figure 15.5 summarises the circulation of nitrogen on a farm.

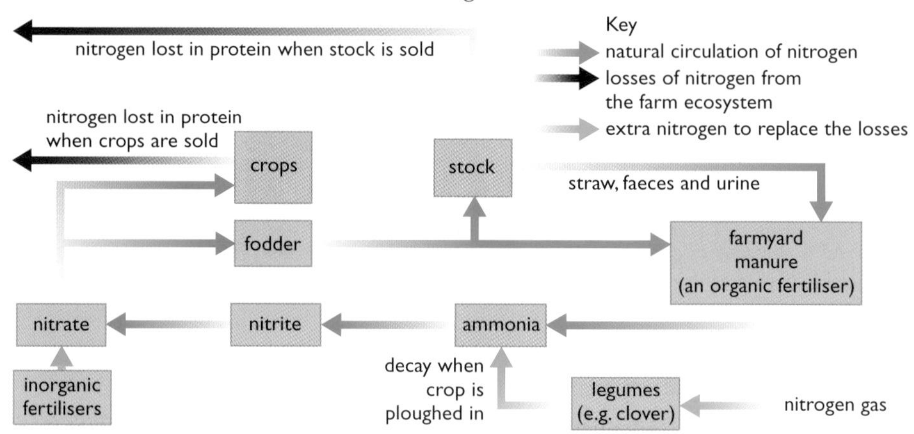

Figure 15.5 *The nitrogen cycle on a farm. The effects of denitrification and lightning (see Chapter 14) have been omitted.*

Fertilisers on the farm

There are two main types of fertilisers – **organic** and **inorganic**. Many organic fertilisers (such as farmyard manure) are made from the faeces of a range of animals mixed with straw. Inorganic fertilisers are simply inorganic compounds (like potassium nitrate or ammonium nitrate), carefully formulated to yield a specific amount of nitrate (or some other ion) when applied according to the manufacturer's instructions.

Adding farmyard manure returns some of the nitrogen to the soil. But, as farmyard manure is made from livestock faeces and indigestible fodder, it can only replace a portion of the lost nitrogen. Most farmers apply inorganic fertilisers to replace the nitrates and other mineral ions lost. Whilst this can replace *all* the lost ions, it can also lead to problems such as **eutrophication**, which we will discuss later in the chapter on page 185. Inorganic fertilisers can also damage soil structure. This is because they do not replace the organic matter lost to soils that is an essential part of the structure.

Another way to replace lost nitrates is to grow a legume crop (such as clover) in a field one year in four. Legumes have nitrogen-fixing bacteria in nodules on their roots (see chapter 14). These bacteria convert nitrogen gas in the soil air to ammonium ions. Some of this is passed to the plants, which use it to make proteins. At the end of the season, the crop is ploughed in and the nitrogen in the proteins is decayed (putrefied) to ammonia. This is then oxidised to nitrate by nitrifying bacteria and is available to next year's crops.

In Britain, about 30% of the potential maize crop is lost to weeds, insects and fungal diseases (Figure 15.6).

Figure 15.6 *Damage to a maize plant by the Corn Earthworm.*

Pests on the farm

Pests are organisms that reduce the yield of crop plants or stock animals. By doing this, they cause economic damage to the farmer. Any type of organism – plants, animals, bacteria, fungi or protoctistans, as well as viruses – can be a pest. Pests can be controlled in a number of ways. Chemicals called **pesticides** can be used to kill them, or their numbers can be reduced by using **biological control**. Some agricultural practices, such as monoculture (see page 178), can encourage the build-up of pests. Avoiding these practices can reduce pest damage.

Pesticides are named according to the type of organism they kill:

- **herbicides** kill plant pests (they are weedkillers)
- **insecticides** kill insects
- **fungicides** kill fungi
- **molluscicides** kill molluscs.

A farmer uses pesticides to kill specific pests and so improve the yield from the crops or livestock. Pests are only a problem when they are present in sufficient numbers to cause economic damage – a few whiteflies in a tomato crop are not a problem, the problem arises when there are millions of them. Because of this, whether or not to use pesticides and how often to use them is largely a financial decision. The increase in income, due to better yields, must be set against the cost of the pesticides.

A weed is a plant that is growing where it is not wanted. Weeds can be controlled mechanically or chemically. Mechanical control involves physically removing the weeds. Chemical control uses chemical weedkillers that are selective.

One problem with using pesticides is that of resistance. Through chance **mutation** and natural selection (see Chapter 19) a population of a pest can become resistant to a pesticide. This makes the existing pesticide useless and another must be found. In addition, the use of pesticides can cause environmental damage, as some persist in the soil and can be accumulated along food chains (see page 187).

ideas
evidence

Another option sometimes open to a farmer is the **biological control** of pests. Biological control uses another organism, rather than a toxic chemical, to reduce the numbers of a pest. We have already mentioned whiteflies as pests of tomatoes. One way of controlling them in large glasshouses is to introduce a parasite that will kill the whiteflies. A tiny wasp called *Encarsia* parasitises and kills their larvae, so reducing the numbers of whitefly.

A feature of biological control is that it never eradicates a pest. If the control organism killed off all the pests, then it, too, would die out as there would be no food. Biological control aims to reduce pest numbers to a level where they no longer cause significant economic damage (Figure 15.7).

Notice that the graph in Figure 15.7 shows the same features as a typical predator–prey graph (Figure 14.16, page 164). This is to be expected as the control agent depends on the pest for its food in the same way a predator depends on its prey.

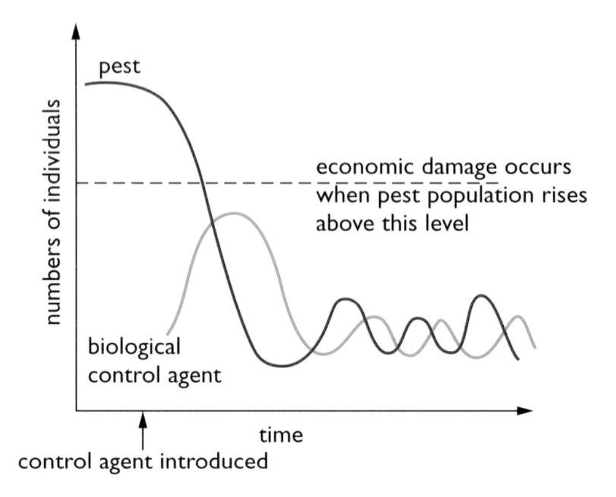

Figure 15.7 *Biological control.*

Methods of biological control include the following:

- Introducing a natural predator – ladybirds can be used to control the populations of aphids in orange groves.
- Introducing a herbivore – a moth was introduced from South America to control the prickly pear cactus that was becoming a serious weed in grazing land in Australia.
- Introducing a parasite – *Encarsia* is used to control whitefly populations in glasshouse tomato crops.
- Introducing a pathogenic (disease-causing) microorganism – the myxomatosis virus was deliberately released in Australia to control the rabbit population.
- Introducing sterile males – these mate with the females but no offspring are produced from these matings, so numbers fall.
- Using pheromones – these animal sex hormones are used to attract the males or females, which are then destroyed, reducing the reproductive potential of the population. Male-attracting pheromones are used to control aphids in plum crops.

Some agricultural practices encourage the build-up and spread of pests, whereas others discourage their spread. **Monoculture** involves giving over vast areas of land to a single crop. In the USA, big agricultural companies plant wheat and maize by the square mile. This makes for efficient harvesting as huge machinery can harvest large amounts of crop in a short period of time. However, it also makes for very efficient spread of pests. If the crop becomes infested, there are millions of crop plants to which the pest can spread. If the same crop is grown year after year in the same field, pests can lie dormant in the soil over winter. The timing of emergence in spring is often closely linked to planting time of the crop and so the pest causes damage from the moment the crop is planted. Over a number of years, the level of the pests builds up, causing increasing economic damage.

Crop rotation involves planting different crops in fields on a rotation basis. Usually, a new crop is grown in each field each year. Figure 15.8 shows a three-year crop rotation. Rotation of crops means that when over-wintering pests emerge, their 'favourite food' is no longer there! This stops the build-up of pests in one particular place.

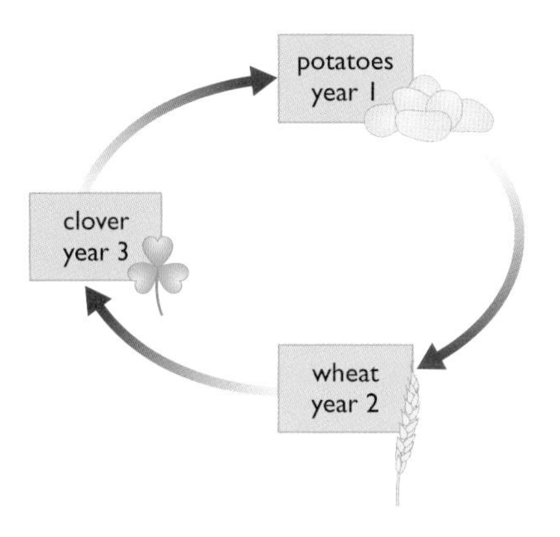

Figure 15.8 *A three-year crop rotation.*

Factory-farming

So far, we have considered only the ways in which a traditional farm produces food. We think of livestock wandering freely across fields to graze. However, nowadays much food is produced by 'factory-farming' methods. In these methods of production, animals are often kept in small, heated enclosures, and are fed a carefully controlled diet. Restricting movement and keeping the animals warm reduce the amount of energy the animal uses. More energy is therefore available for growth.

Many people are concerned about these intensive methods of production. They believe that even animals bred specifically for food production have a right to a reasonable quality of life before they are slaughtered.

In the factory farms the law controls how much space animals are permitted, as well as other conditions that must be adhered to (Figure 15.9). Inspectors check production on farms.

Figure 15.9 *Poultry are 'farmed' intensively under precisely controlled conditions.*

We also eat an increasing amount of fish, most of which traditionally comes from the seas around the UK. However, continued over-fishing of these waters has resulted in stocks of fish being dramatically reduced (Figure 15.10). These waters can no longer supply all our needs. Fishermen must sail further to obtain their catch.

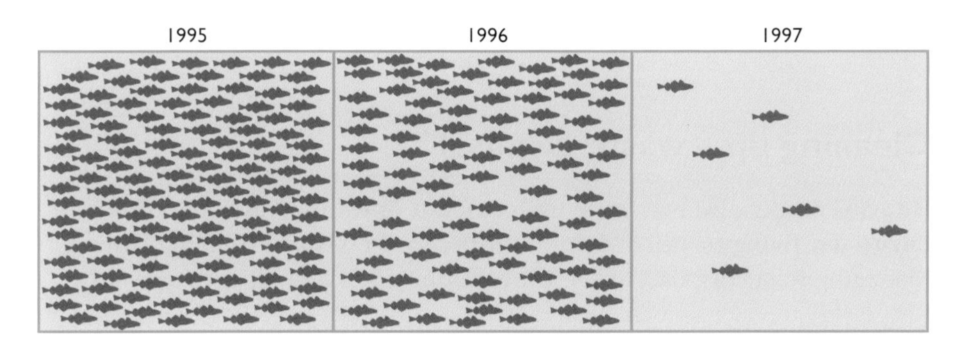

Figure 15.10 *The numbers of fish caught per hour in the North Sea dropped suddenly between 1995 and 1997. Source: The Little Earth Book, James Bruges.*

Fish farming is increasingly meeting the shortfall of fish. The fish are kept in large 'tanks' in which the temperature and oxygenation of the water can be controlled. Their diet is also carefully controlled. Fish farming has many features in common with factory farming of poultry.

In any intensive production system, the potential for the spread of disease is greater than normal because the animals are so close together. Antibiotics are often used to treat disease, but this itself is a cause for concern because the antibiotics may not have been degraded by the time the animals are eaten by humans.

Supplying our building materials and our space

Figure 15.11 *New developments use space and materials from our environment. (a) A complex under construction in Italy. (b) The Winchester bypass carved from the surrounding countryside.*

All the developments shown in Figure 15.11 take up space and need materials from our environment. As our cities extend further into the surrounding countryside, we take space that was once available to plants and other animals. Natural habitats are destroyed and the species richness of these areas decreases. Providing the materials for building also destroys natural habitats. Hillsides are quarried to provide the stone for building and trees are cut down to supply the timber (Figure 15.12).

Figure 15.12 *Wood is a useful resource. (a) Logging in a tropical rainforest. (b) Wood has been used to build this chalet.*

Cleaning up – waste disposal

Humans produce an ever-increasing amount of waste. This comes from our day-to-day living as well as from industrial processes and from agriculture. Managing waste disposal in the UK today involves a number of measures:

- containing the overall production of waste
- reducing the amount of waste going to landfill sites
- recycling 25% of household waste
- ensuring more effective treatment of sewage
- encouraging local authorities to develop schemes to meet the above.

Sewage treatment

Sewage contains suspended organic material, dissolved organic chemicals and inorganic materials. If it is discharged into waterways untreated, the dissolved oxygen levels of these waterways can be significantly reduced. This is because bacteria use up dissolved oxygen in the water as they decay the organic materials. In the UK, there are two main methods of treating sewage: the **percolating** (biological) **filter method** and the **activated sludge method**. These are compared below.

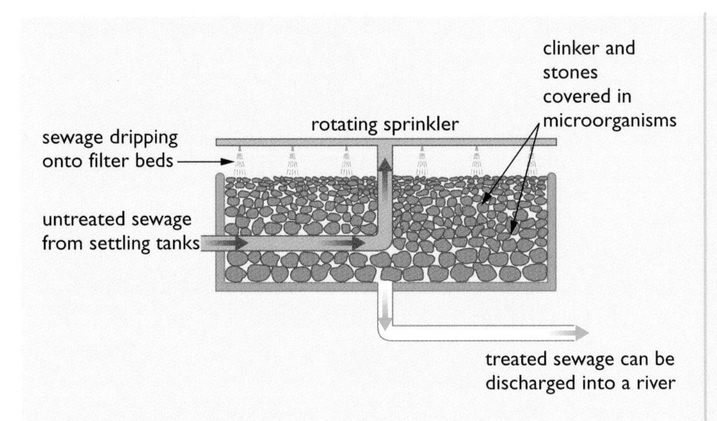

Figure 15.13 *The percolating filter method.*

In this method, when sewage arrives for treatment, it is screened to remove large objects and allowed to stand in large settling tanks to allow other solid material to settle out. It is then pumped through a pipe rotating over the filter bed and allowed to trickle through the filter. Bacteria, fungi and protozoa in the filter oxidise any organic matter. The treated sewage is then discharged into a waterway.

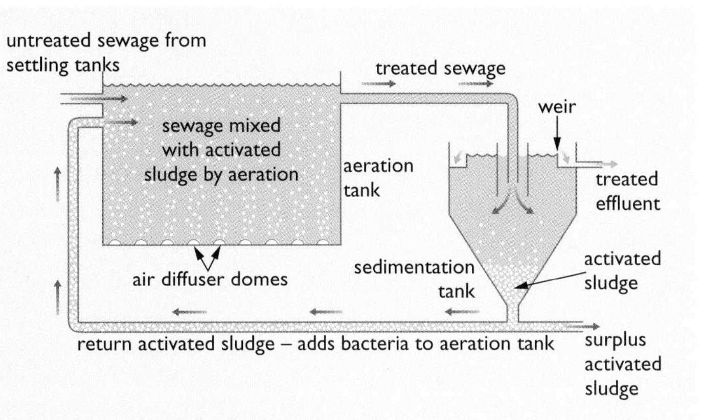

Figure 15.14 *The activated sludge method.*

In this method when sewage arrives for treatment, it is screened to remove large objects and allowed to stand in large settling tanks to allow other solid material to settle out. It is then passed into an aeration tank where the sewage is 'activated' by oxygen being pumped in. Bacteria in the tank oxidise the organic material. From here it is passed to a sedimentation tank where the 'activated sludge' settles out. Some is returned to the aeration tank and the purified effluent is discharged.

In other countries, such as France, other methods of sewage treatment are used as well. Sewage is discharged into large lakes and is oxidised slowly by bacteria. This process is less costly, but takes longer and needs more space.

Household waste

Once you dump your waste into the dustbin, it is disposed of in one of two main ways. It is either buried in a **landfill site**, or it is **incinerated**. Traditionally constructed landfill sites pose health hazards because rubbish is tipped over a large, exposed area over a long period of time, only finally being covered with soil when the entire site is full. Because of this, there is a high risk of transmitting disease. Landfill sites also attract vermin (e.g. rats). In addition, they can easily leak potentially toxic substances into water supplies.

More modern landfill sites use a cellular construction technique (Figure 15.15). These are much more hygienic than the traditional method as the rubbish is contained in smaller cells which are not exposed for as long. They also offer the potential to help local energy schemes, as methane produced by the decomposition of the waste can be extracted and used as a fuel.

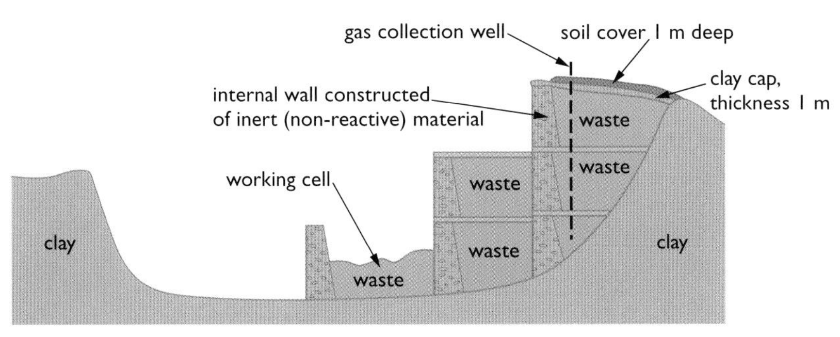

Figure 15.15 *A modern landfill site.*

Both methods of landfill waste disposal can help with land reclamation. Waste can be tipped into an area that has previously been quarried. Once complete, the site can be covered with soil and planted. It could then possibly be used as a recreation area.

Incineration cannot directly help to reclaim land, but it may be preferable in areas where space is at a premium. It has advantages over landfill in that it doesn't use up valuable land, it doesn't attract pests and it doesn't contaminate water supplies. In addition, the heat produced can be used in local heating schemes, although these have not proved to be very efficient. However, incineration does have drawbacks as the plant is costly to build and operate and polluting gases are discharged.

Pollution – the consequences of our actions

Pollution means releasing substances into the environment in amounts that cause harmful effects and which natural biological processes cannot easily remove. A key feature is the *amount*. Small amounts of sulphur dioxide and carbon dioxide would easily be absorbed by the environment and, over time, made harmless. It is the sheer mass of the pollutants that poses the problem.

Modern agricultural practices, building and other industries, as well as individual actions, release many pollutants into the environment. We pollute the air, land and water with a range of chemicals and heat.

Air pollution

We pollute the air with many gases. The main ones are carbon dioxide, carbon monoxide, sulphur dioxide, nitrogen oxides, methane and CFCs (chlorofluorocarbons).

Carbon dioxide

The levels of carbon dioxide have been rising for several hundred years. Over the last 100 years alone, the level of carbon dioxide in the atmosphere has increased by nearly 30%. This recent rise has been due mainly to the increased burning of fossil fuels, including petrol and diesel in vehicle engines. It has been made worse by cutting down large areas of tropical rainforest. These extensive forests have been called 'the lungs of the Earth' because they absorb such vast quantities of carbon dioxide and produce equally large amounts of oxygen. Extensive deforestation means that less carbon dioxide is being absorbed. Figure 15.16 shows changes in the level of carbon dioxide in the atmosphere from 1960 to 1990.

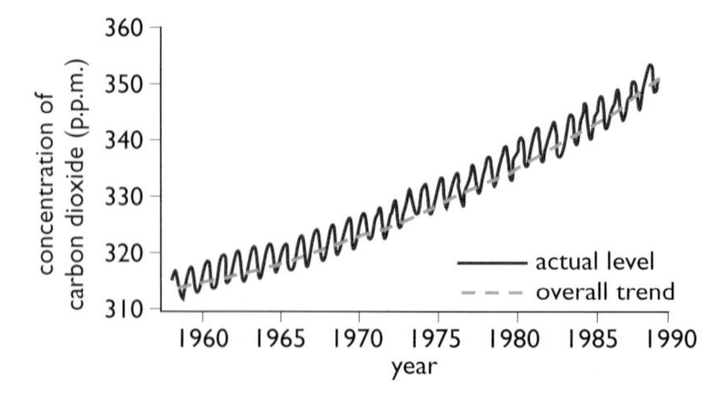

Figure 15.16 *The changes in levels of CO_2 at Mauna Loa, Hawaii from 1960 to 1990.*

In any one year, there is a peak and a trough in the levels of carbon dioxide. This is shown more clearly in Figure 15.17.

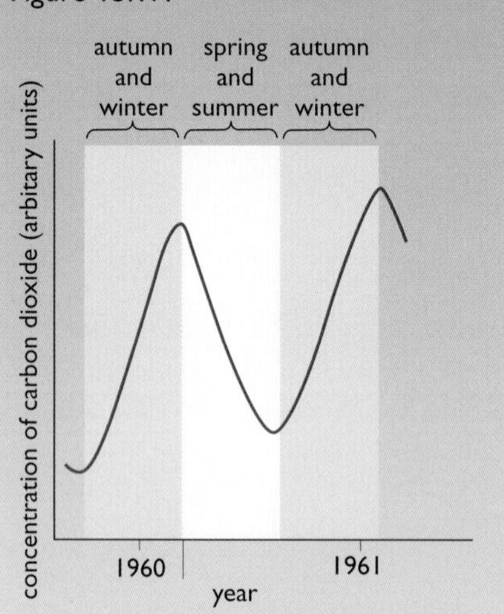

Figure 15.17 *Seasonal fluctuations in carbon dioxide levels.*

In the autumn and winter, trees lose their leaves. They photosynthesise much less and so absorb little carbon dioxide. They still respire, which produces carbon dioxide, so in the winter months, they give out carbon dioxide and the level in the atmosphere rises. In the spring and summer, with new leaves, the trees photosynthesise faster than they respire. As a result, they absorb carbon dioxide from the atmosphere and the level decreases. However, because there are fewer trees overall, it doesn't quite get back to the low level of the previous summer.

The increased levels of carbon dioxide contribute to **global warming**. Carbon dioxide is just one of the so-called 'greenhouse gases' that form a layer in the Earth's atmosphere.

Other greenhouse gases include methane (CH_4), nitrous oxide (N_2O) and chlorofluorocarbons (CFCs).

Short wave radiation from the sun strikes the planet. Some is absorbed and some is reflected as longer wave radiation. The greenhouse gases absorb, then re-emit towards the Earth, some of this long wave radiation, which would otherwise escape into space. This is the '**greenhouse effect**' and is a major factor in global warming (Figure 15.18).

If there were no greenhouse gases and no global warming, the Earth would be the same temperature as the Moon. Life, as we know it, would be impossible.

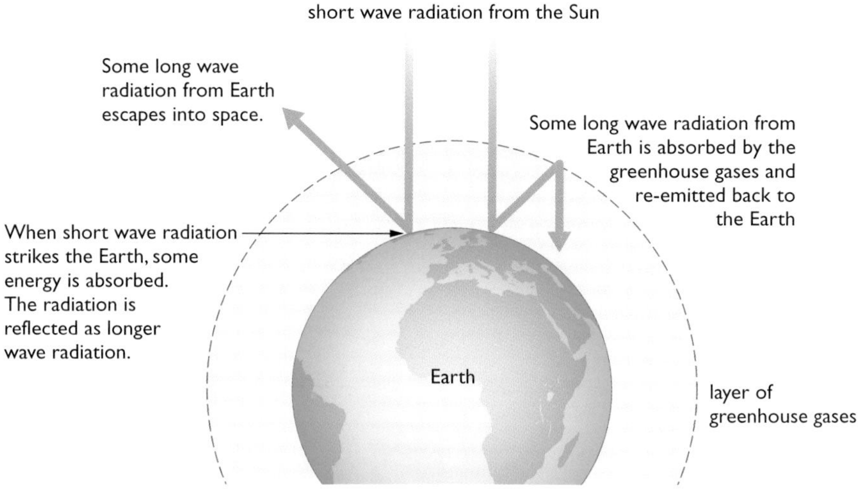

Figure 15.18 *The greenhouse effect.*

A rise in the Earth's temperature of only a few degrees would have many effects:

- Polar ice caps would melt and sea levels would rise.
- A change in the major ocean currents would result in warm water being redirected into previously cooler areas.
- A change in global rainfall patterns could result. With all the extra water in the seas, there would be more evaporation from the surface and so more rainfall in most areas.
- It could change the nature of many ecosystems. If species could not migrate quickly enough to a new, appropriate habitat, or adapt quickly enough to the changed conditions in their current habitat, they could become extinct.
- Changes in agricultural practices would be necessary as some pests became more abundant. Higher temperatures might allow some pests to complete their life cycles more quickly.

Carbon monoxide

When substances containing carbon are burned in a limited supply of oxygen, carbon monoxide (CO) is formed. This happens when petrol and diesel are burned in vehicle engines. Exhaust gases contain significant amounts of carbon monoxide. It is a dangerous pollutant as it is colourless, odourless and tasteless and can cause death by asphyxiation. Haemoglobin binds more strongly with carbon monoxide than with oxygen. If a person

In Europe there are now strict laws controlling the permitted levels of carbon monoxide in the exhaust gases produced by newly designed engines. These levels are lower than those allowed in the M.O.T. test of road-worthiness of vehicles three or more years old.

inhales carbon monoxide for a period of time, more and more haemoglobin becomes bound to carbon monoxide and so cannot bind with oxygen. The person may lose consciousness and, eventually, may die as a result of a lack of oxygen.

Sulphur dioxide

Sulphur dioxide (SO_2) is an important pollutant as it is a major constituent of **acid rain**. It is formed when fossil fuels are burned, and it can be carried hundreds of miles in the atmosphere before finally combining with rainwater to form acid rain.

Some lichens are more tolerant of sulphur dioxide than others. Patterns of lichen growth can, therefore, be used to monitor the level of pollution by sulphur dioxide. The different lichens are called **indicator species** as they 'indicate' different levels of sulphur dioxide pollution (see Figure 15.19).

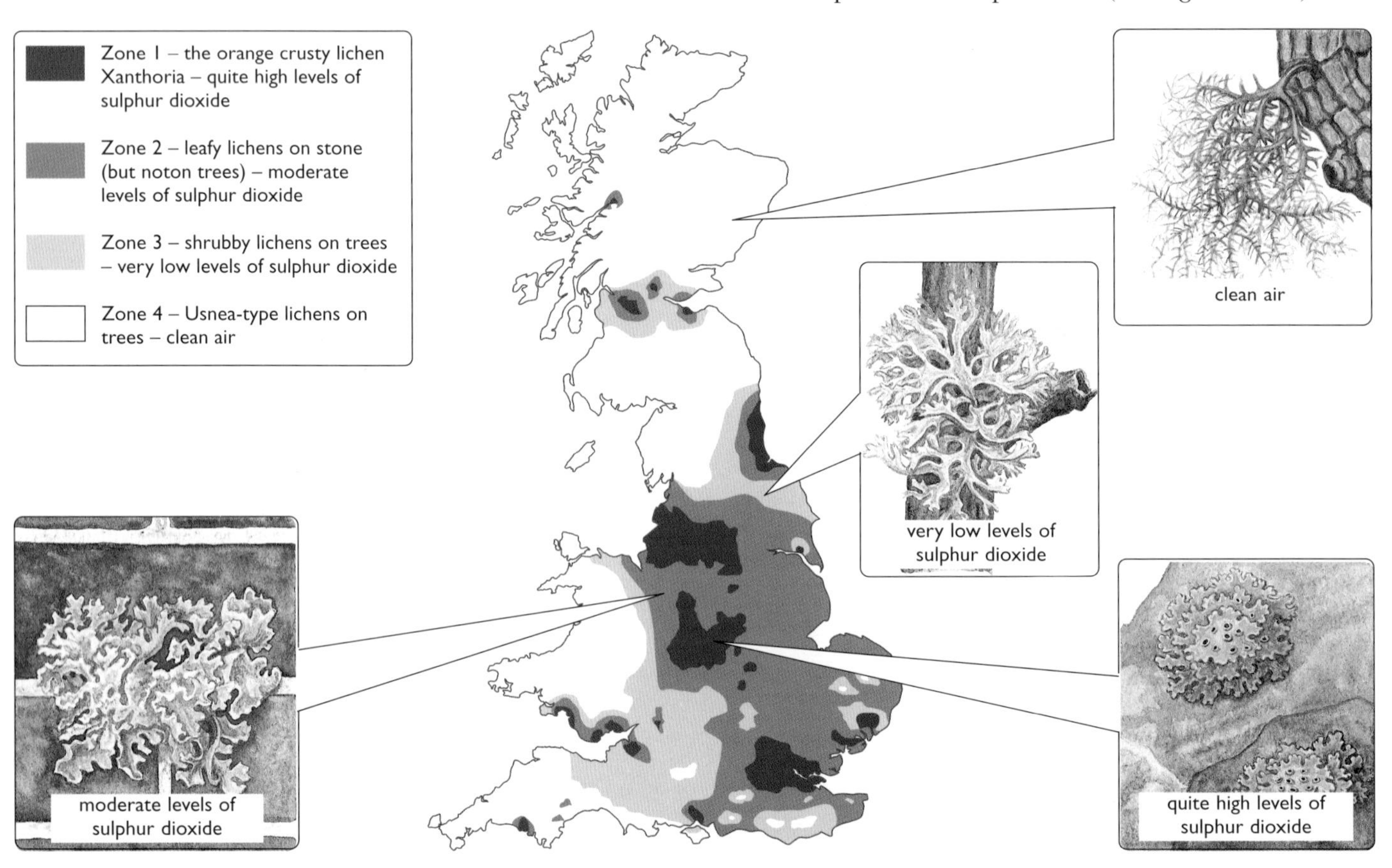

Figure 15.19 *Lichens are sensitive to pollution levels.*

You can make a rough estimate of how much sulphur dioxide is in the air around you by observing which lichens grow on trees and walls. If the only lichens that grow are flat, and pressed close to the surface, this indicates a high level of sulphur dioxide. If the lichens are 'shrubby' in appearance and grow well clear of the surface, there is likely to be less sulphur dioxide polluting the air.

Nitrogen oxides

Nitrogen oxides (NO_x) are also constituents of acid rain. They are formed when petrol and diesel are burned in vehicle engines.

Acid rain

Rain normally has a pH of about 5.5 – it is slightly acidic due to the carbon dioxide dissolved in it. Both sulphur dioxide and nitrogen oxides dissolve in rainwater to form a mixture of acids, including sulphuric acid and nitric acid. As a result, the rainwater is more acidic with a much lower pH than normal rain (Figure 15.20).

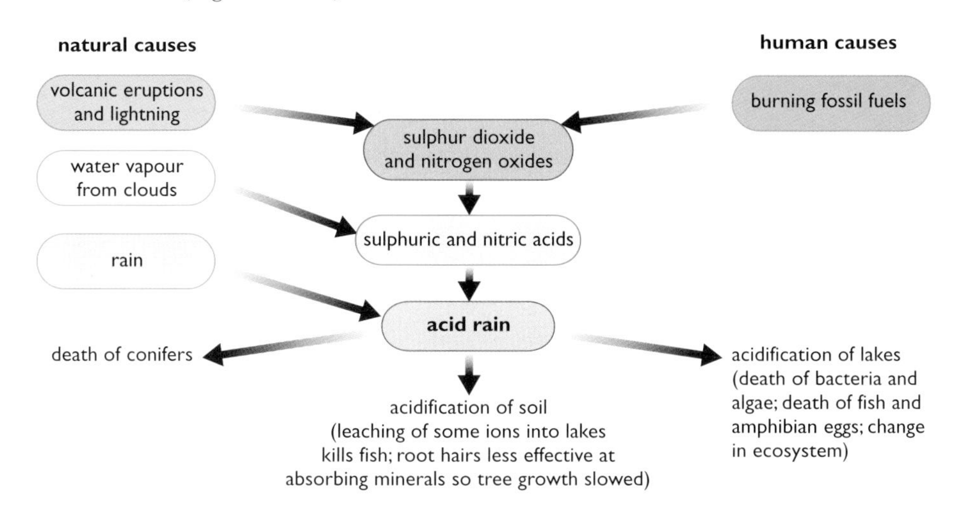

Figure 15.20 *The formation of acid rain and its effects on living organisms.*

Besides its effects on living things, acid rain causes damage to stonework by reacting with carbonates and other ions in the stone (Figure 15.21).

Figure 15.21 *This stone lion outside Leeds Town Hall has been dissolved by acid rain.*

Methane

Methane (CH_4) is an organic gas. It is produced when microorganisms ferment larger organic molecules to release energy. The most significant locations of these microorganisms are:

- decomposition of waste in landfill sites by microorganisms
- fermentation by microorganisms in the rumen of cattle and other ruminants
- fermentation by bacteria in rice paddy fields.

Methane is a greenhouse gas, with effects similar to carbon dioxide. Although there is less methane in the atmosphere than carbon dioxide, each molecule has a bigger greenhouse effect.

Herds of cattle can produce up to 40 dm^3 of methane per animal per hour. This adds up to a lot of methane being belched and farted into the atmosphere!

Freshwater pollution

The three main pollutants of freshwater are nitrates from fertilisers, organic waste and detergents.

Nitrates from fertilisers

ideas evidence

Farmers add inorganic fertilisers to soils to replace mineral ions lost when crops are removed. The ions in these fertilisers (particularly nitrates) are very soluble. As a result they are easily **leached** (carried out with water) from the soils and can enter waterways. The level of nitrates (and other ions) can rise rapidly in these lakes and rivers. This increase in mineral ions is called **eutrophication**, which is a natural process in nearly all waterways (Figure 15.22). What is *not* natural is the speed with which it happens due to leaching of ions in fertilisers from soils. Rapid eutrophication can have disastrous consequences for a waterway.

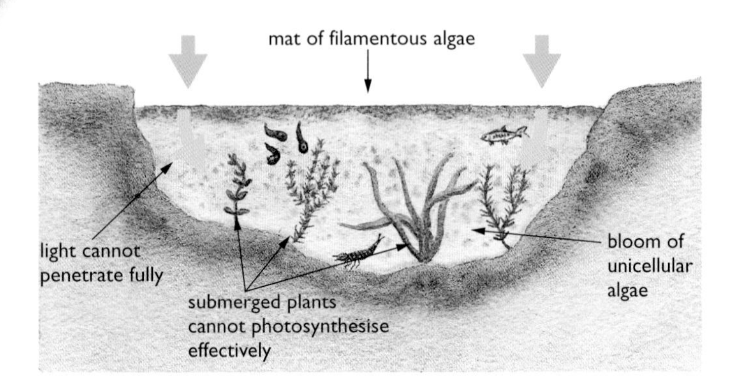

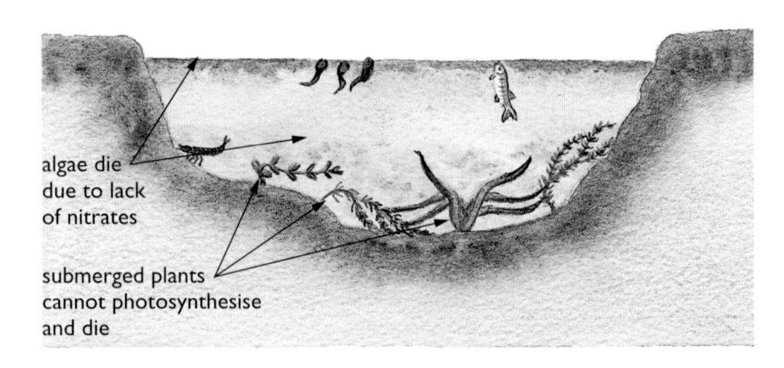

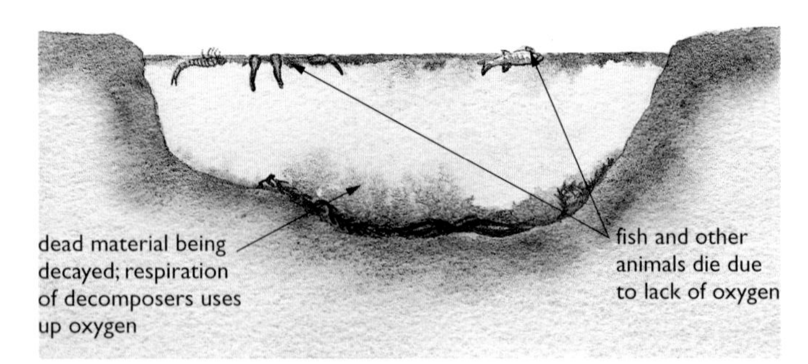

Figure 15.22 *Stages in eutrophication in a pond.*

1 As nitrate levels rise, algae reproduce rapidly. They use the nitrates to make extra proteins for growth, just like the crop plants for which the nitrates were intended.

2 The algae form an **algal bloom** – a kind of algal pea soup if the algae are unicellular, or a mass of filaments if the algae are filamentous.

3 The algae prevent light from penetrating further into the water.

4 Submerged plants cannot photosynthesise and so die.

5 The algae also die as they run out of nitrates.

6 Bacteria decay the dead plants and algae (releasing more nitrates and allowing the cycle to start again).

7 The bacteria reproduce (due to the large amount of dead matter) and their respiration uses up more and more oxygen.

8 The water may become totally **anoxic** (without oxygen) and all life in the water will die.

The problems can be more severe in hot weather because the nitrates can become more concentrated as the heat evaporates water. All the processes are speeded up due to increased enzyme activity. The problems are less severe in moving water because the nitrates are rapidly diluted and the water is continually being re-oxygenated.

Rapid eutrophication is less likely when farmers use organic fertilisers (like manure). The organic nitrogen-containing compounds in manure are less soluble and so are leached less quickly from the soil. However, water can sometimes be polluted with large amounts of organic matter, for example when untreated sewage is released into waterways. Bacteria and fungi decay this and use up oxygen as they respire. The water becomes anoxic in the same way as in eutrophication. Fish and other animals die due to a lack of oxygen.

The level of organic water pollution can be monitored by the 'indicator species' present. Figure 15.23 shows some of these.

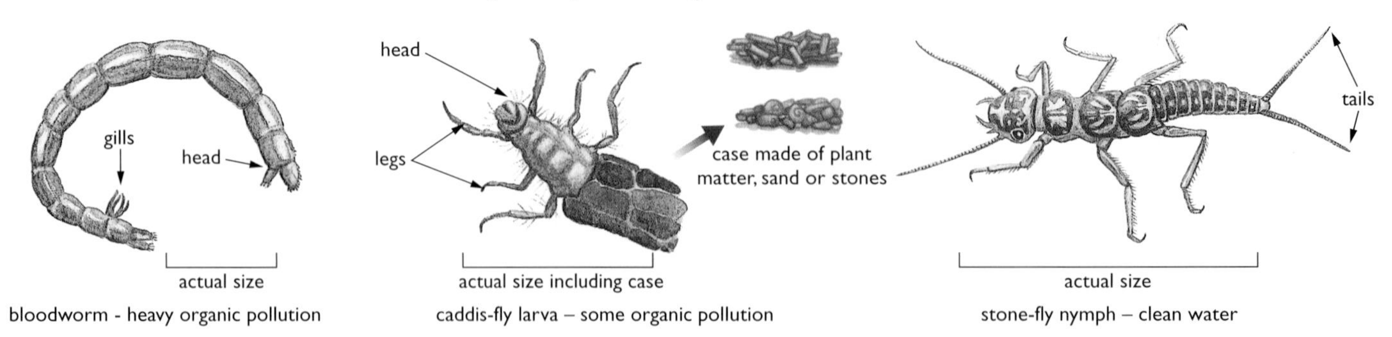

Figure 15.23 *Some freshwater animals will only live in very clean water, while others can survive in very polluted areas.*

When organic matter pollutes moving water, the point of the polluting outlet becomes very low in oxygen, as bacteria decompose the organic material. Only those species adapted to such conditions can survive. As the water moves away from the outlet, it becomes oxygenated again as it mixes with the air at the surface. The increase in oxygen levels allows more species of 'clean water' animals to survive. Figure 15.24 shows the changes in oxygen content, numbers of clean water animals and bloodworms at, and just downstream from, a sewage outlet into a river.

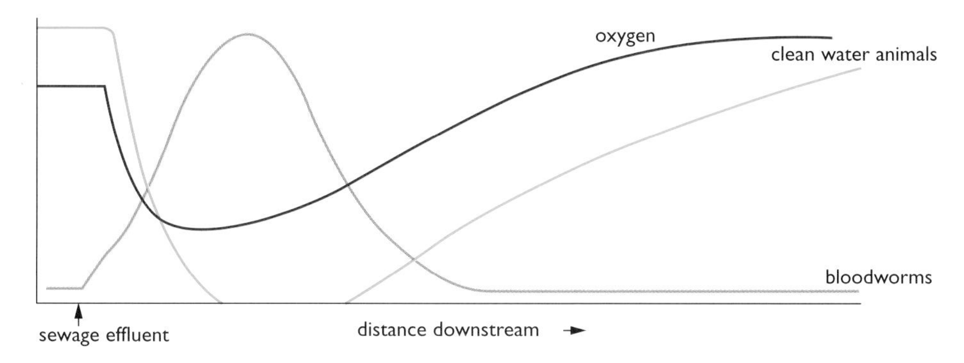

Figure 15.24 *The changes in oxygen levels, numbers of clean water animals and numbers of blood worms around a sewage outlet into a river.*

Other forms of pollution

We pollute our air with poisonous gases and our water with organic pollutants and nitrates. We also pollute our environment in a number of other ways.

Thermal pollution

Water is used as a coolant in power stations and in other industrial plants. As water removes the heat, it becomes warmer and its ability to dissolve oxygen decreases. This can affect the number of animals able to survive in the water. Besides the effects of decreasing oxygen content, the direct effect of changing temperature may also kill animals and plants. Many animals cannot regulate their body temperature (as mammals and birds can). Their temperature changes with that of the environment. There have been cases reported in North America of river water being heated to over 80°C and being completely lifeless. This is thermal pollution of water at its worst.

Pesticide pollution

Farmers frequently use pesticides to control many different types of pests (see page 177). Many of these have no serious side effects, but some are persistent – they are not degraded (decomposed) easily. Traces of some herbicides (weedkillers) can remain on or in the crops that were sprayed, with the risk that they could then be eaten with the food. A single, small 'dose' of herbicide is unlikely to cause any harm, but if the dose is repeated many times then the amount may accumulate and begin to have more serious consequences. Pesticides are sometimes stored in fatty tissue where the amount builds up over a period of time. This is called **bioaccumulation**.

Sometimes the effects of bioaccumulation can be *magnified* as a pesticide is passed along a food chain. The best documented cases of this happening involve the insecticide **DDT**. DDT is an extremely effective insecticide and was widely used after the Second World War. It prevented millions of deaths

An ideal pesticide should:

- control the pest effectively
- be biodegradable, so that no toxic products are left in the soil or on crops
- be specific, so that only the pest is killed
- not accumulate in organisms
- be safe to transport and store
- be easy and safe to apply.

from malaria in some areas by significantly reducing the mosquito population. However, DDT was passed along food chains and accumulated in harmful amounts in the top carnivores. This happened because DDT is extremely persistent (a single application can take over 20 years to degrade) and is fat soluble, so easily stored in living tissue. The food chain in Figure 15.25 shows the extent to which DDT can be accumulated. The amounts in brackets are parts per million of DDT.

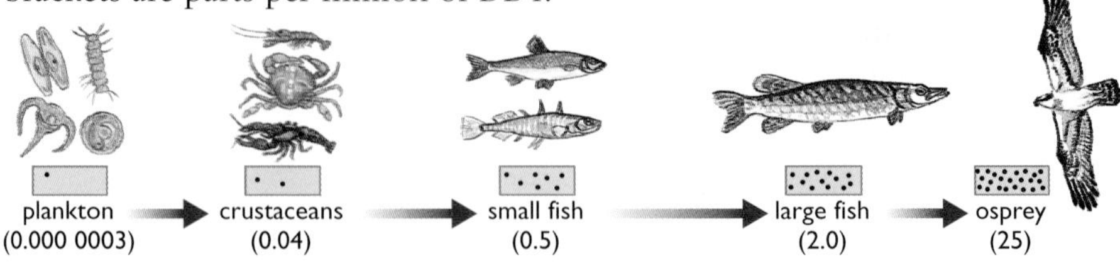

Figure 15.25 *Biomagnification of DDT in a food chain.*

This increase in concentration along a food chain is called **biomagnification**. It happens because each organism in the chain eats many of the preceding organisms and accumulates the DDT (bioaccumulation). The effects of bioaccumulation are therefore magnified at each stage in the food chain.

The use of DDT has been restricted since 1972 because of its ecological effects.

ideas
evidence

Conservation – what can we do about it?

Some people think that conservation means leaving the Earth alone and not using any of its resources. This is not the case. Conservation simply means what the name implies – conserving whatever is present now for future generations to be able to use.

Many agricultural and industrial practices have resulted in the depletion of habitats for wildlife. We have a 'duty of care' for the wildlife of the Earth and should try to ensure that when we take resources from the Earth, we do not needlessly destroy the habitat of another species. Many species are now classified as 'endangered species' and could easily become extinct if their populations do not recover.

As a result of commercial whaling, several species of whales have populations that are only a fraction of what they used to be. Destruction of the habitat of the Bengal tiger has reduced the population to the point where extinction is very likely. Felling tropical rainforest does not just remove trees, but destroys the habitats of a whole range of organisms. Many organisms, including some as yet undiscovered, will become extinct if the tropical rainforests are lost. We will also lose valuable biological knowledge and potential medical products.

What can we do to help? Many people believe that little can be done about the environmental problems that face us, or that only governments working together on a global scale can have any impact. Neither of these is true. Individuals and local authorities can do a great deal to preserve and conserve our natural environment.

Today, whenever any new development is proposed which would disrupt a significant area of land, the developers must prepare an **Environmental**

Impact Statement. This must be submitted to any local authorities from which permission for the development must be obtained.

An environmental impact statement must give information about the following aspects of a development.

1. The nature of the development itself.
- The size of the development. What area of land will be used? How high will the highest building be?
- Will there be any significant pollution risks?
- Will there be any risk of accidents and, if so, what can be done to minimise the risk?
- What resources will be used? Where will they come from?

2. The location of the development.
- The nature of the existing land that will be used; e.g. pasture, woodland, etc.
- Are there any nearby areas that are particularly sensitive to changes in the environment?
- Could the developers have chosen any other site for the development?

3. The likely impact of the development.
- What are likely to be the immediate and long-term effects on surrounding areas?
- Are any of these effects likely to be reversible?

When a development proposal is submitted, committees from the local authority will consider the statement, alongside any objections to the development, before they grant permission for the development to go ahead.

Some important areas of conservation are discussed below.

Reducing the effects of pollution

Table 15.4 shows some possible courses of action to reduce the effects of pollution.

Problem	Possible individual action	Possible government/industrial action
global warming (greenhouse gases, CO_2, CH_4)	use as little energy as possible – less then needs to be generated; reduce use of private transport as far as possible	international agreements to set acceptable levels of CO_2 emissions; reduce deforestation: encourage sustainable felling and replanting schemes; encourage use of more recycled metals
acid rain (SO_2, NO_x)	reduce use of private transport as far as possible; reduce use of electricity where possible	legislation to enforce desulphurisaton of emissions from power stations; encourage use of 'cleaner' fuels such as natural gas (methane) and low-sulphur petrol
ozone depletion (overuse of CFCs)	reduce use of any aerosol containing CFCs	international legislation to restrict the use of CFCs
organic pollution of water	individual farmers can ensure safe storage and use of organic fertiliser (manure)	legislation to enforce treatment of sewage before discharge into waterways
nitrate pollution of water	individual farmers can make more use of 'organic' farming practices – use of crop rotations and organic manure; we can encourage organic farming by buying more organic produce	increase monitoring of waterways; legislation to limit levels of nitrates in water
pesticide pollution of soils	individual farmers can reduce use of pesticides and use biological control and crop rotation to limit pest build up	encourage development of 'safer' pesticides

Table 15.4: *Problems associated with pollution and possible solutions.*

Discouraging removal of hedges

Farmers have tended to remove hedgerows to create larger fields for growing crops. This allows more efficient crop production methods to be used, but it also encourages the build up of pests and erosion of the soil. Removing hedgerows also removes habitats for many birds and small mammals.

Flooding caused by heavy rainfall can be made worse if there are no hedgerows. The roots of hedges and trees help to stabilise the soil and act as a barrier to water movement.

Farmers are now paid a subsidy to leave an area around each field uncultivated to allow wild flowers to return. Wild flowers encourage insects and other animals that find habitats there.

Alternatives to deforestation

If removing hedgerows removes habitats on a small scale, deforestation removes them on a grand scale. The extensive forests that used to cover much of Europe have mostly gone, to be replaced by grazing land, towns, cities and roads. Tropical rainforests are now being cut down at an alarming rate. Besides destroying habitats, deforestation can also:

- reduce the amount of carbon dioxide absorbed from the atmosphere by photosynthesis
- increase the amount of carbon dioxide added to the atmosphere if trees are burned
- allow soil to be eroded more easily as the roots can no longer stabilise the soil
- reduce the nitrogen content of soils – as trees are removed for logging they take the nitrogen with them.

The use of timber need not result in deforestation. There are other options, such as planting young trees as older ones are felled. This maintains the habitats and ensures that the same resource will be there for future generations. Figure 15.26 shows one method of sustainable felling of trees in a tropical rainforest.

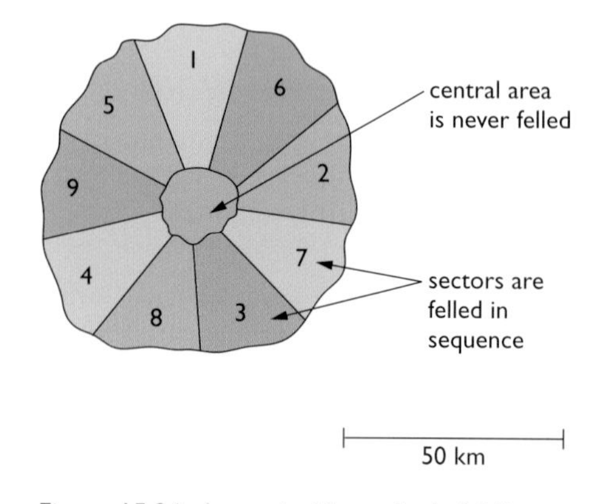

Figure 15.26 *A sustainable method of felling trees in a tropical rainforest.*

It takes about 35 years to complete the felling of one sector, and it will not be felled again until 280 years later. By that time, the trees will have completely re-grown. The animals and plants that find habitats in the trees will also have returned.

We can all help by checking timber products to see if the product comes from an area where replanting is practised.

Feeding the increasing world population

As the world's population has grown, the problem of feeding more people has meant ever-increasing demands for food. Some of the 'solutions' to these problems have created other problems in turn. Tropical rainforest is being felled to create more land for agriculture, with all the negative effects discussed earlier and many seas have been over-fished.

One problem stems from the fact that food chains are an inefficient method of transferring energy (see Chapter 14). Only about 10% of the energy entering one trophic level is passed to the next, and so on. It therefore follows that shorter food chains are more efficient (Figure 15.27).

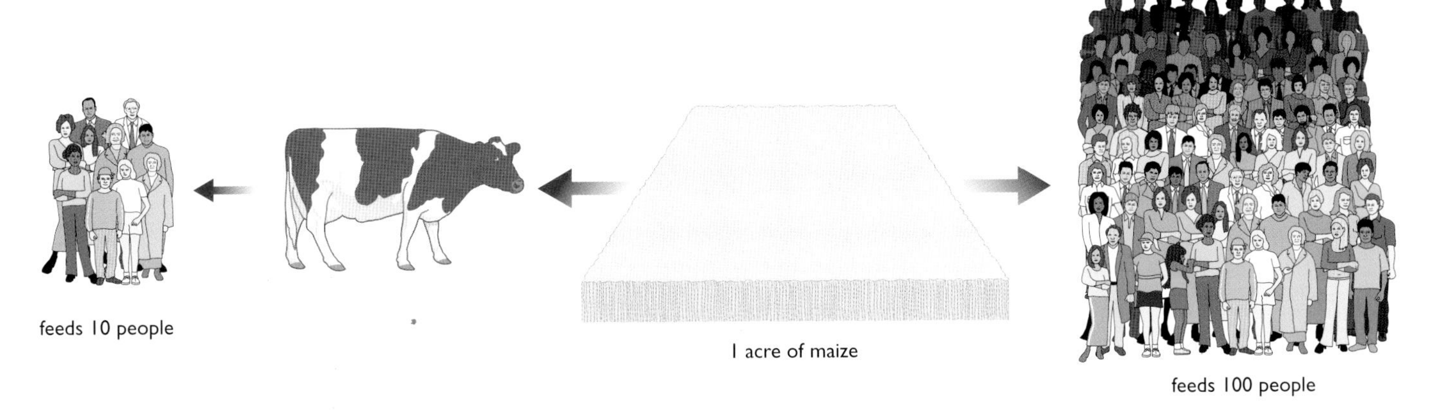

Figure 15.27 *Shorter food chains are more efficient and can feed more people.*

Shortening food chains reduces the number of trophic levels and so reduces the energy losses. Many people in less economically developed countries already obtain a very high proportion of their energy from plant sources. If people in more economically developed countries were to follow their example, it would be easier to supply enough food for everyone.

Saving energy and recycling

The cost of extracting metals from their ores is not just financial; it also causes pollution. The processes usually require a lot of energy, which often comes from burning fossil fuels. This produces pollutants such as sulphur dioxide, nitrogen oxides and carbon dioxide. Recycling these metals and other products generally uses less energy and so reduces pollution levels (Figure 15.28).

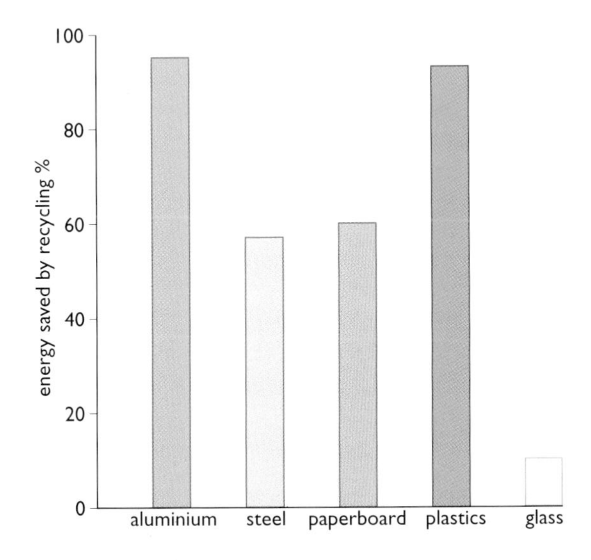

Figure 15.28 *Energy savings from recycling some materials.*

We can all do something to help conserve the resources of our planet for future generations:

- Use as many recycled products as is appropriate. This reduces the energy consumption needed to extract metals from ores and to manufacture new products, so reducing the pollution that goes with energy generation.
- Turn off electrical appliances when they are not needed – again this reduces energy consumption.
- Have cars serviced regularly so that the engine is running efficiently and producing the minimum levels of pollutants.
- Turn down the thermostat on the central heating in your home – less energy used means that less needs to be generated.
- Try to make fewer journeys alone in the car – share transport or use public transport where convenient.
- Don't have endless conversations on your mobile 'phone. The batteries in it are expensive to produce – in energy terms as well as financially!
- Try to encourage wildlife in your garden, if you have one. Planting wild species such as teasels will encourage goldfinches in winter. Buddleias encourage a whole range of butterflies. Construct a small wildlife pond to provide a habitat for frogs and newts.
- Use compost on the garden rather than inorganic fertilisers.

We need to alter our way of thinking so that conserving wildlife, materials and energy becomes a way of life, not just a special event.

End of Chapter Checklist

If you haven't got a copy of your specification, read the introduction on page vi.

You will need to be able to do some or all of the following. Check your Awarding Body's specification (syllabus) to find out exactly what you need to know.

- Explain how the growth in the human population has made increasing demands on our environment.
- Explain how modern agriculture has been able to increase food production.
- Explain how cropping and sale of livestock remove nutrients from the farm ecosystem.
- Describe the effects of pests on yields.
- Explain the role of pesticides, biological control and crop rotation in controlling pests.
- Describe some of the dangers of using pesticides.
- Describe how factory farming and fish farming increase natural productivity.
- Explain the need for sewage treatment and describe the two methods (percolating filter and activated sludge methods) used in the UK.
- Describe how household waste can be disposed of and explain the relative benefits of incineration and landfill tipping.
- Describe the ways in which the air is being polluted and explain the consequences of this pollution.
- Describe the ways in which water is being polluted and explain the consequences of this pollution.
- Explain what is meant by conservation and describe some practical measures that can be undertaken by individuals and groups to conserve our natural resources.

Questions

More questions on human influences on the environment can be found at the end of Section D on page 195.

1 Why are humans having much more of an impact on their environment now than they did 500 years ago?

2 The graph shows the changing concentrations of carbon dioxide at Mauna Loa, Hawaii over a number of years.

a) Describe the overall trend shown by the graph.

b) Explain the trend described in *a)*.

c) In any one year, the level of atmospheric carbon dioxide shows a peak and a trough. Explain why.

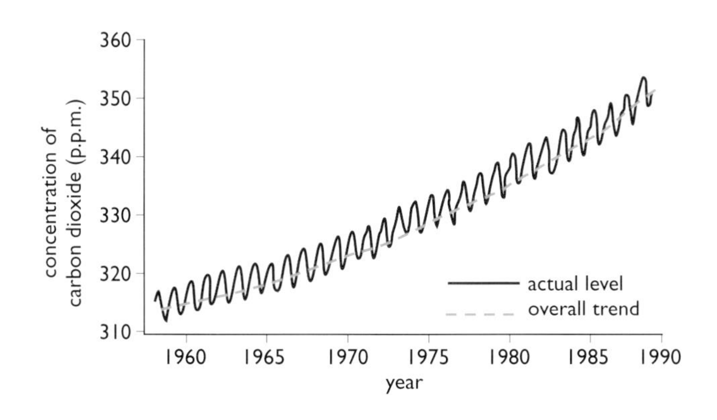

3 The diagram shows how the greenhouse effect is thought to operate.

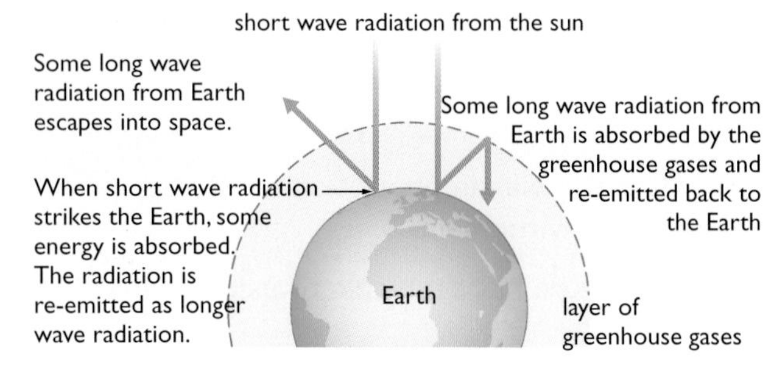

a) Name two greenhouse gases.

b) Explain one benefit to the Earth of the greenhouse effect.

c) Suggest why global warming may lead to malaria becoming more common in Europe.

4 Several factories on an industrial estate burn fossil fuels. As a result, a considerable amount of sulphur dioxide is produced. Describe how you could use the patterns of lichen growth to estimate the changing levels of sulphur dioxide in the air at different distances from the industrial estate. You need not give full practical details.

5 The diagram shows the profile of the ground on a farm either side of a pond.

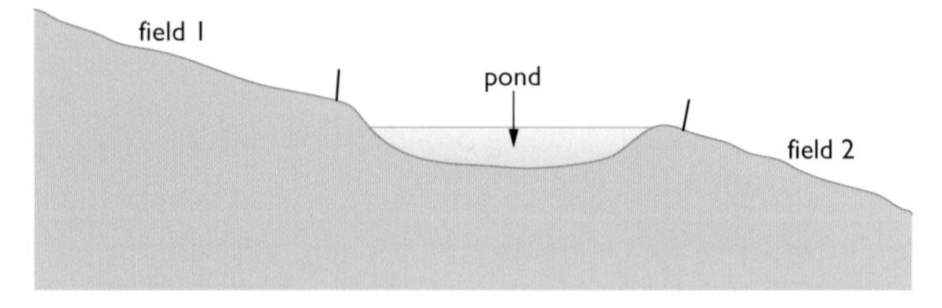

The farmer applies nitrate fertiliser to the two fields in alternate years. When he applies the fertiliser to Field 1, the pond often develops an algal bloom. This does not happen when fertiliser is applied to Field 2.

a) Explain why an algal bloom develops when he applies the fertiliser to Field 1.

b) Explain why no algal bloom develops when he applies the fertiliser to Field 2.

c) Explain why the algal bloom is more pronounced in hot weather.

6 Some untreated sewage is accidentally discharged into a small river. A short time afterwards, a number of dead fish are seen at the spot. Explain, as fully as you can, how the discharge could lead to the death of the fish.

7 Some farmers use pesticides and fertilisers to improve crop yields. Those practising 'organic' farming techniques do not use any artificial products.

a) Describe how the use of pesticides and fertilisers can improve crop yields.

b) Explain how organic farmers can maintain fertile soil and keep their crops pest free.

End of Section Questions

1 The diagram shows a simplified food web of the adult herring.

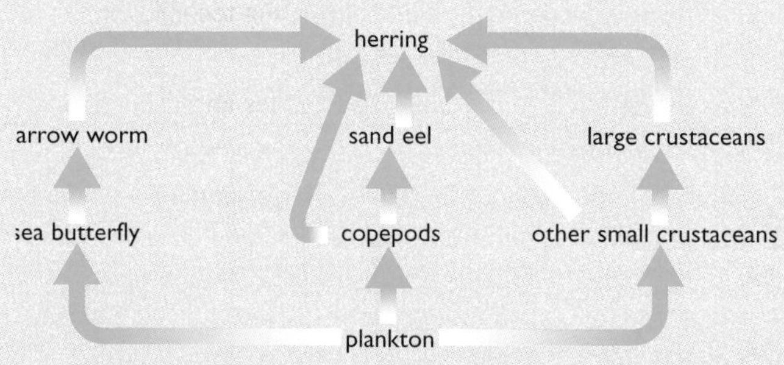

a) *i)* Write out a food chain from the above food web containing four organisms. *(1 mark)*

ii) From your food chain, name the primary consumer and secondary consumer. *(2 marks)*

iii) Name one organism in the web that is both a primary consumer and a secondary consumer. Explain your answer. *(2 marks)*

b) The amount of energy in each trophic level has been provided for the following food chain. The units are $kJ/m^2/yr$.

plankton (8869) → copepod (892) → herring (91)

i) Sketch a pyramid of energy for this food chain. *(1 mark)*

ii) Calculate the percentage of energy entering the plankton that passes to the copepod. *(2 marks)*

iii) Calculate the percentage of energy entering the copepod that passes to the herring. *(2 marks)*

iv) Calculate the amount of energy that enters the food chain per year if the plankton use 0.1% of the available energy. *(2 marks)*

v) Explain two ways in which energy is lost in the transfer from the copepod to the herring. *(2 marks)*

Total 14 marks

2 Animals and plants that live in hot, dry environments show special adaptations.

a) The diagram shows a section through a leaf of a plant that lives in these conditions.

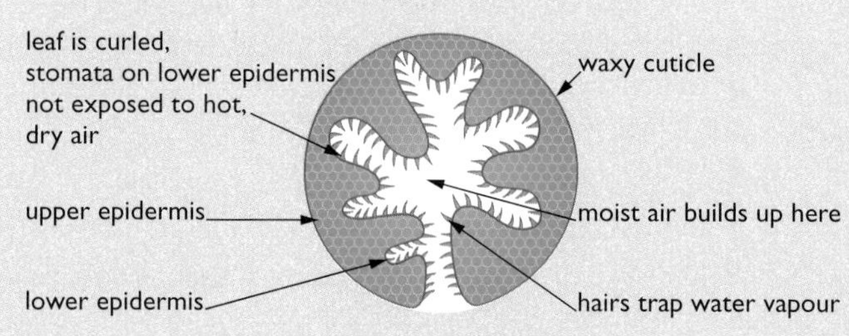

i) Describe two features visible in the diagram that are adaptations to a hot, dry environment. Explain how each feature increases the chance of survival of the plant. *(4 marks)*

ii) Describe two other features that a plant living in a hot, dry environment might possess. Explain how these features increase the chance of survival of the plant. *(4 marks)*

b) Describe three ways in which the camel is adapted to life in the desert. Explain how each feature increases the chance of survival of the camel. *(6 marks)*

c) Describe and explain two ways in which the *behaviour* of a gerbil increases its chances of survival in a desert. *(4 marks)*

Total 18 marks

3 Insecticides are used by farmers to control the populations of insect pests. New insecticides are continually being developed.

a) A new insecticide was trialled over three years to test its effectiveness in controlling an insect pest of potato plants. Three different concentrations of the insecticide were tested. Some results are shown in the table.

Concentration of insecticide	Percentage of insect pest killed each year		
	Year 1	Year 2	Year 3
1 (weakest)	95	72	18
2 (intermediate)	98	90	43
3 (strongest)	99	91	47

i) Describe, and suggest an explanation for, the change in the effectiveness of the insecticide over the three years. *(3 marks)*

ii) Which concentration would a farmer be most likely to choose to apply to potato crops? Explain your answer. *(3 marks)*

b) The trials also showed that there was no significant bioaccumulation of the insecticide.

i) What is bioaccumulation? *(1 mark)*

ii) Give an example of bioaccumulation of an insecticide and describe its consequences. *(2 marks)*

iii) Explain why it is particularly important that there is no bioaccumulation of *this* insecticide. *(1 mark)*

Total 10 marks

4 Carbon is cycled through ecosystems by the actions of plants, animals and decomposers. Humans influence the cycling of carbon more than other animals.

a) Explain the importance of plants in the cycling of carbon through ecosystems. *(2 marks)*

b) Describe two human activities that have significant effects on the global cycling of carbon. *(2 marks)*

c) The graph shows the activity of decomposers acting on the bodies of dead animals under different conditions.

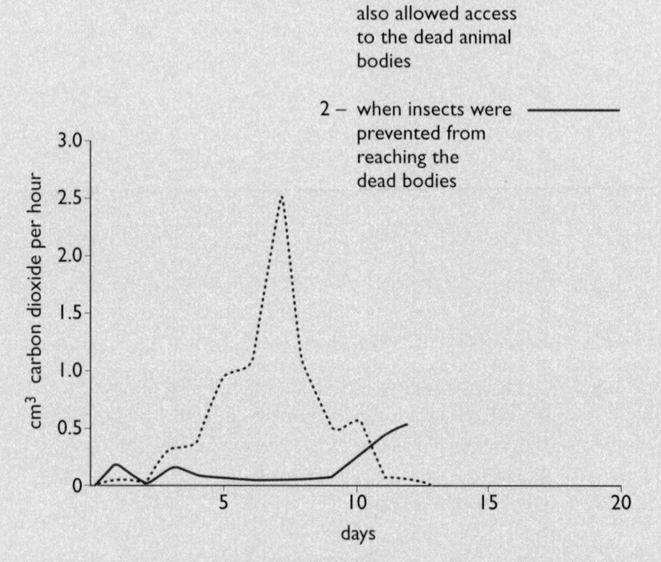

i) Why was carbon dioxide production used as a measure of the activity of the decomposers? *(2 marks)*

ii) Describe and explain the changes in decomposer activity when insects were also allowed access to the dead bodies (1). *(3 marks)*

iii) Describe two differences between curves (1) and (2). Suggest an explanation for the differences you describe. *(4 marks)*

Total 14 marks

5 In natural ecosystems, there is competition between members of the same species as well as between different species.

a) Explain how competition between members of the same species helps to control population growth. *(3 marks)*

b) Crop plants must often compete with weeds for resources. Farmers often control weeds by spraying herbicides (weedkillers).

i) Name two factors that the crop plants and weeds may compete for and explain the importance of each. *(4 marks)*

ii) Farmers usually prefer to spray herbicides on weeds early in the growing season. Suggest why. *(2 marks)*

c) Two species of the flour beetle, *Tribolium*, compete with each other for flour. Both are parasitised by a protozoan. The graphs show the changes in numbers of the two species over 900 days when the parasite is absent and when it is present.

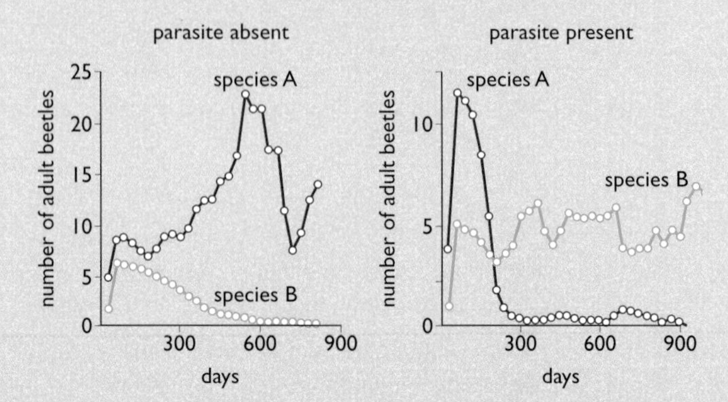

i) Which species is the most successful when the parasite is absent? Justify your answer. *(2 marks)*

ii) What is the effect of the parasite on the relative success of the two beetles? Suggest an explanation for your answer. *(4 marks)*

Total 15 marks

6 The table gives information about the pollutants produced in extracting aluminium from its ore (bauxite) and in recycling aluminium.

Pollutants	**Amount (g per kg aluminium produced)**	
Air	**Extraction from bauxite**	**Recycling aluminium**
sulphur dioxide	88 600	886
nitrogen oxides	139 000	6 760
carbon monoxide	34 600	2 440
Water		
dissolved solids	18 600	575
suspended solids	1600	175

a) Calculate the percentage reduction in sulphur dioxide pollution by recycling aluminium. *(2 marks)*

b) Explain how extraction of aluminium from bauxite may contribute to the acidification of water hundreds of miles from the extraction plant. *(3 marks)*

c) Suggest two reasons why there may be little plant life in water near an extraction. *(4 marks)*

Total 9 marks

Chapter 16: Chromosomes, Genes and DNA

This chapter looks at the structure and organisation of genetic material, namely chromosomes, genes and DNA.

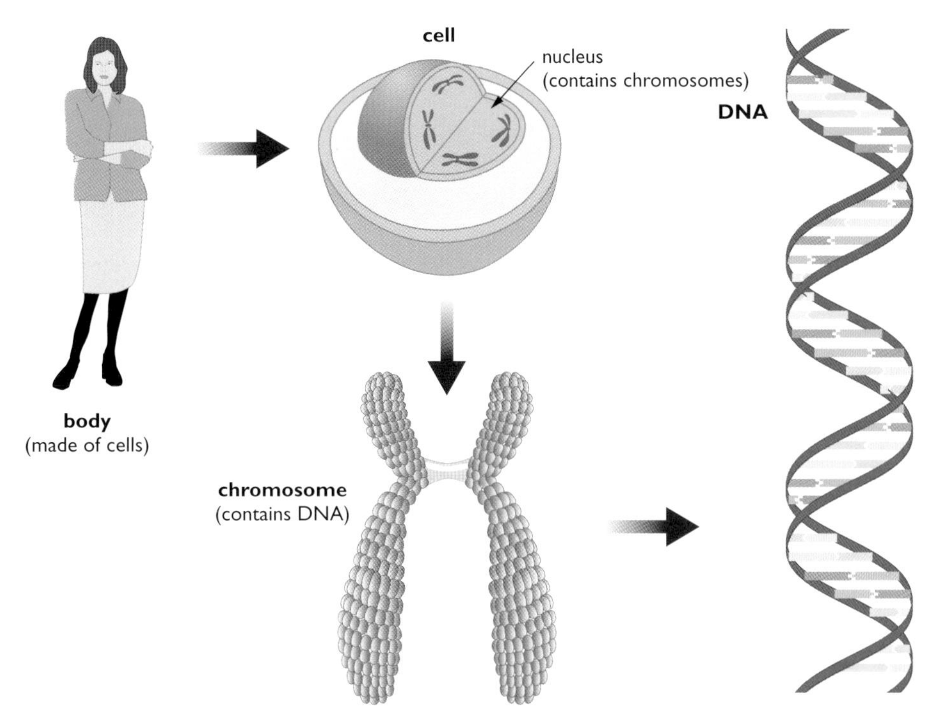

Figure 16.1 *Our genetic make-up.*

Figure 16.2 *(a) Watson and Crick with their double-helix model.*

DNA is short for deoxyribonucleic acid. It gets its 'deoxyribo' name from the sugar in the DNA molecule. This is deoxyribose – a sugar containing five carbon atoms.

The chemical that is the basis of inheritance in nearly all organisms is **DNA**. DNA is usually found in the nucleus of a cell, in the **chromosomes**. A small section of DNA that determines a particular feature is called a **gene**. Genes determine features by instructing cells to produce particular proteins which then lead to the development of the feature. So a gene can also be described as a section of DNA that codes for a particular protein.

DNA can replicate (make an exact copy of) itself. When a cell divides by mitosis (see Chapter 17), each new cell receives exactly the same type and amount of DNA. The cells formed are genetically identical.

ideas evidence

Figure 16.2 *(b) Rosalind Franklin (1920–1958).*

The structure of DNA

Who discovered it?

James Watson and Francis Crick, working at Cambridge University, discovered the structure of the DNA molecule in 1953. Both were awarded the Nobel prize in 1962 for their achievement. However, the story of the first discovery of the structure of DNA goes back much further. Watson and Crick were only able to propose the structure of DNA because of the work of others – Rosalind Franklin had been researching the structure of a number of substances using a technique called X-ray diffraction.

Watson and Crick were able to use her results, together with other material, to propose the now familiar double helix structure for DNA. Rosalind Franklin died of cancer and so was unable to share in the award of the Nobel Prize (it cannot be awarded posthumously).

A molecule of DNA is made from two strands of **nucleotides**, making it a **polynucleotide**. Each nucleotide contains a nitrogenous base (adenine (a), thymine (t), cytosine (c) or guanine (g)), a sugar molecule and a phosphate group (Figure 16.3).

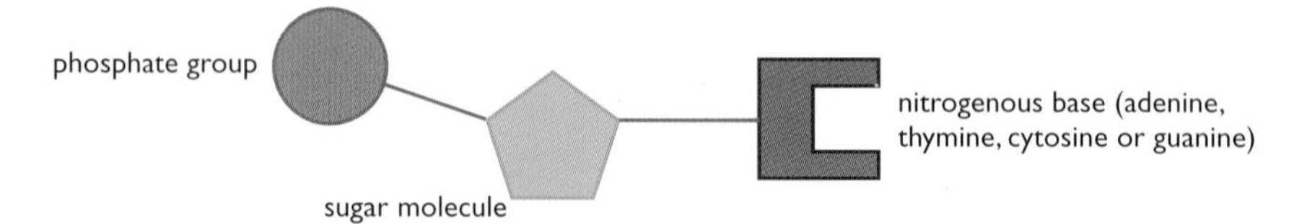

Figure 16.3 *The structure of a single nucleotide.*

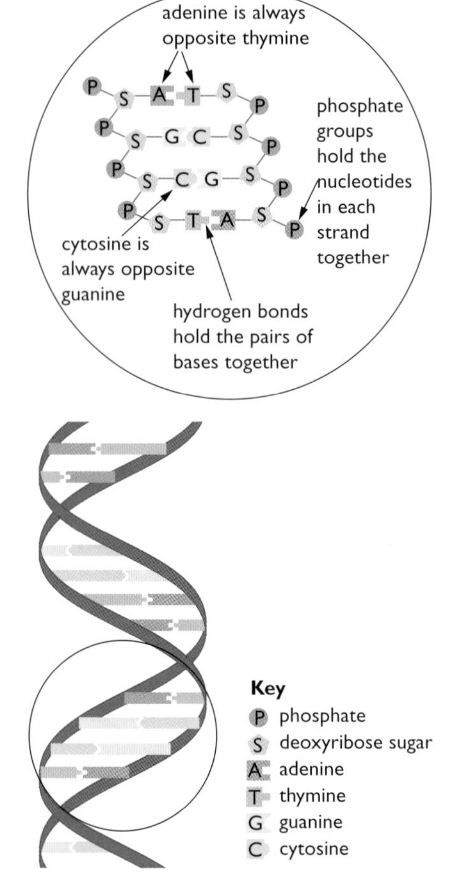

Figure 16.4 *Part of a molecule of DNA.*

Notice that, in the two strands, nucleotides with adenine are always opposite nucleotides with thymine, and cytosine is always opposite guanine. Adenine and thymine are **complementary bases**, as are cytosine and guanine. Complementary bases always bind with each other and never with any other base. This is known as the **base-pairing rule**.

DNA is the only chemical that can replicate itself exactly. Because of this, it is able to pass genetic information from one generation to the next as a 'genetic code'.

One consequence of the base-paring rule is that, in each molecule of DNA, the amounts of adenine and thymine are equal, as are the amounts of cytosine and guanine.

The DNA code

Only one of the strands of a DNA molecule actually codes for the manufacture of proteins in a cell. This strand is called the **sense strand**. The other strand is called the **anti-sense** strand. The proteins manufactured can be **intracellular enzymes** (enzymes that control processes within the cell), **extracellular enzymes** (enzymes that are secreted from the cell to have their effect outside the cell), **structural proteins** (e.g. used to make hair, haemoglobin, muscles, cell membranes) or **hormones**.

Proteins are made of chains of amino acids. A sequence of *three* nucleotides in the sense strand of DNA codes for one amino acid. As the sugar and phosphate are the same in all nucleotides, it is actually the bases that code for the amino acid. For example, the base sequence TGT codes for the amino acid cysteine. Because three bases are needed to code for one amino acid, the DNA code is a **triplet code**. The sequence of bases that codes for *all* the amino acids in a protein is a gene (Figure 16.5).

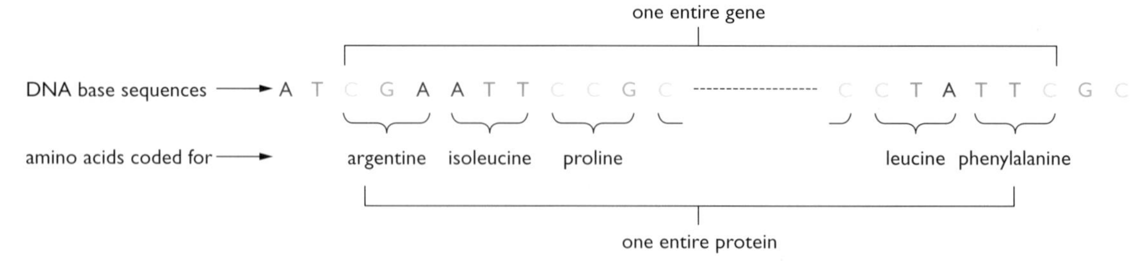

Figure 16.5 *The triplet code.*

The triplets of bases that code for individual amino acids are the same in all organisms. The base sequence TGT codes for the amino acid cysteine in humans, bacteria, bananas, monkfish, or in any other organism you can think of – the DNA code is a **universal code**.

How DNA controls the manufacture of proteins

DNA is found in chromosomes in the nucleus of a cell. Proteins are manufactured by organelles called **ribosomes** outside the nucleus, in the cytoplasm of the cell. To manufacture a protein in a cell, two important events must occur:

- the code must be transferred from DNA to a ribosome
- amino acids must be brought to the ribosome to be assembled into the protein.

Both these functions are carried out by a second nucleic acid called ribonucleic acid or **RNA**.

The RNA that carries the code from DNA to the ribosomes is called **messenger RNA** or just **mRNA**. It is a much smaller molecule than DNA and can pass out of the nucleus through pores in the nuclear membrane. The RNA that brings the amino acids to the ribosome is called **transfer RNA** or **tRNA**. Table 16.1 shows the stages in protein synthesis in a cell.

A triplet code involving four bases has 64 possible combinations and could therefore code for 64 different amino acids. However, organisms only use about 20 amino acids to manufacture the different proteins. This means there is a lot of spare capacity in the system, and some amino acids have more than one code. Because of this spare capacity, we say that the DNA code is **degenerate.**

Code	Amino acid
UGT	cysteine
UGC	cysteine
UGG	trypsin

Table 16.1 *Some codons are very specific but sometimes an amino acid is coded for by more than one codon.*

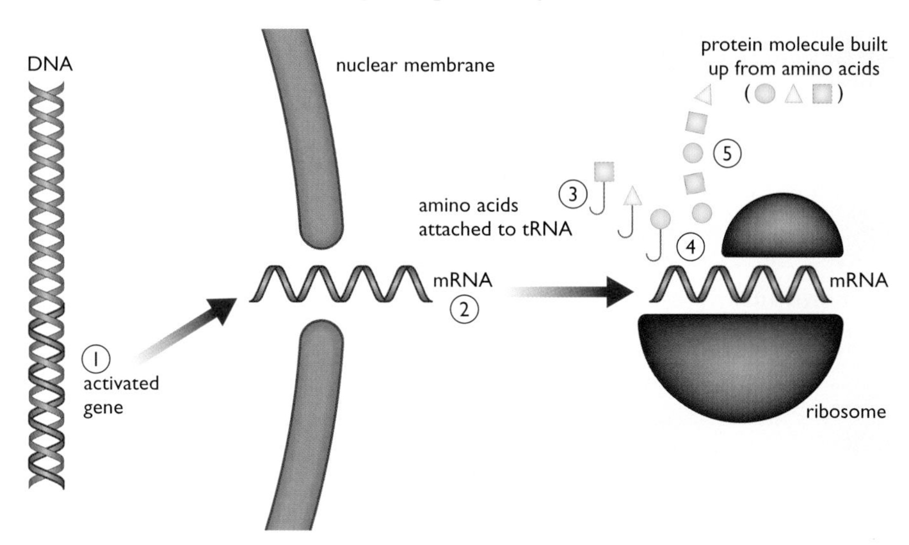

(1) The gene that codes for the protein is 'switched on' or 'activated'. The triplets of bases in the activated gene are transcribed (ie 're-written') as triplets in an mRNA molecule. These triplets are called codons.

(2) The mRNA travels from the nucleus to the ribosomes, carrying the transcribed code.

(3) Each tRNA molecule carries a specific amino acid to the ribosome.
It can recognise the codon on mRNA that codes for its own particular amino acid.

(4) The amino acids are assembled into the protein molecule actually inside the ribosome. A tRNA molecule with its amino acid can only enter a ribosome when it can recognise the codon on the mRNA that is inside the ribosome.

(5) Once the tRNA molecule (with its amino acid) has entered the ribosome, the amino acid is joined onto the growing protein molecule.

The mRNA moves and a different codon is brought into place in the ribosome.
The process continues until all the codons have been 'read' and all the amino acids joined together.

Figure 16.6 *Protein synthesis.*

Not only is the DNA code universal, but the actual DNA in different organisms is very similar. 98% of our DNA is the same as that of a chimpanzee; 50% of it is the same as the DNA of a banana!

Once the amino acids have all been joined in the correct sequence, the protein molecule folds itself into its final shape. The shape of protein molecules is important to their function. For example, an enzyme's active site must fit around its substrate. If it didn't, the enzyme could not bind with the substrate and catalyse the reaction (see chapter 1).

Putting the amino acids together in the correct sequence is crucial. The protein would be altered if just one amino acid was missing or out of place. It would fold itself into a different shape.

DNA replication

When a cell is about to divide (see Mitosis, Chapter 17) it must first make an exact copy of each DNA molecule in the nucleus. This process is called **replication**. As a result, each cell formed receives exactly the same amount and type of DNA. Figure 16.7 summarises this process.

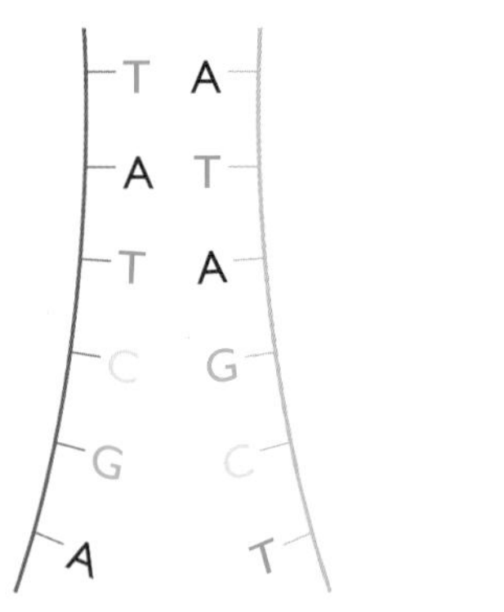

The polynucleotide strands of DNA separate.

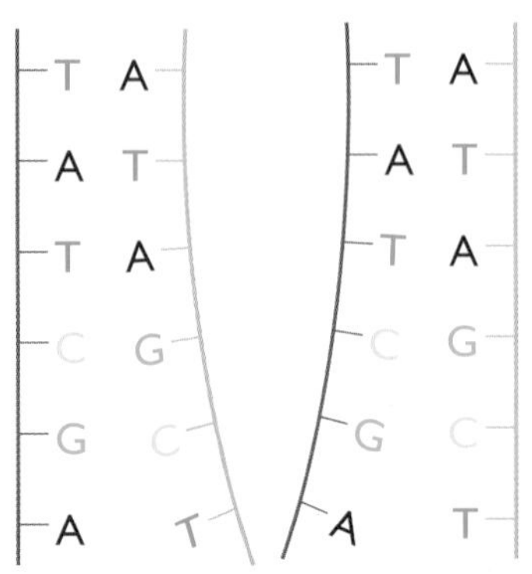

Each strand acts as a template for the formation of a new strand of DNA.

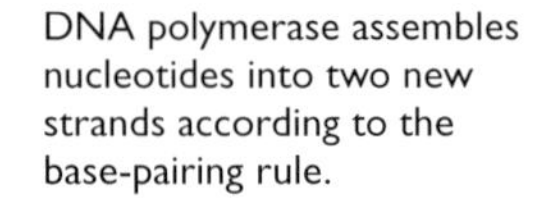

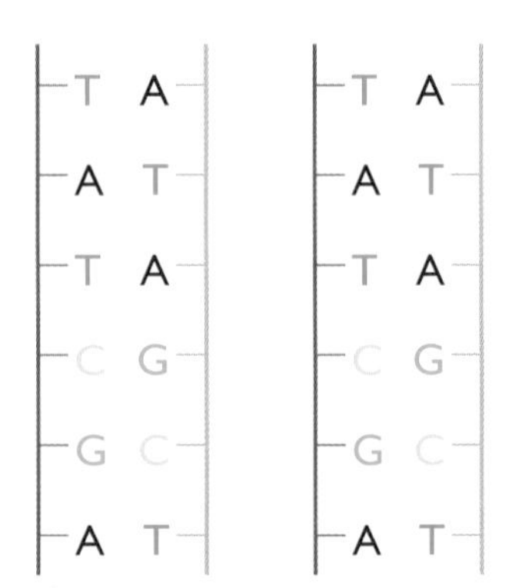

Two identical DNA molecules are formed – each contains a strand from the parent DNA and a new complementary strand.

Figure 16.7 *How DNA replicates itself.*

extra T becomes first base of next triplet

replaced by first base of next triplet

Figure 16.8 Gene mutations (a) duplication, (b) deletion, (c) substitution, (d) inversion.

Gene mutations – when DNA makes mistakes

A **mutation** is a change in the DNA of a cell. It can happen in individual genes or in whole chromosomes. Sometimes, when DNA is replicating, mistakes are made and the wrong nucleotide is used. The result is a **gene mutation** and it can alter the sequence of the bases in a gene. In turn, this can lead to the gene coding for the wrong protein. There are several ways in which gene mutations can occur (Figure 16.8).

In **duplication**, Figure 16.8 (a), the nucleotide is inserted twice instead of once. Notice that the entire base sequence is altered – each triplet after the point where the mutation occurs is changed. The whole gene is different and will now code for an entirely different protein.

In **deletion**, Figure 16.8 (b), a nucleotide is missed out. Again, the entire base sequence is altered. Each triplet after the mutation is changed and the whole gene is different. Again, it will code for an entirely different protein.

In **substitution**, Figure 16.8 (c), a different nucleotide is used. The triplet of bases in which the mutation occurs is changed and it *may* code for a different amino acid. If it does, the structure of the protein molecule will be different. This may be enough to produce a significant alteration in the *functioning* of a protein or a total lack of function. However, the new triplet may not code for a different amino acid as most amino acids have more than one code (see Table 16.1, page 199).

In **inversions**, Figure 16.8 (d), the sequence of the bases in a triplet is reversed. The effects are similar to substitution. Only one triplet is affected and this may or may not result in a different amino acid and altered protein stucture.

ideas
evidence

Mutations that occur in body cells such as those in the heart, intestines or skin, will only affect that particular cell. If they are very harmful, the cell will die and the mutation will be lost. If they do not affect the functioning of the cell in a major way, the cell may not die. If the cell then divides, a group of cells containing the mutant gene is formed. When the organism dies, however, the mutation is lost with it; it is not passed to the offspring. Only mutations in the sex cells or in the cells that divide to form sex cells can be passed on to the next generation. This is how genetic diseases begin. One example of such a gene mutation is that which causes cystic fibrosis (see Chapter 18).

Sometimes a gene mutation can be advantageous to an individual. For example, as a result of random mutations, bacteria can become resistant to antibiotics. Resistant bacteria obviously have an advantage over non-resistant types if an antibiotic is being used. They will survive the antibiotic treatment and reproduce. All their offspring will be resistant and so the proportion of resistant types in the population of bacteria will increase as this happens in each generation. This is an example of **natural selection** (see Chapter 19). Pests can become resistant to pesticides in a similar way.

Gene mutations are random events that occur in all organisms. The rate at which they occur can be increased by a number of agents called **mutagens**. Mutagens include:

- ionising radiation (such as ultraviolet light, X-rays and gamma rays)
- chemicals including mustard gas and nitrous oxide, many of the chemicals in cigarette smoke and the tar from cigarettes, and some of the chemicals formed when food is charred in cooking.

Gene mutations and cancer

Many gene mutations are harmful and cause the death of the cell in which they occur. Others affect the cell differently; they cause the cell to divide in an uncontrolled way and a **tumour** develops. Tumours are not necessarily harmful – some **benign tumours** merely form unwanted 'lumps'. Some, however, go on to become harmful, even fatal cancers (Figure 16.9).

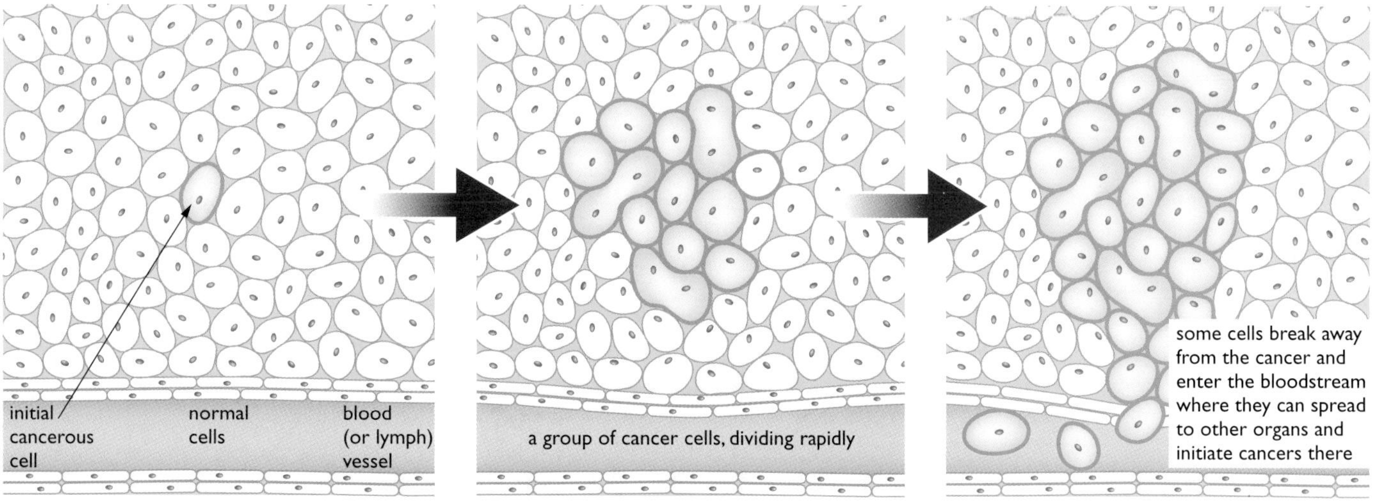

Figure 16.9 *How a cancer forms and can spread.*

If the cancer begins by a mutation in just one cell, it can take from 2.5 to 3 years to form a cancer that can be felt in a routine examination. By this time, the cancer could have a volume of about 1 cm^3 and contain about 5 billion cells. These are *average* values – not all cancers are the same.

The structure of chromosomes

Each chromosome contains one double-stranded DNA molecule. The DNA is folded and coiled so that it can be packed into a small space. Proteins called **histones** surround the DNA (Figure 16.10).

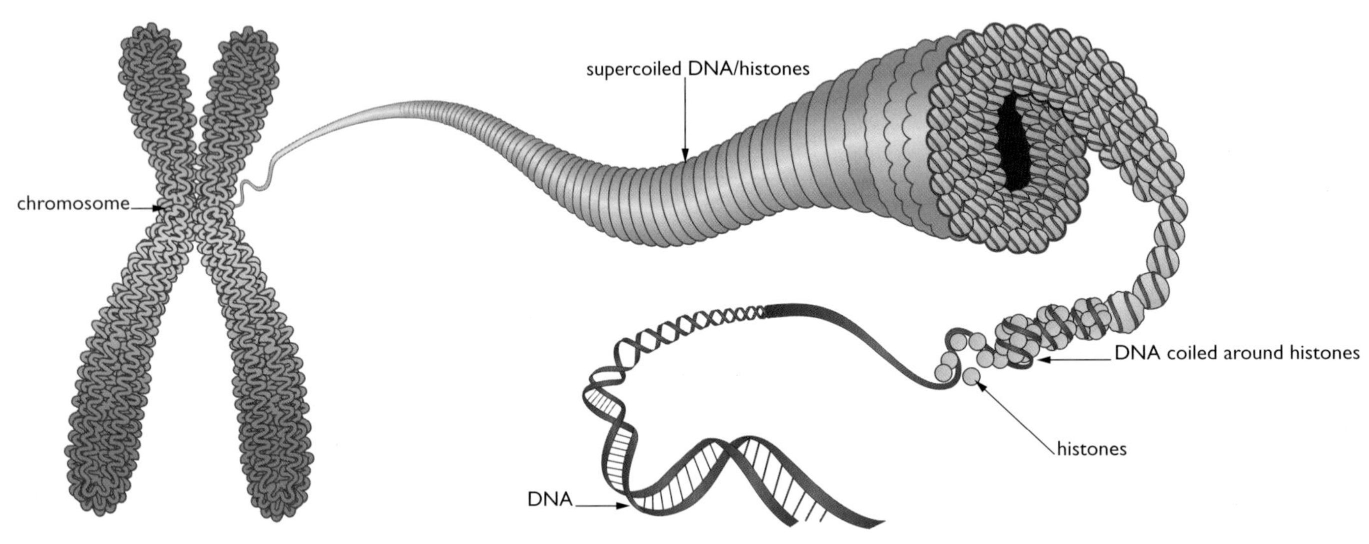

Figure 16.10 *The structure of a chromosome.*

Because a chromosome contains a particular DNA molecule, it will also contain the genes that make up that DNA molecule. Another chromosome will contain a different DNA molecule, and so will contain different genes.

How many chromosomes?

Red blood cells have no nucleus, therefore no chromosomes. This gives them more room for carrying oxygen.

Nearly all human cells contain 46 chromosomes. The photographs in Figure 16.11 show the 46 chromosomes from the body cells of a human male and female.

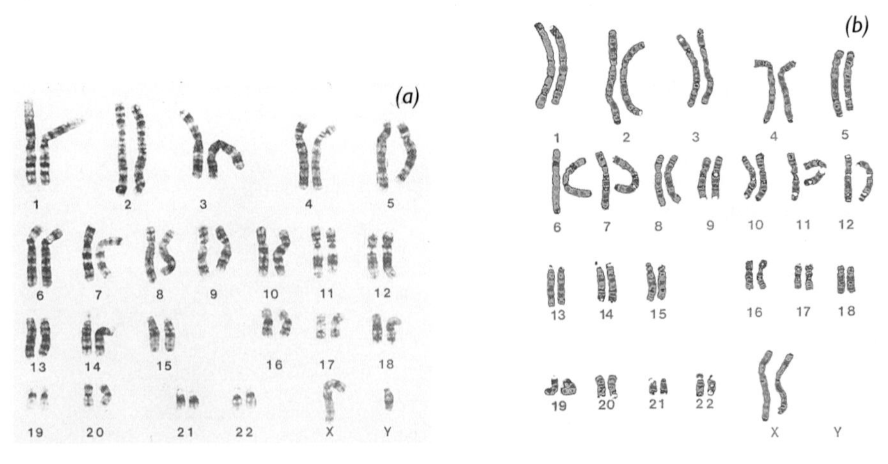

Figure 16.11 *Chromosomes of a human male (a) and female (b). A picture of all the chromosomes in a cell is called a karyotype.*

The chromosomes are not arranged like this in the cell. The original photograph has been cut up and chromosomes of the same size and shape 'paired up'. The cell from the male has 22 pairs of chromosomes and two that do not form a pair – the X and Y chromosomes. A body cell from a female has 23 matching pairs including a pair of X chromosomes.

The X and the Y chromosomes are the **sex chromosomes**. They determine whether a person is male or female.

Pairs of matching chromosomes are called **homologous pairs**. They carry genes for the same features in the same sequence (Figure 16.12). Cells with chromosomes in pairs like this are **diploid** cells.

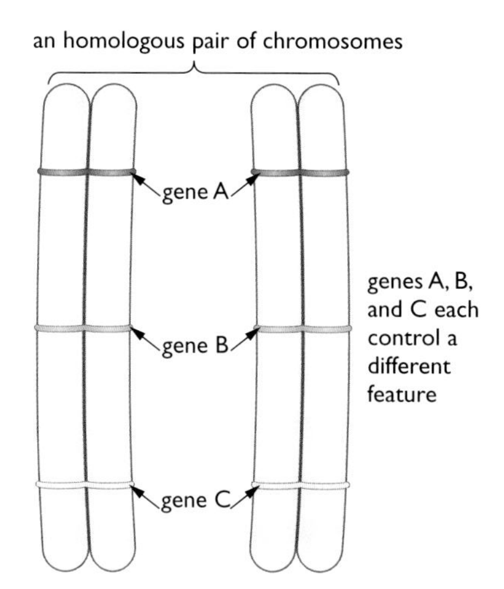

Figure 16.12 *Both chromosomes in an homologous pair have the same sequence of genes.*

Not all human cells have 46 chromosomes. Red blood cells have no nucleus and so have none. Sex cells have only 23 – just half the number of other cells. They are formed by a cell division called **meiosis** (see Chapter 17). Each cell formed has one chromosome from each homologous pair, and one of the sex chromosomes. Cells with only half the normal diploid number of chromosomes, and therefore only half the DNA content of other cells, are **haploid** cells.

When two sex cells fuse in **fertilisation**, the two nuclei join to form a single diploid cell (a **zygote**). This cell has, once again, all its chromosomes in homologous pairs and two copies of every gene. It has the normal DNA content.

Genes and alleles

Genes are sections of DNA that control the production of proteins in a cell. Each protein contributes towards a particular body feature. Sometimes the feature is visible, such as eye colour or skin pigmentation. Sometimes the feature is not visible, such as the type of haemoglobin in red blood cells or the type of blood group antigen on the red blood cells.

Some genes have more than one form. For example, the genes controlling several facial features have alternate forms, which result in alternate forms of the feature (Figure 16.13).

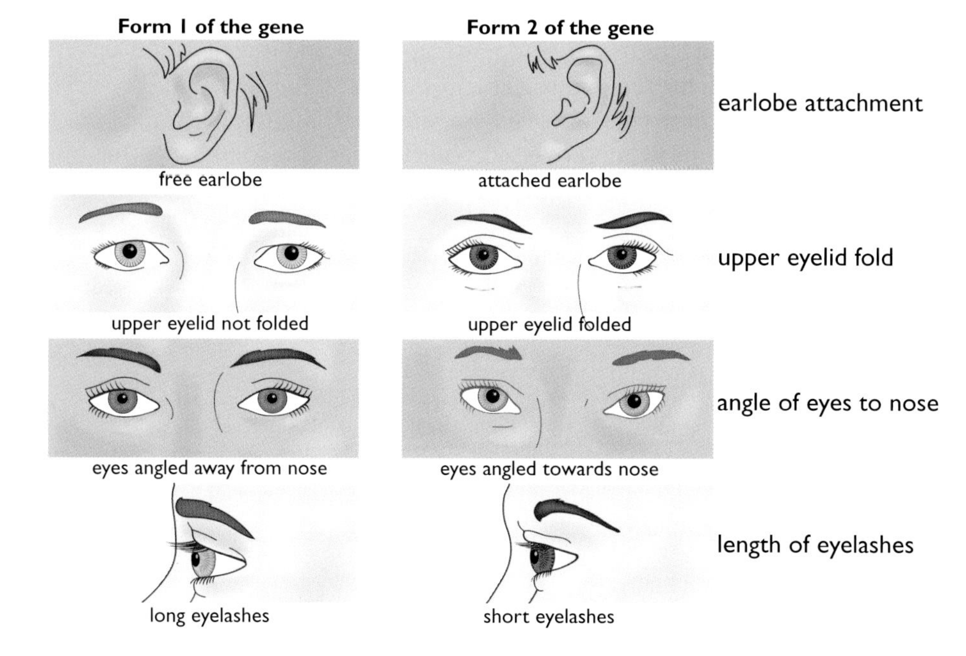

Figure 16.13 *The alternate forms of four facial features.*

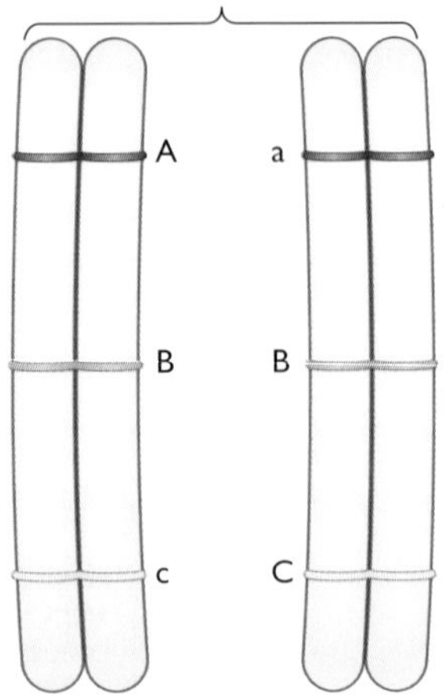

an homologous pair of chromosomes

A a

B B

c C

Figure 16.14 ***A*** *and* ***a***, ***B*** *and* ***b***, ***C*** *and* ***c*** *are different alleles of the same gene. They control the same feature but code for different expressions of that feature.*

The gene for earlobe attachment has the forms 'attached earlobe' and 'free earlobe'. These different forms of the gene are called **alleles**. Homologous chromosomes carry genes for the same features in the same sequence, but the alleles of the genes may not be the same (Figure 16.14). The DNA in the two chromosomes is not quite identical.

Each cell with two copies of a chromosome also has two copies of the genes on those chromosomes. Suppose that, for the gene controlling earlobe attachment, a person has one allele for attached earlobes and one for free earlobes. What happens? Is one ear free and the other attached? Are they both partly attached? Neither. In this case, both earlobes are free. The 'free' allele is **dominant** and 'switches off' the 'attached' allele, which is **recessive**. See Chapter 18 for more detail on how genes are inherited.

Chromosome mutations

When cells divide, they do not always divide properly. Bits of chromosomes can sometimes break off one chromosome and become attached to another. Sometimes one daughter cell ends up with both chromosomes of an homologous pair whilst the other has none. These 'mistakes' are called **chromosome mutations** and usually result in the death of the cells formed.

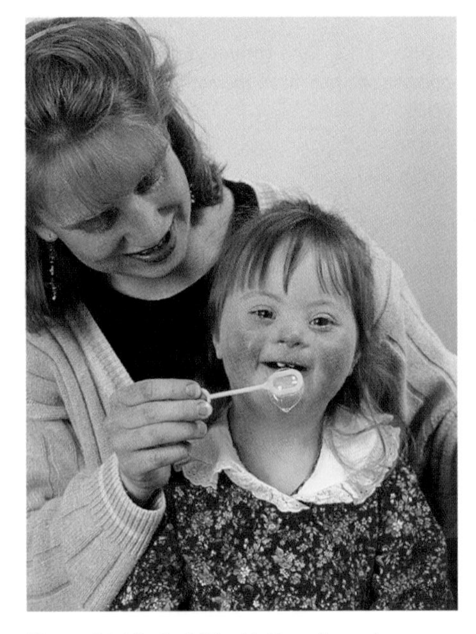

Figure 16.15 *A child with Down's syndrome.*

Sometimes sex cells do not form properly and they contain more (or less) chromosomes than normal. One relatively common chromosome mutation results in ova (female sex cells) containing two copies of chromosome 21. When an ovum like this is fertilised by a normal sperm, the zygote will have three copies of chromosome 21. This is called trisomy (three copies) of chromosome 21. Unlike some other chromosome mutations, the effects of this mutation are usually non-fatal and the condition that results is **Down's syndrome** (Figure 16.15).

Down's syndrome children sometimes die in infancy, as heart and lung defects are relatively common. Those that survive have a near normal life span. Individuals with Down's syndrome can now live much more normal lives than was thought possible just 20 years ago. They require much care and attention during childhood, and particularly in adolescence, but, given this care, they can achieve good social and intellectual growth. Most importantly they achieve personal self-sufficiency. Trisomy of chromosome 21 is more common in women over 40 years of age. As a result, they have more babies with Down's syndrome than younger women.

End of Chapter Checklist

If you haven't got a copy of your specification, read the introduction on page vi.

You will need to be able to do some or all of the following. Check your Awarding Body's specification (syllabus) to find out exactly what you need to know.

- Identify the components of a DNA molecule.
- Describe the role of James Watson, Frances Crick and Rosalind Franklin in discovering the structure of DNA.
- Explain how bases in the DNA molecule code for amino acids.
- Describe how DNA controls protein synthesis in a cell.
- Describe how DNA replicates.
- Describe the nature of gene mutation and explain how different gene mutations result in the formation of abnormal proteins.
- List some factors that can act as mutagens.
- Distinguish between the terms *gene* and *allele*.
- Describe the structure of a chromosome.
- Explain what is meant by the term homologous chromosomes.
- Recall the numbers of chromosomes in: human sex cells, human red blood cells, all other human cells.
- Explain the nature of chromosome mutation and the genetic basis of Down's syndrome.

Questions

More questions on DNA can be found at the end of Section E on page 257.

1 The diagram represents part of a molecule of DNA.

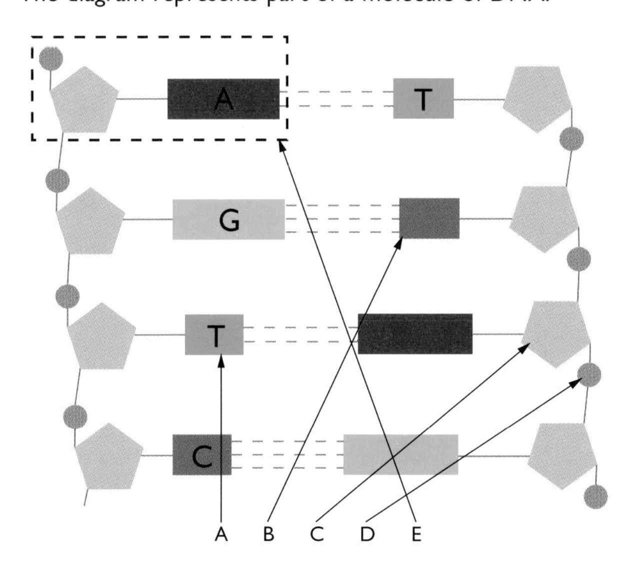

a) Name the parts labelled A, B, C, D and E.

b) What parts did James Watson, Frances Crick and Rosalind Franklin play in discovering the structure of DNA?

c) Use the diagram to explain the base pairing rule.

2 ***a)*** What is:

i) a gene

ii) an allele?

b) Describe the structure of a chromosome.

c) How are the chromosomes in a woman's skin cells:

i) similar to

ii) different from those in a man's skin cells?

3 DNA codes for the production of proteins. The flowchart shows some of the stages in this process.

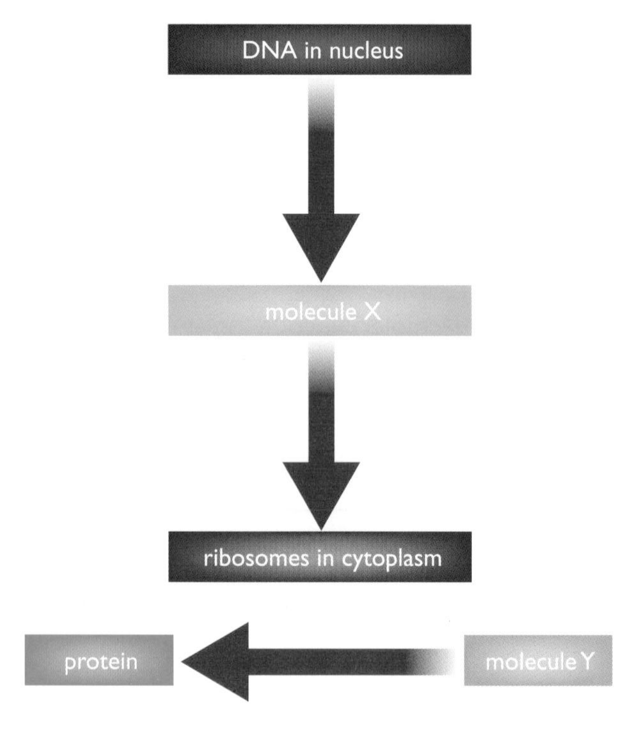

a) Name molecules X and Y.

b) Why is molecule X able to pass out of the nucleus when a molecule of DNA cannot?

c) The DNA code is a universal code. What does this mean?

4 Gene mutation can lead to the formation of tumours.

a) What is a gene mutation?

b) How is a tumour formed?

c) Name three different mutagens (agents that increase the rate of mutation).

5 DNA is the only molecule capable of replicating itself. Sometimes mutations occur during replication.

a) Describe how DNA replicates itself.

b) Explain how a single gene mutation can lead to the formation of a protein in which

i) many of the amino acids are different from those coded for by the non-mutated gene

ii) only one amino acid is different from those coded for by the non-mutated gene.

6 The graph shows the numbers and relative frequency of births of Down's syndrome babies in women aged between 20 and 50.

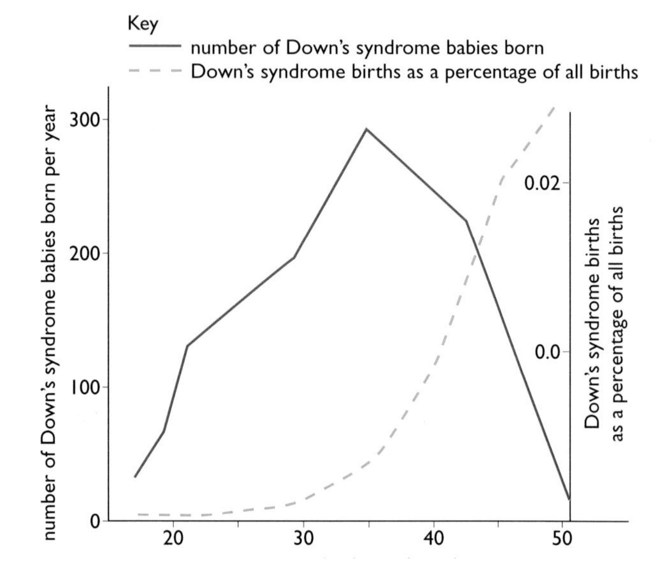

a) What is Down's syndrome?

b) How do the *numbers* of Down's syndrome births change with the age of the mother?

c) Suggest why the trend shown by the frequency of Down's syndrome births is different from that shown by the actual numbers.

Chapter 17: Cell Division

Growth and reproduction are two characteristics of living things. Both involve cell division, which is the subject of this chapter.

In most parts of the body, cells need to divide so that organisms can grow and replace worn out or damaged cells. The cells that are produced in this type of cell division should be exactly the same as the cells they are replacing. This is the most common form of cell division.

Only in the sex organs is cell division different. Here, some cells divide to produce gametes (sex cells), which contain only half the original number of chromosomes. This is so that when male and female gametes fuse together (fertilisation) the resulting cell (zygote) will contain the full complement of chromosomes and can then divide and grow into a new individual.

Human body cells have 46 chromosomes in 23 pairs called homologous pairs. Chromosomes in an homologous pair carry genes for the same features in the same sequence. They do not necessarily have the same *alleles* of every gene (see Chapter 16). These body cells are **diploid** cells – they have *two* copies of each chromosome. The sex cells, with 23 chromosomes,(only one copy of each chromosome) are **haploid** cells.

There are two kind of cell division: **mitosis** and **meiosis**. When cells divide by mitosis, two cells are formed. These have the same number and type of chromosomes as the original cell. Mitosis forms all the cells in our bodies except the sex cells.

Meiosis is sometimes called **reduction division**. This is because it produces cells with only half the number of chromosomes of the original cell.

When cells divide by meiosis, four cells are formed. These have only half the number of chromosomes of the original cell. Meiosis forms sex cells.

Mitosis

When a **parent cell** divides it produces **daughter cells**. Mitosis produces two daughter cells that are genetically identical to the parent cell – both daughter cells have the same number and type of chromosomes as the parent cell. To achieve this, the dividing cell must do two things:

- It must copy each chromosome before it divides. This involves the DNA replicating and more proteins being added to the structure. Each daughter cell will then be able to receive a copy of each chromosome (and each molecule of DNA) when the cell divides.
- It must divide in such a way that each daughter cell receives one copy of every chromosome. If it does not do this, both daughter cells will not contain all the genes.

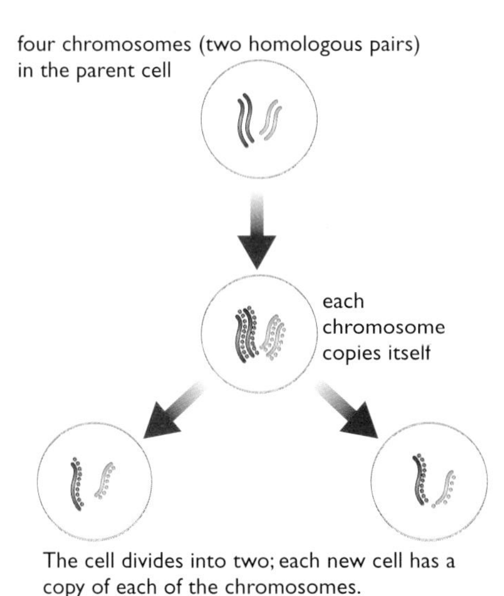

Figure 17.1 *A summary of mitosis.*

These two processes are shown in Figure 17.1.

A number of distinct stages occur when a cell divides by mitosis. These are shown in Figure 17.2. Figure 17.3 is a photograph of some cells from the root tip of an onion. Cells in this region of the root divide by mitosis to allow growth of the root.

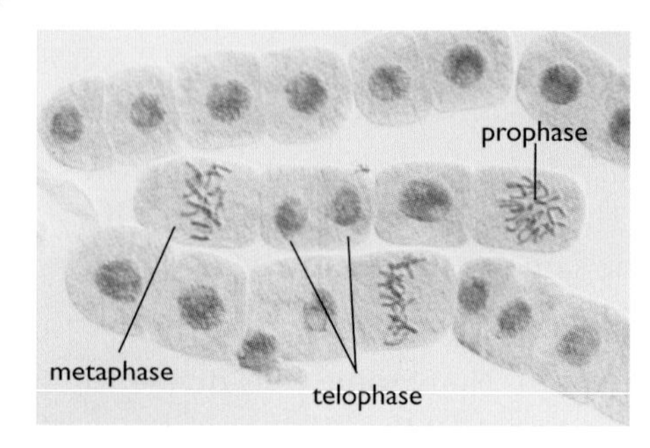

Figure 17.3 *Photomicrograph of cells in the root tip of an onion. Cells in the main stages of mitosis are labelled.*

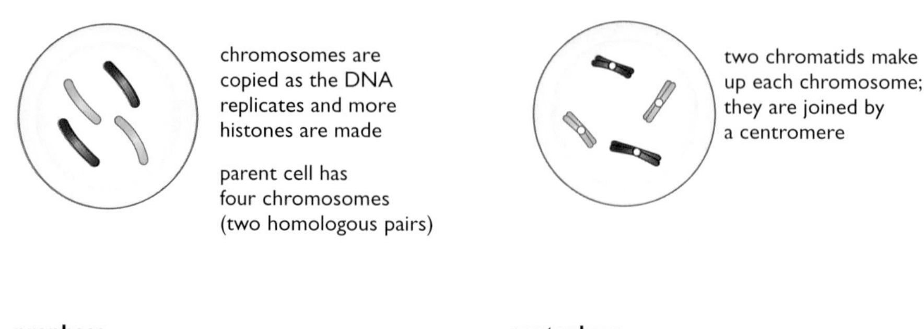

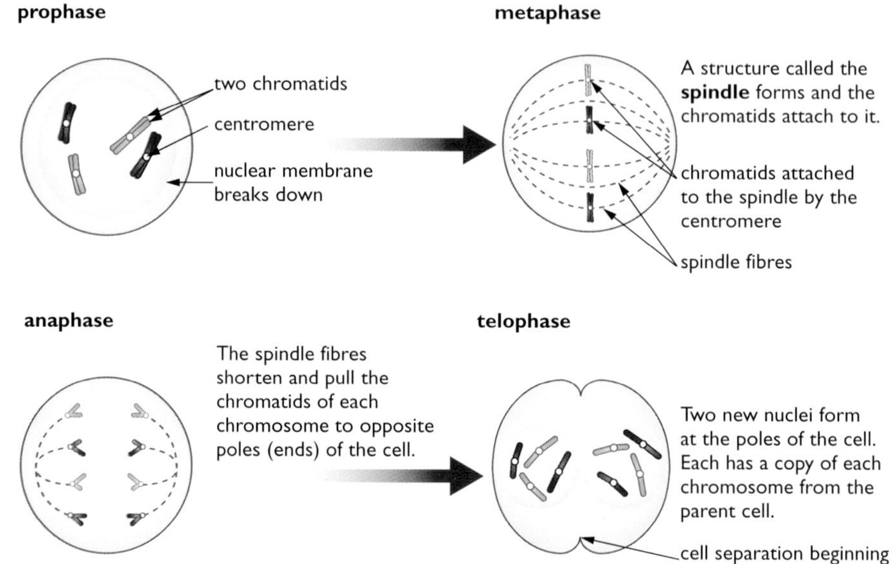

Figure 17.2 *The main stages in mitosis.*

Each daughter cell formed by mitosis receives a copy of every chromosome, and therefore every gene, in the parent cell. Each daughter cell is genetically identical to the others. All the cells in our body (except the sex cells) are formed by mitosis from the zygote (single cell formed at fertilisation). They all, therefore, contain copies of all the chromosomes and genes of that zygote. They are all genetically identical.

Whenever cells need to be replaced in our bodies, cells divide by mitosis to make them. This happens more frequently in some regions than in others.

- The skin loses thousands of cells every time we touch something. This adds up to millions every day that need replacing. A layer of cells beneath the surface is constantly dividing to produce replacements.
- Cells are scraped off the lining of the gut as food passes along. Again, a layer of cells beneath the gut lining is constantly dividing to produce replacement cells.
- Cells in our spleen destroy worn out red blood cells at the rate of 100 000 000 000 per day! These are replaced by cells in the bone marrow dividing by mitosis. In addition, the bone marrow forms all our new white blood cells and platelets (Figure 17.4).
- Cancer cells divide by mitosis. The cells formed are exact copies of the parent cell, including the mutation in the genes that makes the cells divide uncontrollably.

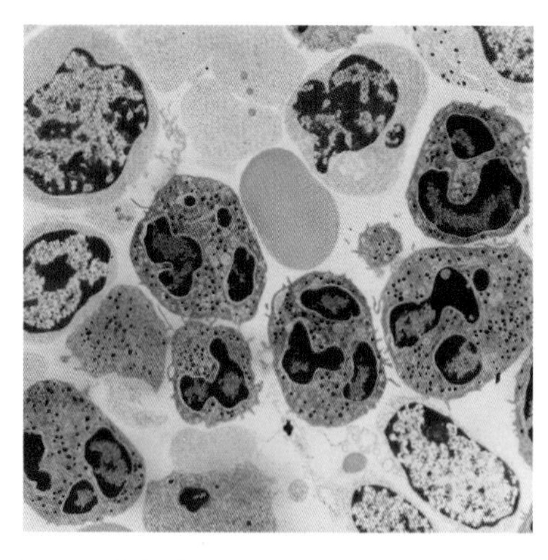

Figure 17.4 *Cells in bone marrow dividing to produce blood cells.*

Meiosis

Meiosis forms only sex cells. It is a more complex process than mitosis involving two cell divisions, but you don't need to know details of all the stages. Meiosis produces four cells that are haploid and not genetically identical. The dividing cell must do two things:

- It must copy each chromosome so that there is enough genetic material to be shared between the four daughter cells.
- It must divide twice, in such a way that each daughter cell receives just one chromosome from each homologous pair.

These processes are summarised in Figure 17.5. Figure 17.6 shows cells in an anther dividing by meiosis.

The sex cells formed by meiosis don't all have the same combinations of alleles – there is **genetic variation** in the cells. During the two cell divisions of meiosis, the chromosomes are divided between the two daughter cells independently of each other. The only 'rules' that operate are:

- During the first division of meiosis, one chromosome from each homologous pair goes into each daughter cell.
- During the second division of meiosis the chromosome separates into two parts. One part goes into each daughter cell. This allows for a lot of variation in the daughter cells (Figure 17.7).

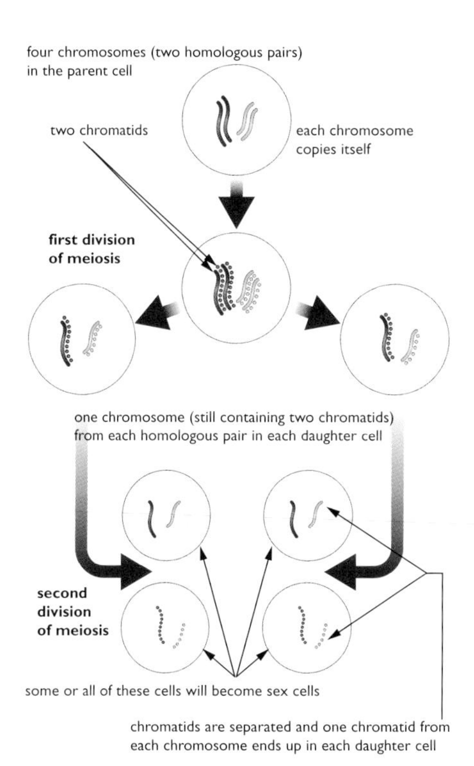

Figure 17.5 *A summary of meiosis.*

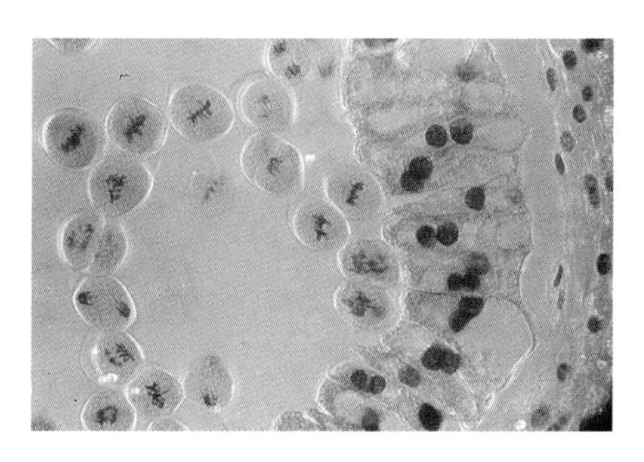

Figure 17.6 *Photomicrograph of an anther showing cells dividing by meiosis.*

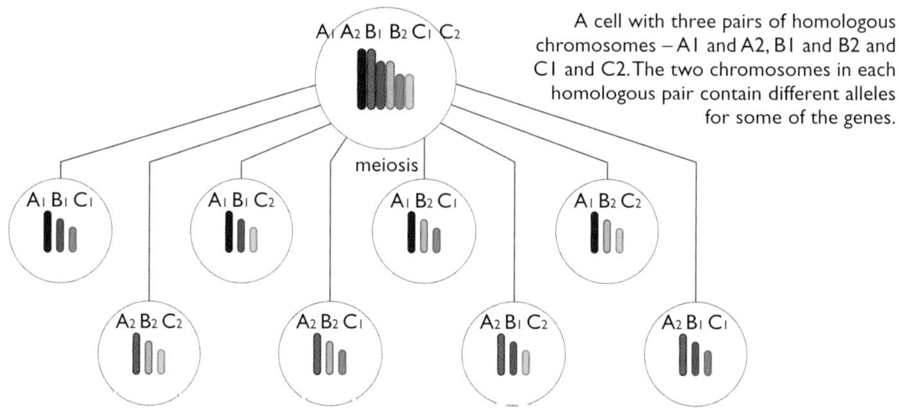

As a result of the two divisions of meiosis, each sex cell formed contains one chromosome from each homologous pair. This gives eight combinations. As A1 and A2 contain different alleles (as do B2 and B2, and C1 and C2) the eight possible sex cells will be genetically different.

Figure 17.7 *How meiosis produces variation.*

There is a mathematical rule for predicting how many combinations of chromosomes there can be. The rule is:

number of possible combinations = 2^n

where n = number of *pairs* of chromosomes.

With two pairs of chromosomes, the number of possible combinations = $2^2 = 4$. With three pairs of chromosomes, the number of possible combinations = $2^3 = 8$. With the 23 pairs of chromosomes in human cells, the number of possible combinations $=2^{23} = 8\,388\,608$!

The features of mitosis and meiosis are show in table 17.1.

Feature of the process	Mitosis	Meiosis
Do the chromosomes duplicate before division begins?	yes	yes
How many cell divisions are there?	one	two
How many cells are formed by the process?	two	four
Are the cells formed haploid or diploid?	diploid	haploid
Is there genetic variation in the cells formed?	no	yes

Table 17.1: *Comparison of meiosis and mitosis.*

Sexual reproduction and variation

Sexual reproduction in any multicellular organism involves the fusion of two sex cells to form a zygote. The offspring from sexual reproduction vary genetically for a number of reasons. One reason is because of the huge variation in the sex cells. The other main reason is because of the random way in which fertilisation takes place. In humans, any one of the billions of sperm formed by a male during his life could, potentially, fertilise any one of the thousands of ova formed by a female.

This variation applies to both male and female sex cells. So, just using our 'low' estimate of about 8.5 million different types of human sex cells means that there can be 8.5 million different types of sperm and 8.5 million different types of ova. When fertilisation takes place, any sperm could fertilise any ovum. The number of possible combinations of chromosomes (and genes) in the zygote is 8.5 million × 8.5 million = 72 trillion! And remember, this is using our 'low' number!

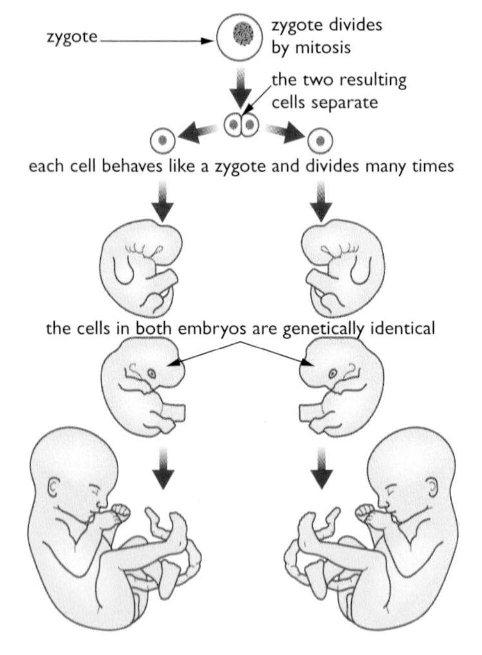

Figure 17.8 *How identical twins are formed.*

Cloning is a process that produces a group of genetically identical offspring (a **clone**) from part of the parent organism. No sex cells are involved.

This means that every individual is likely to be genetically unique. The only exceptions are **identical twins** (and identical triplets and quadruplets). Identical twins are formed from the *same* zygote – they are sometimes called **monozygotic twins.** When the zygote divides by mitosis, the two *genetically identical* cells formed do not 'stay together'. Instead, they separate and each cell behaves as though it were an individual zygote, dividing and developing into an embryo (Figure 17.8). Because they have developed from genetically identical cells (and, originally, from the same zygote), the embryos (and, later, the children and the adults they become) will be genetically identical.

Non-identical twins or **fraternal twins** develop from different zygotes and so are not genetically identical.

Seeds are part of the product of sexual reproduction in plants. Each seed contains an embryo which results from a pollen grain nucleus fusing with an egg cell nucleus. Embryos from the same plant will vary genetically because they are formed by different pollen grains fertilising different egg cells and so contain different combinations of genes.

Plant breeders have known about this variation for a long time. They realised that if a plant had some desirable feature, the best way to get more of that plant was not to collect its seeds, but to **clone** it in some way. Modern plant-breeding techniques allow the production of many thousands of identical plants from just a few cells of the original (see Chapter 20).

Asexual reproduction and cloning

When organisms reproduce asexually, there is no fusion of sex cells. A part of the organism grows and somehow breaks away from the parent organism. The cells it contains were formed by mitosis, so contain exactly the same genes as the parent. Asexual reproduction produces offspring that are genetically identical to the parent, and genetically identical to each other.

Asexual reproduction is common in plants. For example, flower bulbs grow and divide asexually each season to produce more bulbs. Asexual reproduction also occurs in some animals. Figure 17.9 shows *Hydra* (a small freshwater animal) reproducing asexually by 'budding'.

Figure 17.9 Hydra *reproducing asexually by budding.*

Genes and environment both produce variation

Pea plants are either tall or short because of the genes they inherit. There are no 'intermediate height' pea plants. However, all the tall pea plants are not *exactly* the same height and neither are all the short pea plants *exactly* the same height. Figure 17.10 illustrates the different types of variation in pea plants.

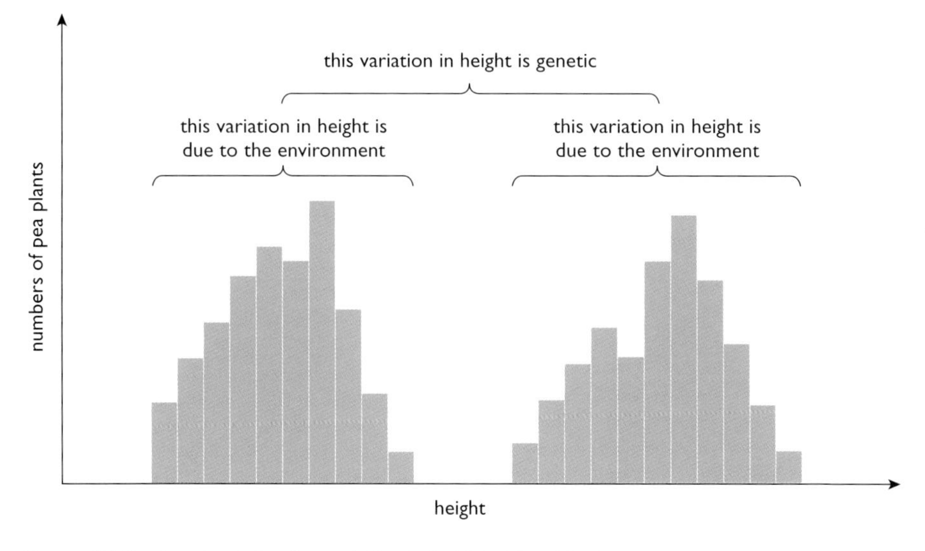

Figure 17.10 *Bar chart showing variation in height of pea plants.*

Several **environmental factors** can influence their height:

- They may not all receive the same amount of light and so some will not photosynthesise as well as others.
- They may not all receive the same amount of water and mineral ions from the soil: this could affect the manufacture of a range of substances in the plant.
- They may not all receive the same amount of carbon dioxide: again, some will not photosynthesise as well as others.

Similar principles apply in humans. Identical twins have the same genes, and often grow up to look very alike (although not quite identical). Also, they often develop similar talents. However, identical twins never look *exactly* the same. This is especially true if, for some reason, they grow up apart. The different environments affect their physical, social and intellectual development in different ways.

ideas evidence

Francis Galton was Charles Darwin's cousin (Figure 17.11). He was interested in many areas of science, including the development of human intelligence. He observed that outstanding talent (intelligence, sport, music) seemed to 'run in families'. He suggested that they were probably inherited from parents and devised a number of methods of studying intelligence in families. He and his colleagues developed a mathematical formula called a **correlation coefficient**. This could be used to see if two factors are always related.

Figure 17.11 *Francis Galton (1822–1911).*

The value of a correlation coefficient can be between +1 and –1. A coefficient of +1 means perfect positive correlation. As the value of one factor *increases*, the value of the other factor also *increases*. A coefficient of –1 means perfect negative correlation. As the value of one factor increases, the value of the other factor *decreases*. A value of 0 means no correlation – there is no relation at all between the values of the two factors. These relationships can be shown in graphs (Figure 17.12).

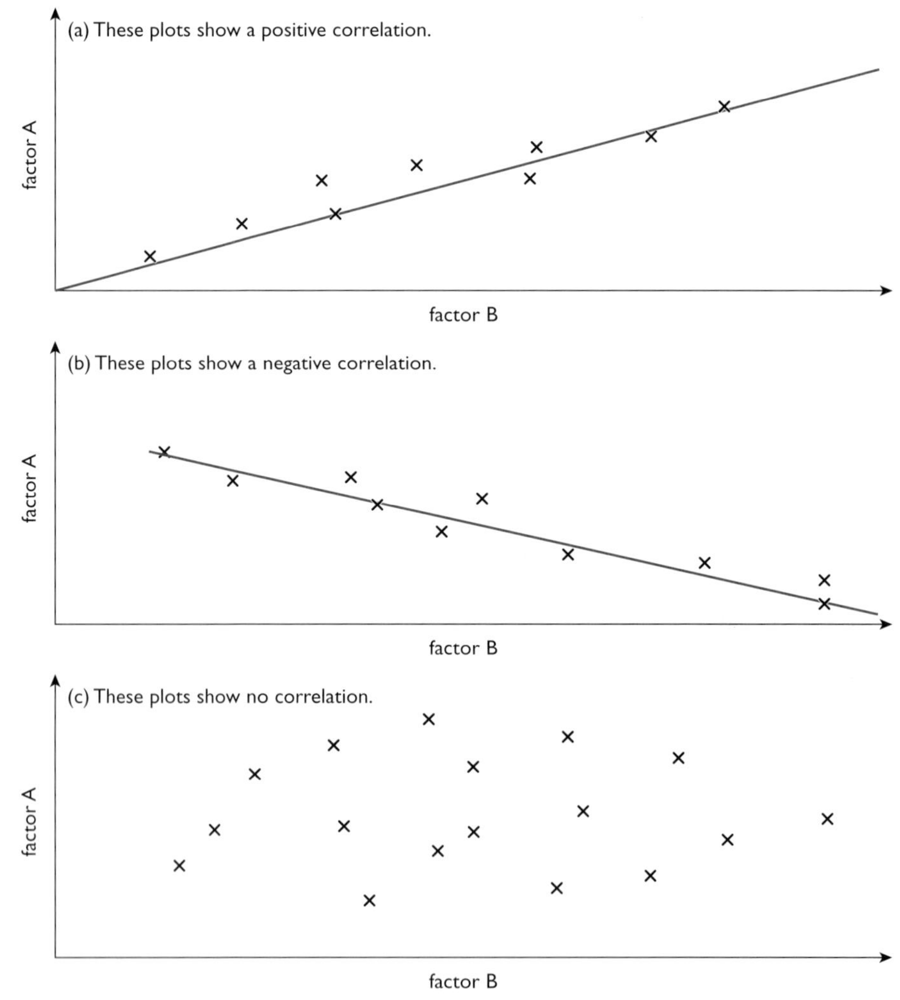

Figure 17.12 *Graphs showing (a) positive correlation, (b) negative correlation and (c) no correlation.*

Experimental psychologists have carried out a lot of research into the effects of inheritance and environment on the development of human intelligence. Intelligence can be measured by intelligence tests and the score expressed as an **intelligence quotient** or **IQ**. The correlations between IQs of different types of family members are shown in Table 17.2.

	Family members	Correlation
1	identical twins reared together	+0.87
2	identical twins reared apart	+0.75
3	fraternal twins reared together	+0.60
4	siblings reared together	+0.47
5	child and biological mother when this parent raises the child	+0.42
6	child and biological mother when this parent does *not* raise the child	+0.31
7	child and unrelated adoptive mother	+0.17

Table 17.2: *Correlation of IQ between family members.*

Siblings are brothers and/or sisters who are not twins.

There is a lot of evidence here to suggest that both heredity *and* environment are involved. For example, the correlation between the IQs of genetically identical twins is the highest. This is true whether they are reared together or apart, so the inherited genes must play an important part. But the correlation is not +1.0 (perfect correlation) so intelligence cannot solely be determined by the genes. Also, the correlation is reduced if they are reared apart. Environment must have some influence.

The relation between biological mother and child when reared together and apart also suggests that both inheritance and environment influence intelligence. When the mother rears the child, the correlation is +0.42 (a quite high positive correlation). When someone else rears the child, the different environment reduces the correlation. However, there is still quite a high correlation between mother and child, which is likely to be due to genes in common.

Biologists and psychologists are both divided as to the parts that inheritance and environment play in the development of intelligence. This debate, called the **nature/nurture** debate (nature meaning inheritance and nurture the environmental effect) seems set to run for some considerable time.

End of Chapter Checklist

If you haven't got a copy of your specification, read the introduction on page vi.

You will need to be able to do some or all of the following. Check your Awarding Body's specification (syllabus) to find out exactly what you need to know.

- Explain the roles of mitosis, meiosis and fertilisation in life cycles.
- Explain the role of mitosis in replacing body cells and the role of meiosis in producing sex cells.
- Describe the process of mitosis and understand why the cells produced by mitosis are genetically identical.
- Describe the process of meiosis and understand why the cells produced by meiosis show genetic variation.
- Explain the reasons for genetic variation in organisms produced sexually.
- Explain why asexual reproduction produces clones.
- Describe and give examples of the interaction between genes and the environment in inheritance.
- Describe and understand the nature/nurture debate about the development of human intelligence.

Questions

More questions on cell division can be found at the end of Section E on page 257.

1 Cells can divide by mitosis or by meiosis.

a) Give one similarity and two differences between the two processes.

b) Do cancer cells divide by mitosis or meiosis? Explain your answer.

c) Why is meiosis sometimes called reduction division?

2 Daffodils reproduce sexually by forming seeds and asexually by forming bulbs. Explain why:

a) the bulbs formed from a single daffodil plant produce plants very similar to each other and to the parent plant

b) the seeds formed by a single daffodil plant produce plants that vary considerably.

3 The diagram shows two cuttings. They were both taken from the same clover plant and planted in identical soil. After a few days, some nitrogen-fixing bacteria were added to the pot labelled 'inoculated'.

a) Why were cuttings from the same plant used rather than seeds from the same plant?

b) What does this experiment suggest about the influence of genes and the environment on variation in the height of clover plants?

4 Some cells divide by mitosis, others divide by meiosis. For each of the following examples, say whether mitosis or meiosis is involved. In each case give a reason for your answers.

   a) Cells in the testes dividing to form sperm.

   b) Cells in the lining of the small intestine dividing to replace cells that have been lost.

   c) Cells in the bone marrow dividing to form red blood cells and white blood cells.

   d) Cells in an anther of a flower dividing to form pollen grains.

   e) A zygote dividing to form an embryo.

5 Identical twins have very similar physical features and there is a high positive correlation between their mental abilities. Fraternal twins can be very different physically, although there is still quite a strong positive correlation between their mental abilities.

   a) *i)* What are identical twins?

   ii) Why do identical twins have very similar physical features?

   b) *i)* Why can fraternal twins be very different physically?

   ii) Explain what is meant by a *strong positive correlation*.

   iii) Explain why fraternal twins who are very different physically can still show a strong positive correlation between their mental abilities.

6 Variation in organisms can be caused by the environment as well as by the genes they inherit. For each of the following examples, state whether the variation described is likely to be genetic, environmental or both. In each case give a reason for your answers.

   a) Humans have brown, blue or green eyes.

   b) Half the human population is male, half is female.

   c) Cuttings of hydrangea plants grown in soils with different pH values develop flowers with slightly different colours.

   d) Some pea plants are tall; others are dwarf. However, the tall plants are not exactly the same height and neither are all the dwarf plants the same height.

   e) People in some families are more at risk of heart disease than people in other families. However, not every member of the 'high risk' families have a heart attack and some members of the 'low risk' families do.

7 In an investigation into mitosis, the distance between a chromosome and the pole (end) of a cell was measured. The graph shows the result of the investigation.

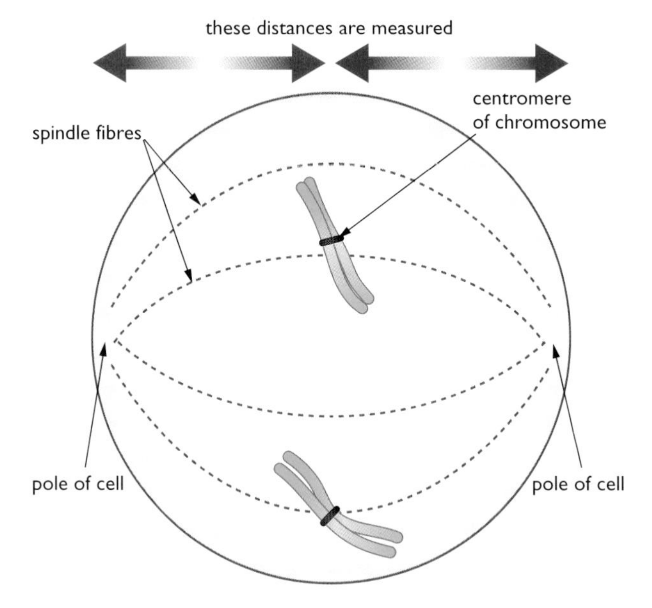

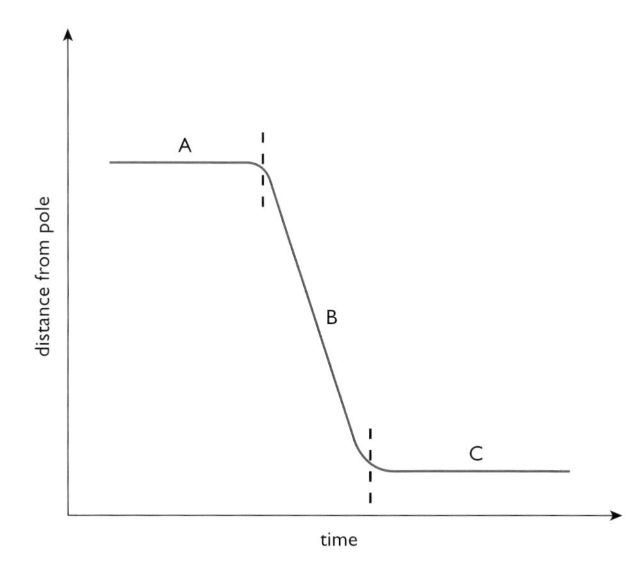

   a) Describe two events that occur during stage A.

   b) Explain what is happening during stage B.

   c) Describe two events that occur during stage C.

8 Write a short essay on the nature/nurture debate on the development of human intelligence. Keep your essay to one side of A4. You could use the Internet to find out more.

Chapter 18: Chromosomes, Genes and Inheritance

How and why do we inherit features from our parents? This chapter answers these questions by looking at the work of Gregor Mendel and how he has helped us to unravel the mysteries of inheritance.

Genes are sections of DNA that determine a particular feature (see Chapter 16) by instructing cells to produce particular proteins. As the DNA is part of a chromosome, we can also define a gene as 'part of a chromosome that determines a particular feature'. The ground-breaking research that uncovered the basic rules of how features are inherited was carried out by Gregor Mendel and published in 1865.

Gregor Mendel

ideas
evidence

Figure 18.1 *Gregor Mendel (1822–1884).*

Gregor Mendel was a monk and lived in a monastery in Brno in what is now the Czech Republic. He became interested in inheritance and his first attempts at controlled breeding experiments were with mice. This was not well received in the monastery and he was advised to use pea plants instead. As a result of the experiments with pea plants, he was able to formulate the basic laws of inheritance.

Mendel established that, for each feature he studied:

- a 'heritable unit' (we now call it a **gene**) is passed from one generation to the next
- the heritable unit (gene) can have alternate forms (we now call these different forms **alleles**)
- each individual must have two alternate forms (alleles) per feature
- the sex cells only have one of the alternate forms (allele) per feature
- one allele can be dominant over the other.

Mendel was able to use his ideas to predict outcomes from breeding certain types of pea plant and then test his predictions by experiment. Mendel published his results and ideas in 1865 but very few people took any notice.

At that time, biologists had little knowledge of chromosomes and cell division, so Mendel's ideas had no physical basis. Also, biology then was very much a descriptive science and biologists of the day were not interested in the mathematical treatment of results. Mendel's work went against the ideas of the time that inheritance resulted from some kind of blending of features. The idea of a distinct 'heritable unit' just did not fit in.

It was not until 1900 that Hugo DeVries and other biologists working on inheritance rediscovered Mendel's work and recognised its importance. In 1903, Walter Sutton pointed out the connection between Mendel's suggested behaviour of genes and the behaviour of chromosomes in meiosis. The science of genetics was well and truly born.

Mendel's experiments on inheritance

Mendel noticed that many of the features of pea plants had just two alternate forms. For example, plants were either tall or dwarf, they either had purple or white flowers; they produced yellow seeds or green seeds.

There were no intermediate forms, no pale purple flowers or green/yellow seeds or intermediate height plants. Figure 18.2 shows some of the contrasting features of pea plants that Mendel used in his breeding experiments.

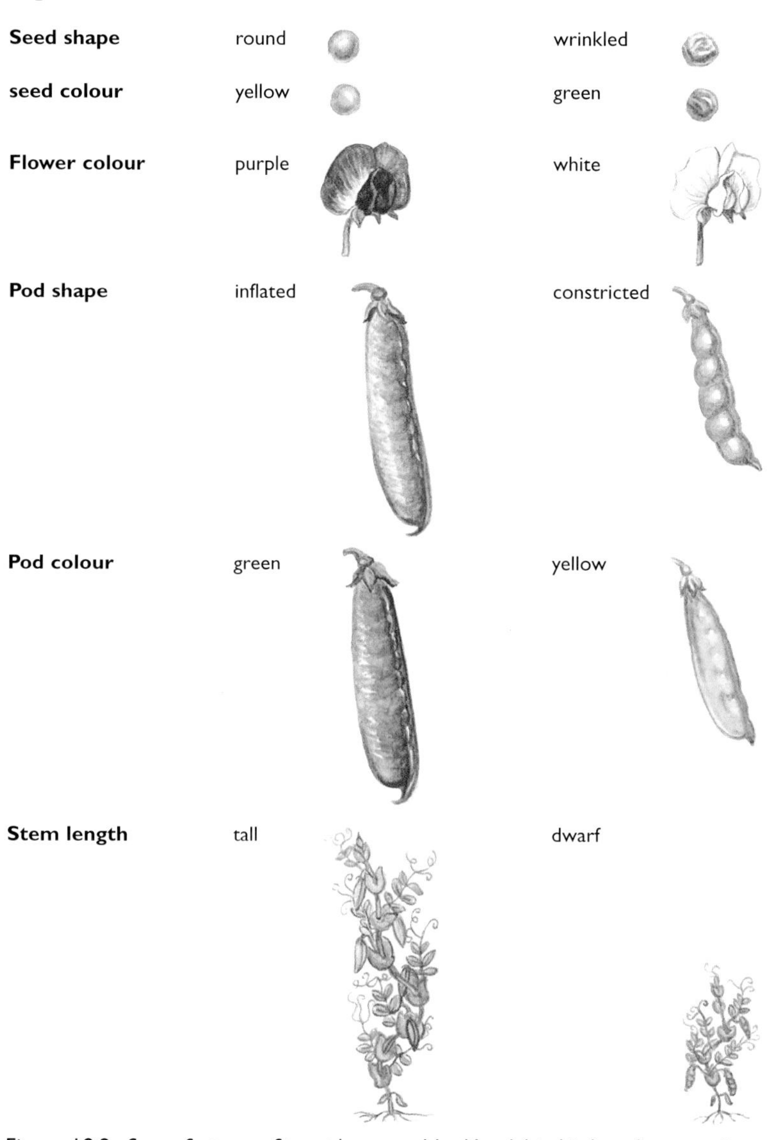

Figure 18.2 *Some features of pea plants used by Mendel in his breeding experiments.*

Mendel decided to investigate systematically the results of cross breeding plants that had contrasting features. These were the 'parent plants', referred to as '**P**' in genetic diagrams. He transferred pollen from one experimental plant to another. He also made sure that the plants could not be self-fertilised.

In his breeding experiments, Mendel initially used only plants that had 'bred true' for several generations. For example, any tall pea plants he used had come from generations of pea plants that had all been tall.

He collected all the seeds formed, grew them and noted the features that each plant developed. These plants were the first generation of offspring, or the 'F_1' generation. He did not cross-pollinate these plants, but allowed them to self-fertilise. Again, he collected the seeds, grew them and noted the features that each plant developed. These plants formed the second generation of offspring or the 'F_2' generation. When Mendel used pure-breeding tall and pure-breeding dwarf plants as his parents, he obtained the results shown in Figure 18.3.

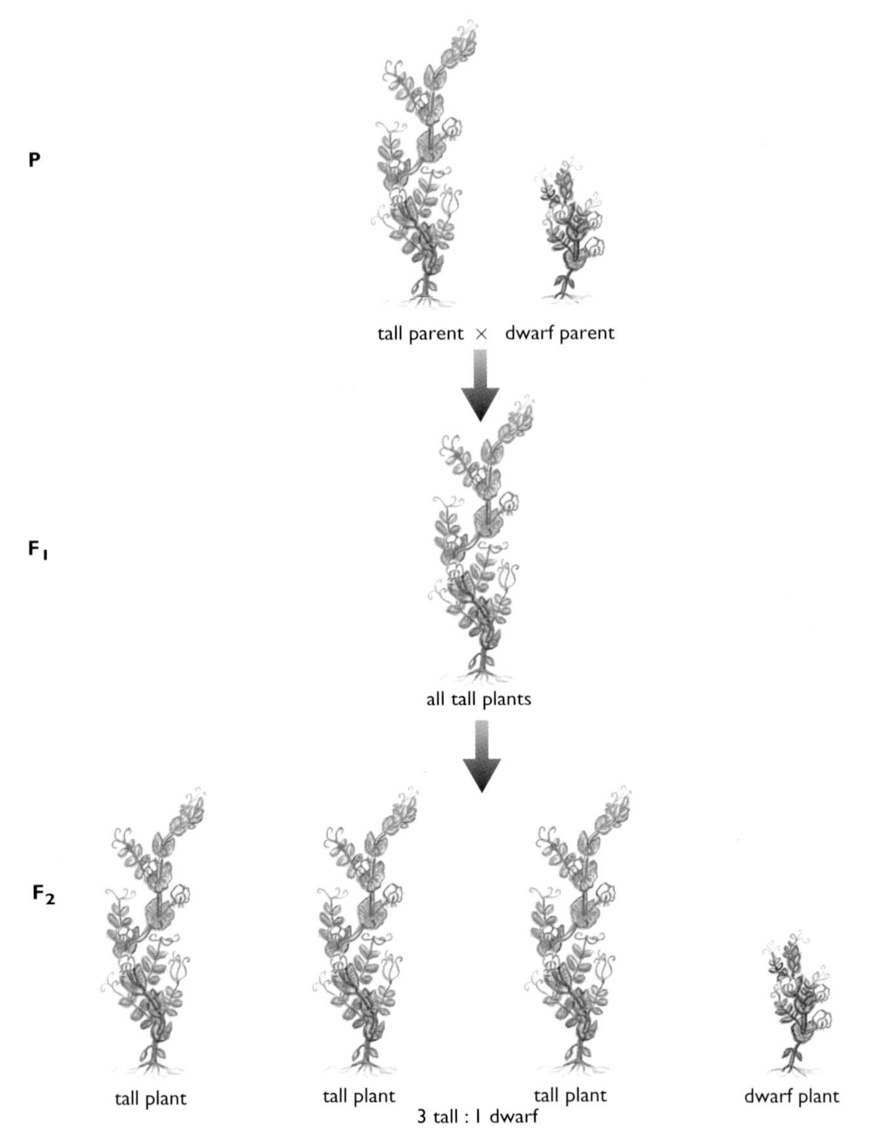

Figure 18.3 *A summary of Mendel's results from breeding tall pea plants with dwarf pea plants.*

Mendel obtained very similar results when he carried out breeding experiments using plants with different pairs of contrasting characters (Figure 18.4). He noticed two things in particular:

- All the plants of the F_1 generation were always of just one type. This type was not a blend of the two parental features, but one or the other. Every time he repeated the experiment with the same feature, it was always the same type that appeared in the F_1 generation. For example, when tall and dwarf parents were cross-bred, the F_1 plants were always all tall.
- There was always a 3:1 ratio of types in the F_2 generation. Three-quarters of the plants in the F_2 generation were of the type that appeared in the F_1 generation. One-quarter showed the other parental feature. For example, when tall and dwarf parents were cross-bred, three-quarters of the F_2 plants were always tall and one-quarter were dwarf.

Mendel was able to use these patterns in his results to work out how features were inherited, without any knowledge of genes and chromosomes.

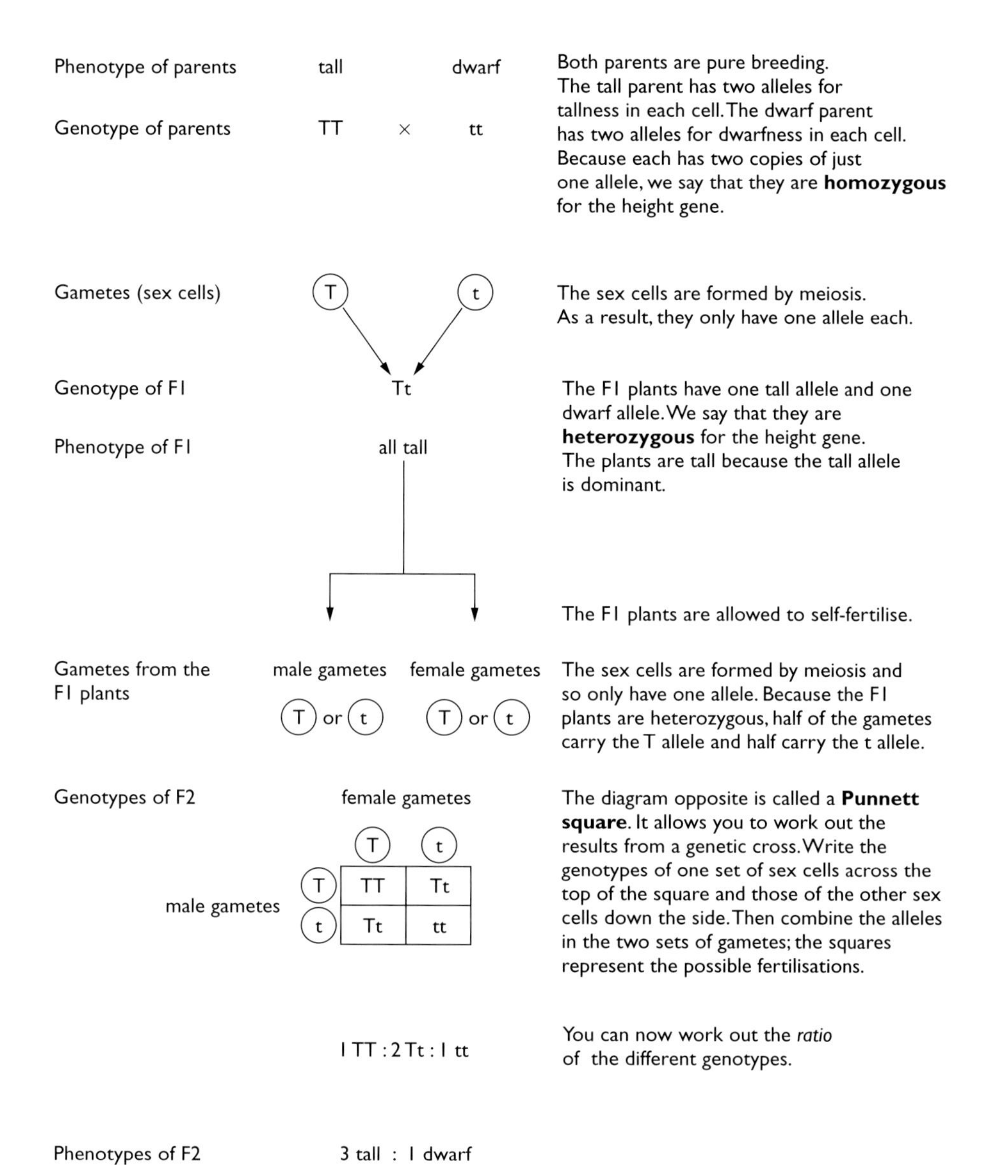

Figure 18.4 *Results of crosses using true-breeding tall and dwarf pea plants.*

Whenever you have to work out a genetic cross, you should choose suitable symbols to represent the dominant and recessive alleles, give a key to explain which symbol is which and write out the cross exactly like the one shown.

Genotype describes the alleles each cell has for a certain feature, e.g. TT. **Phenotype** is the feature that results from the genotype.

Explaining Mendel's results

We can now explain Mendel's results using the ideas of chromosomes, genes, mitosis and meiosis (Chapters 16 and 17).

- Each feature is controlled by a gene, which is found on a chromosome.
- There are two copies of each chromosome and each gene in all body cells, except the sex cells.
- The sex cells have only one copy of each chromosome and each gene
- There are two alleles (forms) of each gene.
- One allele is **dominant** over the other allele, which is **recessive**.
- When two different alleles (one dominant and one recessive) are in the same cell, only the dominant allele is expressed (is allowed to 'work').
- An individual can have two dominant alleles, two recessive alleles or a dominant allele and a recessive allele in each cell.

The genotype of an organism is represented by two letters, each letter representing one allele of the gene that controls the feature. Normally, we use the initial letter of the dominant feature to represent the gene. Writing it as a capital letter indicates the dominant allele, the lower case letter represents the recessive allele. For the feature, height in pea plants, plants can be either tall or dwarf. Using the initial letter of the word 'tall', the alleles are shown as **T** and **t**. **TT** means that a plant has two alleles for tallness.
A plant with two alleles for dwarfness would be represented by **tt**. Although we use pea plants to illustrate Mendel's ideas, the same principles apply to other organisms also.

We can use the cross between tall and dwarf pea plants as an example (Figure 18.4). In pea plants, there are tall and dwarf alleles of the gene for height. We will use the symbol **T** for the tall allele and t for the dwarf allele. The term **genotype** describes the alleles each cell has for a certain feature (e.g. TT). The **phenotype** is the feature that results from the genotype (e.g. a tall plant).

Working out genotypes – the test cross

You cannot tell just by looking at it whether a tall pea plant is homozygous (TT) or heterozygous (Tt). Both these genotypes would appear equally tall because the tall allele is dominant. It would help if you knew the genotypes of its parents. You could then write out a genetic cross and perhaps work out the genotype of your tall plant. If you don't know the genotypes of the parents, the only way you can find out is by carrying out a breeding experiment called a **test cross**.

In a test cross, the factor under investigation is the unknown genotype of an organism showing the dominant feature. A tall pea plant could have the genotype TT or Tt. You must control every other possible variable *including the genotype of the plant you breed it with*. The only genotype you can be *certain* of is the genotype of plants showing the *recessive* feature (in this case dwarf plants). They *must* have the genotype tt.

In a test cross, you breed an organism showing the dominant feature with one showing the recessive feature.

In this example, you must breed the 'unknown' tall pea plant (TT or Tt) with a dwarf pea plant (tt). You can write out a genetic cross for both possibilities (TT × tt and Tt × tt) and *predict* the outcome for each (Figure 18.5). You can then compare the results of the breeding experiment with the predicted outcomes to see which one matches most closely.

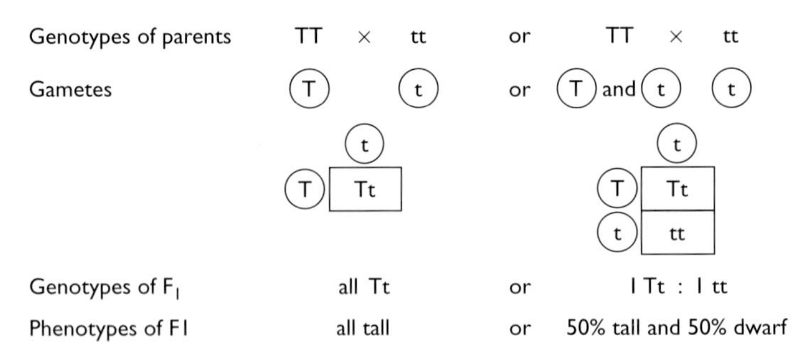

Figure 18.5 *A test cross.*

From our crosses we would expect:

- *all* the offspring to be tall if the tall parent was homozygous (TT)
- *half* the offspring to be tall and *half* to be dwarf if the tall parent was heterozygous (Tt).

Ways of presenting genetic information

Writing out a genetic cross is a useful way of showing how genes are passed through one or two generations, starting with just two parents. To show a proper family history of a genetic condition requires more than this. We use a diagram called a **pedigree**. Polydactyly is an inherited condition in which a person develops an extra digit (finger or toe) on the hands and feet. It is determined by a dominant allele. The recessive allele causes the normal number of digits to develop.

If we use the symbol D for the polydactyly allele and d for the normal-number allele, the possible genotypes and phenotypes are:

- DD – person has polydactyly (has two dominant polydactyly alleles)
- Dd – person has polydactyly (has a dominant polydactyly allele and a recessive normal allele)
- dd – person has the normal number of digits (has two recessive, normal-number alleles).

We don't use P and p to represent the alleles as you would expect, because P and p look very similar and could easily be confused. The pedigree for polydactyly is shown in Figure 18.6.

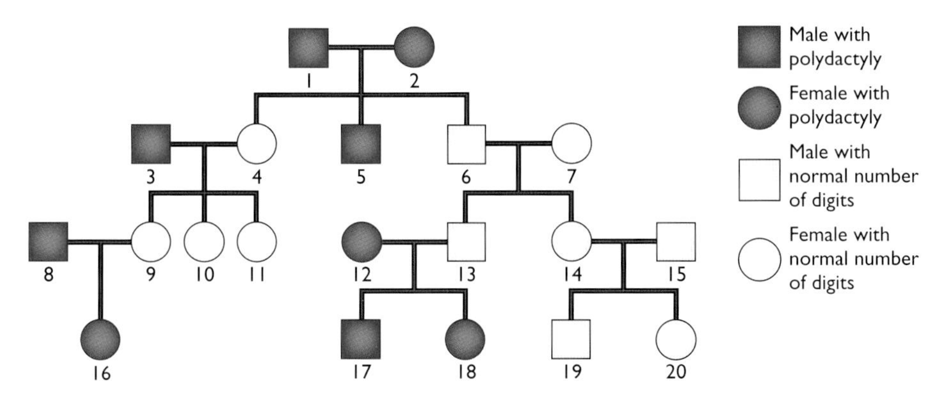

Figure 18.6 *A pedigree showing the inheritance of polydactyly in a family.*

We can extract a lot of information in a pedigree. In this case:

- there are four generations shown (individuals are arranged in four horizontal lines)
- individuals 4, 5 and 6 are children of individuals 1 and 2 (a family line connects each one directly to 1 and 2)
- individual 4 is the first-born child of 1 and 2 (the first-born child is shown to the left, then second born to the right of this, then third born and so on)
- individuals 3 and 7 are not children of 1 and 2 (no family line connects them directly to 1 and 2)
- 3 and 4 are father and mother of the same children – as are 1 and 2, 6 and 7, 8 and 9, 12 and 13, 14 and 15 (a horizontal line joins them).

It is usually possible to work out which allele is dominant from pedigrees. Look for a situation where two parents show the same feature and at least one child shows the contrasting feature. In this pedigree, 1 and 2 both have polydactyly, but children 4 and 6 do not. We can explain this in only one way:

- The normal alleles in 4 and 6 can only have come from their parents – 1 and 2, so 1 and 2 have normal alleles.
- 1 and 2 show polydactyly, so they *must* have polydactyly alleles as well.
- If they have both polydactyly alleles *and* normal alleles but show polydactyly, the polydactyly allele must be the dominant allele.

Now that we know which allele is dominant, we can work out most of the genotypes in the pedigree. All the people with the normal number of digits *must* have the genotype dd (if they had even one D allele, they would show polydactyly). All the people with polydactyly must have *at least one* polydactyly allele (they must be either DD or Dd).

From here, we can begin to work out the genotypes of the people with polydactyly. To do this we need to bear in mind that people with the normal number of digits must inherit one 'normal-number' allele from each parent, and also that people with the normal number of digits will pass on one 'normal-number' allele to each of their children.

From this we can say that any person with polydactyly who has children with the normal number of digits must be heterozygous (the child must have inherited one of their two 'normal-number' alleles from this parent), and also that any person with polydactyly who has one parent with the normal number of digits must also be heterozygous (the normal parent can only have passed on a 'normal-number' allele). Individuals 1, 2, 3, 16, 17 and 18 fall into one or both of these categories and must be heterozygous.

We can now add this genetic information to the pedigree. This is shown in Figure 18.7.

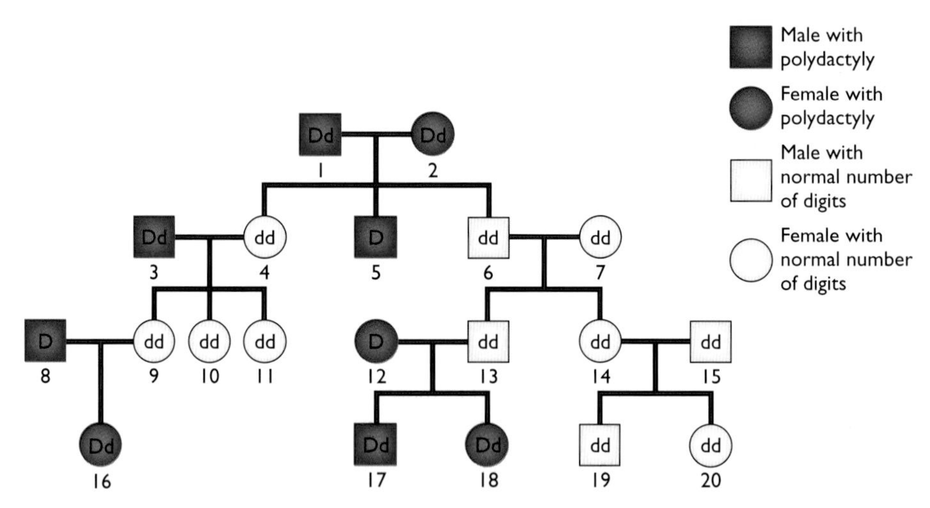

Figure 18.7 *A pedigree showing the inheritance of polydactyly in a family, with details of genotypes added.*

We are still left uncertain about individuals 5, 8 and 12. They could be homozygous or heterozygous. For example, individuals 1 and 2 are both heterozygous. Figure 18.8 shows the possible outcomes from a genetic cross between them. Individual 5 could be any of the outcomes indicated by the shading. It is impossible to distinguish between DD and Dd.

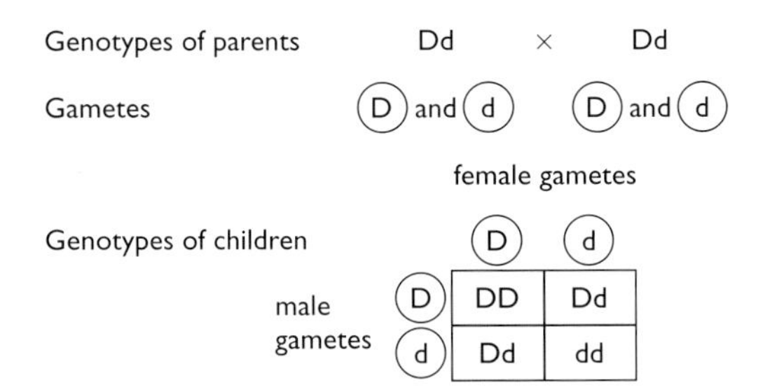

Figure 18.8 *Possible outcomes from a genetic cross between two parents, both heterozygous for polydactyly.*

Inherited diseases

Our genes determine all our features. In some cases they can determine conditions that we recognise as 'genetic diseases'. Three such conditions are Huntington's disease, cystic fibrosis and sickle cell anaemia.

Huntington's disease

This condition is named after the American doctor (George Huntington) who first described the symptoms and development of the disease in 1872. Huntington's disease is caused by a dominant allele on chromosome 4. The condition does not usually develop until middle age, when nerve cells start to degenerate much more rapidly than is normal. The disease is always fatal, with death often occurring within ten years of the first appearance of the symptoms. It is particularly tragic, as, by the time a person becomes aware that they have the condition, they may well already have children. These children may have inherited the dominant allele causing the disease and until recently would have no way of knowing if they would develop Huntington's disease themselves. Now DNA tests can establish whether or not people have the disease-causing allele. Genetic counsellors would advise such people not to have children and so help to eradicate this allele from the population.

The disease is most commonly inherited when one parent is heterozygous for the condition and the other is a non-sufferer (Figure 18.9). There is a 50% chance of a child inheriting the condition in such a relationship.

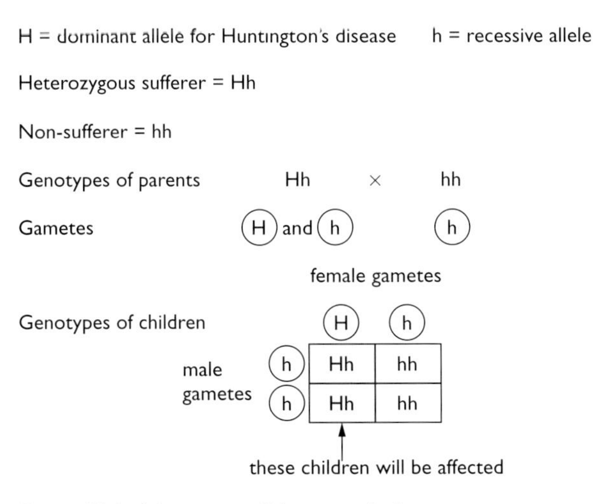

Figure 18.9 *Inheritance of Huntington's disease.*

Cystic fibrosis

This condition is determined by a recessive allele of the gene that controls the production of mucus by cells in glands throughout the body. The gene has a high mutation rate, and if mutations occur in the sex cells (or in the cells that form the sex cells) the mutant allele could be inherited.

The normal, dominant allele results in normal mucus being secreted. The mutated, recessive allele causes the production and secretion of viscous (very thick) mucus. This has several adverse effects:

- It blocks the pancreatic duct so that pancreatic enzymes cannot reach the small intestine. This affects the digestion of carbohydrates, lipids and proteins (see Chapter 3).
- It cannot be easily moved out of the lungs by the cilia (see Chapter 2) as can normal mucus. Gas exchange suffers as a result.

Because of these effects, people suffering from cystic fibrosis often die young. However, treatment is now much better and can extend their life-span. Gene therapy may offer a cure in the future.

To be affected, a person must inherit two alleles (one from each parent), so each parent must carry at least one recessive, cystic fibrosis allele. The disease is most commonly inherited when both parents are heterozygous for the condition (Figure 18.10). There is a 1 in 4 (25%) chance of a child from such a relationship developing cystic fibrosis.

When a recessive allele determines a genetic disease, people who are heterozygous for the condition appear outwardly normal. They can, however, pass on the allele to their children. Because of this they are often called **carriers**.

N = dominant normal allele resulting in production of normal mucus

n = recessive cystic fibrosis resulting in the production of viscous mucus

Heterozygous non-sufferer = Nn

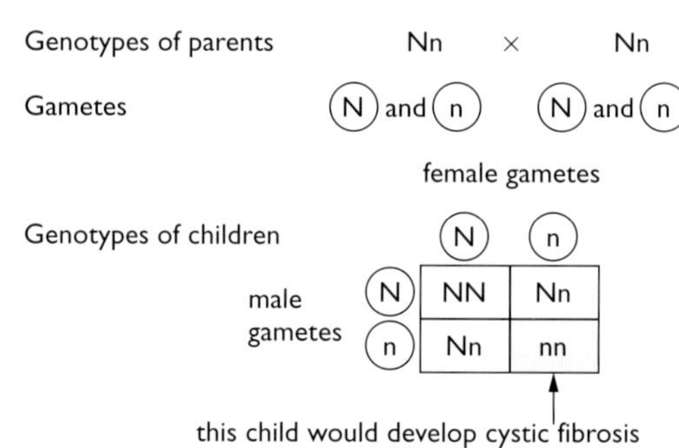

Figure 18.10 *Inheritance of cystic fibrosis.*

A sickle is a crescent-shaped agricultural tool, used for cutting down vegetation.

Sickle cell anaemia

Like cystic fibrosis, this condition is determined by a mutant, recessive allele. It is an allele of the gene that codes for the production of haemoglobin. The haemoglobin produced in sufferers is abnormal. It causes red blood cells to distort when the oxygen concentration of the surroundings is low. Instead of being the usual disc shape, they become sickle shaped (Figure 18.11). This can happen in any active tissues that use up oxygen very rapidly as they respire.

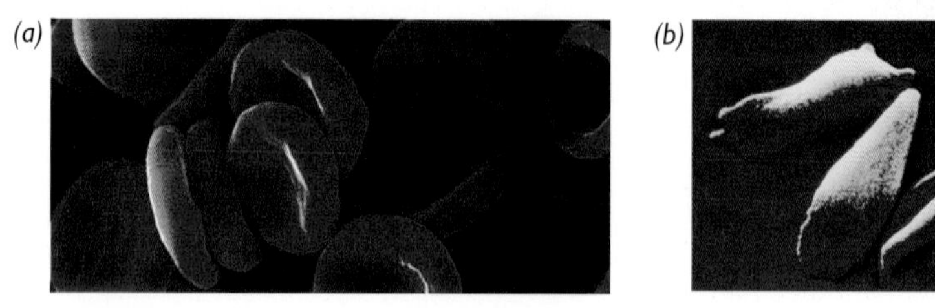

Figure 18.11 *(a) Normal red blood cells and (b) distorted red blood cells from a person suffering from sickle cell anaemia.*

There are two main consequences as a result of the red blood cells becoming sickle shaped:

- The sickle-cells tend to form blood clots that block capillaries, leading to 'sickle cell crisis'. This is a painful condition in which one or more organs are damaged due to a lack of blood. If the blood supply to the brain is affected, a stroke may result.
- The sickle cells are more fragile than normal cells and easily burst. They are also destroyed at a higher rate than normal red blood cells by the spleen. The drastic reduction in the numbers of red blood cells causes the anaemia that gives the disease its name.

Anaemia describes any condition in which the concentration of haemoglobin in the blood falls significantly below normal.

To be fully affected, a person must be homozygous for the sickle cell allele. In other words they must inherit a 'sickle cell' allele from each parent. The condition develops soon after birth and infant mortality is very high, usually due to lack of blood to an organ during a sickle-cell crisis. Sufferers who survive childhood have an increased risk of infection by bacteria that cause a certain type of pneumonia. They also have an increased risk of developing gallstones.

The disease is most commonly inherited when both parents are heterozygous carriers of the condition (Figure 18.12). There is a 1 in 4 (25%) chance of a child from such a relationship developing sickle cell anaemia.

H = normal allele resulting in production of normal haemoglobin

h = allele resulting in the production of abnormal (sickling) haemoglobin

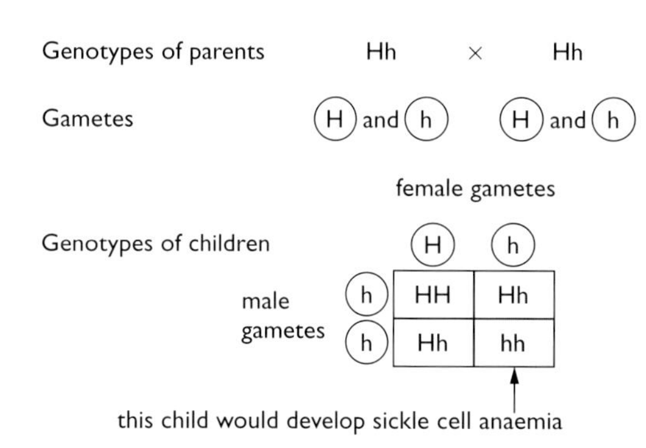

Figure 18.12 *Inheritance of sickle cell anaemia.*

Sickle cell anaemia can be treated by a blood transfusion following a crisis and can be cured by a bone marrow transplant. Bone marrow is the 'production line' for red blood cells. If bone marrow from a non-sufferer is transplanted successfully, it will produce normal red blood cells. The treatment is only available in specialist centres, but the success rate is around 85%. In the future, gene therapy may be used to introduce non-sickling alleles into the bone marrow of sufferers, resulting in the production of normal red blood cells.

Inheriting the sickle cell allele need not be a bad thing. Heterozygous carriers of the allele usually show no symptoms of the disease at all. In fact, they can actually benefit from their condition – being a carrier for sickle cell anaemia gives increased resistance to malaria (see Chapter 19).

The inheritance of sickle cell anaemia is not quite as straightforward as the inheritance of some other characteristics. This is because the normal allele is not actually dominant over the sickle cell allele. The red blood cells of individuals who are heterozygous contain half normal and half abnormal haemoglobin. So *both* the alleles are actually having an effect. Neither is 'switched off'. This type of inheritance is called **codominance**.

Heterozygotes do not suffer sickle cell crises because there is sufficient normal haemoglobin in their red blood cells for them to maintain their normal shape. However, under very low oxygen concentrations, some of the cells will deform into the typical sickle shaped cells. This can be seen under a microscope.

Sex determination

Because the Y chromosome, when present, causes a zygote to develop into a male, some people cannot resist describing it as 'dominant'. This is incorrect: dominant and recessive are terms that are only applied to individual alleles.

Our sex – whether we are male or female – is not under the control of a single gene. It is determined by the X and Y chromosomes – the sex chromosomes. As well as the 44 non-sex chromosomes, there are two X chromosomes in all cells of females (except the egg cells) and one X and one Y chromosome in all cells of males (except the sperm). Our sex is effectively determined by the presence or absence of the Y chromosome. The full chromosome complement of male and female is shown in Figure 16.11 on page 202.

The inheritance of sex follows the pattern shown in Figure 18.13. In any one family, however, this ratio may well not be met. Predicted genetic ratios are usually only met when large numbers are involved. The overall ratio of male and female births in all countries is 1: 1.

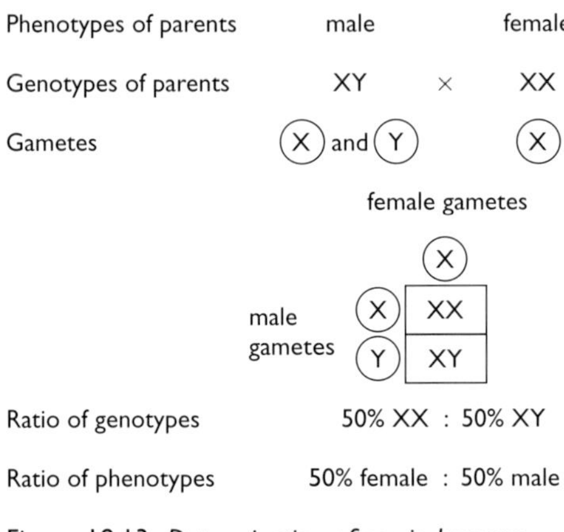

Figure 18.13 *Determination of sex in humans.*

The Human Genome Project

ideas evidence

You can download the first draft of the genome from the Internet and find out more about the Human Genome Project.

The complete sequencing of all the bases in the DNA nucleotides of 24 chromosomes is an extremely complex task. Scientists need to arrange over 3 billion bases in the correct order!

All the alleles of all the genes possessed by a species make up its **genome**. The Human Genome Project aims first of all to identify more than 100 000 genes, and alleles of those genes, on the 24 human chromosomes. It is 24 chromosomes and not 23, because the X and Y sex chromosomes carry different genes. The 'first draft' of this is now complete.

The most important technique in the project is **gene sequencing**. Developed in 1977 by Fred Sanger, gene sequencing allows biologists to determine the sequence of DNA nucleotides (see Chapter 16) in a gene.

The complete identification of all the human alleles brings many potential benefits, but it is not without considerable controversy. Some of the benefits could be:

- a much better understanding of how genes operate
- a better understanding, in particular, of the genes causing genetic diseases and the development of cures for these diseases
- a better understanding of the genetic control of immunity to disease
- development of new anti-cancer treatments.

However, some people are concerned that companies involved in the research are trying to patent the gene sequences so that they own the discoveries. These companies would then control all use of the genes and products of the genes they patent. In addition, there is a risk that if genetic information on individuals becomes freely accessible, those identified as being at increased genetic risk of heart disease (for example) may find it difficult to obtain medical insurance, as the insurance companies would also have access to the genetic database.

Embryos could be screened for signs of potential genetic disorders and only those apparently free from such disorders allowed to develop. Also, individual genetic records could be on file in government controlled organisations as a matter of course.

There are many other issues and there are no easy 'right' and 'wrong' answers.

End of Chapter Checklist

If you haven't got a copy of your specification, read the introduction on page vi.

You will need to be able to do some or all of the following. Check your Awarding Body's specification (syllabus) to find out exactly what you need to know.

- Appreciate the importance of Gregor Mendel's research in laying the foundations for modern genetics.
- Understand and define the following terms: gene, allele, dominant, recessive, genotype, phenotype, homozygous, heterozygous.
- Write out a genetic cross, given the genotypes of parents.
- Deduce genotypes of individuals from data supplied in questions.
- Predict ratios of genotypes and phenotypes in the offspring.
- Interpret the information provided in a pedigree and deduce genotypes of individuals shown.
- Understand the pattern of inheritance of Huntington's disease, sickle cell anaemia and cystic fibrosis.
- Understand why people heterozygous for the sickle-cell allele have an increased resistance to malaria.
- Explain how gender is determined.
- Appreciate the significance of the Human Genome Project (HGP).

Questions

More questions on chromosomes, genes and inheritance can be found at the end of Section E on page 257.

1 Predict the *ratios* of offspring from the following crosses between tall/dwarf pea plants.

a) **TT** × **TT**, *b)* **TT** × **Tt**, *c)* **TT** × **tt**, *d)* **Tt** × **Tt**, *e)* **Tt** × **tt**, *f)* **tt** × **tt**.

2 In cattle, a pair of alleles controls coat colour. The allele for black coat colour is dominant over the allele for red coat colour. The genetic diagram represents a cross between a pure-breeding black bull and a pure-breeding red cow. **B** = dominant allele for black coat colour; **b** = recessive allele for red coat colour.

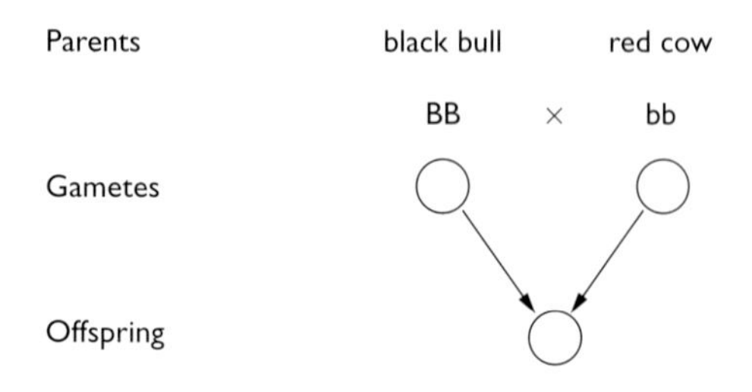

a) *i)* What term describes the genotypes of the pure-breeding parents?

ii) Explain the terms dominant and recessive.

b) *i)* What are the genotypes of the sex cells of each parent?

ii) What is the genotype of the offspring?

c) Cows with the same genotype as the offspring were bred with bulls with the same genotype.

i) What genetic term describes this genotype?

ii) Draw a genetic diagram to work out the ratios of:

- the genotypes of the offspring
- the phenotypes of the offspring.

3 In nasturtiums, a single pair of alleles controls flower colour.

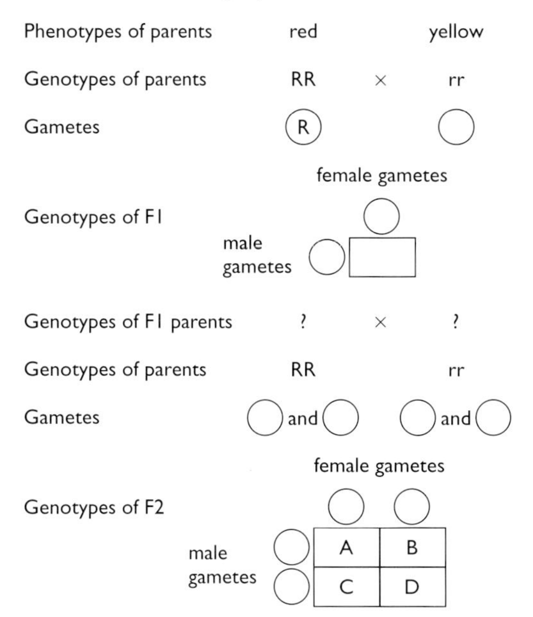

The allele for red flower colour is dominant over the allele for yellow flower colour. The diagram represents the results of a cross between a pure-breeding red-flowered nasturtium and a pure-breeding yellow-flowered nasturtium. **R** = dominant allele for red flower colour; **r** = recessive allele for yellow flower colour.

a) Copy and complete the genetic diagram.

b) What are the colours of the flowers of A, B, C and D?

4 Cystic fibrosis is an inherited condition. The diagram shows the incidence of cystic fibrosis in a family over four generations.

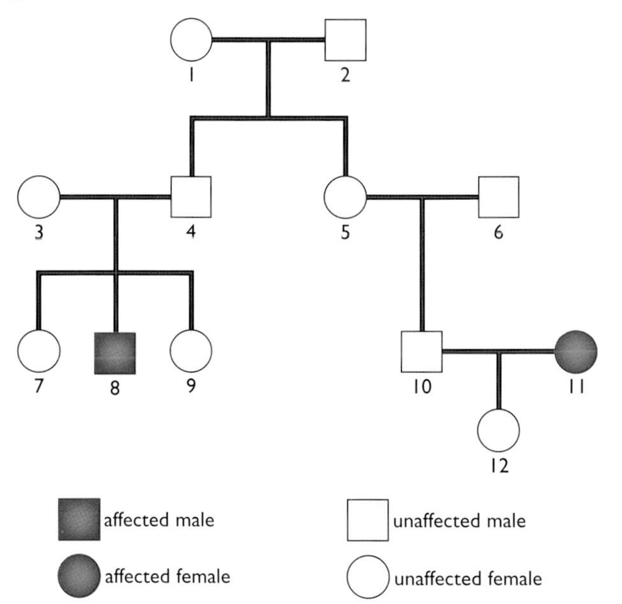

a) What evidence in the pedigree suggests that cystic fibrosis is determined by a recessive allele?

b) What are the genotypes of individuals 3, 4 and 11? Explain your answers.

c) Draw genetic diagrams to work out the probability that the next child born to individuals 10 and 11 will *i)* be male, *ii)* suffer from cystic fibrosis.

5 In guinea pigs, the allele for short hair is dominant to that for long hair.

a) Two short-haired guinea pigs were bred and their offspring included some long-haired guinea pigs. Explain these results.

b) How could you find out if a short-haired guinea pig was homozygous or heterozygous for hair length?

6 Sickle cell anaemia is determined by a single mutant allele. Sufferers are homozygous for this allele. Heterozygotes show virtually no signs of the condition, but have an increased resistance to malaria.

a) What is a sickle cell crisis?

b) Why does a bone marrow transplant often cure sickle cell anaemia?

c) A sickle cell sufferer survived to have children and married a person heterozygous for the condition. What proportion of their children would develop sickle cell anaemia? Use a genetic diagram to explain your answer.

d) Why do people heterozygous for the condition have an increased resistance to malaria?

7 Write an essay about the aims, benefits and concerns of the Human Genome Project (HGP). It should be one side of A4 paper. You could word process it and you could use the Internet to look up more information.

Chapter 19: Natural Selection and Evolution

Over millions of years, life on this planet has evolved from its simple beginnings into the vast range of organisms present today. This has happened by natural selection.

Humans have been asking the question 'Where did we come from?' for thousands of years. The theory of evolution, occurring by natural selection, is the most widely accepted scientific explanation of the answer to this question. The two terms are quite distinct:

- **Evolution** is a gradual change in the range of organisms on the Earth. New species continually arise from species that already exist and other species become extinct.
- **Natural selection** is the *mechanism* by which new species arise. Natural selection 'allows' different forms of a species to survive in different areas. Over time, these different forms become increasingly different and may eventually become different species. If the environment of a species changes and that species is no longer adapted to survive in the new conditions, it may become extinct.

ideas evidence

The work of Charles Darwin

Charles Darwin, the son of a country doctor, did not do particularly well at school. He was unable to settle to prepare himself for any profession. His father is reputed to have said: 'you're good for nothing but shooting guns and rat-catching ... you'll be a disgrace to yourself and all of your family'.

Figure 19.1 *Charles Darwin (1809–1882).*

At the age of 22, Charles Darwin became the ship's naturalist on HMS Beagle, which left England for a five-year voyage in 1831.

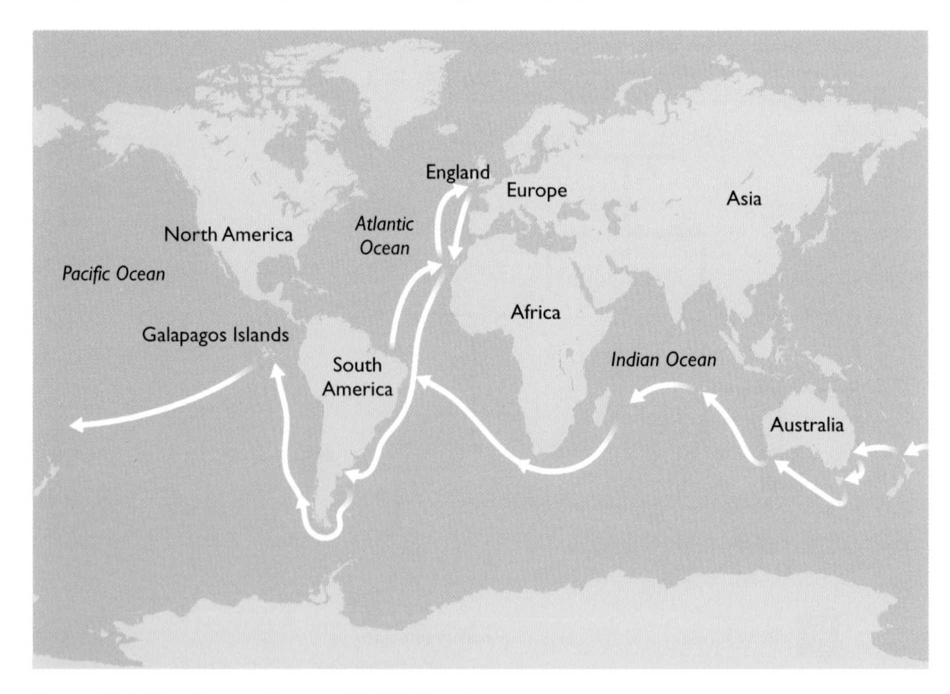

Figure 19.2 *The five-year journey of HMS Beagle.*

During the voyage, Darwin collected hundreds of specimens and made many observations about the variety of organisms and the ways in which they were adapted to their environments. He gained much information, in particular, from the variety of life forms in South America and the Galapagos Islands. Darwin was influenced by the work of Charles Lyell who was, at the time, laying the foundations of modern geology. Lyell was using the evidence of rock layers to suggest that the surface of the Earth was constantly changing. The layers of sediments in rocks represented different time periods. Darwin noticed that the fossils found in successive layers of rocks often changed slightly through the layers. He suggested that life forms were continually changing – evolving. This was in contrast to the religious ideas of special creation, a common belief that all life had been created at one time and had not changed since.

Figure 19.3 *Darwins ideas were unpopular and many newspapers of the time made fun of them.*

On his return to England, Darwin began to evaluate his data and wrote several essays, introducing the ideas of natural selection. He arrived at his theory of natural selection from observations made during his voyage on the Beagle and from deductions made from those observations. Darwin's observations were that:

- organisms tend to produce more offspring than are needed to replace them – a single female salmon can release 5 million eggs per year; a giant puffball fungus produces 40 millions spores
- despite the over-reproduction, stable, established populations of organisms tend to remain the same size – the seas are not overflowing with salmon, and you are not surrounded by piles of giant puffball fungi!
- members of the same species are not identical – living things vary.

He made two important deductions from these observations:

- From the first two observations he deduced that there is a 'struggle for existence'. Many offspring are produced, yet the population stays the same size. There must be competition for resources and many must die.
- From the third observation he deduced that, if some offspring survive whilst others die, those organisms best equipped or best suited to their environment will survive to reproduce. Those less suited will die. This gave rise to the phrase 'survival of the fittest'.

By using the phrase 'survival of the fittest', Darwin was not referring to physical fitness, but to biological fitness. This means how well suited, or well adapted, an organism is to its environment.

Notice a key phrase in the second deduction – the best-suited organisms survive *to reproduce*. This means that those characteristics that give the organism a better chance of surviving will be passed on to the next generation. Those organisms that are less suited to the environment, survive to reproduce in smaller numbers. The next generation will have more of the type that is adapted and fewer of the less well adapted type. This will be repeated in each generation.

Darwin was not aware of genes and how they determine characteristics when he put forward his theory of natural selection. Gregor Mendel had yet to publish his work on inheritance.

Another naturalist, Alfred Russell Wallace, had also studied life forms in South America and had reached the same conclusions as Darwin. Darwin and Wallace published a scientific paper on natural selection jointly, although it was Darwin who went on to develop the ideas further. In 1859 he published his now famous book *'The Origin of Species'*.

This book changed forever the way in which biologists think about how species arise. Darwin went on to suggest that humans could have evolved from ape-like ancestors, for which he was ridiculed, largely by people who had misunderstood his ideas. He also carried out considerable research into plant tropisms (see Chapter 12).

A century before Darwin put forward his theory, the French biologist Jean Baptiste Lamarck put forward a theory of evolution. He, too, thought that living things would become increasingly adapted to their environment. However, the mechanism that Lamarck proposed included the idea that characteristics acquired during a lifetime can be inherited. This is best illustrated in his now famous theory of how giraffes got their long necks (Figure 19.4).

Ancestral giraffes probably had short necks. They fed from leaves on trees and stretched their necks to reach the leaves.

Because the parents stretched their necks, the offspring had longer necks than the parents. The offspring, too, stretched their necks to reach the leaves.

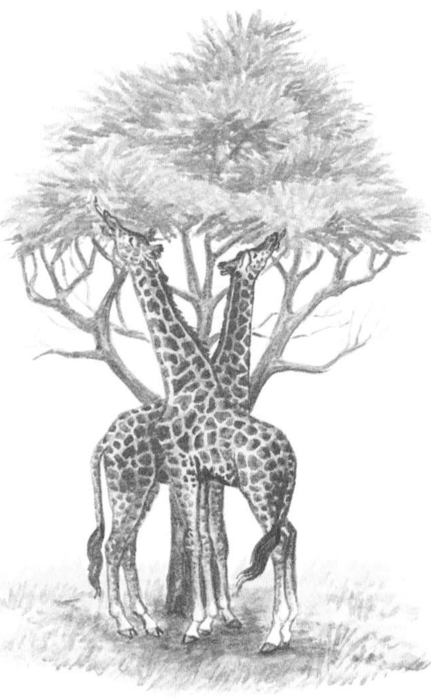

After many generations of stretching their necks and passing this increased neck length on to their offspring, giraffes became the long-necked animals that we now see.

Figure 19.4 *Lamarck's ideas on evolution included the idea that acquired characteristics could be inherited.*

The idea that characteristics acquired during a lifetime can be inherited is now largely discredited. Very few biologists believe that this is likely.

Evidence for natural selection

The theory of natural selection proposes that some factor in the environment 'selects' which forms of a species will survive to reproduce under those conditions. Forms that are not well adapted will not survive. Any evidence for natural selection must show that:

- there is variation within the species
- changing conditions in the environment (a **selection pressure**) favours one particular form of the species (which has a **selective advantage**)
- the frequency of the favoured form increases (it is selected *for*) under these conditions (survival of the fittest)
- the frequency of the less well adapted form decreases under these conditions (it is selected *against*)
- the changes are not due to any other factor.

The peppered moth

The peppered moth has two forms, one is greyish-white with dark markings ('peppered') and one is much darker (See Chapter 13).

Initially, nearly all the moths were of the peppered type. The first record of a dark moth in Manchester was in 1848. By 1895, 98% of the moths in Manchester were of the dark form. What had caused this change? The following pieces of information will give you some clues:

- Peppered moths are food for birds.
- Increasing industrialisation during the nineteenth century killed off many of the lichens growing on tree trunks and also covered the trunks with soot.
- The distribution of the two forms of peppered moth was linked to the degree of industrialisation (Figure 19.5).

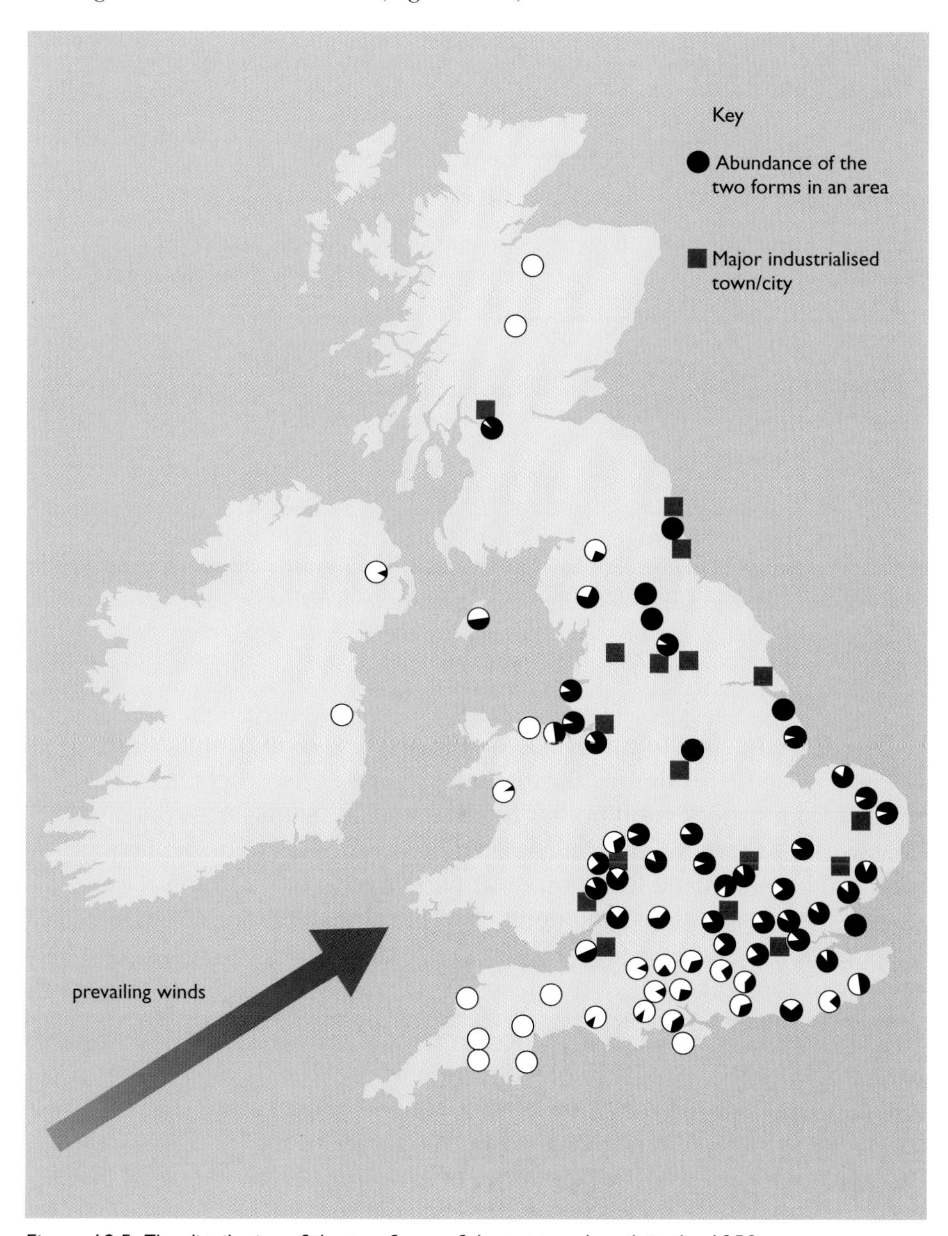

Figure 19.5 *The distribution of the two forms of the peppered moth in the 1950s.*

Natural selection explains the moth distribution in the following way:

- In any area there is an over production of offspring and so there is a struggle for existence. Because of this, the form most suited to its environment will survive.
- In the country, tree trunks are still relatively unpolluted; in the cities, the tree trunks and buildings are covered with soot.
- In the country, the peppered form is camouflaged; in the cities, the dark form is camouflaged.
- Camouflaged moths are less likely to be eaten by birds.
- In the country, more of the peppered form survive to reproduce; in the cities, more of the dark form survive to reproduce.
- Over many generations, the numbers of the dark form increase in the cities, while the numbers of the peppered form remain high in the countryside.

Table 19.1 illustrates this.

Feature of natural selection	**Effect on population of peppered moths in:**	
	Countryside	**City**
selection pressure	predation by birds on moths on clean tree trunks	predation by birds on moths on soot-covered tree trunks and other surfaces
natural variation in the species	some moths are 'peppered', others are dark.	
type with selective advantage	peppered form (camouflaged on clean, lichen-covered tree trunks)	dark form (camouflaged on dark, soot-covered surfaces)
type selected for	peppered form	dark form
type selected against	dark form	peppered form
result of natural selection over many generations	percentage of peppered form increases or remains high, percentage of dark form decreases or remains low	percentage of dark form increases or remains high, percentage of peppered form decreases or remains low

Table 19.1: *Peppered moths as evidence for natural selection.*

Recently, some biologists have questioned the methods by which the peppered moth data was obtained. They do not, however, think that the conclusion about the way in which natural selection is thought to operate is necessarily wrong. They think that it presents a picture that is too clear cut and further data, obtained in a more rigorous manner, is needed to support the conclusion.

Shortly after the data in the map was obtained, a law called the 'Clean Air Act' was passed. This limited the amount of smoke and other pollutants emitted from factories and houses in certain areas. 'Smoke free zones' were established. As time went by, the regulations of the Clean Air Act became more rigid and less and less smoke was permitted in cities. The surfaces of trees became less polluted and buildings were cleaned. More and more peppered forms survived to reproduce as the cleaner surfaces once again gave them camouflage.

This appears to be excellent evidence for natural selection in action. It shows natural selection operating in one direction as the cities became more polluted and then reversing as the cities became cleaner again. The independent variable was the nature of the surface of the tree trunks, the dependent variable was the percentage of each form of the peppered moth. Industrialisation and the Clean Air Act changed the independent variable in the cities. Throughout it all, the countryside acted as a kind of unchanging control experiment. This showed that over the same period, when the independent variable was *not* changed, the percentage of each form of the peppered moth was unaltered. Other factors were not causing the change – it *must* have been the nature of the surfaces offering camouflage to different forms of the moth.

Bacterial resistance to antibiotics

ideas
evidence

Alexander Fleming discovered penicillin, the first antibiotic, in 1929. Since then, other natural antibiotics have been discovered and many more have been synthesised in laboratories. The use of antibiotics has increased dramatically, particularly over the last 20 years. We now almost expect to be given an antibiotic for even the most trivial of ailments. This can be dangerous, as it leads to the development of bacterial resistance to an antibiotic (Figure 19.6).

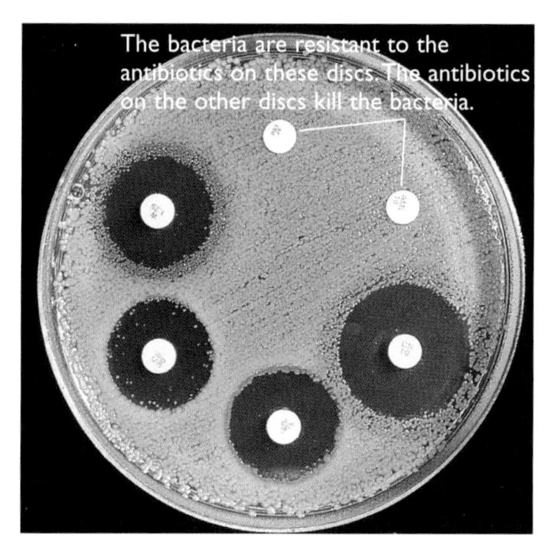

Figure 19.6 *Bacterial resistance to an antibiotic.*

Mutations happen all the time in all living organisms. In bacteria, a chance mutation could give a bacterium resistance to an antibiotic. In a situation where antibiotics are widely used, this new resistant bacterium has an advantage over non-resistant bacteria of the same type. The resistant bacterium will survive and multiply in greater numbers than the non-resistant types. The generation time of a bacterium can be as short as 20 minutes. This means that there could be 72 generations in a single day – the equivalent of about 1500 years of human generation time. The numbers of resistant types would increase with each generation. Very soon a population of bacteria could become almost entirely made up of resistant types. Table 19.2 shows how natural selection can introduce resistance to an antibiotic in a population of bacteria.

Feature of natural selection	Effect on population of non-resistant bacteria
selection pressure	repeated use of antibiotics
natural variation in the species	some are resistant (due to a chance mutation), others are not
type with selective advantage	resistant type – will survive antibiotic treatment
type selected for	resistant type
type selected against	non-resistant type
result of natural selection over many generations	percentage of resistant types in the population increases

Table 19.2: *Bacteria and natural selection.*

Some people talk about bacteria becoming *immune* to antibiotics. This is a misunderstanding. *Individuals* become immune to *microorganisms* that infect them. This happens as a result of the immune response they make to those pathogens and usually takes a few days – a fraction of a lifetime. *Populations* of bacteria become resistant to antibiotics as a result of chance mutations and natural selection over many generations.

Doctors are now more reluctant to prescribe antibiotics. They know that by using them less, the bacteria with resistance have less of an advantage and will not become as widespread.

There is growing concern about so-called 'super-bugs' – bacteria that are resistant to several antibiotics. This multiple resistance is sometimes caused by 'jumping genes'. Bacteria can occasionally transfer genes between different species. A bacterium with a gene that gives resistance to penicillin could transfer this gene to a bacterium with a gene giving resistance to tetracycline (another antibiotic). This would produce a bacterium resistant to both antibiotics. Bacteria with multiple resistance to antibiotics will have a large selective advantage in any situation where those antibiotics are widely used.

Sickle cell anaemia and malaria

Sickle cell anaemia is caused by a mutant allele (see Chapter 18). It affects the formation of haemoglobin in red blood cells. The abnormal haemoglobin causes the red blood cells to become distorted (sickle-shaped) when the oxygen concentration of the surroundings is low. The condition can be fatal in individuals homozygous for the allele.

Homozygous means having two alleles of a gene that are the same (e.g. two alleles for sickle cell or two normal alleles). Heterozygous means having two different alleles of a gene (e.g. one sickle cell allele and one normal allele). People heterozygous for sickle cell anaemia are called 'carriers'.

Heterozygous 'carriers' of the allele usually show no symptoms of the disease at all, although 50% of the haemoglobin in their red blood cells is abnormal. They do have an important benefit, however. They are more resistant to malaria than people with 100% normal haemoglobin (homozygous for the normal allele).

The red blood cells of carriers look normal, but are slightly more fragile than normal red blood cells (because of the 50% abnormal haemoglobin). The malarial parasite is transmitted by the female *Anopheles* mosquito and spends part of its life cycle inside red blood cells (see Chapter 18). When these parasites enter the fragile red blood cells of carriers, the cells often burst before the parasite has time to develop and the parasite dies. The life cycle is broken. Table 19.3 shows how natural selection affects the incidence of sickle cell anaemia in an area where malaria is common.

Feature of natural selection	Effect on incidence of sickle cell anaemia
selection pressure	infection by the malarial parasite
natural variation in the species	carriers (people who are heterozygous for the sickle cell allele) and people homozygous for the normal allele; people homozygous for the sickle cell allele often die at an early age
type with selective advantage	carriers – malarial parasite cannot complete life cycle
type selected for	carriers
type selected against	people with 100% normal haemoglobin (homozygous for normal allele)
result of natural selection over many generations	numbers of heterozygotes in the population are higher than in areas where malaria is absent; numbers of people suffering from the disease are also higher than in other areas

Table 19.3: *Sickle cell anaemia and natural selection.*

The carriers have a selective advantage over those homozygous for the normal allele in areas where malaria is common. However, if two carriers marry, they can produce children who are homozygous for the sickle cell allele. As a result, sickle cell anaemia is more common in these areas also (Figure 19.7).

Figure 19.7 *A map showing areas of the world where sickle cell anaemia and malaria are common.*

Natural selection and the formation of new species

Natural selection favours the survival of individuals with an advantage over others in the population. Consequently, over time, the least well-adapted members of a population do not survive to reproduce, and the population becomes increasingly adapted to its environment. Suppose that there were two populations of the same species in different environments. Different forms would have an advantage in the *different* environments (Figure 19.8).

Biologists define a species as a group of individuals that share common genes and can interbreed to produce fertile offspring (offspring that can also breed and produce offspring).

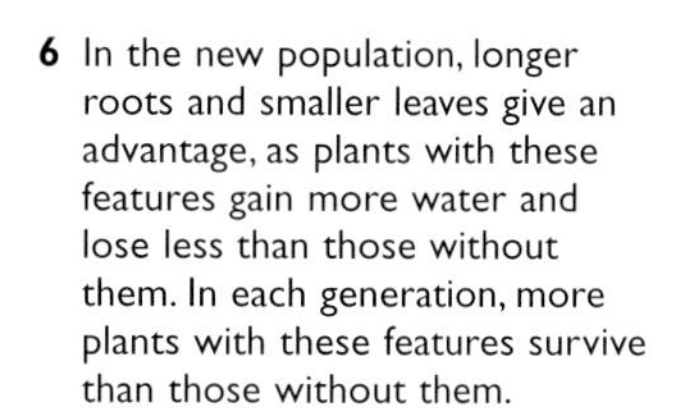

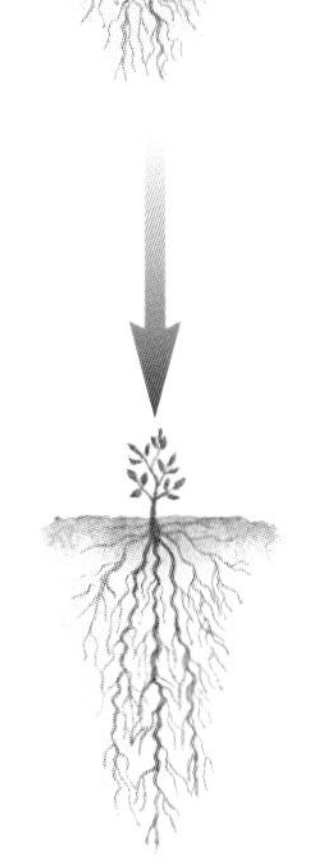

1 A population of plants lives in a fairly normal type of soil, with normal rainfall.

2 Some of these plants colonise a different area, where water is found much deeper in the soil and the rainfall is considerably less. In this new environment, natural selection favours those plants with longer roots (able to reach the soil water) and smaller leaves with fewer stomata (to minimise water loss).

3 The two populations of plants are isolated from each other and cannot interbreed.

4 There is natural variation in these features in both populations as a result of sexual reproduction and gene mutation.

5 In the original population, longer roots and smaller leaves give no advantage and natural selection maintains the original form for as long as the environment remains stable.

6 In the new population, longer roots and smaller leaves give an advantage, as plants with these features gain more water and lose less than those without them. In each generation, more plants with these features survive than those without them.

7 The new plants become more and more different from the original population. Long-rooted and small-leaved forms survive best and the population eventually consists almost entirely of this type.

8 Eventually the two populations are so different that they cannot interbreed. At this point we consider them to be separate species.

Figure 19.8 *How natural selection can lead to the formation of a new species.*

The course of evolution

Many biologists now believe that there is sufficient evidence to suggest that evolution has followed the general course outlined below:

- Life began in water as a result of reactions between chemicals in the early Earth's atmosphere and oceans.
- The first life forms were unicells (single cells), similar to bacterial cells.
- These unicells became more complex as the cells acquired more and more organelles (such as mitochondria and chloroplasts).
- Simple multicellular organisms evolved from the unicells – possibly by unicells undergoing cell division but the two cells formed failing to separate.
- The multicellular organisms became more and more complex, giving rise to plants, animals, fungi and other types of organisms.
- Some of these organisms colonised the land and the evolution of land animals, plants and fungi began (Figure 19.9).

You could carry out an Internet search on the 'endosymbiont' theory as to how early cells acquired their organelles.

A niche is a description of the habitat of an organism and the role of the organism in that habitat. For example, the niche of a lion could be described as 'predator of wildebeest on the Serengeti plains in Africa'. The niche of a camel is that of herbivore in hot deserts.

From time to time catastrophic events caused mass extinctions. Niches that had been filled by one species became 'vacant'. Other species filled these niches. As they occupied the new niches, they evolved into different species as a result of natural selection (see page 230). For example, when the dinosaurs became extinct, the small mammals that existed at the time were able to fill many of the niches left vacant by the dinosaurs and then evolve into the mammals of today.

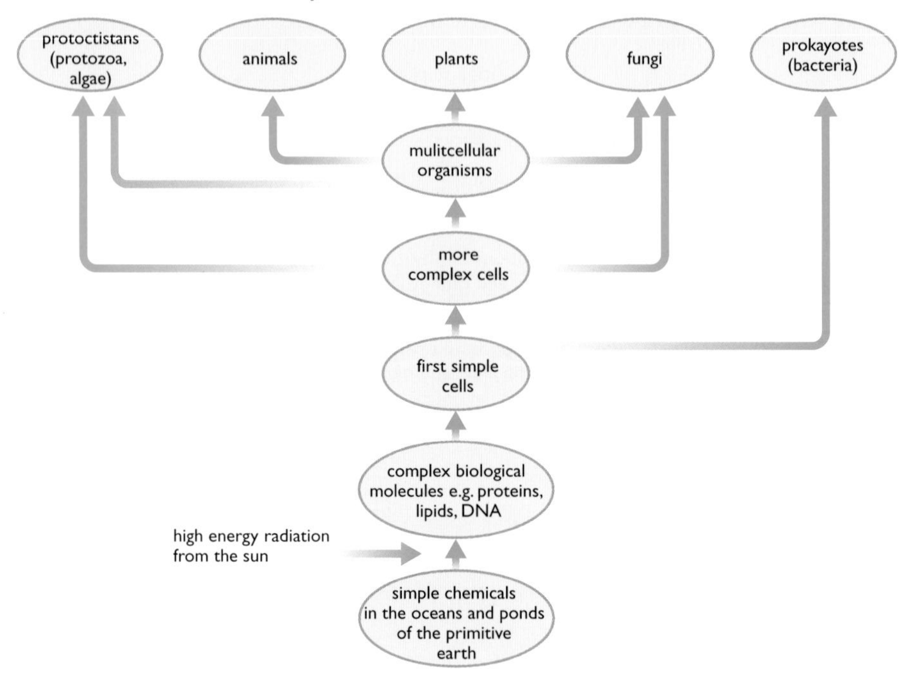

Figure 19.9 *The main stages in evolution.*

ideas evidence

Evidence for evolution

The theory of evolution is now supported by evidence from many areas, the most significant being that from **fossils**.

Whole organisms

Microorganisms that decay dead organisms are called **decomposers**. They help to recycle nutrients through ecosystems (see Chapter 14 for the carbon and nitrogen cycles). To bring about decay, they require a suitable temperature as well as a supply of oxygen and water.

Normally, when an organism dies, it decays and no trace remains. However, if the right conditions are not present, decay will not occur. For example, the freezing conditions of Siberia are far too cold for decay to take place. Woolly mammoths (extinct relatives of elephants) have been found here in almost perfect condition. Some even had undigested grass in their stomachs. Some complete insects have been fossilised inside resin from trees. The resin excludes oxygen, which would normally allow decay to take place.

Hard parts of organisms

Parts of an organism like the skull, teeth and other bones of vertebrates do not decay easily. Sometimes, when whole organisms are not found, it is possible to make deductions from the parts that are found about those that are not there. For example, a femur (thigh-bone) that is one metre long must come from a large organism. The shape and angle of the joint at the top of the bone can give information about how it was joined to the pelvis. This allows biologists to deduce whether the animal stood on two legs or on all four. Hard parts of insects – like the exoskeleton – do not decay easily and are commonly found in fossil-bearing rocks.

Petrified remains and imprints

When an organism dies, it is often covered with some kind of sediment – such as soil. If it does not decay completely, then the remaining parts can be fossilised. If the sediment is compressed into a rock, the remains of the organism become part of that rock. The remains may then stay in their original form, as bones or exoskeletons for example. In some cases the original parts are dissolved away and replaced by other material which takes on the shape of the original organism. This becomes petrified (turned to rock). The original parts may be dissolved away and not replaced, leaving only an imprint of the organism in the rock (Figure 19.10).

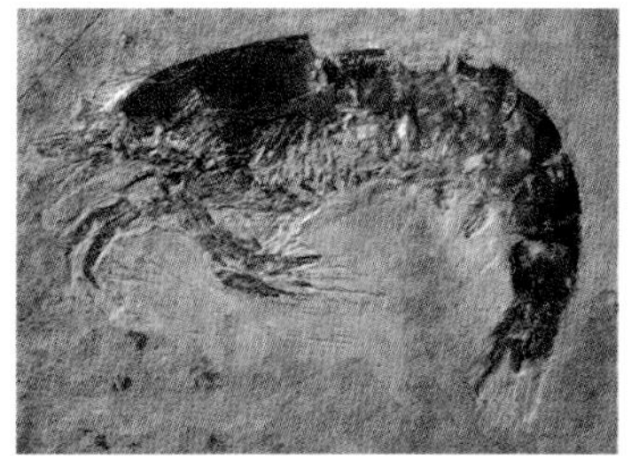

Figure 19.10 *Different types of fossils*

Rocks containing fossils are often found in strata (layers). The oldest rocks are usually in the lower strata whilst the upper strata contain younger rocks. This helps to date any fossils that are present. Dating rocks in this way is called **stratigraphy**. Sometimes fossils in successive layers show gradual changes. This provides evidence of the way in which the organism evolved over the period of time represented by the rocks in which they were found.

Evidence of this type has helped biologists to suggest a possible course of evolution of the modern horse. The fossil record suggests that it evolved from a much smaller animal. This animal had toes on all its feet and teeth similar to those of a monkey. Figure 19.11 shows some of the changes that occurred during the evolution of the modern horse.

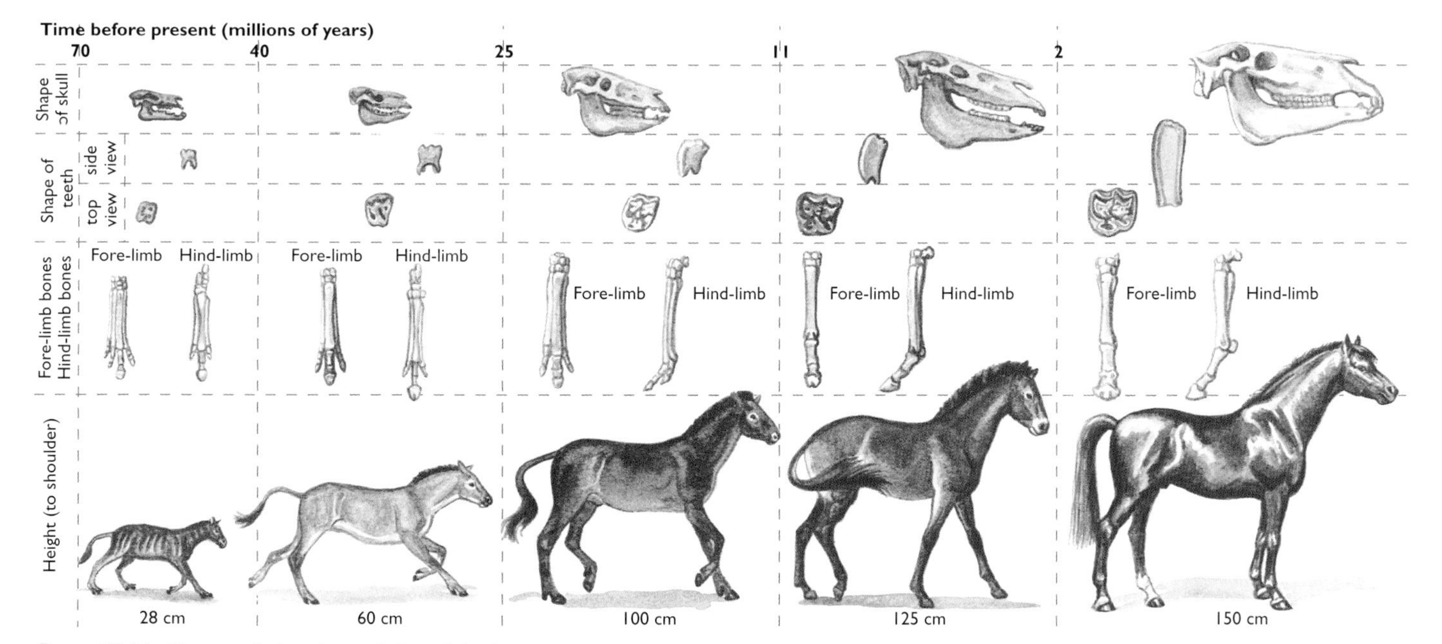

Figure 19.11 Changes during the evolution of the horse.

Biologists do not believe that one form of horse simply evolved into another, then another, with the original animal evolving directly into the modern horse. They believe that several other horse-like animals evolved in different parts of the world. However, none of these could compete successfully with other animals and so became extinct. Other evidence for evolution is described in Table 19.4.

Term	Description	Illustration
comparative anatomy	Biologists compare the anatomy (internal structure) of groups of animals and plants. By comparing the structure of the limbs of vertebrates, a common pattern is found – the **pentadactyl** (five-fingered) **limb**. Surely this did not arise separately in all the vertebrate groups? It must have arisen once and been modified to the various forms shown.	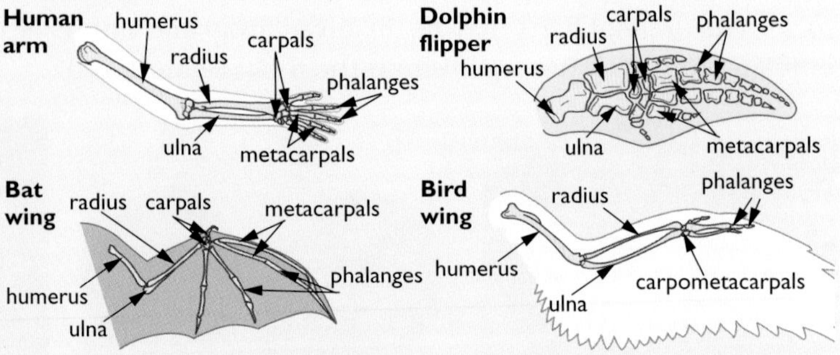 Figure 19.12 *The pentadactyl limb. All these limbs have the same basic plan:* • *a single upper limb bone* • *two lower limb bones* • *a cluster of bones at the wrist* • *a number of digits (fingers), usually five.* *They have been modified for different purposes by natural selection.*
comparative embryology	Comparing the ways in which the embryos of different animals develop can give clues about how closely related they are. It also reveals remarkable common features. Human embryos have structures resembling gills for a brief period in their development.	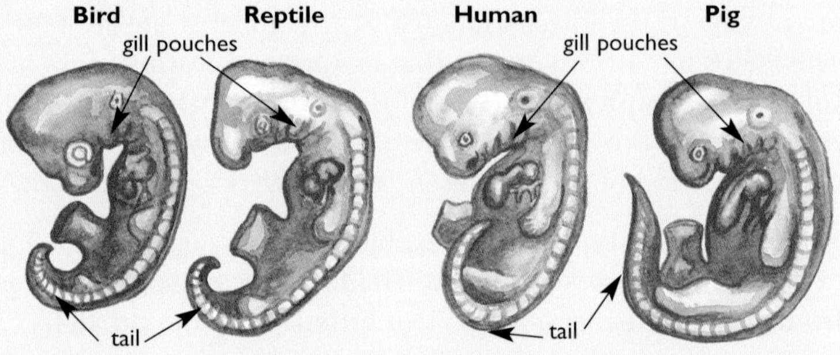 Figure 19.13 *Four different embryos at similar stages of development.*
comparative biochemistry	Comparing the structure of DNA and proteins of different organisms can suggest evolutionary links. 98% of our DNA is the same as that of a chimpanzee. Using this technique, the chimpanzee appears to be our closest relative. Analyses of proteins have suggested that the gorilla is our closest relative and the chimpanzee our second closest.	
parallel evolution	Species of animals from different groups have evolved to occupy similar niches in different parts of the world. For example, marsupial mammals (young develop in a pouch, e.g. the kangaroo) and placental mammals (young develop in the uterus) have both produced species that appear very similar and yet are, genetically, very different indeed. This must be due to natural selection operating in the same way on different animals in similar environments.	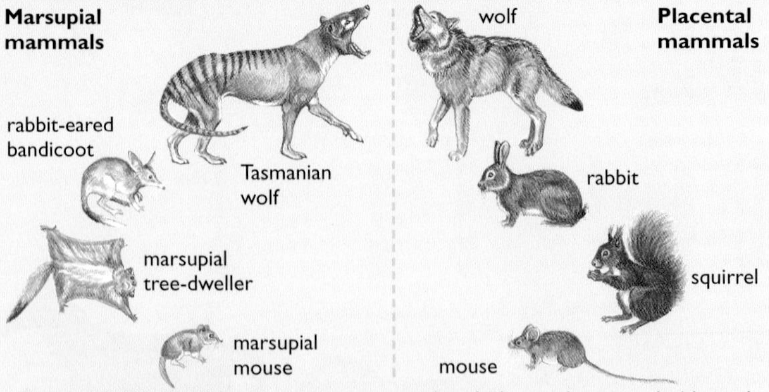 Figure 19.14 *Parallel evolution in marsupial and placental mammals. Natural selection has produced similar forms to fill similar niches in different parts of the world. Some of the marsupial mammals of Australia look remarkably similar to some placental mammals of Europe.*

Table 19.4: *Other evidence for evolution.*

Classification of organisms

Most of us have a natural tendency to 'put things in boxes' to make them easier to find. We often hang all shirts or blouses together, put socks in one drawer and T-shirts in another. This means that when we want our favourite T-shirt, we at least know where to start looking. We have **classified** our clothes into certain groups and placed all the articles of each group together.

For centuries biologists have had the same urge to classify living things. At first, many of the classification systems used were what we now call **artificial classifications**. Today, when classifying organisms, we attempt to take account of evolution and try to produce **natural classifications**. These place organisms that are closely related through evolution into similar groups. For example, all cats (domestic cat, wild cat, lynx) are placed in a group called *Felis*.

In artificial classifications, organisms are placed in groups according to the number of structures that they have in common. This sounds quite logical until you think that a hummingbird, bumblebee and horseshoe bat might all be placed in the same group because they all have wings.

Many early attempts at classification resulted in an organism being given a complex descriptive name that might run to ten or more Latinised words. Few people could make sense of this. Modern systems of classifying and naming organisms are based on a system devised in the eighteenth century by the Swedish biologist Carl von Linne. His family 'Latinised' the name and we know him better as Linnaeus (Figure 19.15).

ideas
evidence

Figure 19.15 *Linnaeus (1707–1778).*

Linnaeus was the first person to devise a systematic method of naming organisms. It was called the **binomial system** and is still used today. Linnaeus' work was primarily in classifying plants, but the system was so successful that it is applied to all organisms, from the simplest bacteria to the most complex mammals. In this system, every organism has a two-part name. For example, all cats have the first name *Felis*. This puts them in the cat group. The second name tells us exactly which cat it is. The common 'moggie' is *Felis domestica* – the domestic cat. Linnaeus classified many thousands of plants and gave them all names using his binomial system. He published his work in a book called *'Systema Naturae'*. The first edition was only 14 pages long. By the time he published the 12th edition in 1766, it had become much longer – 2300 pages! Copies of Linnaeus' work are now owned by the Linnaean Society in London.

A century before Linnaeus, an Englishman, John Ray, had also carried out work on classifying plants. In Ray's classification, the plants in one of his groups had to have many similarities – not just share a few key features as had been the case with previous systems. We still use today many of the plant groupings that Ray established. John Ray also tried to define species in terms of their ancestry rather than their physical appearance. He said that 'two organisms belong to the same species if one is descended from the other, or, if they both share a common ancestor'. We still use some of Ray's ideas in defining a species, but they do not include the idea of being able to produce fertile offspring. However, Ray did not believe that new species were constantly appearing. He still thought that the species that existed then would always exist with very little change. It would be another 200 years before Charles Darwin published *'The Origin of Species'*.

Modern classification

All cats are placed in the group *Felis*. But what of the 'big cats' – the lions, tigers and panthers? They are not cats, yet there are obvious similarities.

Dogs, too, are carnivorous and have some similarity to both groups. Any classification system must take account of both similarities *and* differences. There must be some way of putting dogs, cats and the big cats in the same group (to show similarities) and also in different groups (to take account of differences). The answer is different-sized groups. All mammals that hunt and kill other animals are carnivores. Within that group are bears, cats, panthers, dogs and others. Within the cat group are the different types of cats.

The different sized groups are arranged as shown in Table 19.5.

Group	Description
Kingdom	The largest group. There are only five kingdoms. The **animal kingdom** includes all multicellular organisms which are able to move at some stage in their life cycles and have cells *without* cell walls. The **plant kingdom** is made up of multicellular organisms which have cells with cellulose cell walls and can photosynthesise. The three other kingdoms are: • **prokaryotes** – single celled organisms, nearly all are bacteria • **protoctistans** – a varied group including many unicellular organisms (such as *Amoeba*) and all the algae (the seaweeds and some of the plankton) • **fungi** – plant-like organisms that have no chlorophyll and cannot photosynthesise; cells have non-cellulose cell walls.
Phylum	The next largest grouping. Each kingdom has more than one phylum. There are several phyla within the animal kingdom. A phylum usually has a distinct body plan. **Chordates** all have some kind of supporting column *inside* their bodies – most have a backbone.
Class	A sub-group of a phylum. **Mammals** are a class of chordates that have hair and mammary glands.
Order	A division of a class. The order **Carnivora** includes those mammals that hunt and kill other animals.
Family	A group of essentially similar types. Within the carnivora, all the cat-like mammals are placed in the cat family – the **Felidae**. This includes lions, tigers, panthers, as well as the domestic cat.
Genus	A group of species with a good number of features in common. Within the Felidae, two genera are *Panthera* (the panthers) and *Felis* (the cats).
Species	A group of individuals that share common genes and can interbreed to produce fertile offspring. Within the genus *Panthera*, *Panthera leo* is the lion. In the genus *Felis*, *Felis domestica* is the domestic cat (Figure 19.16).

Table 19.5: *The 5-kingdom system of classification.*

Our species is *Homo sapiens*. 'Homo' tells us that we are part of the genus that includes all humans, living and dead. 'Sapiens' means 'intelligent' – we describe ourselves as 'intelligent humans'.

The species that exist today represent a 'snapshot in time' of the progress of evolution to date. Many of these species may become extinct in the future. Whole classes of present-day animals and plants may become extinct in the next mass extinction event – whenever it occurs.

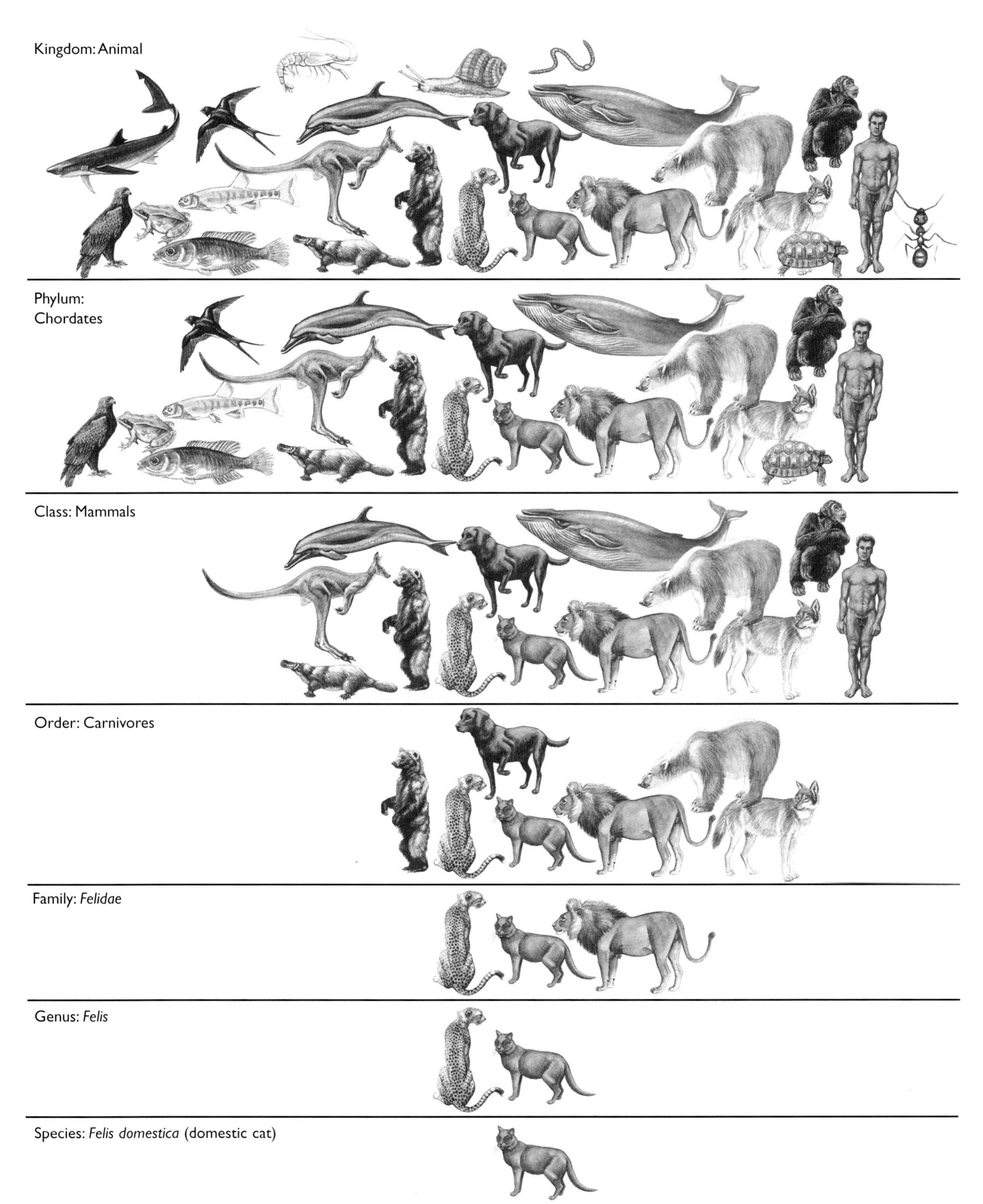

Figure 19.16 *The classification of the domestic cat.*

End of Chapter Checklist

If you haven't got a copy of your specification, read the introduction on page vi.

You will need to be able to do some or all of the following. Check your Awarding Body's specification (syllabus) to find out exactly what you need to know.

- Understand the importance of the work of Charles Darwin and Alfred Russell Wallace in putting forward the theory of natural selection.
- Understand how natural selection operates to produce a population of organisms that is well adapted to its environment.
- Deduce, from information given, the selection pressure operating in the situation described, the forms that will be selected for and those that will be selected against, and the likely outcome if that selection pressure continues to operate for many generations.
- Describe the role of natural selection in the examples quoted in this chapter, for example sickle cell anaemia and the peppered moth.
- Explain the role of natural selection in creating new species.
- Describe the outline course of evolution.
- Describe some of the evidence for evolution, including fossil evidence and dating of fossils, comparative anatomy, comparative embryology, comparative biochemistry and parallel evolution.
- Understand the importance of the work of Linnaeus in establishing the principles of modern classification systems.
- Understand the concept of a species.
- Describe the main features of plants and animals.

Questions

More questions on natural selection and evolution can be found at the end of Section E on page 257.

1 ***a)*** What does the term 'survival of the fittest' mean?

b) Which two biologists arrived at the same idea concerning the 'survival of the fittest' at the same time?

c) Why were the ideas of natural selection controversial when they first appeared in the nineteenth century?

2 The changes in the distribution of the peppered moth as areas became industrialised is often used as evidence of natural selection. There are two forms of the peppered moth, the peppered form and the dark form.

a) Explain why the dark form of the moth became much more common in industrialised cities.

b) Explain why the dark form of the moth remained extremely rare in the countryside during the same period.

c) Suggest why the two types of the peppered moth have not become separate species.

3 Copy and complete the table showing the classification of human beings.

Classification group	Human example
kingdom	
phylum	chordates
	mammals
	primates
family	hominidae
genus	
species	

4 Warfarin is a pesticide that was developed to kill rats. When it was first used in 1950, it was very effective. Some rats, however, had a mutant allele that made them resistant to warfarin. Nowadays the pesticide is much less effective.

   a) Use the ideas of natural selection to explain why warfarin is much less effective than it used to be.

   b) Suggest what might happen to the number of rats carrying the warfarin resistance allele if warfarin were no longer used. Explain your answer.

5 In the Galapagos Islands, Charles Darwin identified a number of species of finch. He found evidence to suggest that they had all evolved from one ancestral type, which had colonised the islands from South America. The main differences between the finches was in their beaks. The diagram shows some of the beak types and that of the likely ancestral finch.

   a) Explain how the seed-eating finches are adapted to their environment.

   b) Explain how the finches that eat insects and live in woodland are adapted to their environment.

   c) Use the information in the diagram to help you explain how the common ancestor could have evolved into the different type of finches.

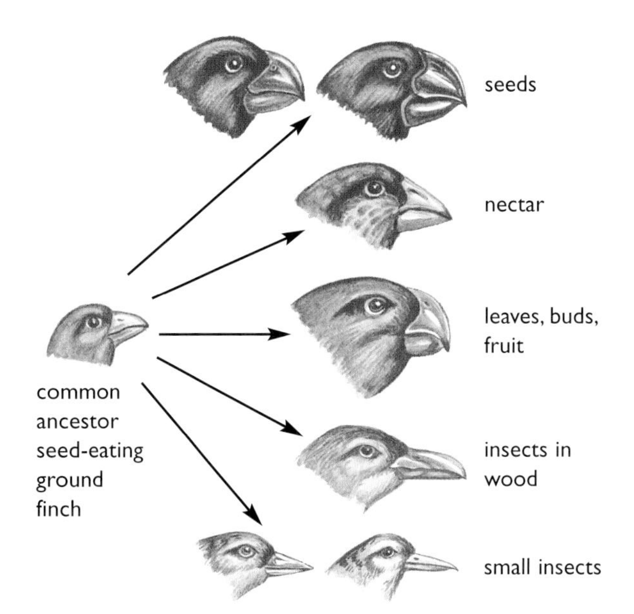

6 Write an essay outlining how a biologist who accepts evolution as a fact might try to convince a sceptical person (one who does not necessarily accept evolution as a fact). In your essay, you should refer to possible doubts or objections by the sceptical person and evidence that could possibly answer these objections.

Chapter 20: Selective Breeding

Humans have been selectively breeding animals and plants ever since they first became farmers. This chapter looks at how the process has changed over the years and highlights some moral and ethical issues linked with current practices.

About 12 000 years ago, the human way of life changed significantly. Humans began to grow plants and keep animals for milk and meat. They became farmers rather than hunters. This change took place first in the Middle East. Similar changes took place a little later in the Americas (where potatoes and maize were being grown) and in the Far East (where rice was first cultivated).

In the Middle East, humans first grew the cereal plants wheat and barley, and domesticated sheep and goats. Later, they domesticated cattle and pigs. Cultivating crops and keeping stock animals made it possible for permanent settlements to appear – human village life began. Because of the more certain food supply, there was spare time, for the first time ever, for some people to do things other than hunt for food.

Ever since the cultivation of the first wheat and barley and the domestication of the first stock animals, humans have tried to obtain bigger yields from them. They cross-bred different maize plants (and barley plants) to obtain strains that produced more grain. They bred sheep and goats to give more milk and meat – selective breeding had begun. Today, animals and plants are bred for much more than food. They are bred to produce a range of medicines, and for research into spare-part surgery and the action of drugs.

Selective breeding is best described as the breeding of only those individuals with desirable features. It is sometimes called '**artificial selection**', as human choice, rather than environmental factors, is providing the **selection pressure** (see Chapter 19).

The methods used today for selective breeding are vastly different from those used only 50 years ago. Modern gene technology makes it possible to create a new strain of plant within weeks, rather than years. These new techniques raise serious moral and ethical questions which will be discussed later.

The production of modern bread wheats by selective breeding is probably one of the earliest examples of producing genetically modified food. Each original wild wheat species had 14 chromosomes per cell. The wild emmer hybrid had 28 chromosomes per cell. Modern bread wheat has 42 chromosomes per cell. Selective breeding has modified the genetic make-up of wheat.

Traditional selective breeding

Plants

Traditionally, farmers have bred crop plants of all kinds to obtain increased yields. Probably the earliest example of selective breeding was the cross-breeding of strains of wild wheat. The aim was to produce wheat with a much increased yield of grain (Figure 20.1). This wheat was used to make bread.

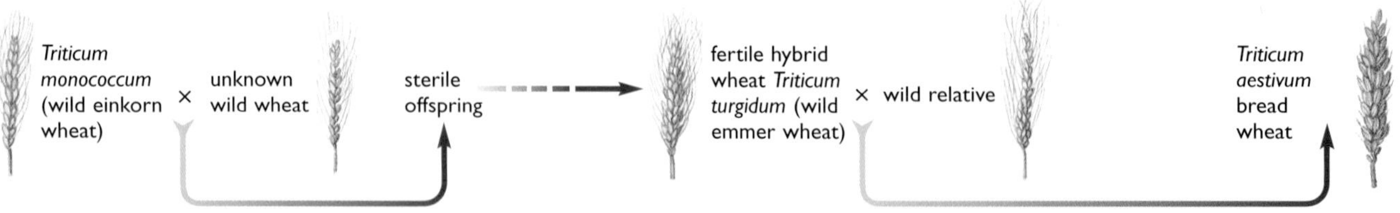

1 About 11 000 years ago, two strains of wild wheat were cultivated by farmers. Initially, all attempts at crossbreeding to produce wheats with a better yield gave only sterile offspring.

2 About 8000 years ago, a fertile hybrid wheat appeared from these two wild wheats. This was called emmer wheat and had a much higher yield than either of the original wheats.

3 The emmer wheat was cross-bred with another wild wheat to produce wheat very similar to the wheats used today to make bread. This new wheat had an even bigger yield and was much easier to 'process' to make flour.

Figure 20.1 *Modern wheat is the result of selective breeding by early farmers.*

Other plants have been selectively bred for certain characteristics. *Brassica* is a genus of cabbage-like plants. One species of wild brassica (*Brassica olera*) was selectively bred to give several strains, each with specific features. Some of the strains had large leaves, others had large flower heads, and others produced large buds.

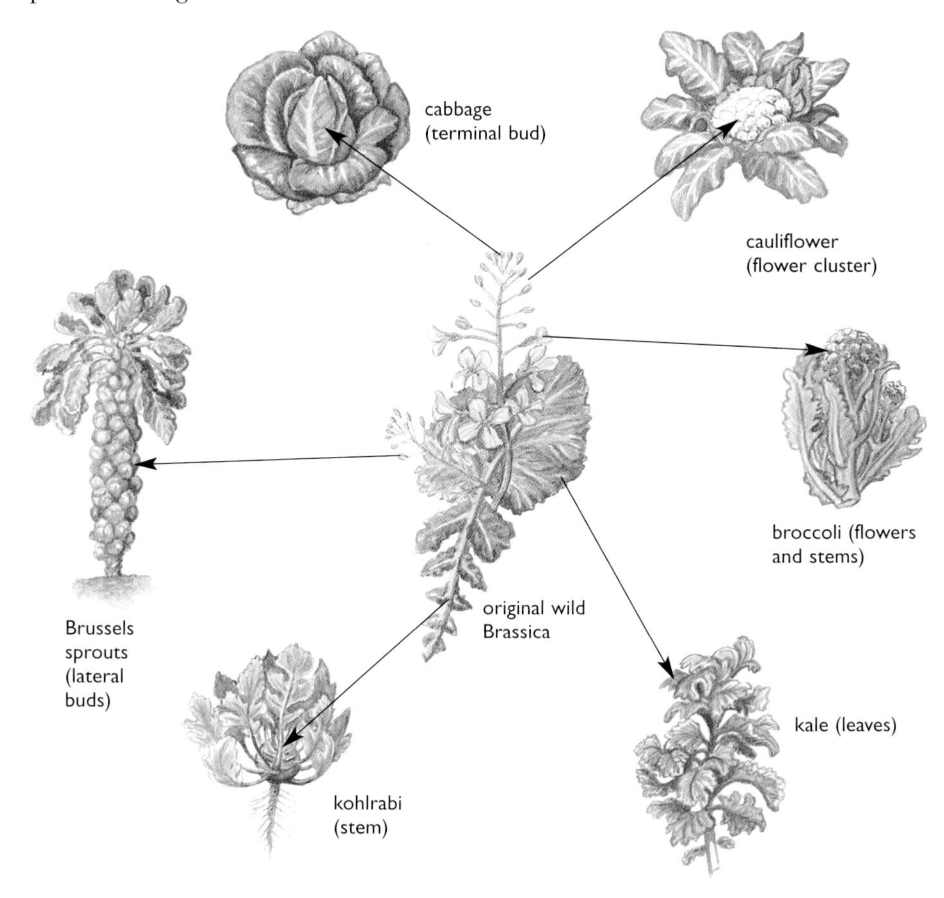

Figure 20.2 *Selectively breeding the original wild brassica plants to enhance certain features has produced several familiar vegetables.*

Selective breeding has produced many familiar vegetables. Besides the ones produced from *Brassica*, selective breeding of wild *Solanum* plants has produced the many strains of potatoes that are eaten today. Carrots and parsnips are also the result of selective breeding programmes.

Crop plants are bred to produce strains that:

- give higher yields
- are resistant to certain diseases (the diseases would reduce the yields)
- are resistant to certain insect pest damage (the damage would reduce the yield)
- are hardier (so that they survive in harsher climates or are productive for longer periods of the year)
- have a better balance of nutrients in the crop (for example, plants that contain more of the types of amino acids needed by humans).

Figure 20.3 shows a field of potato plants. Some have been bred to be resistant to insect pests, while others were not selectively bred in this way.

Plant breeders have not just bred plants for food. Nearly all garden flowers are the result of selective breeding. Breeders have selected flowers of a particular size, shape, colour and fragrance. Roses and orchids are among the most selectively bred of our garden plants.

Figure 20.3 *Selective breeding can reduce damage by pests. The plants in area A are bred to be resistant to a pest. Plants in area B have not been bred to be resistant.*

Animals

Farmers have bred stock animals for similar reasons to the breeding of crops. They have selected for animals that:

- produce more meat, milk or eggs
- produce more fur or better quality fur
- produce more offspring
- show increased resistance to diseases and parasites.

Again, like crop breeding, breeding animals for increased productivity has been practised for thousands of years. Figure 20.4 shows a picture of a stone tablet found in Iran and dated at over 5000 years old. It appears to record the results of breeding domesticated donkeys.

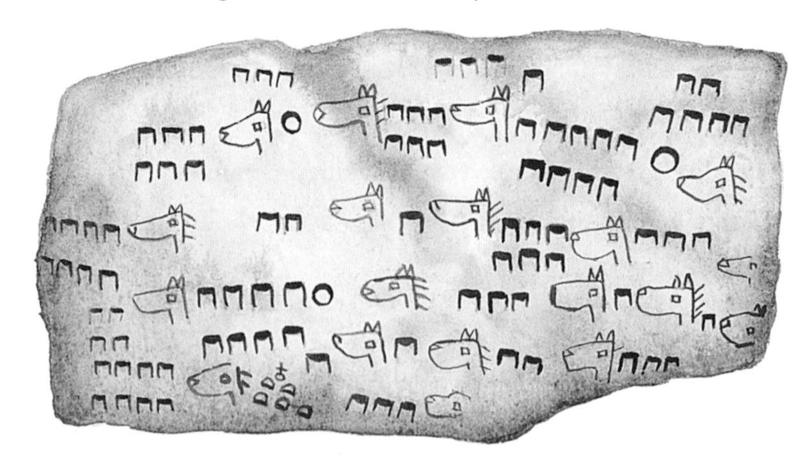

Figure 20.4 *The tablet of stone with these markings is over 5000 years old. Does it show a breeding programme for domestic donkeys?*

For many thousands of years, the only way to improve livestock was to mate a male and a female with the features that were desired in the offspring. In cattle, milk yield is an important factor and so high yielding cows would be bred with bulls from other high yielding cows.

Since the Second World War, the technique of **artificial insemination** (**AI**) has become widely available. Bulls with many desirable features ('superior bulls') have been purchased by the Milk Marketing Board and kept in special centres. Semen obtained from these bulls is diluted, frozen and stored. Farmers can purchase quantities of this semen to inseminate their cows. AI makes it possible for the semen from one bull to be used to inseminate many thousands of cows.

Modern sheep are domesticated wild sheep, and pigs have been derived from wild boars. Just think of all the varieties of dogs that now exist. All these have been derived from one ancestral type. This original 'dog' was a domesticated wolf (Figure 20.5). In domesticating the wolf, humans gained an animal that was capable of herding stock animals. The sheepdog has all the same instincts as the wolf except the instinct to kill. This has been selectively 'bred out'.

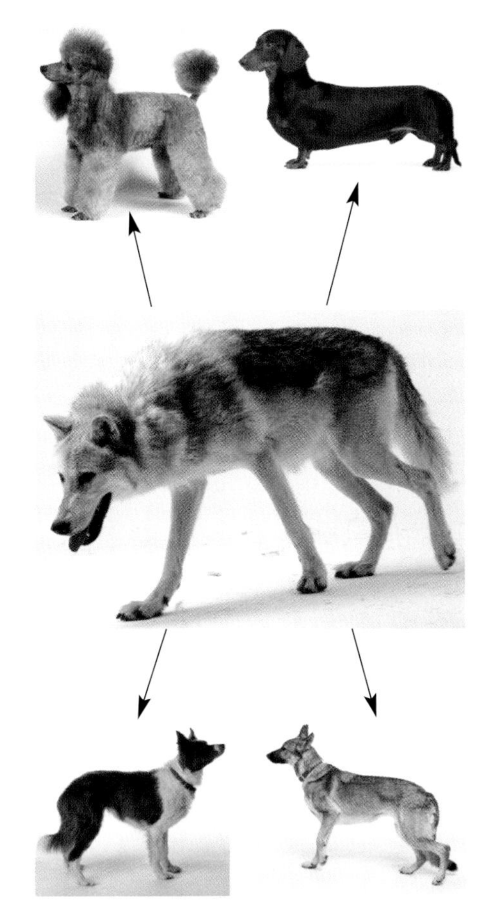

Figure 20.5 *The many different breeds of dog all originate from a common ancestor – the wolf.*

Modern selective breeding

Plants

If you visit any major department store just before Mother's day, you will see hundreds of seemingly identical pot plants on show. They seem identical because they are identical, genetically at least. They have been **cloned** from one individual plant that has the necessary features.

The term cloning describes any procedure that produces genetically identical offspring. Taking cuttings of plants and growing them is a traditional cloning technique (Figure 20.6).

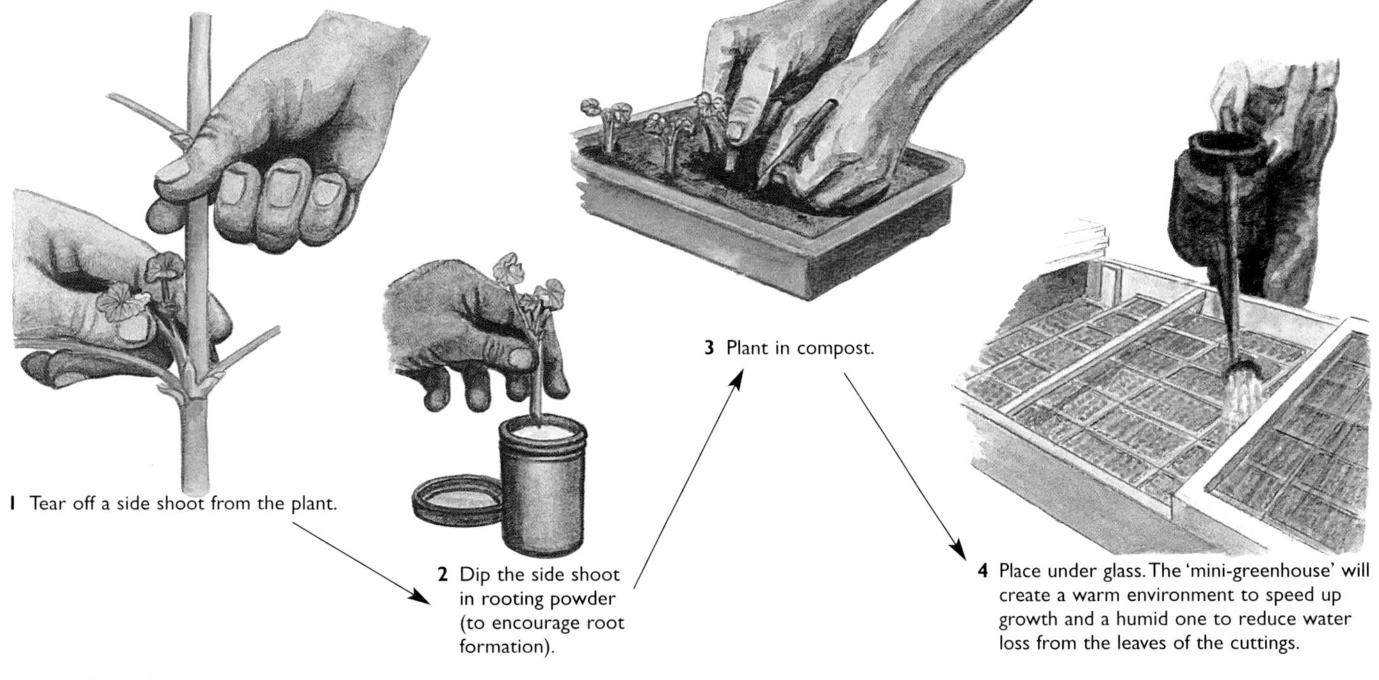

Figure 20.6 *Taking stem cuttings.*

All the cuttings contain identical genes as they are all parts of the same parent plant. As they grow, they form new cells by mitosis, copying the genes in the existing cells exactly. The cuttings develop into a group of genetically identical plants – a **clone**. Any differences will be due to the environment. Many garden flowers have traditionally been propagated this way.

Some modern cloning techniques are essentially the same as taking cuttings – removing pieces of a plant and growing them into new individuals. The technology, however, is much more sophisticated. Using the technique of **micropropagation**, thousands of plants can quickly be produced from one original (Table 20.1).

Stages	Illustrations
The tips of the stems and side shoots are removed from the plant to be cloned. These parts are called **explants**. The explants are trimmed to a size of about 0.5 – 1mm. They are then placed in an agar medium that contains nutrients and plant hormones to encourage growth. More explants can be taken from the new shoots that form on the original ones. This can be repeated until there are enough to supply the demand.	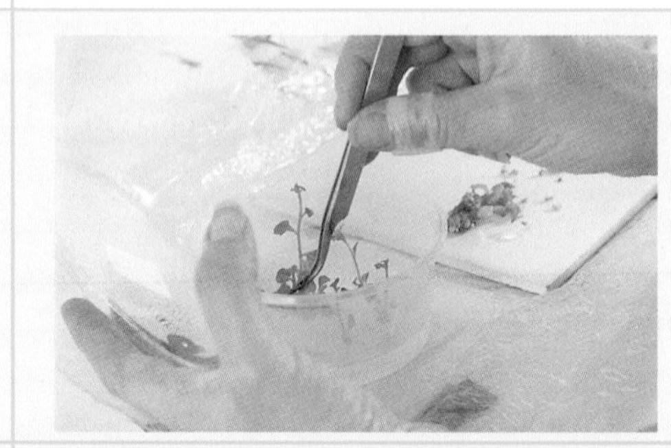 Figure 20.7 *Explants growing in a culture medium.*
The explants with shoots are transferred to another culture medium containing a different balance of plant hormones to induce root formation.	 Figure 20.8 *Explants forming roots.*
When the explants have grown roots, they are transferred to greenhouses and transplanted into compost. They are then gradually acclimatised to normal growing conditions. The atmosphere in the greenhouse is kept very moist to reduce water loss from the young plants. Because of the amount of water vapour in the air, they are often called 'fogging greenhouses'.	 Figure 20.9 *Young plants being grown in compost in a greenhouse.*

Table 20.1: *The main stages in micropropagation.*

There are many advantages to propagating plants in this way:

- Large numbers of genetically identical plants can be produced rapidly.
- Species that are difficult to grow from seed or from cuttings can be propagated by this method.
- Plants can be produced at any time of the year.
- Large numbers of plants can be stored easily (many can be kept in cold storage at the early stages of production and then developed as required).
- Genetic modifications can be introduced into thousands of plants quickly after modifying only a few plants.

Many strains of bananas are infertile. They are now commonly reproduced by micropropagation. Other plants produced this way include lilies, orchids and agave plants (used to make the drink tequila).

Animals

Artificial insemination (AI) was the first step to speeding up the genetic improvement of stock animals. AI allows the sperm from one superior bull to be used to inseminate many cows. It does not, however, allow a superior cow to produce more than a few calves. This was only made possible by the techniques of ***in vitro* fertilisation** (**IVF**) and **embryo transplantation**. IVF involves fertilising ova (egg cells) in the laboratory.

In vitro means in glass.

Using IVF, sperm form a superior bull can be used to fertilise many ova from a superior cow. The steps in the procedure are described in Figure 20.10.

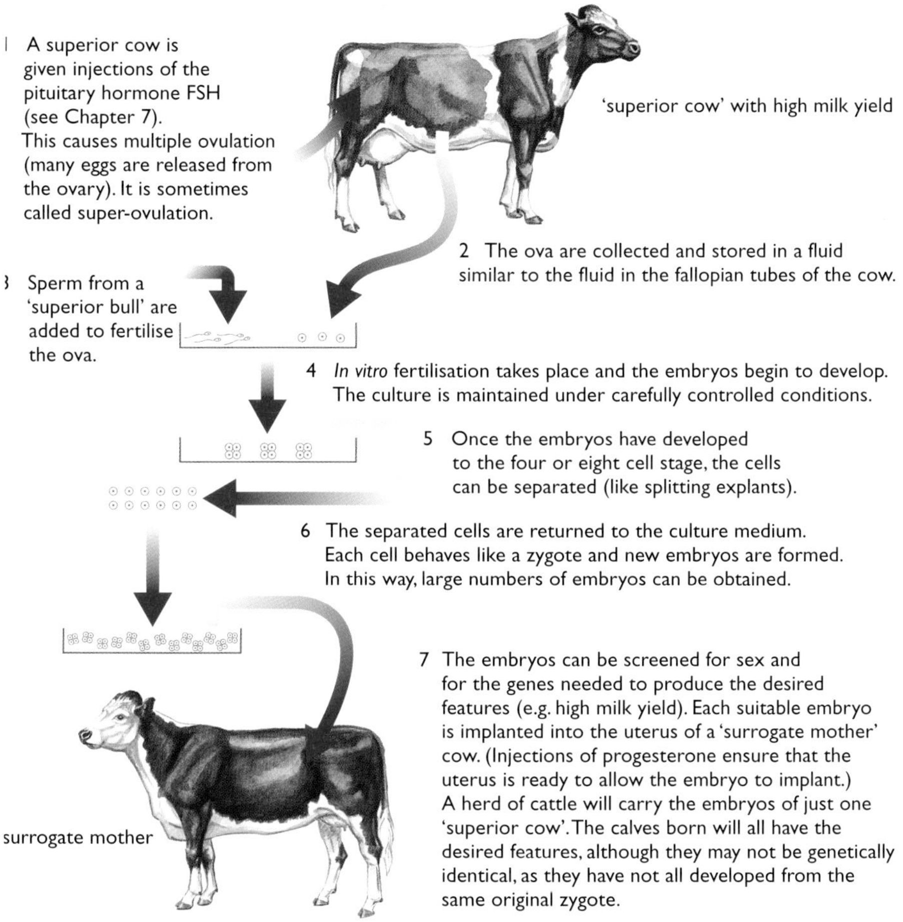

1 A superior cow is given injections of the pituitary hormone FSH (see Chapter 7). This causes multiple ovulation (many eggs are released from the ovary). It is sometimes called super-ovulation.

'superior cow' with high milk yield

2 The ova are collected and stored in a fluid similar to the fluid in the fallopian tubes of the cow.

3 Sperm from a 'superior bull' are added to fertilise the ova.

4 *In vitro* fertilisation takes place and the embryos begin to develop. The culture is maintained under carefully controlled conditions.

5 Once the embryos have developed to the four or eight cell stage, the cells can be separated (like splitting explants).

6 The separated cells are returned to the culture medium. Each cell behaves like a zygote and new embryos are formed. In this way, large numbers of embryos can be obtained.

7 The embryos can be screened for sex and for the genes needed to produce the desired features (e.g. high milk yield). Each suitable embryo is implanted into the uterus of a 'surrogate mother' cow. (Injections of progesterone ensure that the uterus is ready to allow the embryo to implant.) A herd of cattle will carry the embryos of just one 'superior cow'. The calves born will all have the desired features, although they may not be genetically identical, as they have not all developed from the same original zygote.

surrogate mother

Figure 20.10 *The main stages of in vitro fertilisation (IVF).*

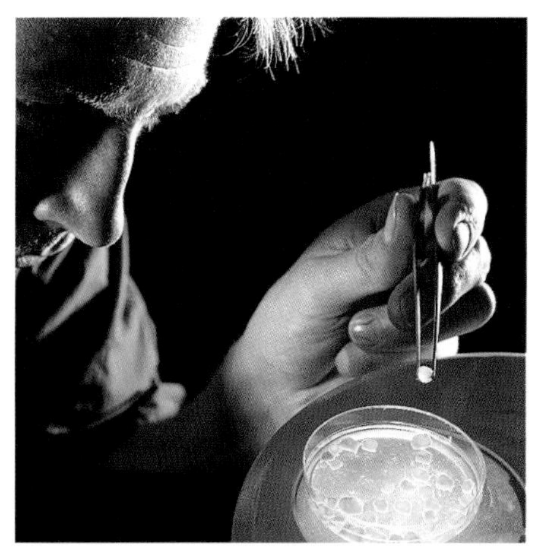

Figure 20.11 *Removal of ovarian cow follicles as part of IVF technique.*

The zygotes formed at Stage 3 will not be identical as different sperm fertilise different ova.

The new 'zygotes' formed by splitting an embryo (at Stage 5) will be genetically identical. All the cells of the embryo were formed by mitosis from the single zygote.

Cloning animals

We have been able to clone plants by taking cuttings for thousands of years. It is now possible to make genetically identical copies of animals. The first, and best-known, example of this is the famous cloned sheep, Dolly.

Dolly was produced by persuading one of her mother's ova (egg cells) to develop into a new individual without being fertilised by a sperm. The nucleus of the ovum was removed and replaced with a cell in the udder of another sheep. The cell that was formed had the same genetic information as all the cells in the donor and so developed into an exact genetic copy. The stages in the procedure are shown in Figure 20.12. Figure 20.13 shows how an udder cell is inserted into an egg cell that has had its nucleus removed.

The nucleus of an ovum is haploid (see Chapter 18). It cannot develop into a new individual because it only has half the chromosomes of normal body cells. An ordinary diploid body cell, even though it has all the chromosomes, is too specialised. Transferring a diploid nucleus into an egg cell that has had its nucleus removed creates a cell that is capable of developing into a new individual. In practice, it is easier to transfer a small whole cell rather than attempt to transfer just the nucleus, as the nucleus alone could too easily be damaged.

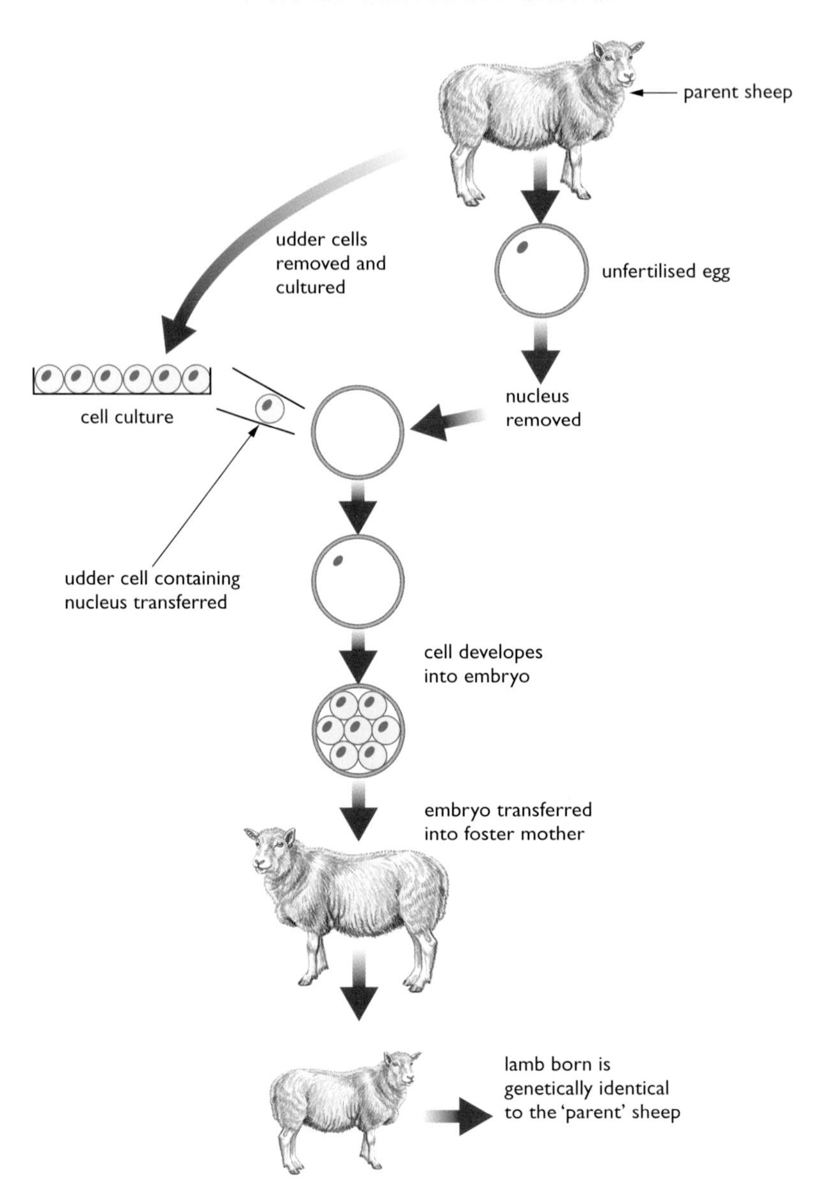

Figure 20.12 *How 'Dolly' was produced.*

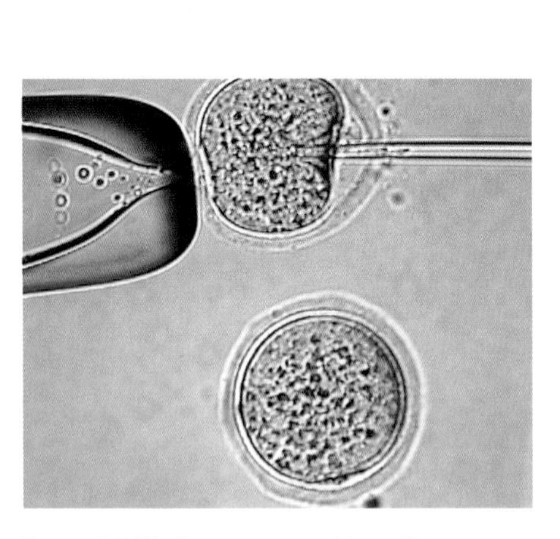

Figure 20.13 *Inserting an udder cell into an egg cell that has had its nucleus removed.*

Dolly was only produced after many unsuccessful attempts. Since then, the procedure has been repeated using other sheep as well as rats, mice and pigs. Some of the animals produced are born deformed. Some do not survive to birth. Biologists believe that these problems occur because the genes that are transferred to the egg are 'old genes'. These genes came from an animal that had already lived for several years and from cells specialised to do things other than produce sex cells. It will take much more research to make the technique reliable.

Cloning humans

Although sheep are very different from humans, both are mammals and so there are also many similarities. The techniques used to produce Dolly could be used to clone humans. A group of doctors in Italy claim to be ready to carry out the first human cloning. There has been no shortage of women volunteers, despite the problems evident in cloning. Nearly all are women who are infertile and see no prospect of having children in any other way.

Down on the 'pharm'

Cloning animals has special value if the animal produces some important product. Sheep have been genetically modified to produce several human proteins. One of these is used to treat conditions such as emphysema and cystic fibrosis. The genetically modified sheep secrete the protein in their milk. Cloning sheep like these would allow production of much more of this valuable protein. Genetically modifying and cloning mammals in this way has been nicknamed 'pharming'! Polly, the first cloned, genetically modified sheep, was born a year after Dolly.

Animals that have had genes transferred from other species are called **transgenic animals**.

'Pharming' could produce not just whole animals, but individual organs, such as kidneys, livers and hearts as well. Research is currently underway into pharming pigs to produce organs that could be used in transplant operations. The pigs must be genetically modified so that the organs will not be rejected by the human immune system (see Chapter 23 for more details).

Some moral and ethical concerns of modern selective breeding

Morality is our personal sense of what is right, or acceptable, and what is wrong. It is not necessarily linked to legality. For example, an individual may think it unacceptable to hunt foxes (currently legal) and quite acceptable to snort cocaine (currently illegal). Another person may think the opposite. The morals of the two are different and it would be difficult for either to persuade the other that he or she is wrong. Morals are often shaped by religion and family upbringing.

Ethics have a sense of right and wrong also, but they are not an individual opinion. Ethics represent the 'code' adopted by a particular group to govern its way of life. They are the 'rules' and, as such, can often be linked to legality. Medical ethics are the rules that the medical profession adopts to define acceptable behaviour in researching diseases and treating patients.

Morals and ethics can sometimes be in conflict. It is currently illegal in England to attempt to clone humans. This makes it *ethically* unacceptable to the medical profession. An individual doctor, however, may think that it is *morally* acceptable to carry out human cloning, for a number of very positive reasons.

Issues	Concerns about selective breeding of plants
Selective breeding of plants resistant to herbicides (weedkillers) may lead to overuse of herbicides, which may accumulate in the environment and pass along food chains. People believe that we should not produce such plants.	Many environmental biologists find this unacceptable. They feel that accumulation of herbicides may reduce the numbers of wild animals. They consider the practice to be an ethical issue. Many ordinary consumers also feel strongly about this. They consider it to be dangerous and would prefer to spend a little extra to purchase vegetables and fruits guaranteed not to have been treated with pesticides. The market for 'organic foods' is increasing all the time.
The release of many different strains of selectively bred plants means that, by the laws of chance, some are bound to be unsafe.	Many people find this morally wrong. They perceive it as an unnecessary risk to generate profit for the companies involved. However, many more people, who have difficulties in growing sufficient crops to feed themselves, do not have such concerns. They perceive it as morally wrong not to produce any new crop that might help to reduce their problem.
By only breeding from certain strains of a plant (for example only using high cropping strains) other genes may be lost from the gene pool of that species.	This is an ethical issue for many biologists. They feel that we should not allow genes to be lost unnecessarily. These genes could be of benefit to future generations and so should be retained.

Table 20.2: *Some concerns about selective breeding of plants.*

The gene pool of a species is the total of all the genes in all the populations of that species.

Some of the concerns about the selective breeding of animals are the same as those regarding selective breeding of plants. Concerns about ecological effects and reduction of the gene pool apply to both plants and animals. There are some concerns that only apply to the selective breeding of animals (Table 20.3).

Issues	**Concerns about selective breeding of animals**
If we selectively breed animals it may be possible to do this to humans. The result would be a 'super race'.	Most people think that this would be morally *and* ethically wrong. However, some biologists think that it would be morally acceptable to breed a race of humans that did not suffer from many of the genetic diseases that currently afflict us. It is not a black and white situation.
Animals should not be bred just to provide a supply of drugs for humans.	Most people who think this have few problems with breeding plants for this purpose. They feel that animals have brains and so may have 'souls' or individuality: 'Because they are capable of thought, they are very like humans and we wouldn't treat ourselves this way'. Clearly, this is a moral issue for many individuals.
Animals should not be bred just to provide a supply of human organs.	Many people find this morally unacceptable. Some could not accept an animal organ transplant on religious grounds. Others just could not stand the thought of an organ from another species inside them and some consider that it degrades the animals. Many biologists believe that other animals have the same right to exist on the planet as us and think that such animal transplants are ethically unacceptable. Others find little difference between breeding pigs to supply bacon and breeding pigs to supply a heart.
Cloning of humans (an extreme form of selective animal breeding) doesn't create a new individual: it merely creates a copy of one that already exists, so who is the new clone?	Many people from all professions find this morally unacceptable. The parent and child would have exactly the same genes, so is the child really just the parent reborn? Would the parent, therefore, insist that the child behave exactly like himself/herself as, genetically, they are the same? However, other people ask 'What is the difference between this situation and that of identical twins?'.

Table 20.3: *Some concerns about the selective breeding of animals.*

End of Chapter Checklist

If you haven't got a copy of your specification, read the introduction on page vi.

You will need to be able to do some or all of the following. Check your Awarding Body's specification (syllabus) to find out exactly what you need to know.

- Explain the term 'selective breeding'.
- Understand the importance of early humans cultivating crops and domesticating animals.
- Describe how modern bread wheats arose from lower-yielding wild wheats.
- Give examples of how selective breeding of wild plants has produced modern crop plants and garden plants.
- Give examples of how selective breeding of wild animals has produced the domestic animals of today.
- Describe how stem cuttings can be taken and explain why the plants produced are genetically identical.
- Describe the technique of micropropagation and explain the benefits of this technique.
- Explain how the techniques of *in vitro* fertilisation and embryo transplantation have allowed a 'superior cow' to produce many offspring with her desirable features.
- Describe how Dolly, the first cloned sheep, was produced.
- Understand why cloning animals often produces deformed offspring.
- Appreciate the moral and ethical concerns that surround modern selective breeding of animals and plants.

Questions

More questions on selective breeding can be found at the end of Section E on page 257.

1 Selective breeding is sometimes called 'artificial selection'.

a) How is selective breeding similar to natural selection?

b) How is selective breeding different from natural selection?

2 Selective breeding of crop plants often aims to increase the yield of the crop.

a) Describe, and explain the reasons for, three other aims of selective breeding programmes in crop plants.

b) Describe two advantages of micropropagation over the more traditional technique of taking cuttings.

c) Explain why plants produced by micropropagation will be genetically identical to each other and to the parent plant.

3 The diagram shows some of the features of a cow that might be used as a basis for a breeding programme.

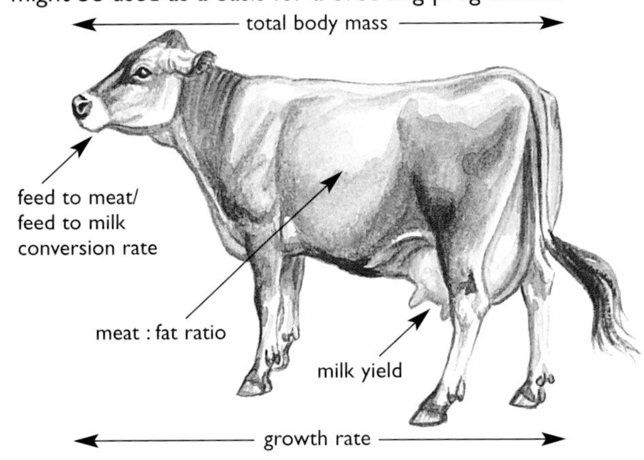

a) Which features would you consider important in a breeding programme for dairy cattle?

b) Assume that you had all the techniques of modern selective breeding available to you. Describe how you would set about producing a herd of high-yielding beef cattle.

4 The diagram shows the results of a breeding programme to improve the yield of maize (sweetcorn).

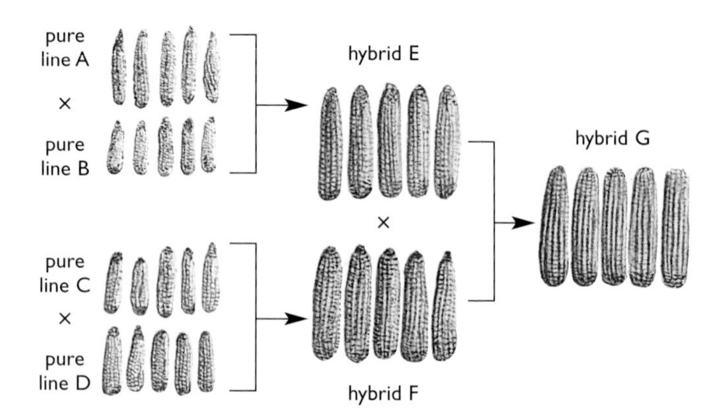

a) Describe the breeding procedure used to produce hybrid G.

b) Describe three differences between the corn cobs of hybrid G and those of hybrid C.

c) How could you show that the differences between hybrid G and hybrid C are genetic?

5 Write an essay about the benefits and concerns of selective breeding of animals. You should produce about one side of A4 word-processed work. Use books and the Internet to find out more information.

End of Section Questions

1 For natural selection to operate, some factor has to exert a 'selection pressure'. In each of the following situations identify both the selection pressure and the likely result of this selection pressure.

a) Near old copper mines, the soil becomes polluted with copper ions that are toxic to most plants. *(2 marks)*

b) In the Serengeti of Africa, wildebeest are hunted by lions. *(2 marks)*

c) A farmer uses a pesticide to try to eliminate pests of a potato crop. *(2 marks)*

Total 6 marks

2 Micropropagation produces thousands of genetically identical plants. Small 'explants' from the parent plant are grown in culture media.

a) Outline the main stages in micropropagation. *(4 marks)*

b) Explain why the plants formed by micropropagation are genetically identical. *(2 marks)*

c) In some cases, the explants used contain only a few cells with neither roots nor shoots. Plant hormones are added to the culture media to encourage root and shoot formation. Two of these hormones are called kinetin and auxin. The diagram shows the effects of using different concentrations of the two hormones on root and shoot growth of the explants.

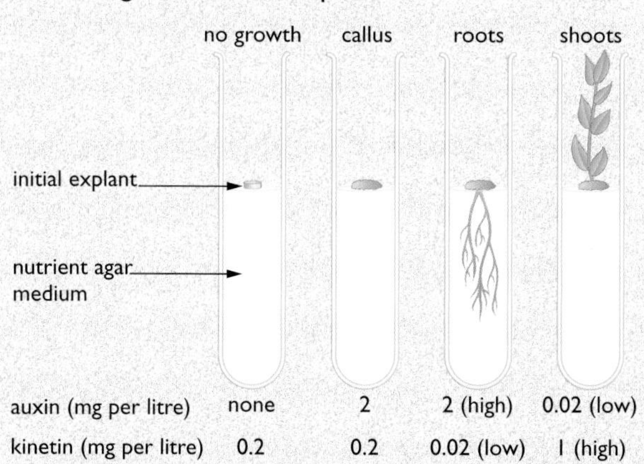

i) What is the effect of adding kinetin or auxin without any other hormone? *(2 marks)*

ii) Describe how you would treat these explants to produce first shoots and then roots. *(3 marks)*

d) Explain one advantage and one disadvantage of micropropagation. *(2 marks)*

Total 13 marks

3 The diagram shows the inheritance of PTC tasting in a family. Although PTC has a very bitter taste, some people cannot taste it.

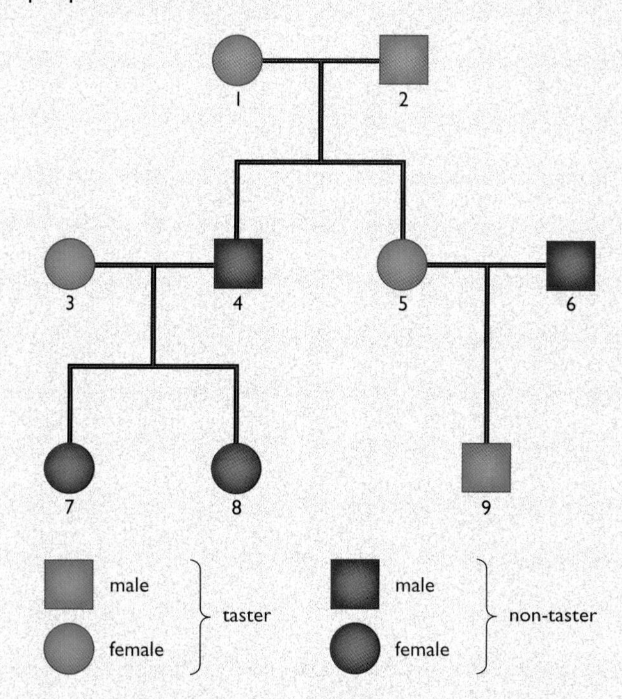

a) What evidence in the diagram suggests that the allele for PTC tasting is dominant? *(2 marks)*

b) Using **T** to represent the tasting allele and **t** to represent the non-tasting allele, give the genotypes of individuals 3 and 7. Explain how you arrived at your answers. *(4 marks)*

c) Why can we not be sure of the genotype of individual 5? *(2 marks)*

d) If individuals 3 and 4 had another child, what is the chance that the child would be able to taste PTC? Construct a genetic diagram to show how you arrived at your answer. *(4 marks)*

Total 12 marks

4 The Human Genome Project has produced the first draft of the complete human genome. This is now freely available on the Internet to anyone who wishes to download a copy. As a result of this project, it will become possible to compile a database of the complete genotype of individuals.

a) What is the human genome? *(2 marks)*

b) Suggest an advantage of a genetic database being freely available. *(2 marks)*

c) Suggest two disadvantages of a genetic database being freely available. *(4 marks)*

Total 8 marks

5 The graph shows some estimates of brain size of *Australopithecus*, *Homo erectus*, *Homo habilis* and *Homo sapiens*.

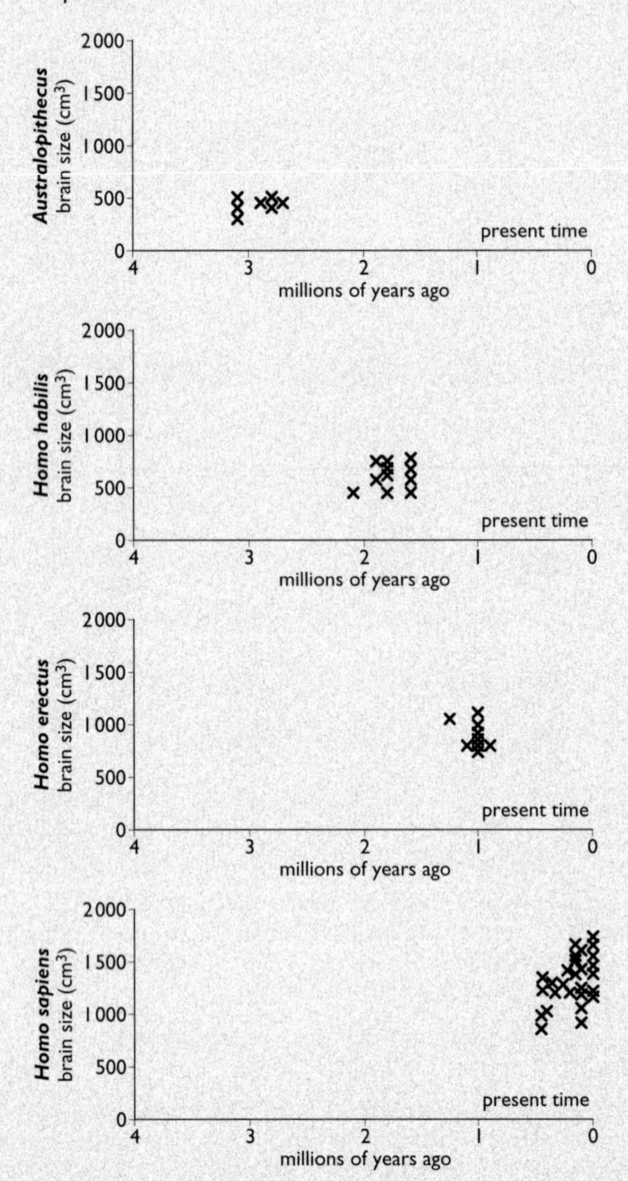

a) What is the range of estimates of brain sizes of *Australopithecus*? *(2 marks)*

b) *i)* What is the age and brain size of the oldest *Homo sapiens* specimen? *(2 marks)*

ii) Compare this specimen to the most recent *Homo sapiens* specimen. By what percentage has the brain size increased? *(3 marks)*

c) Suggest a possible selective advantage of the increase in brain size during the evolution of *Homo sapiens*. *(3 marks)*

Total 10 marks

6 The diagrams A to F show an animal cell during cell division.

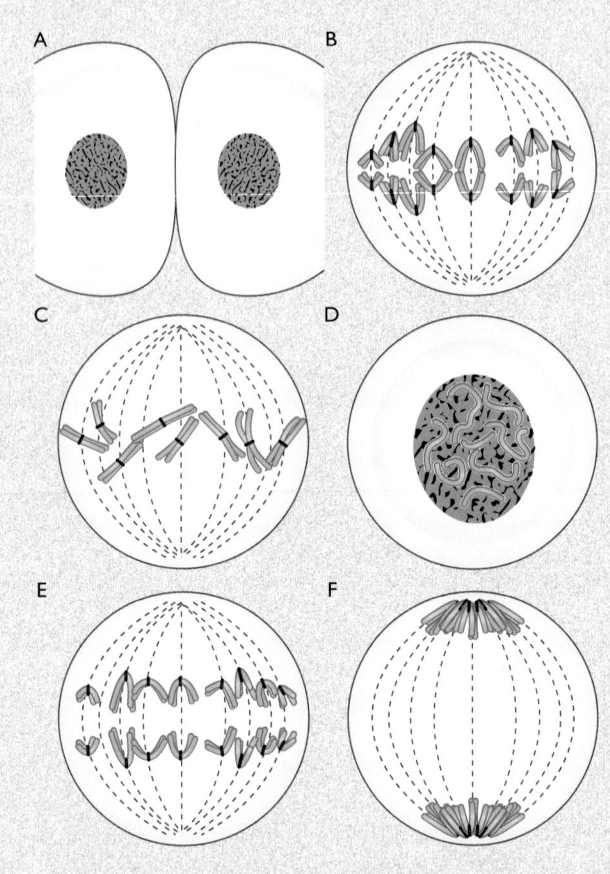

a) Put the pictures in the correct order. *(3 marks)*

c) Is the cell going through mitosis or meiosis? Explain your answer. *(2 marks)*

c) This cell has eight chromosomes which is its diploid number. How many chromosomes would a diploid human cell have? *(1 mark)*

d) Describe two differences between mitosis and meiosis. *(2 marks)*

Total 8 marks

Chapter 21: Microorganisms

Most microorganisms are invisible to the naked eye, but they are the most numerous living things on Earth. In this chapter we will look at the structure of some microorganisms, how they reproduce, and how some are the cause of disease.

What are microorganisms?

Microorganisms are living things that you can only see with the help of a microscope. The 'bodies' of most microorganisms are made of a single cell, although sometimes millions of cells are gathered together to form a **colony**. The colony of cells may then be visible to the naked eye.

Microorganisms have critical roles to play in recycling the waste products of organisms, as well as recycling the organisms themselves when they die. Many types of microorganisms are studied because they cause disease in animals and plants. On the other hand, humans have harnessed the great reproductive capacity of microorganisms to make useful products, such as food, drink and medicines.

There are several groups that we call microorganisms, including protozoa and algae, some fungi, bacteria and viruses:

- **Protozoa** are made of single cells that have features like an animal cell. *Amoeba*, which lives in pond water, is an example. Some single-celled organisms have chloroplasts, and are more like plants. These belong to a group called **algae**.
- Many **fungi** are microorganisms, including single-celled **yeasts**, as well as **moulds**, which consist of thread-like filaments of cells.
- **Bacteria** are single-celled organisms made of very small, simple cells.
- **Viruses** are much smaller than bacteria, and are not cells at all. In fact they can be thought of as half-way between a living organism and a chemical.

Figure 21.1 shows just a few examples of the many types of microorganisms.

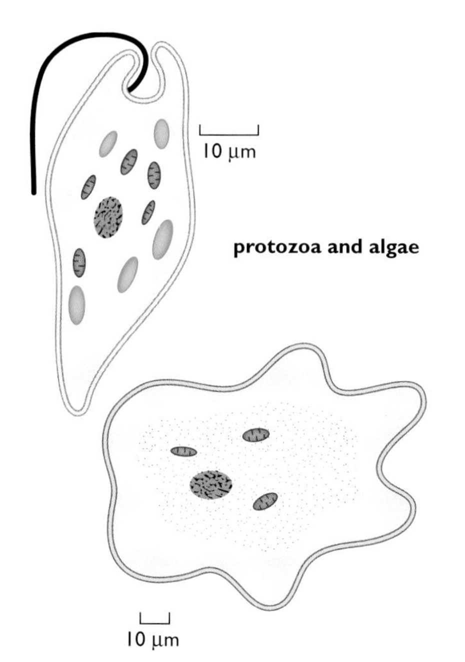

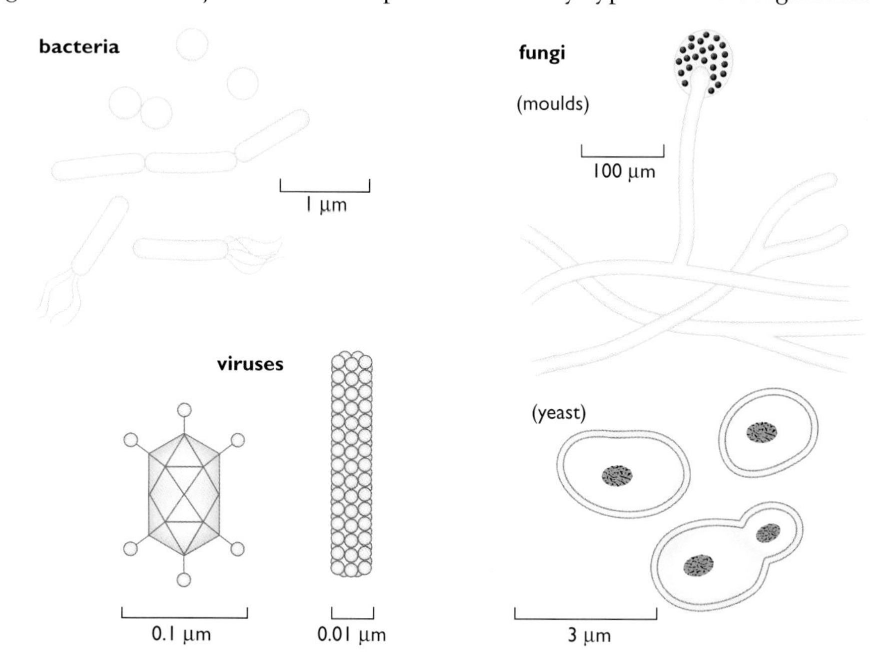

Figure 21.1 *Some examples of microorganisms. They are not drawn to the same scale. Notice the range of size, as shown by the scale bar alongside each organism. One micrometre (1 μm) is a millionth of a metre, or a thousandth of a millimetre.*

Microorganisms: friend or foe?

Microorganisms live in all of the Earth's habitats. Many bacteria and fungi are important **decomposers** (see Chapter 14), recycling dead organisms and waste products in the soil and elsewhere. Protozoa live in wet environments, including soil, ponds, rivers, lakes and the sea. Some species of yeasts live in soil, and others in parts of plants where they can get nutrients, such as on the surface of fruits. These 'free-living' microorganisms usually cause no direct harm to other organisms.

However, some species of microorganisms have adapted to living in or on other organisms, where they cause disease. Any organism that causes a disease is called a **pathogen**, and any disease caused by a microorganism is called an **infectious** disease. Most groups of microorganisms have members that have evolved a pathogenic way of life. You will be most familiar with two of these – some of the bacteria, and all of the viruses. As you will see, viruses can only live as **parasites** inside other cells, where they are responsible for causing diseases. The common cold and influenza (flu) are both caused by viruses.

A parasite is an organism that lives in or on another living organism (its **host**), gaining food from the host.

Bacteria also cause many diseases. Some diseases are **contagious**, which means that they can be passed from person to person, such as whooping cough and tuberculosis (TB). Other diseases are not contagious, such as food poisoning caused by the bacterium *Salmonella*, or tetanus, which is caused by a bacterium in soil.

A contagious disease is one that can be passed from person to person. This can happen by several different routes, such as through the air by sneezing, via skin contact, or through sexual activities.

A few species of protozoa cause disease, such as the parasite that produces malaria. Moulds and yeasts can also be pathogens. Many fungi are parasites of plants, gaining food from the plant and causing it damage. A few species of fungi can even cause disease in animals, such as the fungus that produces 'athlete's foot'. You will learn more about the causes of infectious disease in Chapter 24.

The structure and reproduction of mould fungi

The fungi that most people recognise are mushrooms and toadstools (Figure 21.2). In fact, these are the reproductive structures, or 'fruiting bodies' of various species of fungi. Their function is to produce **spores**, that are spread to new habitats and grow into more fungi. When a mushroom is growing, much of the organism is hidden below the surface of the soil or compost, and consists of a network of fine threads or **hyphae** (pronounced *high-fee*). The hyphae grow through the compost, absorbing nutrients from it.

A mould is rather like a mushroom without the fruiting body. It just consists of the network of hyphae. The whole network is called a **mycelium** (pronounced *my-sea-lee-um*). Moulds feed by absorbing nutrients from dead (or sometimes living) material, so they are found wherever this is present, for example in soil, rotting leaves, decaying fruit and so on. In fact moulds and other fungi, along with bacteria, are the two main types of **decomposers** (see Chapter 14).

Figure 21.2 *Toadstools growing on a rotten log.*

If you leave a piece of bread or fruit exposed to the air for a few days, it will soon become mouldy. Mould spores carried in the air land on the food and grow into a mycelium of hyphae. One common mould that grows in this way is the 'pin mould' called *Mucor* (Figure 21.3).

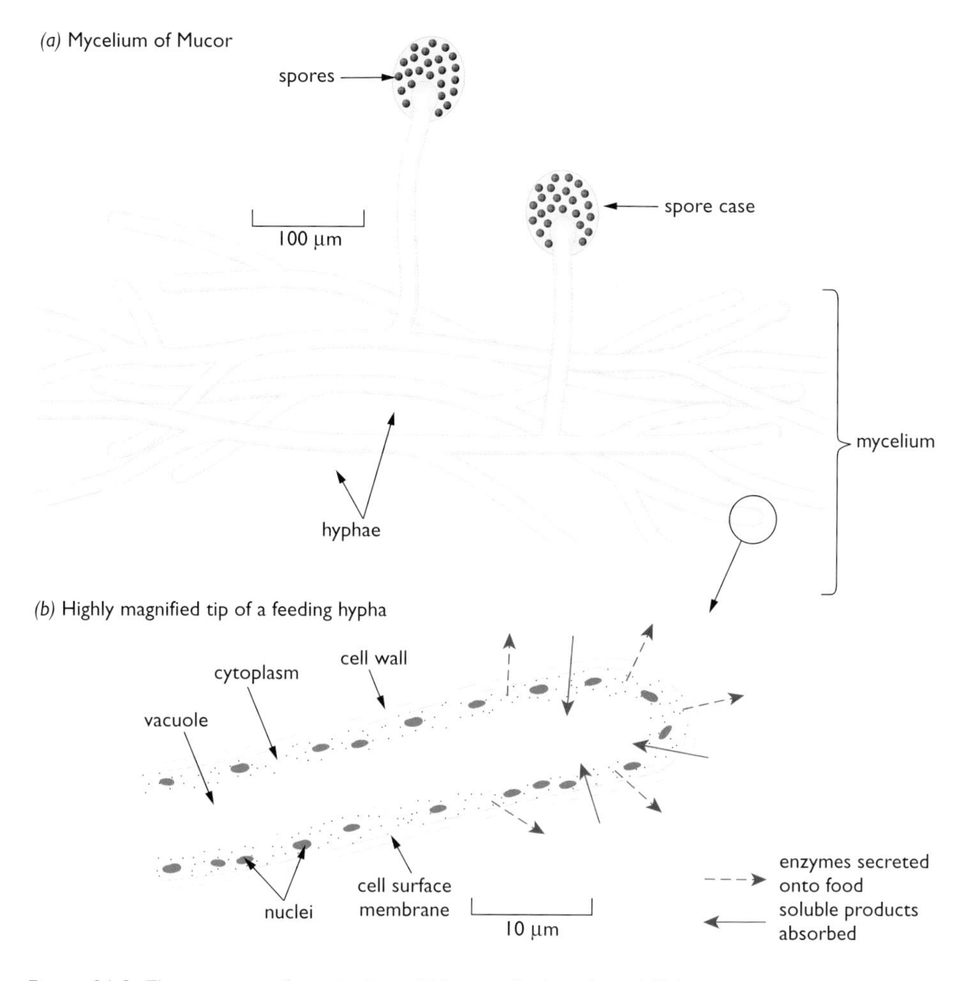

Figure 21.3 *The structure of a typical mould fungus, the bread mould* Mucor.

Because cells of fungi have a wall around the cell surface membrane, fungi were once thought to be plants that had lost their chlorophyll. We now know that their cell wall is not made of cellulose as in plants, but a different chemical called **chitin** (the same material that makes up the outside skeleton of insects). Fungi are quite different from plants in many ways (the most obvious is that they do not photosynthesise) and they are not closely related to plants at all.

Another well-known mould is *Penicillium*, because it produces the **antibiotic** called **penicillin** (see Chapter 26). Antibiotics are chemicals which kill bacteria. *Penicillium* uses this chemical to kill bacteria which might otherwise compete with it for the food source. Humans use penicillin, extracted from *Penicillium*, to kill some bacteria that cause disease.

The thread-like hyphae of *Mucor* have cell walls surrounding their cytoplasm. The cytoplasm contains many nuclei, in other words the hyphae are not divided up into separate cells.

When a spore from *Mucor* lands on food, a hypha grows out from it. The hypha grows and branches again and again, until the mycelium covers the surface of the food. The hyphae secrete digestive enzymes on to the food, breaking it down into soluble substances such as sugars, which are then absorbed by the mould. Eventually, the food is used up and the mould must infect another source of food. Hyphae grow upwards from the mycelium, and the tips of these hyphae develop into spore cases containing hundreds of spores. These spores are released into the air and carried away on air currents. Some of the thousands of spores formed in this way will land on a new source of food to continue the mould's life cycle. Spore production is an example of **asexual** reproduction. Asexual reproduction is described in more detail in Chapter 17.

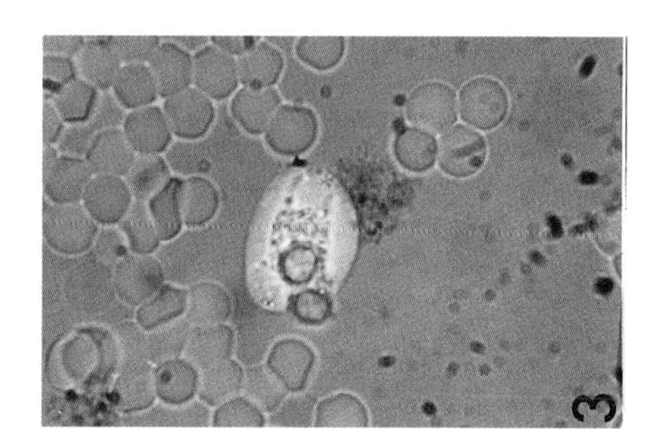

Figure 21.4 *Yeast cells seen through a microscope, greatly magnified.*

The structure and reproduction of yeasts

Yeasts are single-celled fungi that do not form hyphae. Different species of yeasts live everywhere – on the surface of fruits, the nectaries of flowers, in soil, water, and even on dust in the air (Figure 21.4).

A nectary is the part of the flower that makes nectar, a sugary liquid which attracts pollinating insects. Some yeasts can grow in nectaries, using the nectar as a food source.

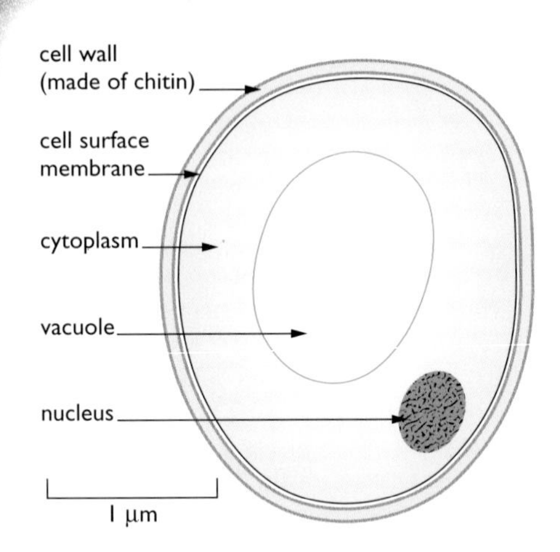

Figure 21.5 *Structure of a yeast cell.*

Like moulds, a yeast cell has a cell wall surrounding the cell membrane and cytoplasm, and a central vacuole. Each cell has a single nucleus (Figure 21.5).

One species of yeast that is very important to humans is called *Saccharomyces*. This is the yeast which is used for baking bread and making wine and beer (see Chapter 22). It reproduces asexually by a process called **budding** (Figure 21.6).

When a yeast cell buds, it forms a small bulge on one side of the cell. The nucleus of the cell divides in two by mitosis (see Chapter 17). The bulge grows steadily bigger and bigger until it is about the same size as the 'parent' cell. The bud containing the new nucleus then pinches off at the base, and breaks away to form a new cell. When yeast cells have a good supply of food, this process can happen very quickly, so that in a few hours thousands more yeast cells are formed.

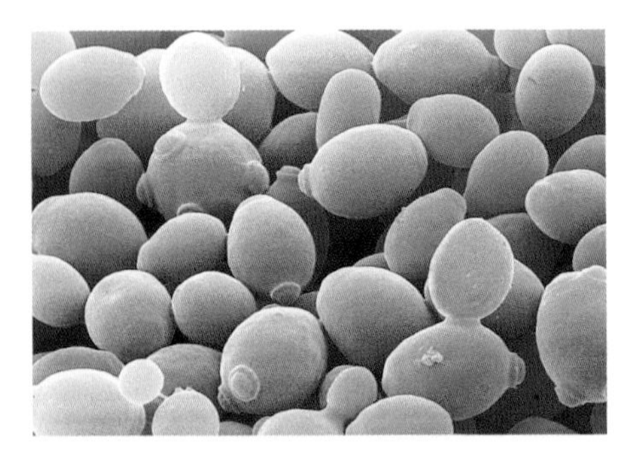

Figure 21.6 *Some of these yeast cells can be seen budding. The buds are different sizes. Eventually each bud will grow big enough to break away and form a new yeast cell.*

The structure and reproduction of bacteria

Bacteria are very small single-celled organisms. To give you some idea of their size, a typical animal cell might be 10 to 50 µm in diameter (1 µm, or one micrometre, is a millionth of a metre, or a thousandth of a millimetre). Compared with this a typical bacterium is only 1 to 5 µm in length (Figure 21.7) and its volume can be thousands of times less than the larger cell.

There are three basic shapes of bacteria: spheres, rods and spirals, but they all have a similar internal structure (Figure 21.8).

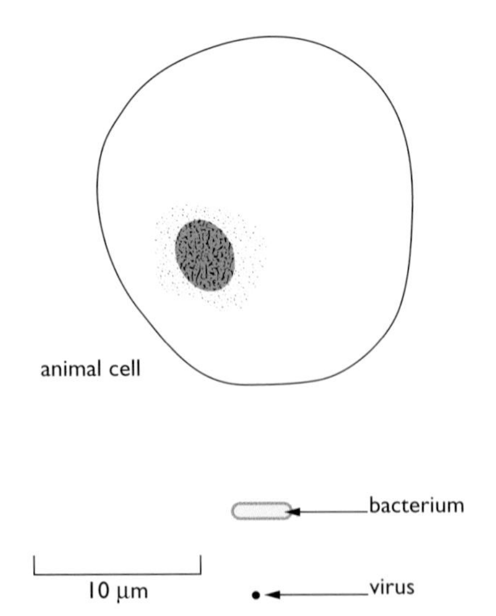

Figure 21.7 *A bacterium is much smaller than an animal cell. The relative size of a virus is also shown.*

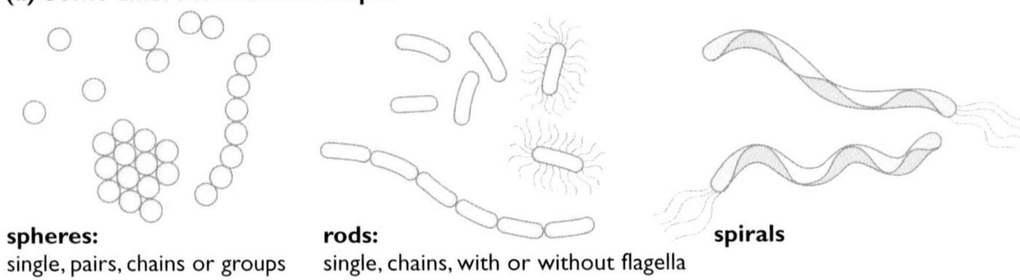

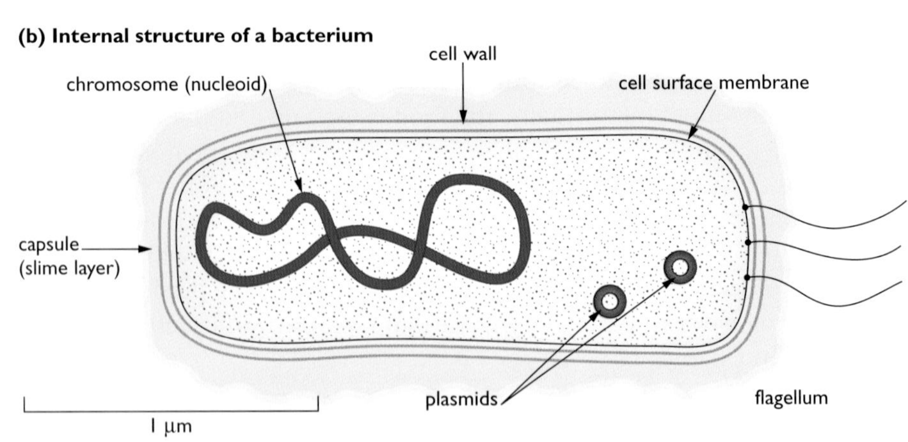

Figure 21.8 *Structure of bacteria.*

All bacteria are surrounded by a **cell wall**, that protects the bacterium and keeps the shape of the cell. Whereas the cell wall of a plant cell is made of cellulose, and cell walls of fungi are made of chitin, bacterial cell walls contain neither of these two substances. Instead, they are composed of complex chemicals made of polysaccharides and proteins. Some species have another layer outside this wall, called a **capsule** or **slime layer** (these are made of similar chemicals, but a capsule is denser than a slime layer). Both give the bacterium extra protection. Underneath the cell wall is the **cell membrane**, as in other cells. The middle of the cell is made of **cytoplasm**. One major difference between a bacterial cell and the more complex cells of animals and plants is that the bacterium has no nucleus. Instead, its genetic material (DNA) is in a **single chromosome** or **nucleoid**, loose in the cytoplasm, forming a circular loop.

Some bacteria can swim, and are propelled through water by corkscrew-like movements of structures called **flagella** (a single one of these is called a flagellum). However, many bacteria do not have flagella and cannot move by themselves. Other structures present in the cytoplasm include the **plasmids**. These are small circular rings of DNA. They carry genes which give the bacterium protection against antibiotics (antibiotic **resistance**), or allow it to use a wider range of nutrients. Not all bacteria contain plasmids, although about three quarters of all known species do. Plasmids have very important uses in **genetic engineering** (see Chapter 23).

Despite the relatively simple structure of the bacterial cell, it is still a living cell that carries out the normal 'processes of life', such as respiration, feeding, excretion, growth and reproduction. As you have seen, some bacteria can move, and they can also respond to a range of stimuli. For example, they may move towards a source of food, or away from a poisonous chemical. You should think about these features when you compare bacteria with the much simpler viruses.

Bacteria reproduce asexually by a process called **binary fission**, which just means 'splitting into two'. First of all, a cell grows until it has doubled its length. The chromosome is now copied in a process called **replication**. The two copies separate to opposite ends of the cell and the cell membrane folds inwards to form a double layer across the middle of the cell. Two new cell walls are then made in between the two membranes (Figure 21.9). Plasmids also replicate themselves, separately from the replication of the bacterial chromosome. Each daughter cell receives copies of the plasmids too.

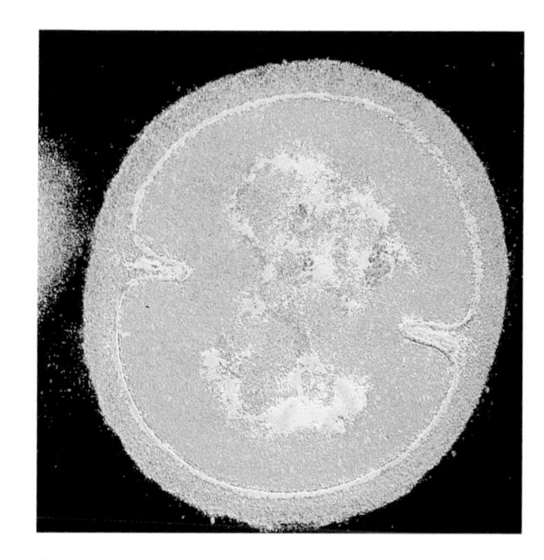

Figure 21.9 *A bacterium reproducing by binary fission.*

The two 'daughter' cells then separate and grow until they are ready to divide again. This process can happen very quickly. Under good conditions, bacteria can divide in two every 20 minutes. In 24 hours, one cell can produce many billions of cells (see Question 7 at the end of this chapter).

Bacterial spores

Like fungi, bacteria can also form spores. However, these are not for reproduction, but for survival. Often bacteria will be exposed to unfavourable conditions that will stop them growing and reproducing. For example:

- very high or low temperatures
- ultraviolet light
- changes in pH of their surroundings
- lack of water
- harmful chemicals, such as disinfectants.

Most bacteria flourish in conditions where there is oxygen, water and moderate temperatures.

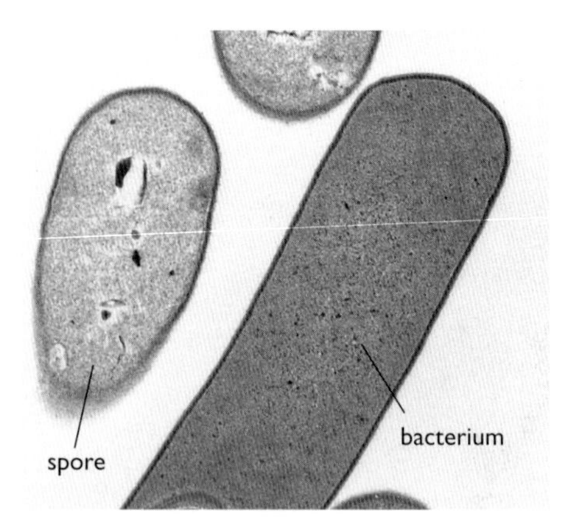

Figure 21.10 *A bacterial spore is for survival, not reproduction.*

In 1998 in a police museum in Norway, a small glass bottle was found which contained deadly contents. Inside were two sugar lumps, each with a small glass tube in the middle. Inside these tubes were spores of the bacterium which causes a disease called anthrax. Anthrax is fatal to many animals, including sheep, horses and humans. The bottle had once belonged to a German spy in the First World War, who was planning to use the anthrax to kill horses that the British used for transport. What is interesting here from a biological point of view is that scientists in 1998 found that the spores were still alive, and could grow into colonies of anthrax bacteria, more than 80 years after they had been sealed in their tubes!

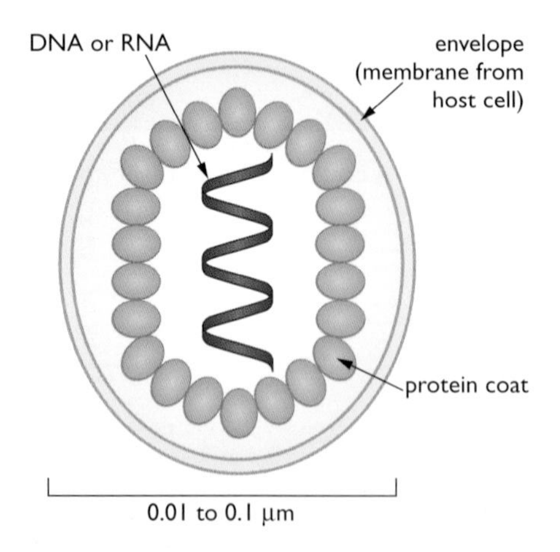

Figure 21.11 *The structure of a virus.*

Under conditions like these, many species of bacteria can survive by forming protective spores. The spore consists of a very thick outer coat surrounding the living bacterial contents (Figure 21.10). It develops inside the bacterial cell, which then bursts open to release it. Protected inside this coat, a bacterium can survive for many years, until conditions are good enough for growth and reproduction again.

The structure and reproduction of viruses

All viruses are parasites, and can only reproduce inside living cells. The cell in which the virus lives is called the host. There are many different types of viruses. Some live in the cells of animals or plants, and there are even viruses that infect bacteria. Viruses are much smaller than bacterial cells: most are between 0.01 and 0.1 μm in diameter (see Figure 21.7).

Notice that we say 'types' of virus, and not 'species'. This is because viruses are not made of cells. A virus particle is very simple. It has no nucleus or cytoplasm, and is composed of a core of genetic material surrounded by a protein coat (Figure 21.11). The genetic material can be either DNA, or a similar chemical called RNA (see Chapter 16 for details of the structure of DNA. Chapter 23 looks at the functions of DNA and RNA in more detail). In either case, the genetic material makes up just a few genes – all that is needed for the virus to reproduce inside its host cell.

Sometimes a membrane called an **envelope** may surround a virus particle, but the virus does not make this. Instead it is 'stolen' from the surface membrane of the host cell.

Viruses do not feed, respire, excrete, move, grow or respond to their surroundings. They do not have any of the normal 'characteristics' of living things except reproduction, and they can only do this parasitically. This is why some scientists think of viruses as being on the border between a living organism and a non-living chemical.

A virus reproduces by entering the host cell and taking over the host's genetic machinery to make more virus particles. After many virus particles have been made, the host cell dies and the particles are released to infect more cells. Many human diseases are caused in this way, including colds, influenza, measles, mumps, polio and rubella (German measles). Of course the reproduction process does not go on forever. Usually the body's immune system destroys the virus and the person recovers (see Chapter 25). Sometimes, however, a virus cannot be destroyed by the **immune system** quickly enough, and it may cause permanent damage or death. With other infections, the virus may attack cells of the immune system itself. This is the case with **HIV** (Human Immunodeficiency Virus) which eventually causes the disease **AIDS** (Acquired Immune Deficiency Syndrome).

Different viruses reproduce in slightly different ways inside their host cell, but you can get a good idea of the process if you consider one example, the virus that causes influenza (flu). The influenza virus is spread from person to person in droplets carried in coughs and sneezes (Figure 21.12). It is also transmitted through the skin in sweat, by direct contact. It infects the cells of the respiratory passages, causing the person to have a high temperature, aches and pains and other symptoms. Reproduction of this virus is shown in Figure 21.13. It is a virus that carries its genes as RNA. These genes, when they enter the host cell, direct the cell's enzymes to make more viral RNA and protein coats. These are then assembled into virus particles, which are released to infect other host cells.

Some DNA viruses 'hide' their genes inside the DNA of the host cell chromosomes. Here, the virus can remain dormant (inactive) for long periods. Eventually it becomes active, reproducing virus particles and causing disease. Viruses like this cause some kinds of cancer.

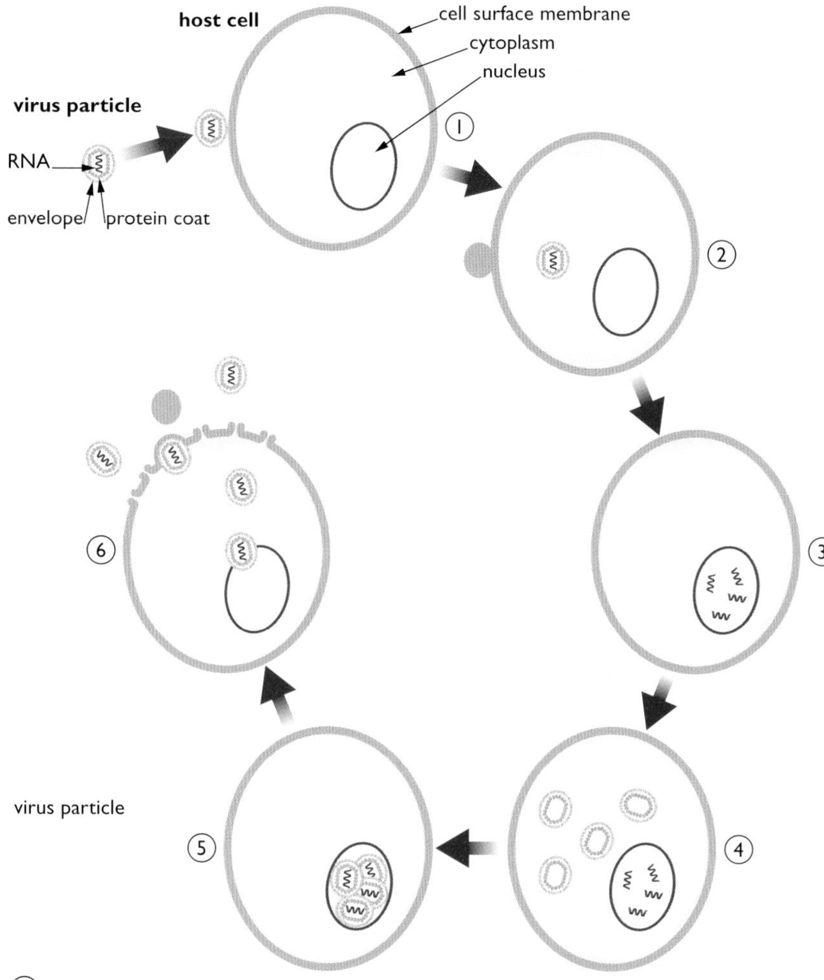

host cell
cell surface membrane
cytoplasm
nucleus
virus particle
RNA
envelope
protein coat
virus particle

1. Virus particle attaches to host cell membrane using the envelope from previous host cell.
2. Virus enters host cell and travels to nucleus.
3. Protein coat of virus breaks down. Copies of RNA made.
4. RNA directs manufacture of new protein coats in the cytoplasm.
5. Protein coats enter nucleus and enclose copies of RNA.
6. Virus particles break out of nucleus and cell, becoming enclosed in a new envelope as they leave the host cell

Figure 21.13 *Reproduction of the influenza virus inside human cells.*

Figure 21.12 *Coughs and sneezes carry thousands of tiny fluid droplets. Each may contain pathogens such as the influenza virus. If they are inhaled by another person this can give them the flu by 'droplet infection'.*

A group of viruses called **bacteriophages**, or just **phages**, only infect bacterial cells (Figure 21.14). Their outer protein coat has a complex shape, with a 'head' that contains DNA and a 'tail' that is used to attach the virus to a bacterium. The bacteriophage reproduces in a similar way to the influenza virus, but it uses its tail to 'inject' the DNA into the host cell, rather like using a syringe. As before, the viral DNA then controls the production of many more virus particles, causing the death of the host cell (Figure 21.15). Some bacteriophages are used in genetic engineering to transfer genes to bacteria (see Chapter 23).

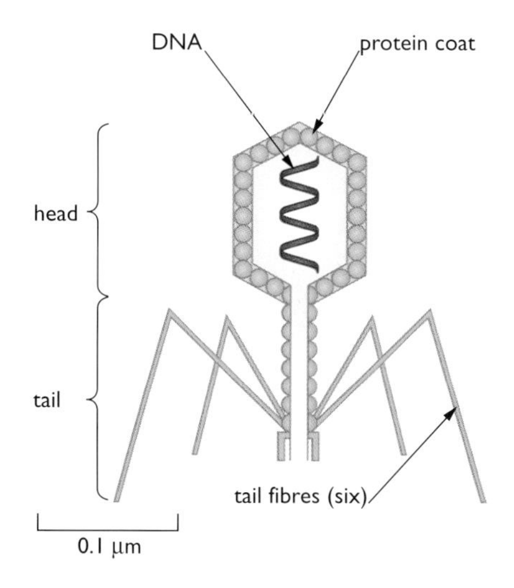

Figure 21.14 *A section through a bacterial virus (bacteriophage). The four tail fibres behind the section are shown.*

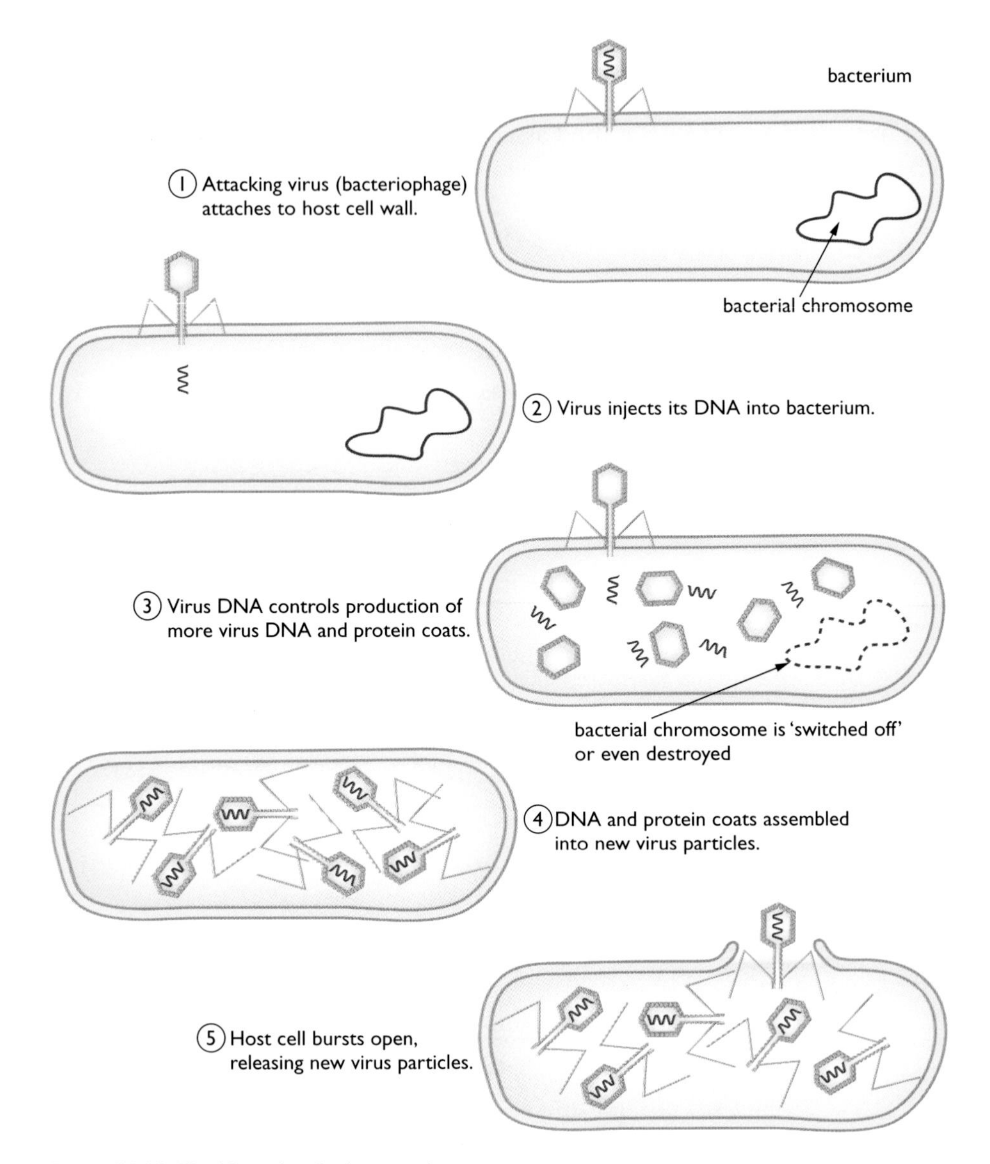

Figure 21.15 *The life cycle of a bacteriophage.*

How do pathogens cause disease?

Pathogens cause disease in a number of ways. Viruses cause cells to be damaged, because they enter the host cell and interfere with the normal processes going on in the cell that are needed to keep it alive. They also destroy the cell when they burst out through the cell membrane.

Some species of pathogenic bacteria also enter cells, but most species remain outside the cells, in the tissue fluid or blood. In either case, there are many ways that they can be harmful. Often, they produce poisonous waste products called **toxins**. Some of these are extremely poisonous. For example, the bacterium that causes cholera makes a toxin that affects the cells lining the gut. It causes them to pump out ions and water by active transport (see Chapter 1) so that the infected person dies of dehydration.

Other bacteria produce enzymes that break down cells, producing dead areas of tissue. Some bacteria damage the immune system, for example by killing phagocytic cells.

Growth curves of microorganisms

Microorganisms can reproduce very quickly. A bacterium placed in a solution containing suitable food and oxygen, called a **nutrient broth**, will soon produce millions of cells by binary fission. We can count the numbers of cells in the solution and plot a graph of the numbers against time, to produce a **growth curve** (Figure 21.16).

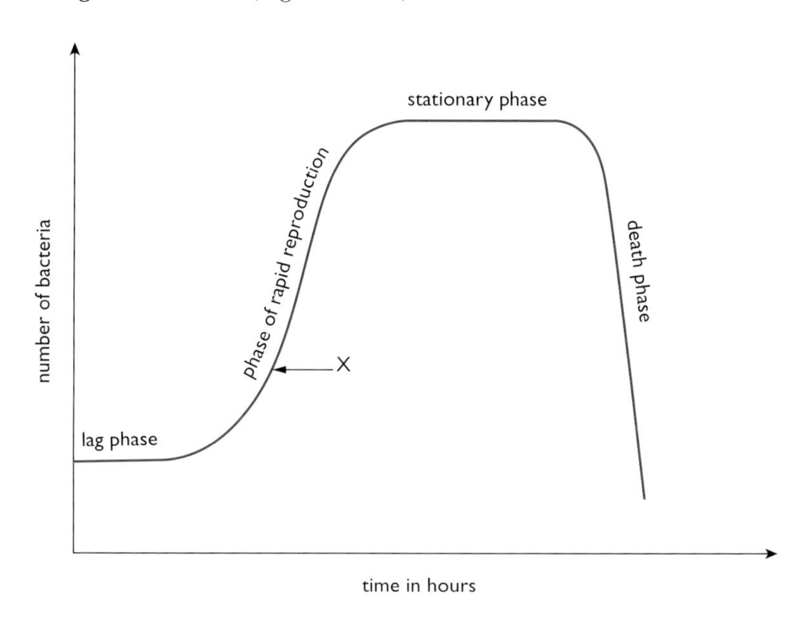

Figure 21.16 *Growth curve for a population of bacteria in a nutrient broth.*

The growth curve doesn't keep on rising forever. Eventually the curve reaches a peak and then drops off again. In fact you can see four phases in the growth curve:

- The **lag phase** is the time just after the first bacteria have been added to the broth. The bacteria cells don't usually start dividing straight away. Instead, they go through a period when they get used to their new food supply. For example, they may have to make new enzymes to break it down.
- When the lag phase is over, the cells start dividing in equal time intervals, producing a **phase of rapid reproduction**.
- Eventually, their reproduction rate slows down, and cells begin to die. When the number of cells dying is equal to the number of new cells being produced, the curve levels off, reaching the **stationary phase**.
- Finally, the number of living cells in the broth decreases, in the **death phase**.

In exam questions, you will need to recognise that the slope of the growth curve shows the rate of change of the population size. The fastest rate of growth is where the gradient is steepest (at point 'X' on the curve).

What causes the stationary and death phases? There are several possible **limiting factors**. As they grow and reproduce, the bacteria may use up the food or oxygen in the broth, or they may produce poisonous waste products. Lack of nutrients or build-up of waste will slow down the rate of reproduction and cause the death of more cells.

Other microorganisms such as yeast and moulds produce similar growth curves. We can't draw a growth curve for viruses, because they only reproduce inside other cells, not in a broth.

The idea of limiting factors is discussed in the section on photosynthesis (Chapter 10). Strictly, the definition of a limiting factor is one which slows down a process, because it is in 'shortest supply'. However, you may see a build-up of poisonous waste described as a 'limiting factor', because it also limits the rate of growth.

Some bacterial spores can even survive boiling at 100 °C for many minutes. When glass apparatus is sterilised in microbiology laboratories, it is heated to 121 °C in an **autoclave**, to make sure all spores are killed (Figure 21.17). An autoclave heats water at a high pressure, so that it boils above 100 °C. You may have a 'pressure cooker' at home which works on the same principle.

Figure 21.17 *An autoclave used to sterilise apparatus.*

The rate of growth of the microorganisms also depends on the temperature of the broth. Most bacteria reproduce fastest between 20 °C and 40 °C. At lower temperatures, their rate of cell division slows down, because their enzymes work more slowly (see Chapter 1, page 00). That is why food stored in a fridge at 4 °C takes longer to go bad than food at room temperature. At very high temperatures (e.g. 100 °C, the boiling point of water) enzymes are destroyed by denaturing, so this kills most bacteria.

The history of microbiology and the work of Louis Pasteur

The first person to see microorganisms was the Dutchman Anton van Leeuwenhoek in 1676. His hobby was making lenses and optical instruments, and he used a simple microscope to see bacteria and other organisms, which he then drew.

At this time, many people thought that food went bad because microorganisms, and even small animals like maggots, arose 'spontaneously' from non-living material. This idea was called **spontaneous generation**, meaning that they appeared out of the food material itself. Other people like Van Leeuwenhoek, who had seen the minute moving creatures through his microscope, believed that they infected food by being carried in the air.

In the late seventeenth century an Italian scientist called Francesco Redi proved that larger organisms such as maggots were the larvae of flies. He covered pieces of meat with a piece of gauze, and stopped flies from reaching the meat. But Redi's gauze could not stop microorganisms entering, so the meat still decayed.

Spallanzani's experiments led to the invention of tinned (canned) food. The Emperor Napoleon needed a way of keeping food fresh so that his armies would have a steady supply. In 1795, he offered a prize for the best method of preserving food. A chef called François Appert won the prize by copying Spallanzani's method of heating food and then sealing it from the air. This is the same principle used today to make a tin of baked beans.

In the middle of the eighteenth century, another Italian called Lazzaro Spallanzani found a way to stop a nutrient broth from going bad. He boiled the broth in a flask for a long time, so that any microorganisms in the broth or the air above it were killed. Then he sealed the flask. The contents of the flask did not decay, unless they were opened to the air.

Some scientists still continued to believe in spontaneous generation. It wasn't until 1861 that a French chemist (later to become one of the most famous microbiologists of all time) finally proved once and for all that microorganisms in food did not arise by spontaneous generation. He was Louis Pasteur (Figure 21.18), who has given his name to a method we use to kill pathogens in milk, called **pasteurisation**.

Pasteur was a remarkable scientist, who was responsible for many important discoveries in the field of microbiology. He got a job in Lille, in northern France, where he experimented with **fermentation** (see Chapter 22), the process used to make beer and wine. People had seen yeasts in samples from wine and beer vats, but no one realised that they actually *caused* fermentation. Pasteur showed that if the yeast cells were killed by boiling, fermentation stopped.

ideas evidence

The winemakers and brewers of Lille had a problem that no one could solve. From time to time, their wine and beer turned sour while it was maturing in the vats. This resulted in a great loss of income, and in 1856 they turned to

Pasteur for help. He used his microscope to investigate the difference between good samples of beer and wine, and others which had gone sour. He showed that the good samples contained only yeast cells. However, when he looked at the sour samples, they contained, as well as yeast, other elongated cells. These were bacteria, which caused the souring by producing acids. Pasteur came up with a remedy for the problem. He suggested to the winemakers and brewers that after the wine or beer was made, they should gently heat it to 55 °C. This killed the bacteria so that the souring was prevented, without altering the taste of the drink. It was the first use of pasteurisation.

Figure 21.18 *Louis Pasteur (1822–95).*

The wine and beer producers of Lille were so grateful to Pasteur that they set up a scientific research institute for him in Paris. Here Pasteur and other scientists made many other important discoveries.

Pasteur disproved the theory of spontaneous generation, by showing that microorganisms that contaminated food came from the air. His most important experiment used a specially designed piece of apparatus called a swan-necked flask (Figure 21.19).

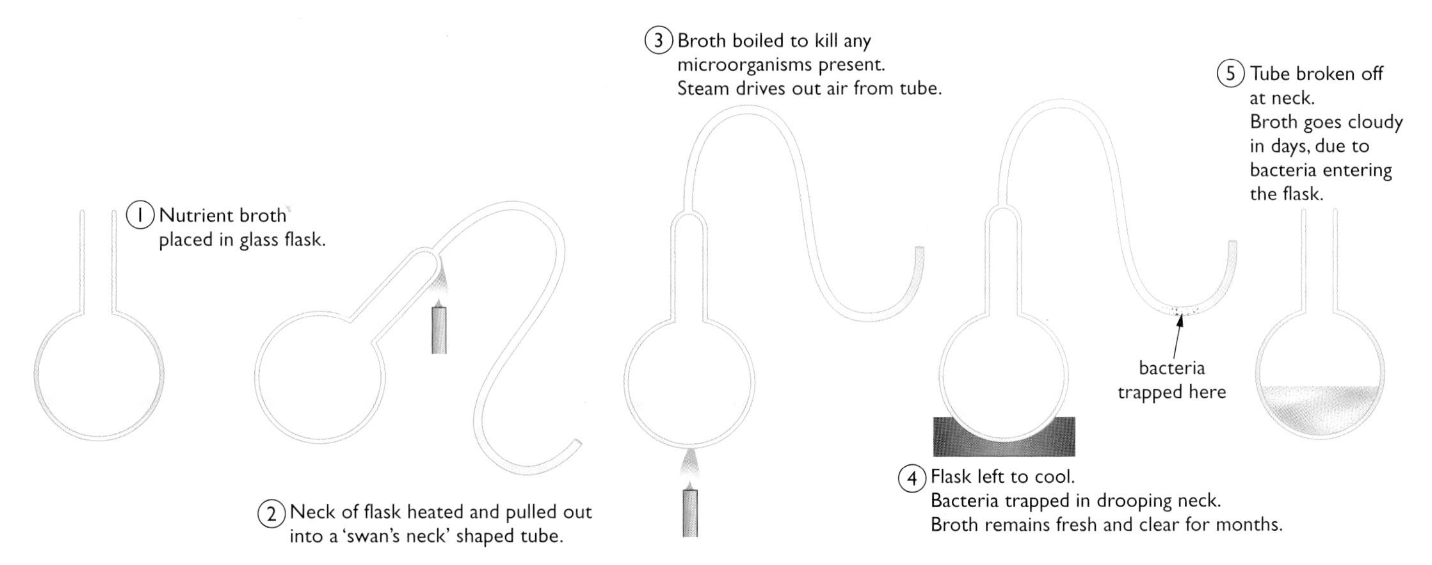

Figure 21.19 *Pasteur's experiments with swan-necked flasks finally disproved the theory of spontaneous generation.*

A nutrient broth was boiled in the flask. This killed all microorganisms in the flask, including the neck. As the broth cooled down, microorganisms were drawn into the neck as air was sucked in, but they settled on the moist walls of the neck. The broth did not decay. If Pasteur tipped the flask, so that the broth ran into the neck, the broth became contaminated with the microorganisms and decayed. If the neck was broken, this also caused the broth to decay, since the microorganisms could get in from the air. Other scientists in France and England repeated his experiments and found the same thing – living organisms only came from other living organisms.

Louis Pasteur contributed greatly to early developments in microbiology. You have only been able to read a brief summary of some of his achievements here. You could carry out an Internet search to find out more about him and his work, or about any of the other scientists mentioned here.

Pasteur went on to study how microorganisms were involved in causing disease. He discovered that a disease of silkworms, that was damaging the silk industry, was caused by an infectious microorganism. He developed a method of vaccinating chickens against cholera, and another procedure for vaccinating against the killer disease anthrax, that affects farm animals and humans. He also produced a successful vaccine against the virus that causes rabies.

End of Chapter Checklist

If you haven't got a copy of your specification, read the introduction on page vi.

You will need to be able to do some or all of the following. Check your Awarding Body's specification (syllabus) to find out exactly what you need to know.

- Understand that microorganisms include protozoa, fungi, bacteria and viruses, and that all viruses, and some bacteria, fungi and protozoa can be pathogens.
- Recall the structure of mould fungi as a mycelium of hyphae that do not contain separate cells. Know that hyphae contain cytoplasm and many nuclei, and have a surface membrane and a cell wall. Know that reproduction is asexual, by spore production.
- Recall the structure of yeast and know that it reproduces by budding.
- Recall the structure of a bacterial cell, including the single chromosome (nucleoid), cytoplasm, surface membrane, cell wall and plasmids.
- Understand that bacteria reproduce by cell division (binary fission) and that some bacteria form spores when unfavourable conditions arise. Know that spores are resistant to low and high temperatures, change in pH, drying out and the harmful effects of chemicals such as disinfectants, and that the spores can grow into new cells if conditions improve.
- Recall the structure of a virus, including the DNA or RNA core and protein coat, and understand that viruses lack some of the characteristics of living organisms.
- Understand how viruses reproduce by insertion of viral DNA/RNA into the host cell, control of protein manufacture by viral DNA, production of new virus particles, destruction of the host cell and release of new virus particles to attack new cells.
- Recognise that microorganisms can reproduce very rapidly. Interpret data from growth curves. Understand the effect of temperature, food supply and the build-up of toxic waste products on the rate of reproduction of bacteria and fungi.
- Describe Louis Pasteur's discoveries in microbiology, and his evidence that microorganisms can cause decay and disease.

Questions

More questions on microorganisms can be found at the end of Section F on page 332.

1 Tuberculosis (TB) is a contagious disease caused by a pathogenic microorganism living as a parasite in the human body. Explain the terms:

   a) contagious

   b) pathogenic

   c) parasite.

2 Describe how the influenza virus infects human cells and reproduces more virus particles. Your description should include one or two simple diagrams and about half a page of writing.

Organism	Most species are made of single cells	Cells have a nucleus	Cells are surrounded by a cell wall	Organism reproduces by forming spores	Organism always causes disease
yeast					
mould		many nuclei			
bacterium					
virus	not made of cells				

3 Copy the table above. Put a tick (✓) in the boxes that are correct and a cross (✗) in those that are incorrect.

4 Explain the meaning of each of the following:

a) hyphae

b) mycelium

c) budding (in yeast)

d) nucleoid

e) plasmid

f) binary fission

g) bacteriophage.

5 The diagram shows a growth curve of a bacterium grown in a nutrient broth over a number of hours.

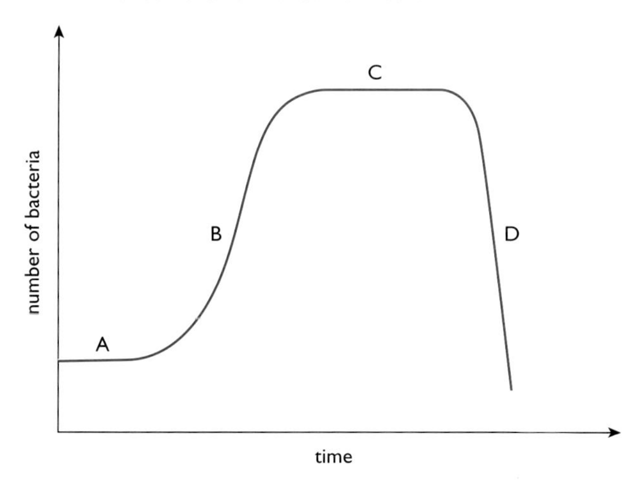

a) Identify each of the phases of growth labelled A to D. Choose from: death phase, lag phase, stationary phase, phase of rapid reproduction.

b) Sketch the graph on a piece of paper. Mark the point on the curve when the bacterial population is increasing at its fastest rate with an 'X'.

c) Write down three limiting factors which could be the cause of the growth curve entering phase C.

6 A student carried out an investigation to try to repeat some of the experiments of Louis Pasteur. She used a nutrient broth in test tubes. The diagram shows the apparatus that she used. She heated all four tubes at 121 °C in a pressure cooker for 20 minutes, which killed any bacteria present in the tubes. She then left them for a few days. If bacterial colonies grew in the broth, it turned from clear to cloudy.

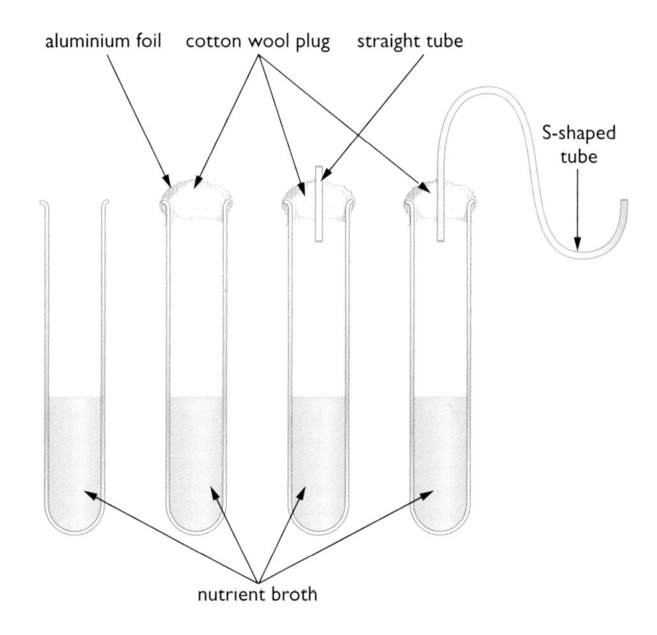

For each of tubes 1 to 4, predict whether you would expect the broth to turn cloudy, and explain the reasoning behind your prediction.

7 *a)* Use a spreadsheet to calculate how many bacteria are produced if one bacterial cell divides every 20 minutes for 24 hours.

b) Use the data in the spreadsheet to plot a graph of the numbers of bacteria against time, for the first 12 generations (4 hours).

c) How does your curve compare with the graph shown on page 267?

d) How many hours does it take to produce

i) a million *ii)* ten million *iii)* a billion bacteria?

Chapter 22: Growing Useful Microorganisms

Some microorganisms make products that are useful to humans, such as food and medicines. Their rapid rates of growth mean that cultured microorganisms can be very important sources of these products.

The previous chapter described how microorganisms such as bacteria and moulds can reproduce very rapidly. Because of this they are easy to grow under controlled conditions, as long as they are given the nutrients that they need and the correct physical conditions. A controlled growth like this is called a **culture**.

Culture media

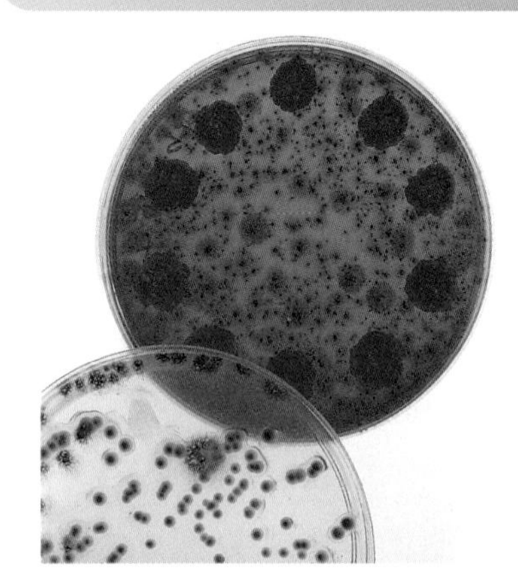

Figure 22.1 *Colonies of microorganisms growing on agar plates.*

Both solid and liquid materials are used for growing microorganisms. They are called **culture media**. The commonest solid culture medium is **agar**. This is a jelly made from a species of seaweed. It can be sterilised by heat and poured into Petri dishes or glass bottles while still molten, so that it sets to form a surface on which the microorganisms will grow (Figure 22.1). Petri dishes containing agar are often called **agar plates**.

Agar has one great advantage – there are very few microorganisms that can digest it, so the jelly does not get eaten away as the microorganisms grow. To supply the microorganisms with food, the agar has to have nutrients added to it when it is prepared, and is known as 'nutrient agar'. Different species need different nutrients in the agar, but these usually include carbohydrates such as sugars as an energy source, and mineral ions such as nitrate. Sometimes other nutrients needed for growth are added, such as proteins and vitamins.

Microorganisms can also be grown in a liquid medium called a **broth**. Liquid cultures vary enormously in size. In a laboratory, small cultures of bacteria or fungi may be grown in flasks containing only a few hundred cm^3 of liquid medium. As the microorganisms reproduce by binary fission (see Chapter 21) the broth culture becomes cloudy. Microorganisms are also grown in massive vessels containing hundreds of thousands of litres of liquid. These are called **fermenters**. They are used to make many products that are useful to humans, a branch of biology called **biotechnology**.

Sterile techniques

Usually, a microbiologist or biotechnologist will only want to grow one species of microorganism in a culture. 'Cross-contamination' refers to when an unwanted species grows as well, or when a harmful organism escapes from the culture.

To grow colonies of microorganisms successfully, it is essential that microbiologists follow procedures called **sterile techniques**. These methods try to ensure that colonies do not become contaminated with unwanted organisms and that organisms in a culture are not allowed to escape out of the culture vessel. Sterile techniques use apparatus that is sterilised before use, and methods that minimise cross-contamination. This is very important when dealing with colonies of pathogens, for example in the microbiology laboratory of a hospital. A colony of bacteria in a Petri dish could contain billions of dangerous organisms able to infect the laboratory workers. Sterile practices are also very important in running industrial fermenters. To make useful products in a fermenter, the culture must be uncontaminated. A contaminating species may kill the wanted organism, or it may produce waste materials which 'pollute' the product.

Imagine that a microbiologist has a Petri dish containing a number of colonies of different species of bacteria, and wants to isolate one of these species. One way to do this, using sterile techniques, is shown in Figure 22.2.

After the agar plate has been inoculated, the wire loop is 'flamed' again to sterilise it. The lid of the Petri dish is then fixed with sticky tape, to make sure it does not accidentally come off after the bacterial colony has grown. This prevents microorganisms in the air from contaminating the culture, and stops any 'escape' of possible pathogens.

The Petri dish must be labelled with the date and source of the bacterium, and placed upside-down in an incubator. In a microbiology laboratory, the incubator will be set at the temperature that is best for the particular bacterium. Usually this will be between 20 and 40°C. In a school laboratory, bacterial colonies should be incubated at 25°C. This reduces the likelihood of growing pathogenic species, which reproduce best at body temperature (37°C).

The Petri dish is incubated for two or three days, until a colony of the bacterium grows and can be examined.

Glass bottles and glass Petri dishes are sterilised by heating them at 121°C in an autoclave (a laboratory pressure cooker). Nowadays though, most Petri dishes are of the disposable plastic kind. They are supplied in sealed bags and have been sterilised by gamma radiation. After use they are destroyed, along with their contents.

Sometimes pure cultures of bacteria are provided, growing on agar in glass tubes. These can be transferred to an agar plate by using a wire loop (Figure 22.3). After collecting bacteria on the loop, the agar plate can be inoculated as shown in Figure 22.2.

1 The wire loop is heated in a Bunsen burner flame, as in Figure 22.2.

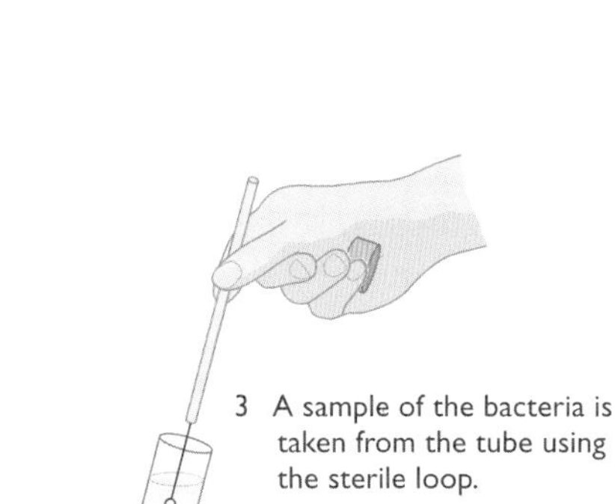

2 Still holding the loop in the right hand, the tube is uncapped with the same hand. Using the left hand, the neck of the tube is passed quickly through the Bunsen burner flame killing any bacteria from outside which might be around the neck of the tube.

3 A sample of the bacteria is taken from the tube using the sterile loop.

4 The neck of the tube is passed through the flame again, to kill any bacteria which might have landed on the neck. The cap is replaced.

Figure 22.3 *Transferring bacteria from a tube to an agar plate. The instructions are for a right-handed person.*

The dish is placed upside-down so that any drops of water that condense inside it don't flood the colonies.

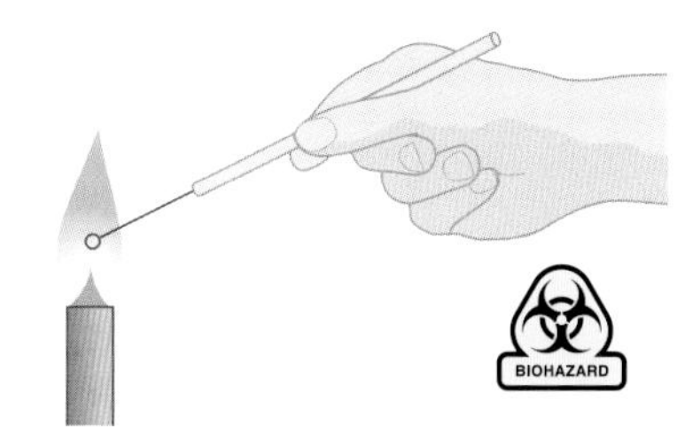

1 A wire loop is placed in a hot Bunsen burner flame for a few seconds, until it glows red-hot. This sterilises the loop. It is then allowed to cool in the air.

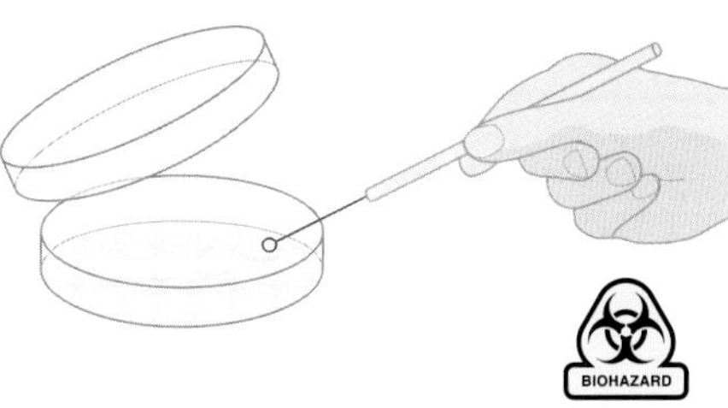

2 The lid of the Petri dish containing the mixture of colonies is lifted to an angle of about 45°, just enough to allow the wire loop to be used to collect a sample of the bacterium of interest. The lid is then replaced.

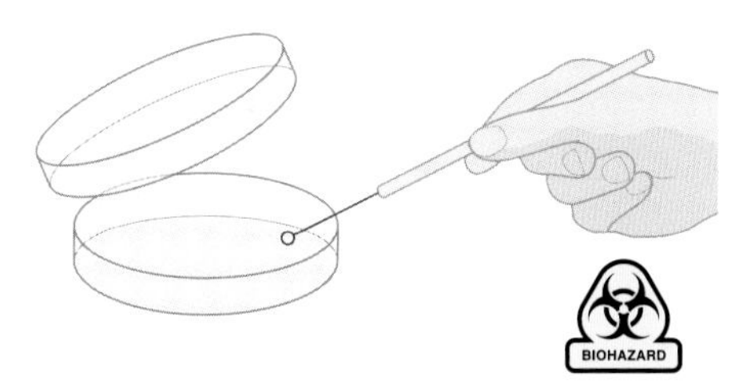

3 The sample on the loop is transferred to a fresh sterile agar plate. The loop is moved in a zigzag pattern on the surface of the agar, just touching the jelly, but without letting the loop dig in. This is called 'inoculating a plate'. The lid of the dish is replaced.

Figure 22.2 *Isolating a species of bacterium from a mixture of different colonies, using sterile techniques.*

It may be possible for you to try inoculating a plate in this way, or you may be able to use a wire loop to collect samples of microorganisms from surfaces in the laboratory and grow them on agar plates. You must always take great care when dealing with potentially dangerous microorganisms. Always use sterile techniques and follow any safety instructions you are given, including washing your hands afterwards.

Fermentation and biotechnology

Many microorganisms use an external food source from their growth medium to obtain energy. In doing this, they change substances in the medium. This is the modern meaning of fermentation. It is used by humans to make many important products. The use of microorganisms to make products useful to humans is called **biotechnology**.

Many microorganisms obtain their energy by anaerobic respiration, which does not need oxygen (see Chapter 1). The original meaning of the word **fermentation** was simply anaerobic respiration. For example, anaerobic respiration of sugars by yeast makes ethanol. This is called 'alcoholic fermentation'. Louis Pasteur, who studied this subject a great deal, described fermentation as 'life without air'. Nowadays, however, many other processes carried out by microorganisms are called fermentation, and many of them are aerobic (use oxygen).

The word 'biotechnology' may be new, but humans have used some biotechnology processes for thousands of years. Since ancient times, fermentation by yeast has been used to make wine and beer, and to produce bread. Other fermented products do not use yeast. Yoghurt is made by the action of bacteria on milk, and other bacteria and moulds are used in cheese manufacture. Another bacterium is used to convert the ethanol in wine into vinegar. You will read about how these and other products are made later in this chapter.

Our ancestors used biotechnology to make products like wine, beer and cheese, but they did not understand *how* they were made, and had no idea of the existence of microorganisms. Nowadays we understand what is happening when fermentation takes place, and can use biotechnology to produce not just foods, but a vast range of products, from medicines like penicillin to chemicals such as enzymes and fuels.

Many products are made by altering the genes of microorganisms, so that they code for new products. This is called **genetic engineering** and makes **genetically modified organisms**, or **GMOs**. It is a topic that you will read about in Chapter 23. The rest of this chapter is mainly concerned with the uses of biotechnology for food production, but also deals with manufacture of the antibiotic penicillin, and production of fuels.

ideas evidence

There are several advantages to using microorganisms as a source of food. They are easy to keep in fermenters, where they will reproduce without being affected by the surrounding conditions, such as climate. They grow very rapidly and can produce a great deal of their products in a short space of time. Microbes can usually be fed with very simple, cheap nutrients. Sometimes they are even fed with waste products of other industrial processes, that would otherwise be difficult to get rid of. There is great potential for biotechnology to help deal with the world's food shortage.

Industrial fermenters

A fermenter is any vessel that is used to grow microorganisms used for fermentation. A glass jar used to make wine at home is a fermenter, and even a baking tray containing a ball of dough could be called a fermenter.

Industrial fermenters are large tanks that can hold up to 200 000 dm^3 of a liquid culture (Figure 22.4). They enable the environmental conditions such as temperature, oxygen and carbon dioxide concentrations, pH and nutrient supply to be carefully controlled so that the microorganisms will yield their product most efficiently. A simplified diagram of the inside of a fermenter is shown in Figure 22.5.

Figure 22.4 *An industrial fermenter holds hundreds of thousands of* dm^3 *of a liquid culture.*

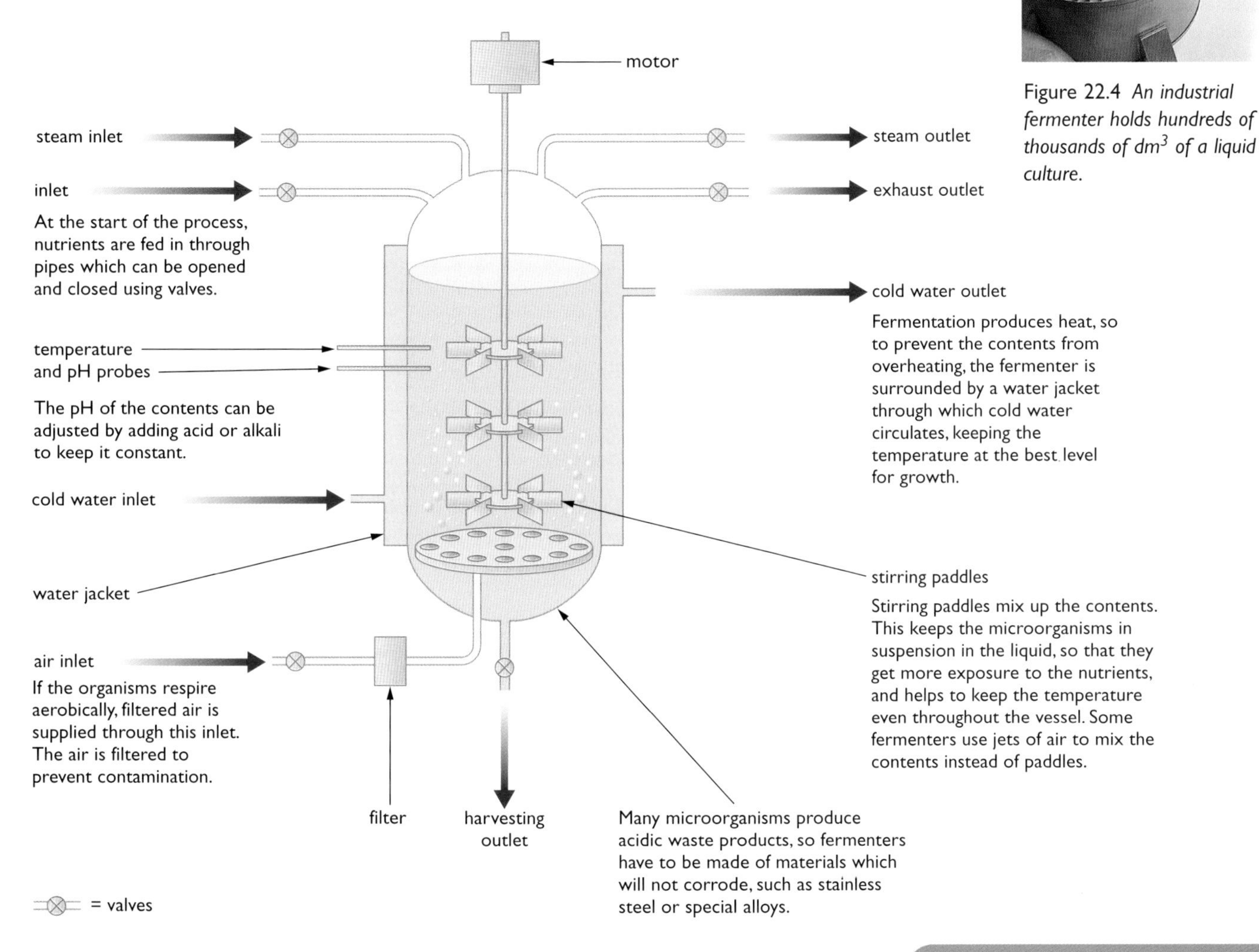

Figure 22.5 *An industrial fermenter. Fermenters like this are used to make many products, such as the antibiotic penicillin.*

When fermentation is finished, the products are collected through an outlet pipe. Before the fermenter is filled with new nutrients and culture, the inside of the tank and all the pipes must be cleaned and sterilised. This is usually done with very hot steam under high pressure. If the inside of the fermenter and the new nutrients are not sterile, two problems are likely to develop. Firstly, any bacteria or fungi that managed to get in would compete with the organism in the culture, reducing the yield of product. Secondly, the product would become contaminated with waste products or cells of the 'foreign' organism.

It is important to realise that the running of the fermenter is highly automated. Its contents are monitored by special probes which record temperature, pH, oxygen and carbon dioxide levels and so on to provide the best environment for growth of the microorganisms. The data from these probes is fed to a computer which controls the internal environment of the fermenter. For example, if the temperature starts to rise, cold water is passed more quickly through the water jacket.

> The method of production of penicillin described is known as a **batch culture**. The microorganism is grown in the fermenter, the product collected, and then the fermenter is shut down and cleaned out, ready to start all over again. A different type of fermenter produces a **continuous culture**. In this process, once the fermentation has started, more nutrients and cells are added, and the product is removed, or 'harvested' continuously. This allows the contents of the fermenter to be held at a particular point in the growth curve (see Chapter 21). Most fermenters are of the batch type. A few processes use continuous fermenters, for example the production of **mycoprotein**.

Production of penicillin

The first step in the manufacture of the antibiotic penicillin is to make a broth of spores of the mould *Penicillium*. This 'starter culture' is used in a fermenter of the type described on page 275. The culture solution contains sugar and other nutrients, and oxygen is supplied to allow the mould to respire. Most culture media for *Penicillium* contain a waste product of the starch industry, called 'corn steep liquor'. It contains sugars such as lactose (milk sugar) which the mould uses as an energy source, as well as other nutrients needed for growth.

The contents of the fermenter are kept at a steady 24°C. The spores develop into filaments of cells which multiply rapidly, doubling their mass every six hours. After about 40 hours, as they start to use up the nutrients in the broth, the cells begin to produce penicillin, and continue to produce it for several more days. The penicillin is secreted out of the cells, so that at the end of fermentation the fermenter contains a dense broth of cells, unused nutrients and penicillin solution.

The broth is filtered to remove the cells, and the penicillin is then extracted from the watery solution by using organic solvents. It is re-dissolved back into water and made to crystallise out as pure penicillin. The crystals are collected and chemically treated to produce a range of penicillins for treating different kinds of bacterial infection.

Production of mycoprotein

Mycoprotein is a meat substitute that is sold under the brand name Quorn™. It is used in a range of pies, burgers, 'ready meals' and other foods (Figure 22.6).

The mycoprotein is made from the mycelium of a mould fungus called *Fusarium*. The mould is grown in very tall (40 metres high) continuous loop fermenters (Figure 22.7). The fermenters are run for about six weeks, all the time more nutrients being added while the mould is harvested.

outlet for waste gases
rising air bubbles cause broth to circulate
denser broth sinks
40 m
rising air bubbles
inlet for air and ammonia
inlet for nutrients
outlet for harvesting
cooling system

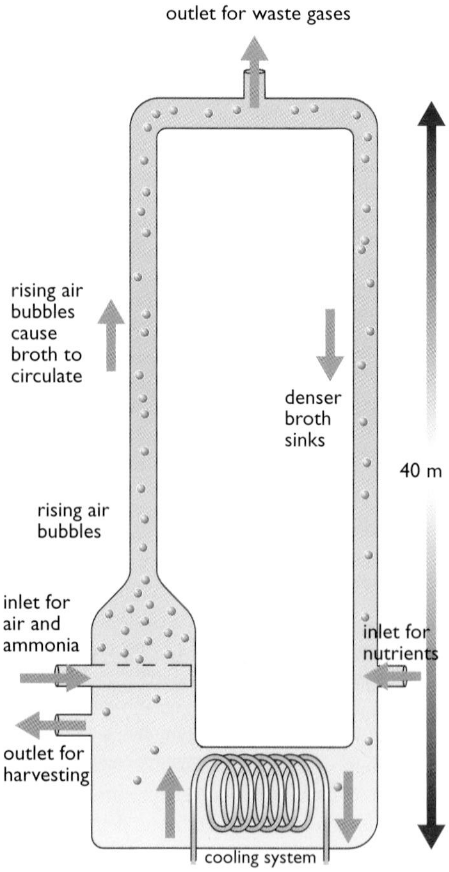

Figure 22.7 *Continuous fermenter used to grow the mould Fusarium to make Quorn™.*

Figure 22.6 *A variety of pies and other products containing Quorn™ mycoprotein.*

Instead of using paddles to stir the broth, this fermenter uses a stream of compressed air to circulate the medium and cells. The mould is fed glucose syrup as its source of carbohydrate, and minerals, including ammonia, to supply its nitrogen needs. The contents of the fermenter are kept at a steady 30°C and a pH of 6.

The harvested mycoprotein contains too much of the nucleic acid called RNA to be healthy for human consumption. This is broken down by heat treatment, and the final product is filtered and dried. It is a mat of interwoven fungal hyphae looking rather like pastry. It has very little taste, so has flavourings added to make it taste like chicken or beef. Because it is made of fibrous hyphae, it is naturally chewy and has a texture similar to that of meat.

Mycoprotein has a very high nutritional value because it contains large amounts of protein, large amounts of fibre, no animal fat and little total fat, and no cholesterol.

The lack of fats, including cholesterol, make mycoprotein a healthier alternative to meat, lowering the risk of heart disease. Fibre is a necessary part of a balanced diet (see Chapter 3) while the protein content of mycoprotein is still quite high. Look at Table 22.1, which compares some of the nutrients present in a mycoprotein burger with the same nutrients in a traditional beef burger.

Nutrient	**Total (per 100 g)**	
	Mycoprotein burger	**Beef burger**
protein	12.8 g	15.0 g
fats	4.6 g	23.8 g
cholesterol	0 g	0.02 g
fibre	4.1 g	0.4 g
carbohydrate	5.8 g	3.5 g
Total energy content	490 kJ	1 192 kJ

Table 22.1: *Comparison of nutrients in mycoprotein and beef burgers.*

Quorn™ is one successful example of a protein-containing food made from a microorganism. During the 1950s to 1970s there was a lot of research into developing other high-protein foods from microorganisms. These were given the general name **single cell proteins** (**SCP**) and were developed as food for both humans and domestic animals, as a replacement for traditional sources of protein, such as meat and fish. SCPs were thought to be a possible answer to the world shortage of dietary protein.

Producing traditional protein-rich food, such as beef, involves large losses of energy at each step in the food chain (see Chapter 14). Growing SCP does not involve these energy losses, so it is a very efficient way of producing food. However, a number of problems with SCPs have been found, including:

- high development and production costs
- competition from traditional foods
- safety of the food products – some were found to contain toxic substances
- acceptability to the public – it is difficult to convince people to eat bacteria or moulds!

As well as Quorn™, several other SCPs have been produced. These include Pekilo™, made from a fungus called *Paecilomyces*, and an animal feed called Pruteen™, made from a bacterium, *Methylophilus*. This organism feeds on methanol from North Sea gas! However, eating microorganisms is not a new idea. Since the 1500s, people in Mexico have made cakes from a slime found in lakes, made up of a photosynthetic bacterium called *Spirulina*. You could find out more about these and other SCPs by doing an Internet search.

Yeast cells normally respire *aerobically* when oxygen (air) is available, producing carbon dioxide and water. They only respire anaerobically if there is little or no oxygen available, as is the case in a wine vat or a dough mixture.

Figure 22.8 *Grapes being harvested at a vineyard in Chile.*

Figure 22.9 *Wine is fermented in large vats.*

Figure 22.10 *Froth forming on the surface of the beer as yeast ferments the sugars to alcohol and carbon dioxide.*

'Traditional' biotechnology – wine, beer and bread

The production of wine, beer and bread all involve the respiration of yeast. When yeast cells are deprived of oxygen, they respire anaerobically, breaking sugar down into ethanol and carbon dioxide:

glucose → ethanol + carbon dioxide

This ethanol is the alcohol in alcoholic drinks like wine or beer. In bread making, carbon dioxide from the yeast produces gas bubbles which expand when the dough is baked, making the bread 'light'.

Making wine

In wine making, the sugars in the grape juice are fermented by the yeast. The grapes are grown in vineyards (Figure 22.8). They are picked and crushed to produce a juice called a **must**. 'Wild' yeasts live on the surface of the grapes and can start the fermentation of the must, although usually sulphur dioxide is added to kill unwanted microorganisms, and 'cultivated' yeast is added. Fermentation of the must is carried out in large vats that stop air reaching the wine (Figure 22.9). The actual fermentation only takes a few days, during which time the alcohol concentration in the wine builds up, killing the yeast cells when it rises above 14%. The fermented pulp is passed through a press, to remove seeds and skins, and then filtered, heated and 'aged' in more vats or barrels; a process which can take years for a good red wine. Wines contain between about 6 and 14% ethanol, depending on the type of wine.

Making beer

Beer is made from **barley**. Unlike grapes, barley does not contain sugars, but starch. The starch first has to be broken down into sugar so that the yeast cells can ferment it to ethanol. When barley seeds germinate, they produce the enzyme **amylase**, which breaks down starch into maltose, or malt sugar. This can be used by the yeast as an energy source in fermentation, so the first step in beer production is to get the barley seeds to germinate. This is done by soaking them in water and laying them out on a flat surface in a **malthouse**. When the seeds have started to germinate, they are killed by heating, without destroying the enzymes. This produces a dried product called **malt**, which can be stored.

To turn malt into beer, the malt is ground up and mixed with hot water in a large vessel called a **mash tun**. The enzymes in the mash now act on the starch, breaking it down into maltose, producing a sweet liquid. This liquid is boiled to stop the enzymes working, and filtered. At this point **hops** are added. Hops give the beer a bitter flavour and stop bacteria growing. Yeast is then added. After a while, the yeast uses up the oxygen in the mixture, and starts to respire the sugars anaerobically. Fermentation carries on for several days (Figure 22.10). To make different types of beer, different species of yeast are used. Brewer's yeast (*Saccharomyces cerevisiae*) is used to make ales, and *Saccharomyces carlsbergensis* for lagers. When fermentation is finished, the beer is centrifuged, filtered and sometimes pasteurised. It is finally put into modern aluminium casks or traditional wooden barrels. The process is summarised in Figure 22.11.

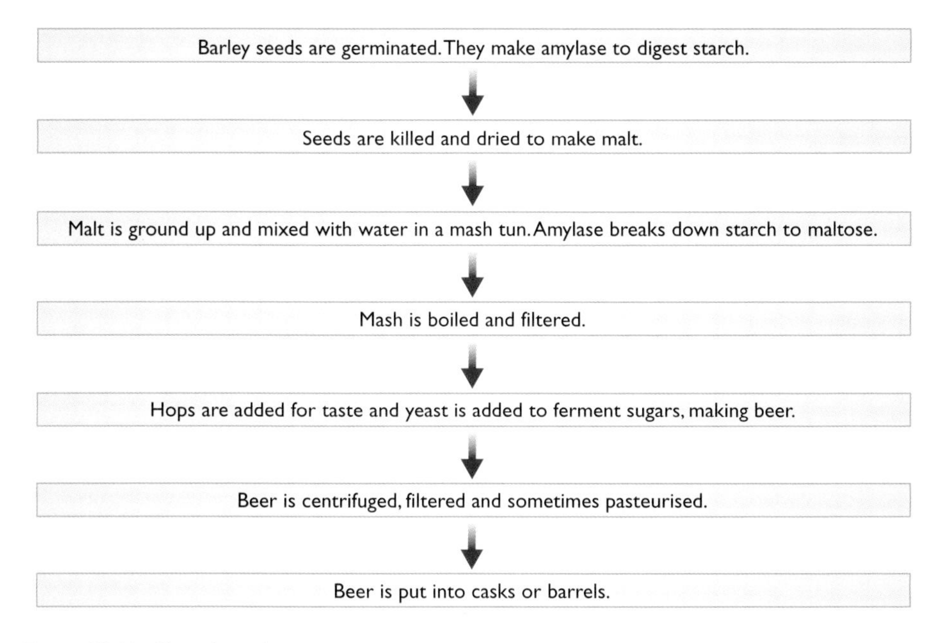

Figure 22.11 *Flow chart showing the stages in beer production.*

Remember – you often hear people say that beer is 'made from hops'. Hops are only used to *flavour* beer. The source of sugar for beer production is the starch in barley seeds (Figure 22.12).

Figure 22.12 *Malting barley seeds (top) are the source of sugar for fermentation. Hop flowers (bottom) add a bitter flavour to the beer.*

Making bread

Brewer's yeast (*Saccharomyces cerevisiae*) is also used to make bread. Flour and water are mixed together, and yeast added, forming the dough. Enzymes from the original cereal grains break down starch to sugars, which are fermented by the yeast. Extra sugar can be added at this stage. Fermentation produces carbon dioxide gas, which makes the dough rise. This is speeded up if the dough is left at a warm temperature. When the dough is later baked in the oven, the gas bubbles expand, giving the bread a light, cellular texture (Figure 22.13). Baking also kills the yeast cells and evaporates any ethanol made by the fermentation.

Figure 22.13 *The 'air' spaces in this bread were produced by bubbles of carbon dioxide released from the respiration of the yeast.*

Figure 22.14 *Bread that is made without using yeast is called* ***unleavened*** *bread. What is the difference in texture between leavened and unleavened bread?*

Fermentation to produce dairy products – cheese and yoghurt

Both cheese and yoghurt are produced by a group of organisms called **lactic acid bacteria**, which ferment milk. However, these foods are produced in rather different ways.

Making cheese

Rennet used to be made from calves' stomachs, but is now more often made by genetically engineered yeasts and moulds. The advantage of this is that the cheese can be eaten by vegetarians.

Cheese making started as a way of preserving the nutrients in milk. It was probably discovered by accident when milk, transported in bags made from the stomachs of sheep or cows, turned into cheese. Cheese is made by the action of lactic acid bacteria, which live naturally in milk and cause it to go sour. A modern cheese such as cheddar is made as follows. Firstly the milk is pasteurised, to kill any natural bacteria that it contains. Next, a starter culture of a controlled mixture of lactic acid bacteria is added to the milk in a cheese vat. These bacteria convert the milk sugar called lactose into lactic acid. This lowers the pH of the milk, which starts to make the milk proteins coagulate (clot into solid lumps called **curds**). More curds are made by adding a substance called **rennet**, which contains more milk-clotting enzymes.

Figure 22.15 *A selection of different cheeses. The blue colour seen in some cheeses is due to the growth of a blue mould.*

The solid curds are then separated from the liquid **whey** and allowed to ripen and mature. This is a very complicated process, and the details vary a great deal, depending on the type of cheese being made. Hard cheeses like cheddar just use the action of lactic acid bacteria. Some soft cheeses are matured through the action of moulds and yeasts that grow on the surface of the cheese. Blue cheeses are given their particular taste and colour by a strain of *Penicillium* that grows throughout the cheese. There are many different ways that cheeses are produced and matured, giving a wide range of tastes and textures (Figure 22.15).

Making yoghurt

Two types of bacteria commonly used in both yoghurt and cheese production are different species of *Lactobacillus* and *Streptococcus*. Both ferment lactose in milk to lactic acid. Lactic acid causes physical and chemical changes in milk, such as coagulating the milk protein and turning the mixture acidic.

Milk can also be fermented to produce yoghurt. This is also a very ancient process, and again uses lactic acid bacteria, but they are different species from those used to make cheese.

Milk is first pasteurised at 85–95°C for 15–30 minutes, to kill any natural bacteria that it contains, then **homogenised** to disperse the fat globules. It is then cooled to 40–45°C and inoculated with a starter culture of two species of bacteria. As with cheese production, these bacteria produce lactic acid, as well as starting to digest the milk proteins. The culture is kept at this temperature for several hours, while the pH falls to about 4.4 (these are the optimum conditions for the bacteria). The mixture thickens as the drop in pH causes the milk proteins to coagulate.

When fermentation is finished, the yoghurt is stirred and cooled to 5°C. Flavourings, colourants and fruit may then be added, before it is packaged for sale.

The drop in pH as the yoghurt forms gradually reduces the reproduction of the lactic acid bacteria (although it doesn't kill them). It also helps to prevent the growth of other microorganisms, which, as with cheese, preserves the nutrients of the milk. The steps in yoghurt production are summarised in the flow chart (Figure 22.16).

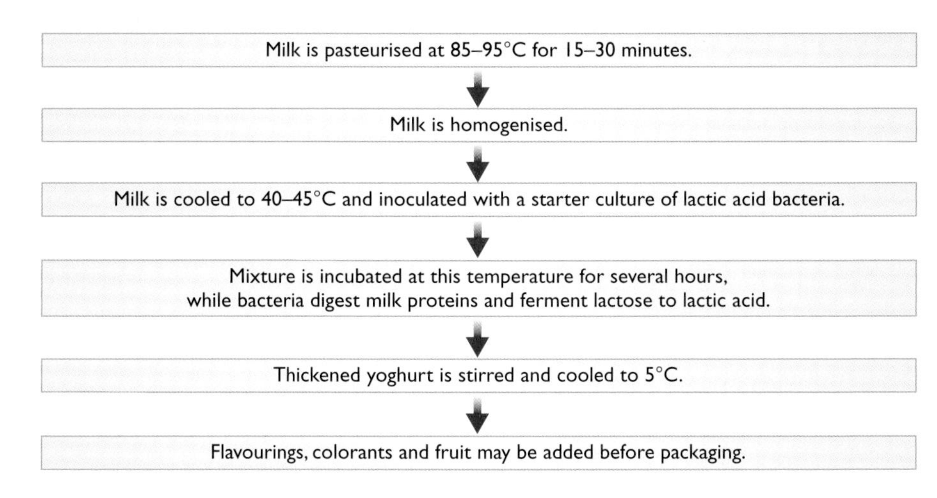

Figure 22.16 *Flow chart showing the stages in yoghurt production.*

Making vinegar

Vinegar is a dilute solution of **ethanoic acid** along with other compounds that give vinegar its taste. As well as being used to flavour food, it is a useful **preservative**. It will keep **pickled** foods fresh for long periods.

The usual way to make vinegar is to ferment beer, wine or cider a second time, using cultures of a bacterium called *Acetobacter* (the old name for ethanoic acid was acetic acid). This organism respires aerobically, oxidising the ethanol to ethanoic acid.

The fermenters used to make vinegar are of the continuous kind. The bacteria need high concentrations of oxygen, so the wine or beer is slowly trickled over a finely divided material, traditionally wood shavings or birch twigs carrying a coating of the bacteria, while air is blown through the fermenter (Figure 22.17). After a few days the ethanol is converted to ethanoic acid and the raw vinegar can then be drawn off from the bottom of the fermenter, as more ethanol is added at the top.

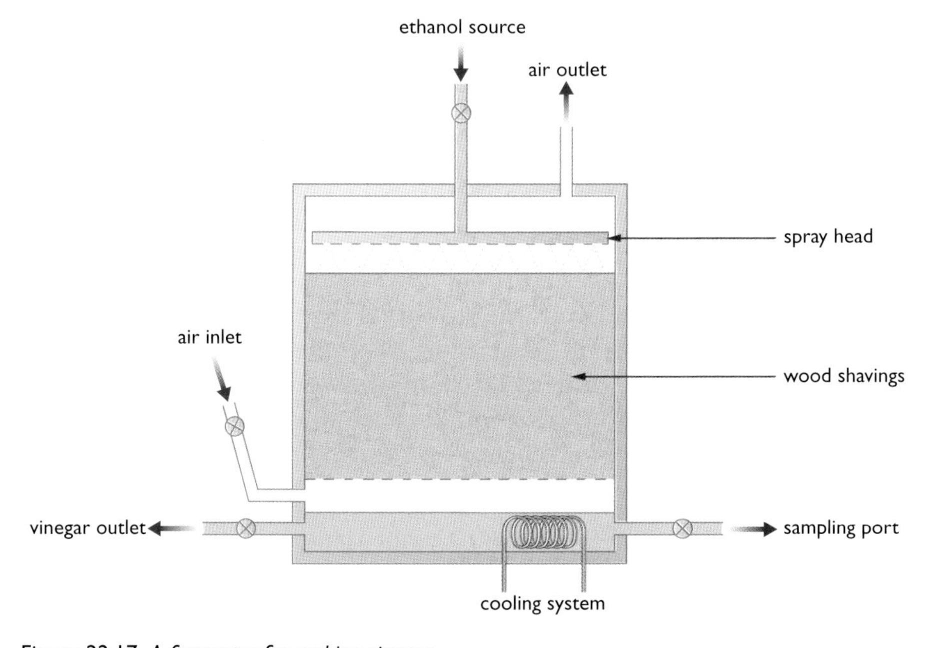

Figure 22.17 *A fermenter for making vinegar.*

Figure 22.18 *Soy sauce is used to flavour many oriental dishes.*

Making soy sauce

Soy sauce is the most widely used fermented food in China, Japan and much of the rest of Asia (Figure 22.18). It is a very important ingredient in oriental cookery, used as a flavouring in many fish, rice and bean dishes that would otherwise taste rather bland. Soy sauce is acidic, salty, slightly alcoholic and rich in amino acids.

The raw materials for making soy sauce are soya beans and roasted wheat. These are converted into the sauce by the action of moulds, bacteria and yeasts in two main fermentation processes. The steps in soy sauce production are easiest to follow as a flow chart (Figure 22.19).

Soya beans are soaked and boiled in a pressure cooker, and wheat is roasted and crushed. The soya bean mash is mixed with flour from the roasted wheat.

Fermentation Stage 1: A fungus called *Aspergillus oryzae* is added to the mixture, which is incubated for 3 days at 30°C, under aerobic conditions. The fungus makes the enzyme amylase, which breaks down starch into sugar molecules. Proteins aere also digested to amino acids.

The mixture is heavily salted.

Fermentation Stage 2: The product of the first fermentation stage is incubated anaerobically in deep tanks for 3–6 months at 15°C, increasing to 25°C. Here, lactic acid bacteria (*Lactobacillus spp.*) produce lactic acid, which causes the pH to fall. Yeasts in the mixture make ethanol.

The mixture is left to 'age' for several months, filtered, and the liquid collected. This liquid is finally pasteurised at 75°C, filtered again, and bottled.

Figure 22.19 *Flow chart showing the stages in soy sauce production.*

Fermentation to produce fuels

When you light a Bunsen burner, the gas that is burning is **methane**, which we get from gas deposits in the rocks under the North Sea. Most of our gas is supplied from this source. However, methane can also be made by fermentation of materials such as sewage or animal waste. The methane is made from the activities of microorganisms, which break down the organic matter – another fermentation process.

ideas evidence

In Britain, farmers are increasingly using fermentation to produce fuel in this way. The fuel is known as **biogas**, and as well as supplying energy, is a useful way to get rid of farming wastes such as farmyard manure or chicken litter.

Biogas is made in large tanks called **digesters** (Figure 22.20), which contain several different types of microorganisms, including anaerobic bacteria called **methanobacteria**. They use carbohydrates in the waste as a food source, breaking it down into methane and some carbon dioxide. The biogas burns well, and is used as an extra source of energy on the farm, for heating or powering electrical generators. However, the digesters can leak gas. This makes them smelly and they have been known to explode if a naked flame is exposed near the gas.

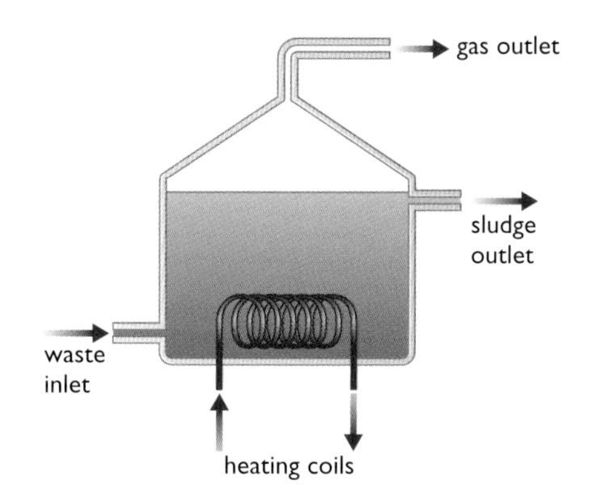

Figure 22.20 *Biogas is made in large tanks called digesters.*

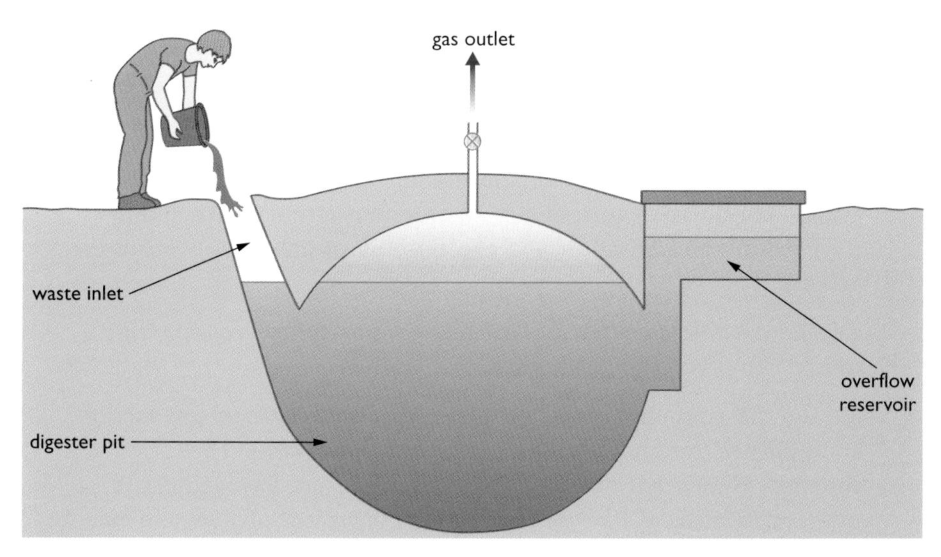

Figure 22.21 *An underground biogas digester. Building the digester underground gives it support and insulation. It must be covered to stop oxygen reaching the contents and to allow the gas to be collected.*

Biogas digesters are very common in some countries. It is estimated that in China there are about 5 million simple digesters like the one shown in Figure 22.21. They are common in rural areas, where they are used for small-scale production of biogas from materials collected locally, such as dung and vegetable waste. The same sort of digester is used in several other less economically developed countries in Asia, Africa and South America. In the future, it may be possible to use digesters in more economically developed countries on a much larger scale, for production of biogas from sewage or waste from sugar factories.

You have seen above how yeast can ferment sugars to produce alcoholic drinks such as wine and beer. Of course, pure ethanol will also burn well, and makes a very good fuel. If it can be produced cheaply enough by fermentation, it can be used as an alternative to fossil fuels.

ideas
evidence

A fuel called **gasohol** is made in certain countries, especially Brazil. Gasohol is a mixture of about 80–90% petrol and 10–20% ethanol. The ethanol is made by anaerobic fermentation of sugar by yeast, and then distilled to produce pure ethanol. The sugar is supplied by sugar cane plantations as raw (unrefined) sugar or cane waste. It can also be made from maize starch by the action of carbohydrase enzymes. If there are sugar plantations nearby, the gasohol can be made cheaply enough to be a viable renewable source of energy. There are problems though. When the world price of petrol falls, it may not be economic to make gasohol. Also, cultivating sugar cane or maize to produce ethanol uses up a lot of land that could be used for growing food crops.

End of Chapter Checklist

If you haven't got a copy of your specification, read the introduction on page vi.

You will need to be able to do some or all of the following. Check your Awarding Body's specification (syllabus) to find out exactly what you need to know.

- Understand the use of solid and liquid culture media and the principles of sterile technique used to prepare and maintain cultures of microorganisms in the laboratory.
- Recall the definition of fermentation: microorganisms use an external food source to obtain energy, changing substances in the medium.
- Understand the workings of a fermenter and explain the need for aseptic conditions, nutrient supply, optimum temperature and pH, oxygenation and agitation, and explain how optimum conditions are achieved.
- Recall how the antibiotic penicillin can be made by growing the mould *Penicillium* in a fermenter.
- Explain the advantages of using microorganisms for food production, and evaluate the potential of biotechnology in relation to world food shortage. Know that different microorganisms are involved in the production of wine, beer, bread, yoghurt, single cell protein and vinegar.
- Remember how mycoprotein (single cell protein) is produced by the fungus *Fusarium*, and understand the advantages of using mycoprotein as a food source, in terms of nutrition and economics.
- Understand that anaerobic respiration by yeast to produce ethanol and carbon dioxide is an example of fermentation, and is used in making wine, beer and bread.
- Describe in outline the stages in beer, wine and bread making, particularly the biological significance of these stages.
- Explain the role of bacteria in the production of cheese from milk and of bacteria and moulds in the ripening of cheese.
- Explain the role of bacteria in the production of yoghurt from milk by the conversion of lactose to lactic acid.
- Understand the stages in the commercial production of soy sauce, including fermentation of a mixture of cooked soya beans and roasted wheat using *Aspergillus*, further fermentation using yeasts and then *Lactobacillus*, filtration, pasteurisation and sterile bottling.
- Know the composition of biogas and how it is made by the fermentation of plant products or waste material containing carbohydrates, on a small or large scale. Evaluate the advantages and disadvantages of different kinds of biogas generator.
- Know how ethanol-based fuels can be made from anaerobic fermentation of sugar cane juices or glucose from maize starch, followed by distillation. Know that ethanol can be used in motor vehicle fuels, and interpret economic and environmental data relating

Questions

More questions on growing useful organisms can be found at the end of Section F on page 332.

1 The diagram shows an industrial fermenter which is used to make the antibiotic penicillin.

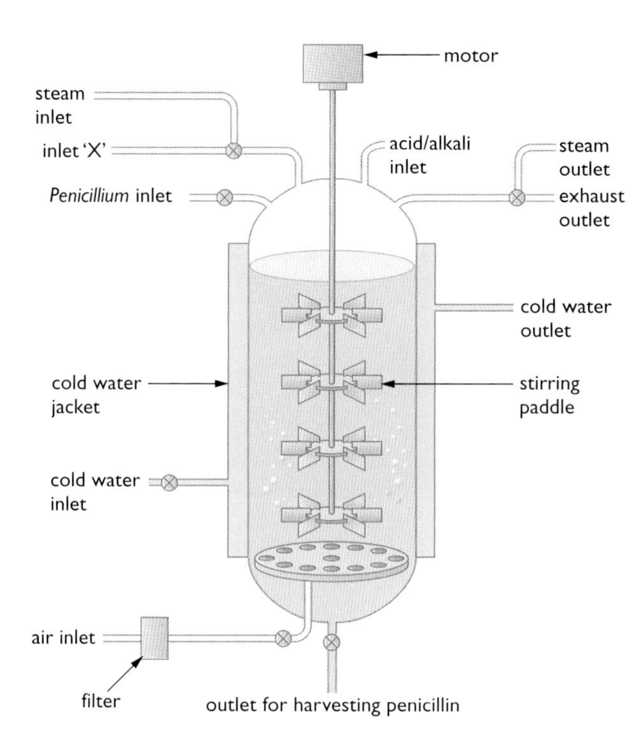

a) Explain how the fermenter is sterilised before use.

b) Why does air need to be pumped through the fermenter? Why is the air filtered?

c) Explain how a steady temperature is maintained in the fermenter.

d) What is supplied through the inlet marked 'X'?

e) Explain what would happen to the growth of *Penicillium* in the fermenter if the paddles stopped working.

2 Which of the following fermentations are aerobic (need oxygen) and which are anaerobic (don't use oxygen)?

making beer, making vinegar, formation of methane in a biogas generator, production of penicillin from *Penicillium*

3 The diagram shows a fermentation flask used to make wine at home.

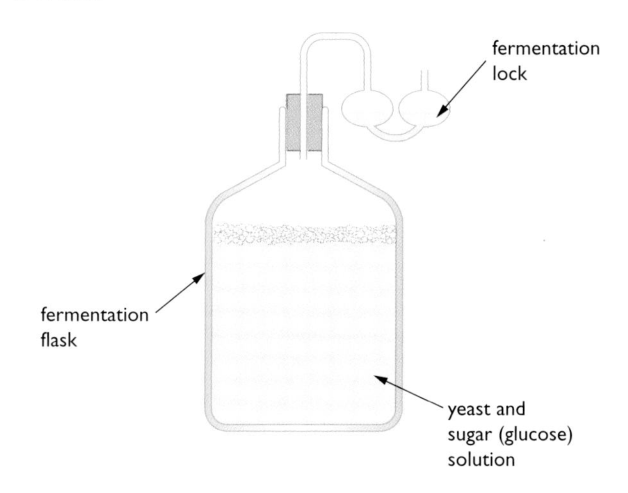

a) The yeast cells in the flask respire anaerobically. Write a word equation for this process.

b) Yeast cells can also respire aerobically. How is this prevented in this flask?

c) Home winemaking flasks like this are often kept in a warm airing cupboard. Explain why this is done.

d) Wine usually contains no more than about 14% ethanol (alcohol). Explain why.

4 Answer these questions about making yoghurt. Try at first to answer them without looking back to the section on page 280!

a) Why is the milk pasteurised at the start of the process?

b) Why is the mixture of milk and bacteria incubated at 45°C?

c) What causes the milk to thicken?

d) Why does fermentation eventually stop?

e) Explain how making yoghurt is a way of preserving the nutrients from milk.

5 Use the information about mycoprotein and beef burgers in Table 22.1 (page 277) to answer the following questions:

a) Write a paragraph explaining why a mycoprotein burger is thought to be more healthy than a beef burger.

b) Suggest why a mycoprotein burger contains more energy than a beef burger.

c) Name the fungus used to produce mycoprotein.

d) Describe the conditions maintained in the fermenter to help this fungus grow.

6 Look at the flow chart showing the production of soy sauce on page 282. Use it to answer the following questions:

a) What will happen to the fungus *Aspergillus oryzae* during the second fermentation stage? Explain your answer.

b) From information in the flow chart, suggest three reasons why soy sauce does not go bad during storage.

7 It is often easier to learn the steps in a complex process from a flow chart. Construct a flow chart for cheddar cheese production similar to the other flow charts in this chapter for beer, yoghurt and soy sauce. An outline of cheese production is given on page 280, but since each cheese uses a slightly different process, you will need to look up the details for cheddar, using other books or an Internet search.

8 Explain the differences between biogas and gasohol.

9 Bacteria, moulds and their spores are everywhere. Describe how you could inoculate some agar plates with bacteria and moulds from a laboratory sink. You can use any of the following apparatus and materials: Petri dishes containing nutrient agar, Bunsen burner, wire loop, sticky tape, wax pencil, incubator, thermometer.
You should write about 15 lines, stressing sterile techniques.

10 The picture shows a selection of foods and drink that are all made using microorganisms. For each product, state which organism(s) is needed.

Chapter 23: Gene Technology

In this chapter, we will look at ways in which it is now possible to manipulate genes and produce **genetically modified organisms** or **transgenic organisms.** This is the science of 'genetic engineering'. But first, we must remind ourselves what genes are made of and how they work.

Ever since Gregor Mendel first identified his 'heritable factors', biologists have tried to find out more and more about genes. We now know what genes are made of and we have a good idea of how they actually work. We have just produced the first ever gene map of all the human genes in the Human Genome Project.

DNA – the stuff of genes

A gene is a section of one strand of a DNA molecule that codes for the production of a protein. Each sequence of three bases (a triplet) in the DNA strand codes for one amino acid. Different genes produce different proteins because each has a unique sequence of bases that codes for a unique sequence of amino acids – that results in a unique protein (Figure 23.1).

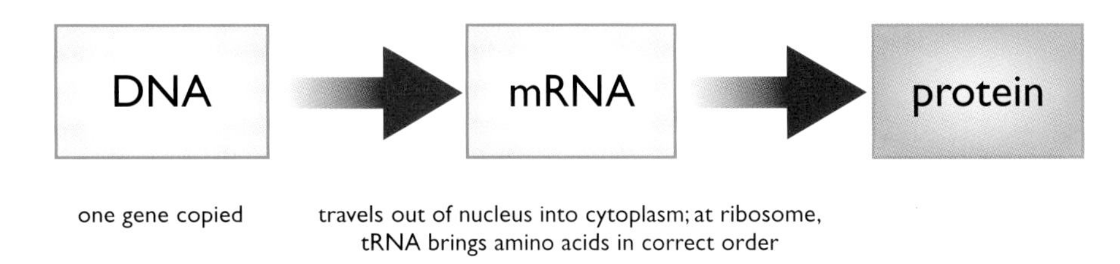

Figure 23.1 *The role of DNA in protein synthesis.*

The protein that is produced could be:

- an enzyme that controls a particular reaction inside a cell or in the digestive system
- a structural protein like the keratin in hair, collagen in skin or one of the many proteins found in the membranes of cells
- a hormone
- a protein with a specific function such as haemoglobin or an antibody.

Recombinant DNA

A transgenic organism is one that contains a gene or several genes from another species. For example, some bacteria have had human genes transferred to them that allow them to make human insulin. Some sheep secrete AAT in their milk because they have the human gene that directs the manufacture of this substance. Because they contain 'foreign' genes, they are no longer quite the same organisms. They are transgenic.

Producing recombinant DNA is the basis of gene technology or genetic engineering. A section of DNA – a gene – is snipped out of the DNA of one species and inserted into the DNA of another. This new DNA is called **recombinant** DNA, as the DNA from two different organisms has been 'recombined'. The organism that receives the new gene is a transgenic organism.

The organism receiving the new gene now has an added capability. It will manufacture the protein its new gene codes for. For example, a bacterium receiving the gene from a human that codes for insulin production will make human insulin. If these transgenic bacteria are cultured by the billion in a fermenter, they become a human insulin factory.

A bacterial chromosome is not like a human chromosome. It is a continuous loop of DNA rather than a strand. Also, the DNA is 'naked' – there are no proteins in it – only DNA.

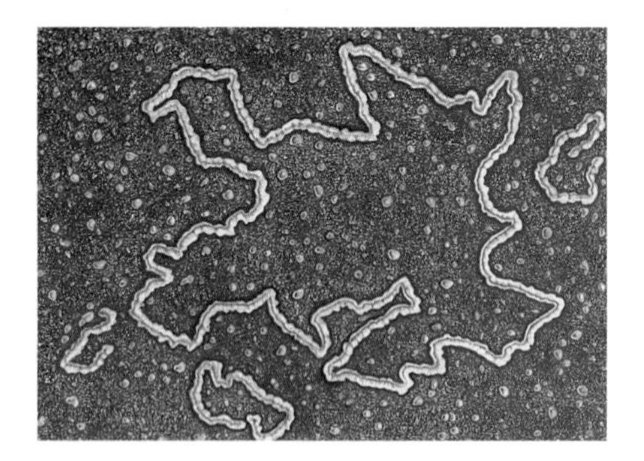

Figure 23.2 *Bacterial DNA.*

Producing genetically modified (transgenic) bacteria

The breakthrough in being able to transfer DNA from cell to cell came when it was found that bacteria have two sorts of DNA – the DNA found in their bacterial 'chromosome' and much smaller circular pieces of DNA called **plasmids** (Figure 23.2).

Bacteria naturally 'swap' plasmids, and biologists found ways of transferring plasmids from one bacterium to another. The next stage was to find molecular 'scissors' and molecular 'glue' that could snip out genes from one molecule of DNA and then stick them back into another. Further research found enzymes that were able to do this:

- **Restriction endonucleases** are enzymes that cut DNA molecules at specific points. Different restriction enzymes cut DNA at different places. They can be used to cut out specific genes from a molecule of DNA.
- **DNA ligases** are enzymes that join cut ends of DNA molecules.

Each restriction endonuclease recognises a certain base sequence in a DNA strand. Wherever it encounters that sequence, it will cut the DNA molecule. Suppose a restriction enzyme recognises the base sequence G-A-A-T-T-C. It will only cut the DNA molecule if it can 'see' the base sequence on both strands. Figure 23.3 illustrates this.

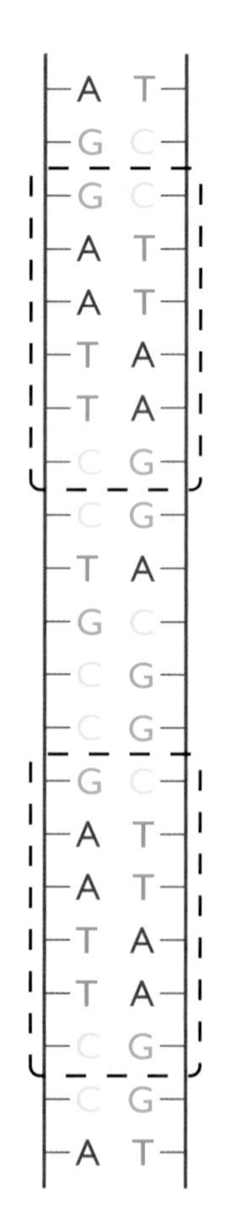

Figure 23.3 *Part of a DNA molecule containing the base sequence G-A-A-T-T-C. Notice that the sequence is present on both strands, but running in opposite directions.*

Some restriction enzymes make a straight cut and the fragments of DNA they produce are said to have 'blunt ends'. Other restriction enzymes make a staggered cut. These produce fragments of DNA with overlapping ends with complementary bases. These overlapping ends are often called 'sticky ends' because fragments of DNA with exposed bases are more easily joined together by ligase enzymes. This is shown in Figure 23.4.

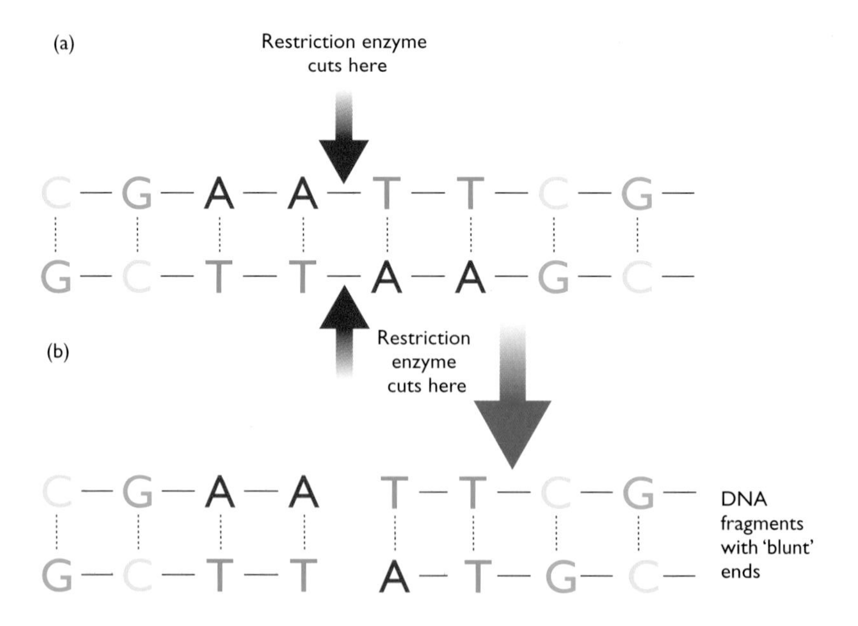

Figure 23.4 *How restriction enzymes cut DNA.*

Biologists now had a method of transferring a gene from any cell into a bacterium. They could insert the gene into a plasmid and then transfer the plasmid into a bacterium. The plasmid is called a **vector** because it is the means of transferring the gene. The main processes involved in producing a transgenic bacterium are shown in Figure 23.5.

ideas evidence

There is a lot more to producing recombinant DNA and transgenic bacteria than is shown here. You could carry out an Internet search or search appropriate CD-encyclopaedias to find out more.

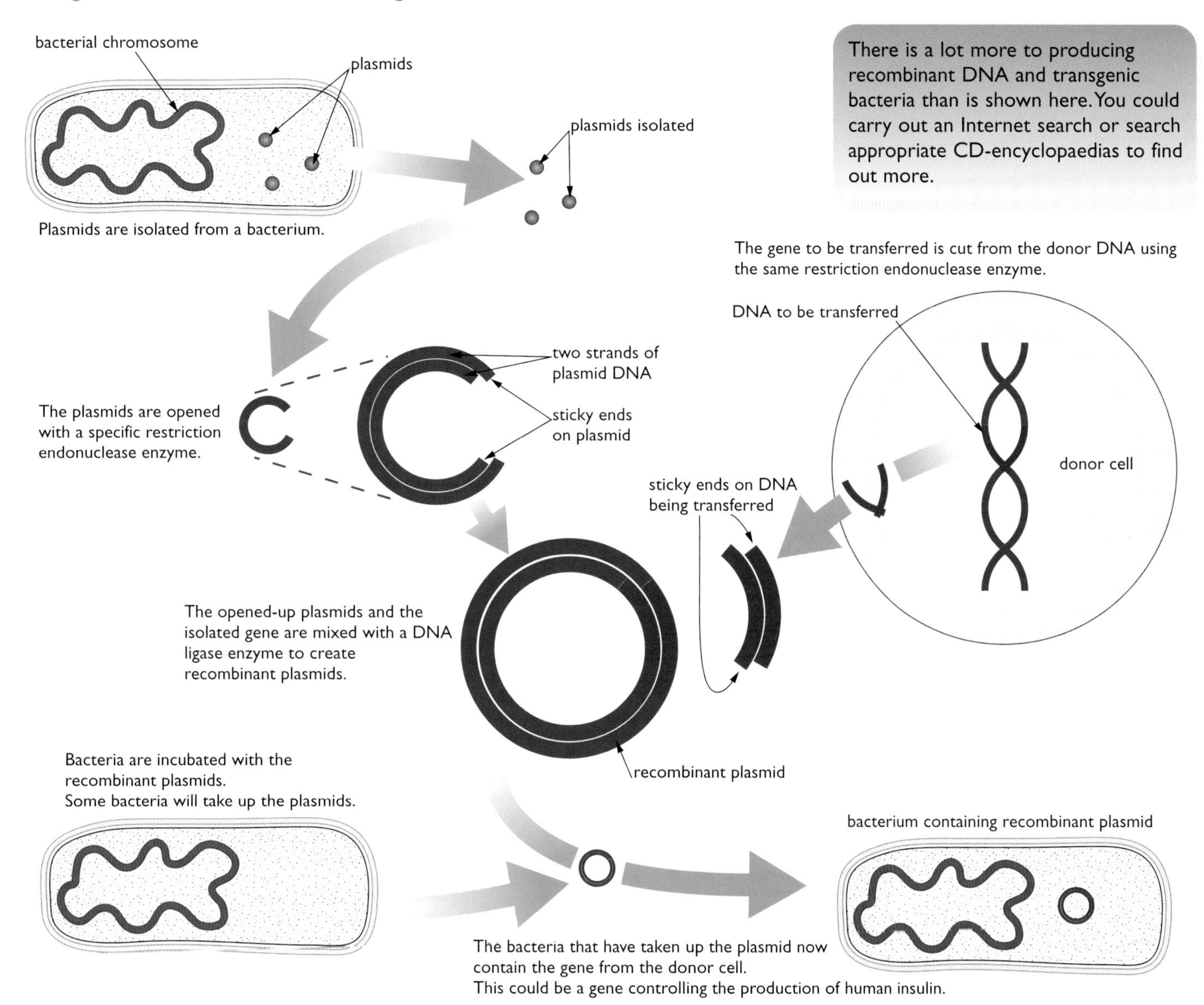

Figure 23.5 *Stages in producing a transgenic bacterium.*

Making use of genetically modified bacteria

Different bacteria have been genetically modified to manufacture a range of products. Once they have been genetically modified, they are cultured in fermenters to produce large amounts of the product (see Chapter 22). Some examples are described here.

- **Human insulin** People suffering from diabetes need a reliable source of insulin. Before the advent of genetic engineering, the only insulin available came from other animals. This is not quite the same as human insulin and does not give quite the same control of blood glucose levels.

More insulin is required every year because the number of diabetics increases world-wide each year and diabetics now have longer life spans.

- **Enzymes for washing powders** Many stains on clothing are biological. Blood stains are largely proteins, grease marks are largely lipids. Enzymes can digest these large, insoluble molecules into smaller, soluble ones. These then dissolve in the water. Amylases digest starch, proteases digest proteins and lipases digest lipids. Bacteria have been genetically engineered to produce enzymes that work at higher temperatures, allowing even faster and more effective action.
- **Enzymes in the food industry** One bacterial enzyme used in the food industry is **glucose isomerase**. This enzyme turns glucose into a similar sugar called fructose. Fructose is much sweeter than glucose and so less is needed to sweeten foods. This has two advantages – it saves money (less is used) and it means that the food contains less sugar and is healthier.
- **Human growth hormone** The pituitary gland of some children does not produce sufficient quantities of this hormone and their growth is retarded. Injections of growth hormone from genetically modified bacteria restore normal growth patterns.

Before human growth hormone from genetically modified bacteria was available, the only source of the hormone was from human corpses. This was a rather gruesome procedure and had health risks. A number of children treated in this way developed Creutzfeld–Jacob disease (the human form of 'mad cow' disease). When this became apparent, the treatment was withdrawn.

- **Bovine somatotrophin (BST)** (a growth hormone in cattle) This hormone increases the milk yield of cows and increases the muscle (meat) production of bulls. Giving injections of BST to dairy cattle can increase the milk yield by up to 10 kg per day. To do this they need more food, but this increased cost is more than offset by the increased income from the increased milk yield (Table 23.1).

	Feed (kg/day)	**Milk output (kg/day)**	**Milk to feed ratio**
without BST	34.1	27.9	0.82
with BST	37.8	37.3	0.99

Table 23.1: *Effects of BST on milk yield.*

- **Human vaccines** Bacteria have been genetically modified to produce the antigens of the Hepatitis B virus. This is used in the vaccine against Hepatitis B. The body makes antibodies against the antigens but there is no risk of contracting the actual disease from the vaccination.

Since the basic technique of transferring genes was worked out, many unicellular organisms have been genetically modified to produce useful products. Also, other techniques for transferring genes into larger organisms have been developed.

Producing genetically modified plants

The gene technology described so far can transfer DNA from one cell to another cell. In the case of bacteria, this is fine – a bacterium only has one cell. But plants have billions of cells and to genetically modify a plant, each cell must receive the new gene. So, any procedure for genetically modifying plants has two main stages:

- introducing the new gene or genes into plant cells
- producing whole plants from just a few cells.

Biologists initially had problems in inserting genes into plant cells. They then discovered a bacterium called *Agrobacterium*, which regularly inserts plasmids into plant cells. Now that a vector had been found, the rest became possible. Figure 23.6 outlines one procedure that uses *Agrobacterium* as a vector.

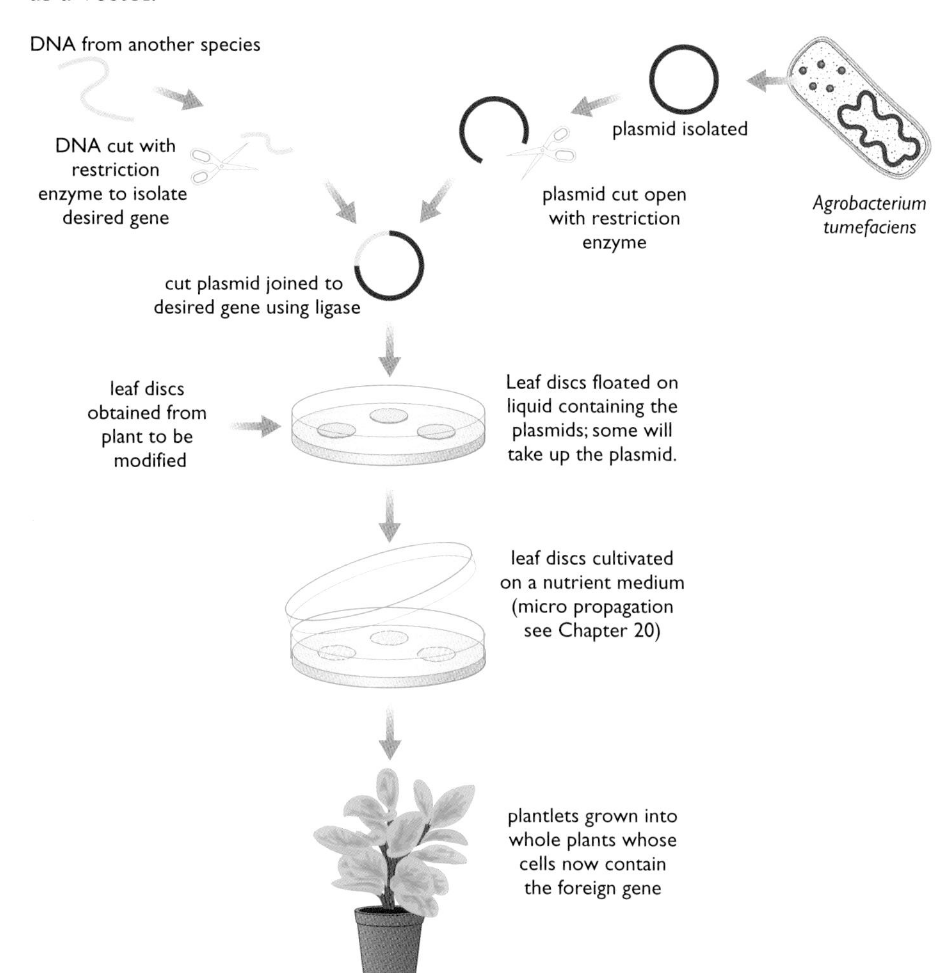

Figure 23.6 *Genetically modifying plants using* Agrobacterium.

Figure 23.7 *The gene gun.*

This technique cannot be used on all plants. *Agrobacterium* will not infect cereals and so another technique was needed for these. Enter the gene gun! This is, quite literally, a gun that fires a golden bullet (Figure 23.7). Tiny pellets of gold are coated with DNA that contains the desired gene. These are then 'fired' directly into plant tissue. Research has shown that if young, delicate tissue is used, there is a good uptake of the DNA. The genetically modified tissue can then be grown into new plants using the same micropropagation techniques as those used in the *Agrobacterium* procedure. The gene gun has made it possible to genetically modify many cereal plants as well as tobacco, carrot, soybean, apple, oilseed rape, cotton and many others.

Making use of genetically modified plants

Large numbers of genetically modified plants are already available to plant growers and farmers. There have been examples of fruit and vegetables with extended shelf lives.

Some plants have been modified to be resistant to herbicides (weedkillers). This allows farmers to spray herbicides at times when they will have maximum effect on the weeds, without affecting the crop plant. There are concerns that this will encourage farmers to be less careful in their use of herbicides. In another example, genes from Arctic fish that code for an 'anti-freeze' in their blood have been transferred to some plants to make them frost resistant.

Figure 23.8 *Golden rice.*

The gene gun has allowed biologists to produce genetically modified rice called 'golden rice'. This rice has had three genes added to its normal DNA content. Two of these come from daffodils and one from a bacterium. Together, these genes allow the rice to make beta-carotene – the chemical that gives carrots their colour. It also colours the rice, hence the name 'golden rice'. More importantly, the beta-carotene is converted to vitamin A when eaten. This could save the eyesight of millions of children in less economically developed countries who go blind because they have no source of vitamin A in their diet.

Genetically modified plants are also helping humans to resist infection. Biologists have succeeded in modifying tobacco plants and soybeans to produce antibodies against a range of infectious diseases. If these can be produced on a large scale, they could be given to people who are failing to produce their own antibodies. Other modified tobacco plants produce the hepatitis B antigens that could be used as the basis for a vaccine. There is always a risk with a vaccine containing viruses that they may somehow become infectious again. This could not happen with a vaccine containing only plant-produced antigens.

What if you could receive a 'vaccination' every time you ate a banana? Scientists are researching the possibility of transferring the genes that produce the antigens for several diseases to bananas. If they succeed, then when you eat the banana, the antigens will stimulate an immune response (see Chapter 25). There will be no risk of you catching the disease – and no need to have needles stuck in you!

Antibodies and antigens produced in this way are sometimes called 'plantibodies' and 'plantigens'! You can read more about conventional antibodies and antigens in Chapter 25.

Besides the specific examples given, research into the genetic modification of plants hopes to provide (or provides already) plants with:

- increased resistance to a range of pests and pathogens
- increased heat and drought tolerance
- increased salt tolerance
- a better balance of proteins, carbohydrates, lipids, vitamins and minerals – more nutritious crop plants.

In addition, some genetically modified oilseed rape plants will be used in large-scale production of biodegradable plastics and anti-coagulants.

One of the biggest achievements would be to modify crop plants like cereals and potatoes to allow nodules of nitrogen fixing bacteria to form on their roots. At the moment only legumes (peas, beans and other plants with seeds in 'pods') can do this (see Figure 14.13 on page 168). Biologists know that the ability is genetically controlled. However, they cannot transfer these genes to other plants yet. If they could, vast areas of infertile soil would be able to yield good crops of cereals without the need to use large quantities of fertilisers.

The bacteria in the root nodules would obtain nitrogen from the air in the soil and 'fix' it in a more usable form (usually ammonia). By doing this, they would make a supply of usable nitrogen available to the plants. The plants would convert this into plant protein and use the protein for growth. The cost of producing these crops would decrease dramatically.

Producing genetically modified animals

Producing genetically modified animals poses some of the same problems as those connected with modifying plants. Animals, like plants, are multicellular. It is not enough simply to transfer a gene to a cell. That cell must then grow into a whole organism. The plasmid technology used to create genetically modified plants depends on the modified cells being grown into whole plants using micropropagation. No such micropropagation techniques exist for animals.

Scientists researching the production of genetically modified animals had to find other techniques. The most successful involves injecting DNA directly into a newly fertilised egg cell. This develops into an embryo, then an adult (Figure 23.9).

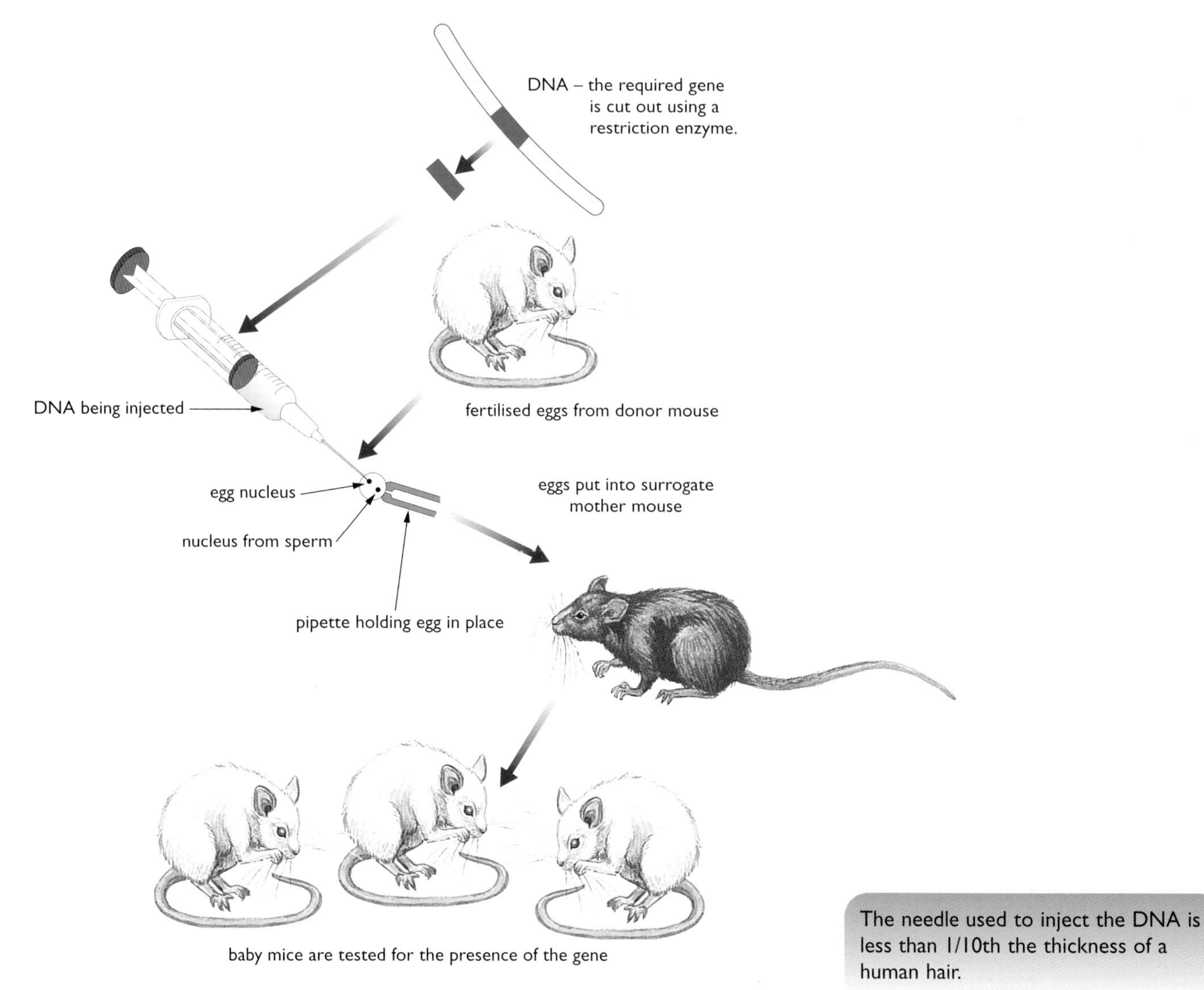

Figure 23.9 The procedures used in producing genetically modified animals.

The needle used to inject the DNA is less than 1/10th the thickness of a human hair.

Xenotransplantation means transplanting organs from other animals into humans. Transgenic pigs have been produced with genes that code for the main human 'marker antigens'. The cells of the pig's organs therefore have these human antigens on their surface and the organs would be less likely to be rejected by a recipient. If this became possible on a large scale, it could help to overcome the shortage of donor organs for transplantation.

Research of this kind can produce beneficial results similar to those achieved by genetically modifying plants:

- increased production of a particular product, e.g. higher milk yield
- increased resistance to disease and other parasites
- manufacture of human antibodies
- manufacture of specific medicinal products.

In addition, genetically modified animals may produce low cholesterol milk and organs for **xenotransplantation**.

Gene technology in action

Identification of genetic diseases

Genetic diseases develop when specific genes become active. Identifying the genes involved helps to identify people at risk. If a person is found to have the gene for one of these diseases, a genetic counsellor can advise the person of the risk of passing on the condition to any children. This is particularly important with diseases that do not develop until after reproductive age as by that time, the person could already have passed on the disease to their children.

A PCR (polymerase chain reaction) machine makes DNA replicate itself in almost the same way as it does in cell division. Because DNA copies itself exactly, the amount of DNA in a sample can be increased until there is enough for the tests.

Gene probes are single strands of DNA. The base sequence of the DNA of a gene probe is **complementary** to the sequence of the gene it tests for. Strands of DNA with complementary base sequences can bind to each other. For a fuller explanation of what complementary means, see Chapter 16.

Testing people for genes that cause disease is called **genetic screening**. It requires a sample of DNA from the person being tested. If there is not enough DNA in the sample, it can be made to copy itself in a **PCR** machine.

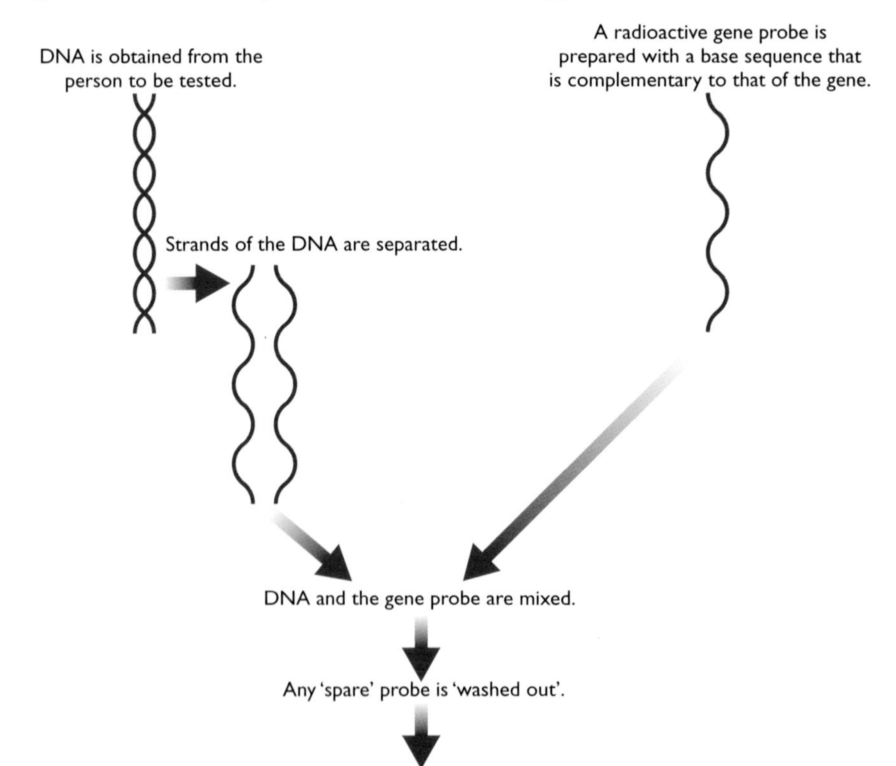

Figure 23.10 *The main stages in identifying a gene.*

The DNA sample is tested with a **gene probe** to see if the person is carrying the disease-causing gene. A gene probe is a short section of manufactured DNA that will bind to a specific gene – and only that gene. Gene probes are made to be either fluorescent or radioactive, so that they can be detected easily afterwards. Figure 23.10 summarises the procedure.

The technique could also be used for genetic screening of human fetuses. This sort of genetic screening raises important moral and ethical issues. What if the fetus *is* carrying the gene for a genetic disorder? Does the mother have the right to opt for an abortion?

A tool in crime detection

With the possible exception of identical twins, each person's DNA is unique. Therefore if a particular sample of DNA found at the scene of a crime matches that of a suspect, it is likely that the suspect was at least present. The gene technology involved in comparing DNA samples like this is called **genetic fingerprinting**. The procedure is shown in Figure 23.11. Figure 23.13 shows how genetic fingerprinting is used to identify suspects present at the scene of a crime.

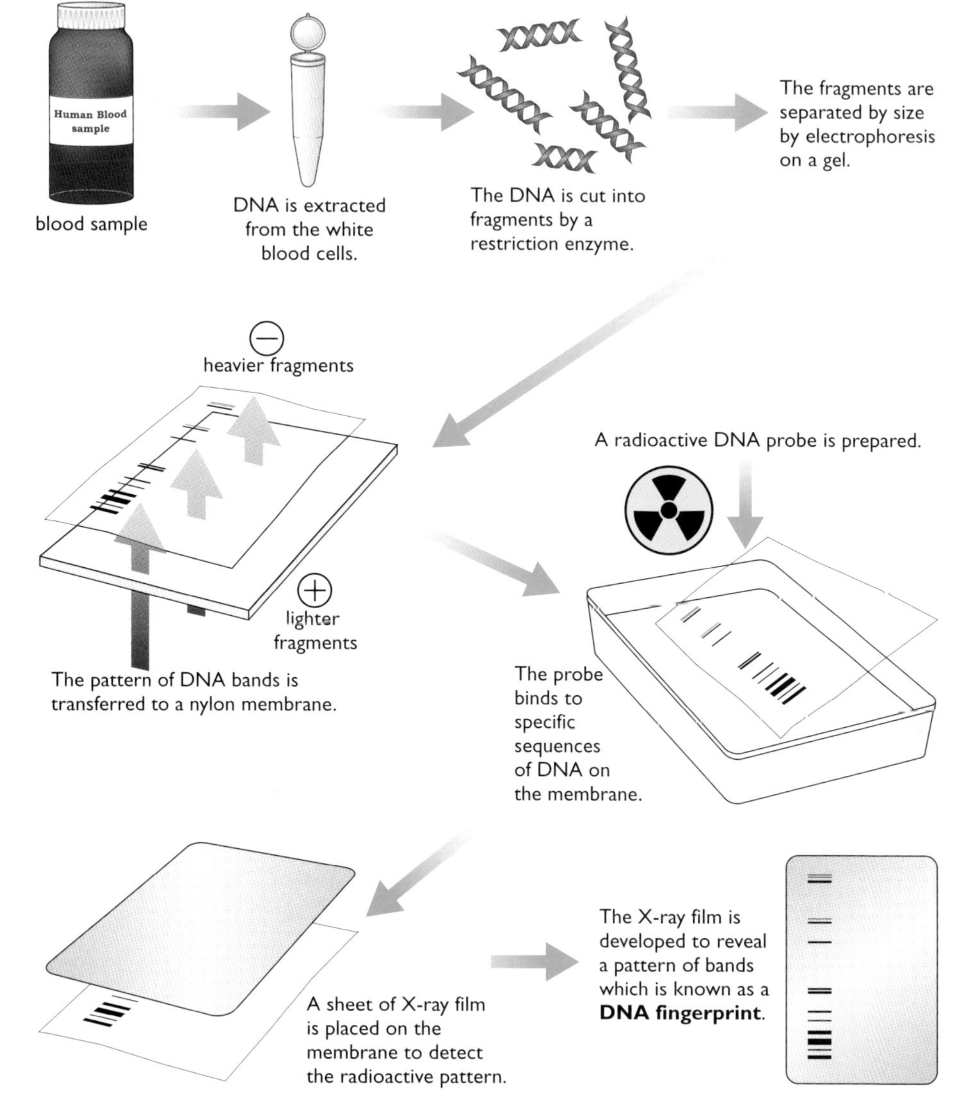

Figure 23.11 *The main stages in producing a genetic fingerprint.*

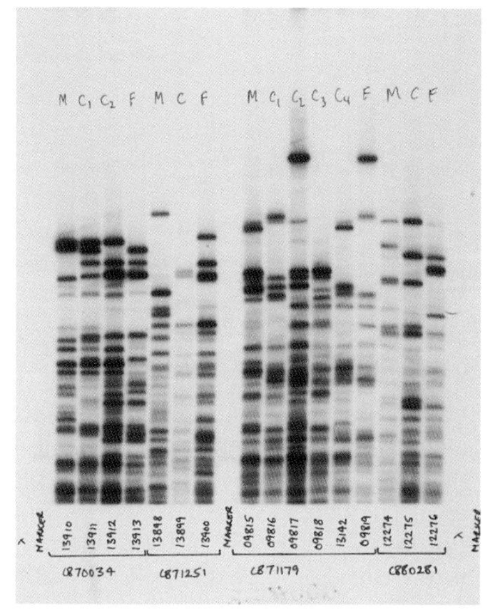

Figure 23.12 *A DNA fingerprint.*

Electrophoresis uses a difference in electrical potential to move charged particles through a gel. DNA fragments are negatively charged and so move towards the positively charged end of the gel. Lighter fragments move further than heavier fragments.

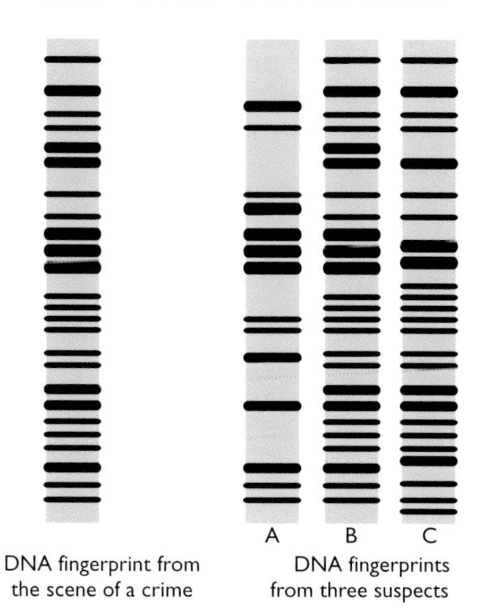

Figure 23.13 *Genetic fingerprints from the scene of a crime and from three suspects. Clearly, the banding pattern in the genetic fingerprint of suspect B is the same as that from the scene of the crime. Suspect B must have been there. The other two are different and so the 'fingerprints' give no proof that they were there.*

Some moral, ethical and practical concerns about genetic engineering

Morality is our personal sense of what is right, or acceptable, and what is wrong. It is not necessarily linked to legality. **Ethics** have a sense of right and wrong also, but they are the not individual opinions. Ethics represent the 'code' adopted by a particular group to govern its way of life (see Chapter 20 for a fuller discussion of morality and ethics). Some of the concerns of genetic engineering are discussed in Table 23.2.

Issues	**Concerns about genetic engineering**
Genes transferred from one species to another species could not normally have got there.	This is a moral issue. Many people feel that a species should not be altered in any way.
We do not know enough about the long-term ecological effects of introducing genetically modified organisms into fields. They may out-compete wild plants and take over an area. They may be toxic to wild animals. Because we do not know, we should not, therefore, introduce such plants into fields. Some people believe that we should not even carry out trials to find out.	Again, this is a moral view. We could never know this with any new breed of plant. The early farmers who crossbred wild wheats had no idea of the impact their new plants would have. Does this make it wrong?
When genetically modified plants are planted the 'new' genes may 'jump' into other species of plants. For instance, genes for herbicide (weedkiller) resistance may transfer into the weeds.	This is not a moral issue, but a purely practical one. Is it likely that resistance genes could be transferred into weeds in sufficient numbers to make a real difference? We can only find this out by conducting field trials.
Using genetic fingerprinting to combat crime will only really be useful if there is a genetic database – a file of the genetic fingerprints of everyone in the country. Once the Human Genome Project finally identifies all the genes, who will have access to this information?	There are concerns that a genetic database would be subject to misuse, that evidence could be manufactured. Also, if insurance companies had access to the genetic database, they may refuse car insurance, or charge higher premiums, to a person with an increased risk of heart disease for example. Employers could (secretly) refuse employment to a person because their 'genetic profile' did not meet the requirements.
Is it acceptable to genetically modify pigs so that they can be bred to provide organs for humans?	Again, this is a moral issue. While some people find this unacceptable, their opinion does not necessarily make it wrong. Other people find this just as acceptable as breeding pigs to produce more or leaner bacon for human consumption. This is especially true of many people on transplant waiting lists.
Gene technology may give doctors the ability to create designer babies. They may be able to obtain a newly fertilised human egg, determine its genotype and ask the parents which genes they would like to modify. They might start only by replacing genes that actually cause disease, but they may then be led into replacing other genes.	Most doctors would find this morally *and* ethically unacceptable. They may consider replacing genes that cause disease, but not replacing genes just to improve a child's image in the eyes of its parents. However, if and when such practices become possible, who will define what is ethically acceptable for doctors? What will be the dividing line between 'cosmetic gene therapy' and 'medical gene therapy'?

Table 23.2: *Some concerns about genetic engineering.*

End of Chapter Checklist

If you haven't got a copy of your specification, read the introduction on page vi.

You will need to be able to do some or all of the following. Check your Awarding Body's specification (syllabus) to find out exactly what you need to know.

- Describe how genes are transferred from one cell to another and name the enzymes used in the procedure.
- Describe some of the ways in which genetically modified bacteria are used.
- Describe how genetically modified plants are produced.
- Describe some of the uses of genetically modified pants.
- Describe how genetically modified animals are produced.
- Describe some of the uses of genetically modified animals.
- Explain the role of gene technology in screening for genetic diseases.
- Explain the role of DNA fingerprinting in crime detection.
- Appreciate some of the concerns surrounding gene technology.

Questions

More questions on gene technology can be found at the end of Section F on page 332.

1 The diagram shows the main stages in transferring the human insulin gene to a bacterium.

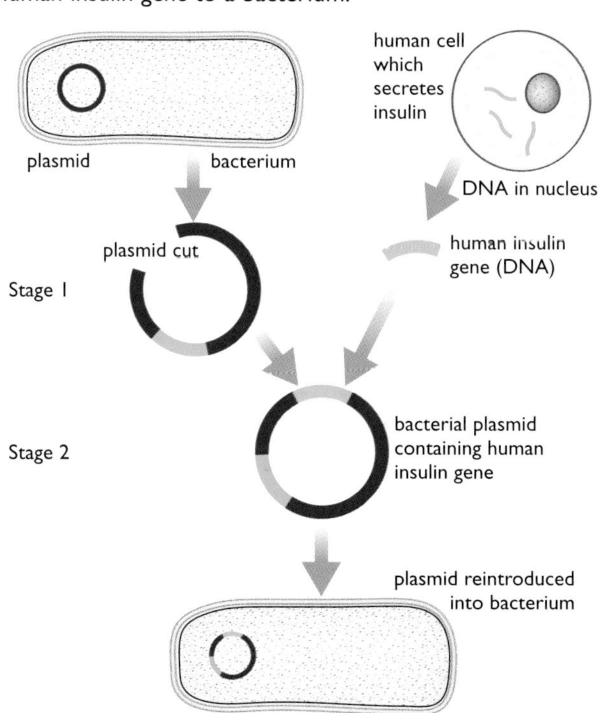

a) Name the enzymes used at stages 1 and 2.

b) What is the role of the plasmid in this procedure?

c) How would the insulin-producing bacteria be used to produce significant amounts of insulin?

d) Why is the insulin produced this way preferred to insulin extracted from other animals?

2 Blood groups and genetic fingerprinting can both be used as evidence in a criminal prosecution.

a) Describe how a genetic fingerprint is produced.

b) Explain why a genetic fingerprint can indicate guilt more strongly than a blood group.

3 Producing genetically modified plants and animals is more complex than producing genetically modified bacteria.

a) Describe two ways in which genes can be introduced into plant cells.

b) How are these genetically modified cells used to produce whole organisms?

c) What sort of animal cell is genetically modified and then used to produce the whole organism?

4 Write an essay about the importance of genetic engineering. In your essay you should make reference to:

a) important potential benefits resulting from genetic engineering in animals and plants

b) concerns about the risks resulting from genetic engineering in animals and plants.

Chapter 24: Disease

Although all organisms can suffer from disease in some way, we shall concentrate on disease in humans in this chapter.

The World Health Organisation (WHO) defines health as 'a state of complete physical, mental and social well-being'. Disease is less easy to define. It does not just mean the absence of health – just because we are less fit than we might be, or are feeling depressed at the thought of revising for an examination, does not necessarily mean that we have a disease. A useful definition of disease might be 'a condition with a specific cause in which part or all of a body is made to function in an abnormal and less efficient manner'. This definition could include diseases of all organisms – including plants. It could also include physical, mental and social aspects of disease in humans.

Types of disease in humans

Antibiotics often kill organisms by interfering with protein synthesis in a cell or by interfering with cell division. See Chapter 26 for a fuller explanation.

Some diseases are **infectious** – they are caused when organisms infect our bodies. Other diseases are caused by other factors and are **non-infectious**.

Organisms that cause infectious diseases

Many types of organisms cause infectious diseases, including bacteria, fungi, viruses and protozoa. Table 24.1 and Figure 24.1 give details of some of these.

Type of microorganism	How the microorganism causes disease	Examples of diseases caused
bacteria	Bacteria release toxins (poisons) as they multiply. These toxins affect cells in the region of the infection and sometimes in other regions of the body as well. Bacteria can be treated with antibiotics as each is a true cell with its own 'metabolic systems' and is capable of cell division.	tuberculosis (TB), pneumonia, cholera
viruses	Viruses enter living cells and disrupt the metabolic systems of that cell. The genetic material of the virus takes over the cell and instructs the cell to produce more viruses. Viruses cannot be treated with antibiotics as they are not true cells and are only active inside other living cells.	influenza ('flu), AIDS, measles, common cold
fungi	When fungi grow in or on living organisms, their **hyphae** (see Chapter 21) secrete digestive enzymes onto the tissues and the digested substances are absorbed. Growth of hyphae also physically damages the tissue. Some fungi secrete toxins while others can cause an allergic reaction (e.g. farmer's lung).	athlete's foot, farmer's lung
protozoa	There is no set pattern as to how protozoa cause disease.	malaria, sleeping sickness

Table 24.1: *The main types of disease-causing organisms.*

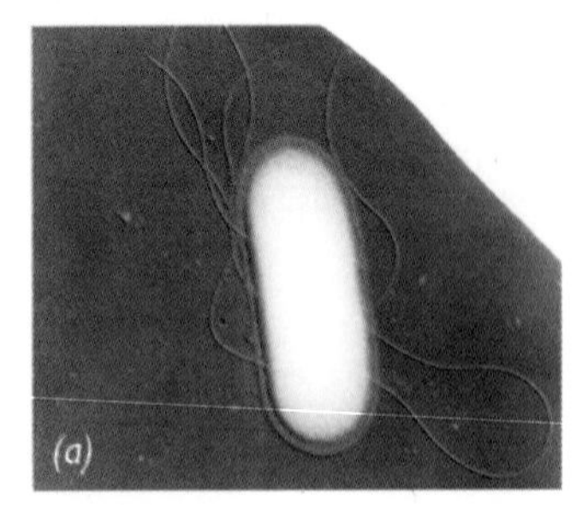

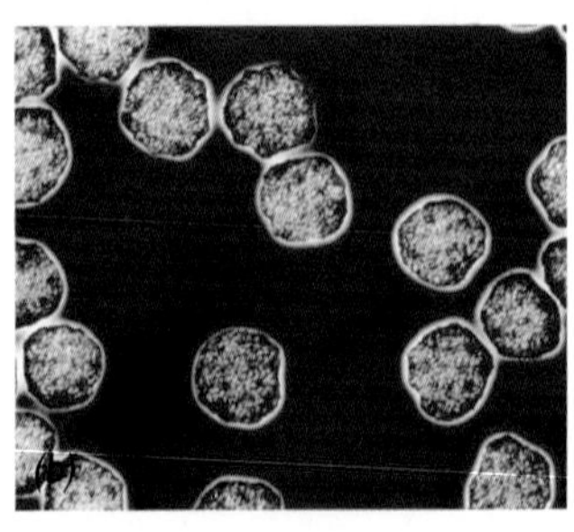

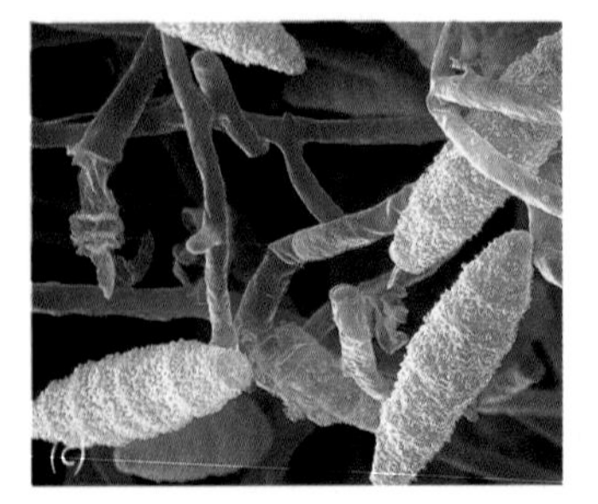

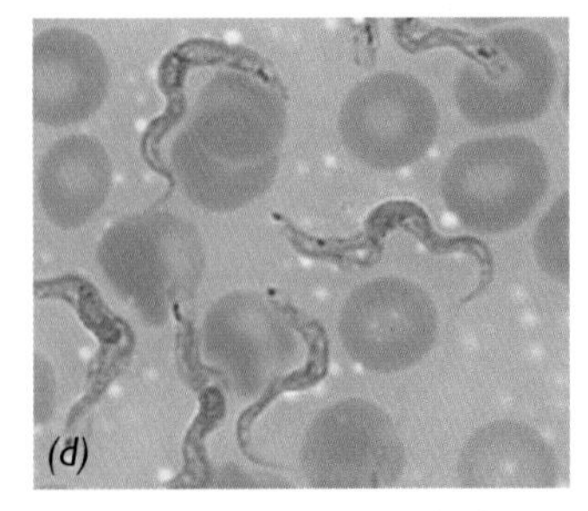

Figure 24.1 *Some disease-causing microorganisms. (a)* Salmonella bacterium *(b) Rubella viruses cause German measles (c) the fungus responsible for ringworm (d)* Trypanosoma gambiense *– the cause of sleeping sickness.*

Transmission of disease

Not all diseases are spread by air as shown in Figure 24.2. However, for a person to be infected, the disease-causing organism must be transmitted (carried) to that person in some way. The main methods of transmission of disease are described in Table 24.2.

Method of transmission	How the transmission route works	Examples of diseases
droplet infection	Many of these diseases are 'respiratory diseases' – diseases affecting the airways of the lungs. The organisms are carried in tiny droplets through the air when an infected person coughs or sneezes. They are inhaled by other people.	common cold, 'flu, pneumonia
drinking contaminated water	The microorganisms transmitted in this way often infect regions of the gut. When unclean water containing the organisms is drunk, they colonise a suitable area of the gut and reproduce. They are passed out with faeces and find their way back into the water.	cholera, typhoid fever
eating contaminated food	Most food poisoning is bacterial, but some viruses are transmitted this way. The organisms initially infect a region of the gut.	salmonellosis, typhoid fever, listeriosis, botulism
direct contact	Many skin infections, such as athlete's foot, are spread by direct contact with an infected person or contact with a surface carrying the organism.	athlete's foot, ringworm
sexual intercourse	Organisms infecting the sex organs can be passed from one sexual partner to another during intercourse. Some are transmitted by direct body contact, such as the fungus that causes candidiasis (thrush). Others are transmitted in semen or vaginal secretions, such as the AIDS virus. Some can be transmitted in saliva, such as syphilis.	candidiasis, syphilis, AIDS, gonorrhoea
blood to blood contact	Many of the sexually transmitted diseases can also be transmitted by blood-to-blood contact. Drug users sharing an infected needle can transmit AIDS.	AIDS, hepatitis B
animal vectors	Many diseases are spread through the bites of insects. Mosquitoes spread malaria and tsetse flies spread sleeping sickness. In both cases, the disease-causing organism is transmitted when the insect bites humans in order to suck blood. Flies can carry microorganisms from faeces onto food.	malaria, sleeping sickness typhoid fever, salmonellosis

Table 24.2: *Methods of transmission of disease.*

Figure 24.2 *A poster from the 1940s, encouraging people to reduce the spread of airborne diseases.*

Figure 24.3 *A fly feeding on human food. The fly releases saliva onto the food as it feeds. The saliva may contain disease-causing organisms.*

Factors affecting the transmission of disease

Transmission of disease is often related to the conditions in the area in which the disease is normally found. A disease is said to be **endemic** to an area if it is common in that area, but quite uncommon in other areas. The extent of transmission can be affected by environmental, social and economic factors. Social and economic factors include standard of living, education, income and better nutrition.

Modern, intensive methods of rearing livestock, such as 'factory farming' of chickens, brings many animals into close contact. Unless hygiene standards are very high, it is easy for *Salmonella* bacteria in the faeces of one animal to be passed to another. Also, once animals are infected, the bacteria can reproduce rapidly and produce toxins, which may remain in the animal after slaughter. If you study the life cycle in Figure 24.4 carefully, you will see that the cycle can be broken and transmission prevented by:

- sensible hygiene measures, such as always washing hands and utensils thoroughly (to remove any bacteria on hands and utensils and prevent transmission)
- cooking food thoroughly (to kill any bacteria present)
- proper hygiene measures in food production (to prevent cross-contamination).

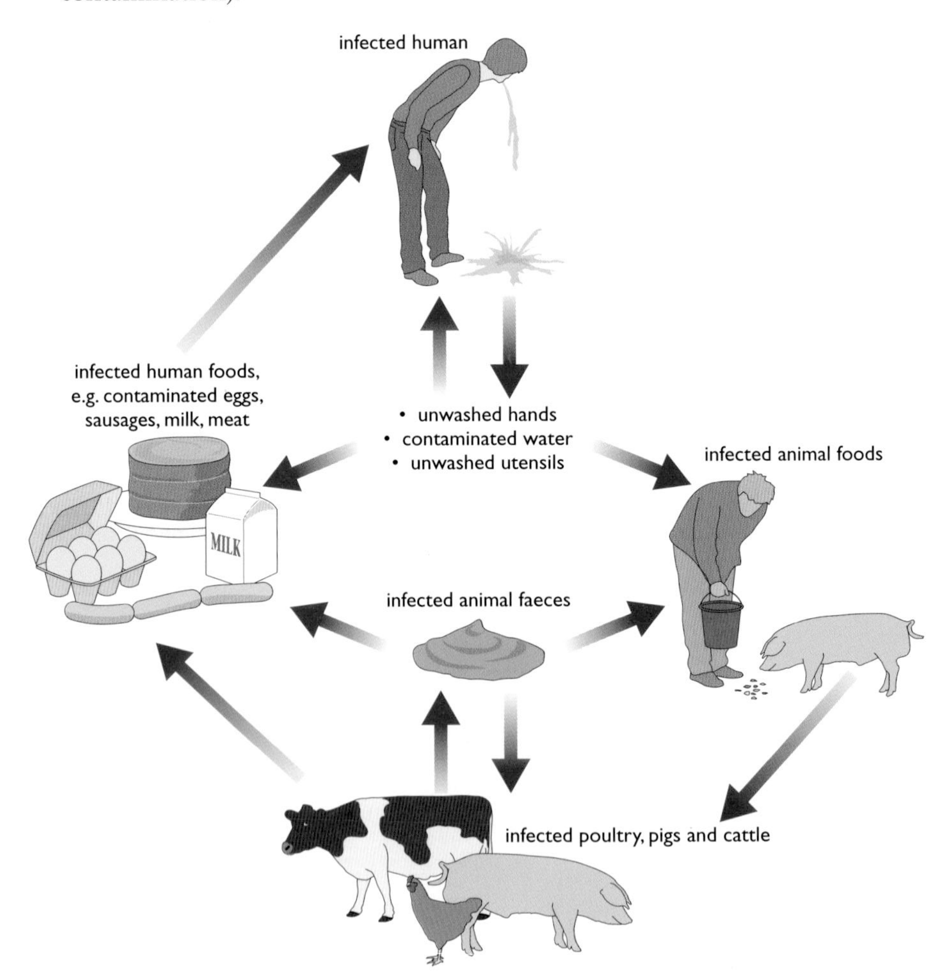

Figure 24.4 *The ways in which* Salmonella *bacteria may be transmitted and cause salmonellosis.*

Cholera is still endemic in countries where sanitation is poor. People drink untreated, contaminated water with the result that parasites in the water easily infect people and complete their life cycles. In more economically developed countries, waterborne diseases like this are rare. Contamination of drinking water is much less common, as untreated sewage is not discharged into water supplies. Also, drinking water is treated to a very high standard of purity in water treatment plants, and chlorine is added to the water to kill microorganisms.

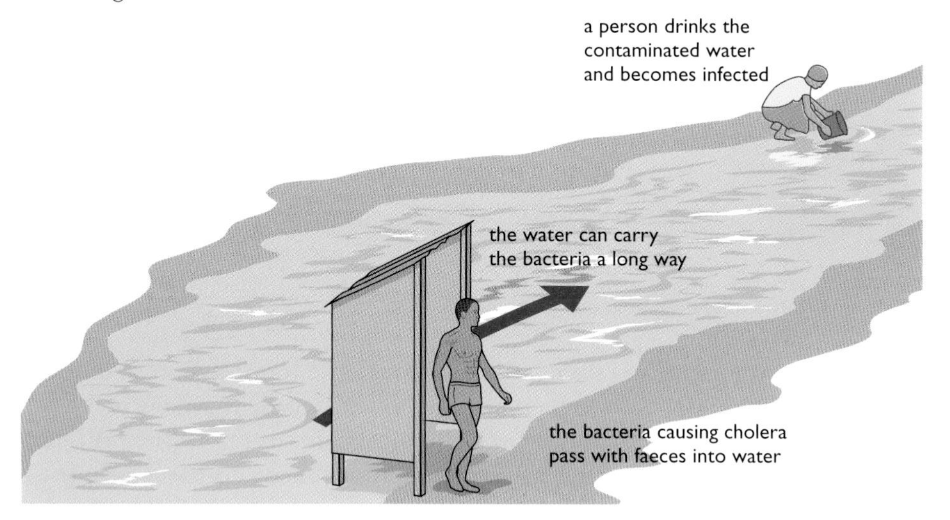

Figure 24.5(a) *How cholera is spread.*

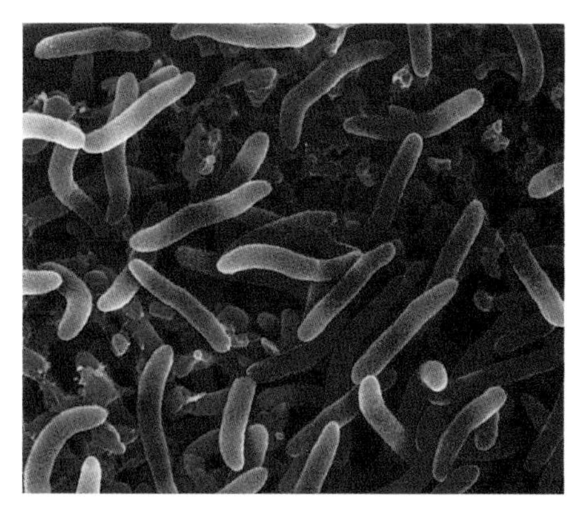

Figure 24.5(b) The bacteria that cause cholera.

Airborne diseases are less common in small villages than they are where the population is denser. In crowded cities we are all in much closer contact. We continually breathe in air that other people have breathed out. If that air contains droplets bearing the common cold virus, we stand a very good chance of contracting yet another cold.

ideas
evidence

Poverty is a key factor in the transmission of disease. If people cannot afford to eat a balanced diet, they risk suffering from deficiency diseases such as scurvy and kwashiorkor. In addition, their immune system may not develop fully and they may be more at risk of contracting infectious diseases. The standard of housing and living conditions in general is linked to poverty. Poor housing with poor sanitation creates conditions in which microorganisms can easily be passed from one person to another.

Education has played an important role in reducing the transmission of disease. People have become more aware of how factors such as personal hygiene and a balanced diet can reduce the transmission of disease. They therefore take more care and so are at less risk of infection.

In some cases, improvements in these 'socio-economic' factors seem to have had more effect on the transmission of disease than advances in medical treatments. Figure 24.6 shows the changes in death rates per million people from measles and from diphtheria.

In the case of diphtheria, the decline in death rates is largely due to advances in medical treatments. Death rates from measles, however, had fallen dramatically long before vaccination began. The decline may be due to improvements in health and living standards.

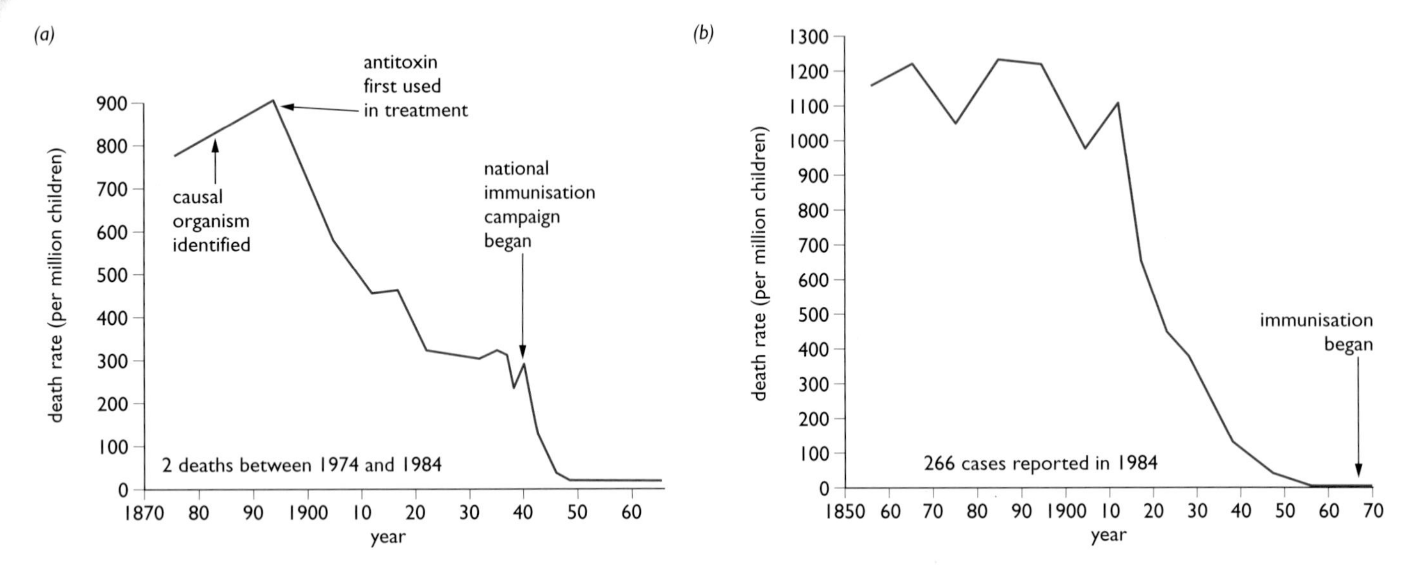

Figure 24.6 *Death rates of children under 15 in England and Wales from (a) diphtheria and (b) measles.*

Changes in social attitudes do not always reduce the transmission of disease. In the 1960s there was a change in attitude to sex outside marriage in Europe and America. At the same time chemical contraception (the pill) was introduced. More people were having sexual intercourse and fewer people used condoms. There was a large increase in the numbers of people contracting sexually transmitted diseases such as syphilis and gonorrhoea.

Distribution of diseases

Some diseases are only found in certain parts of the world. For example, malaria is common in tropical and sub-tropical countries. Sleeping sickness is common in Africa. These diseases are endemic to these regions (Figure 24.7).

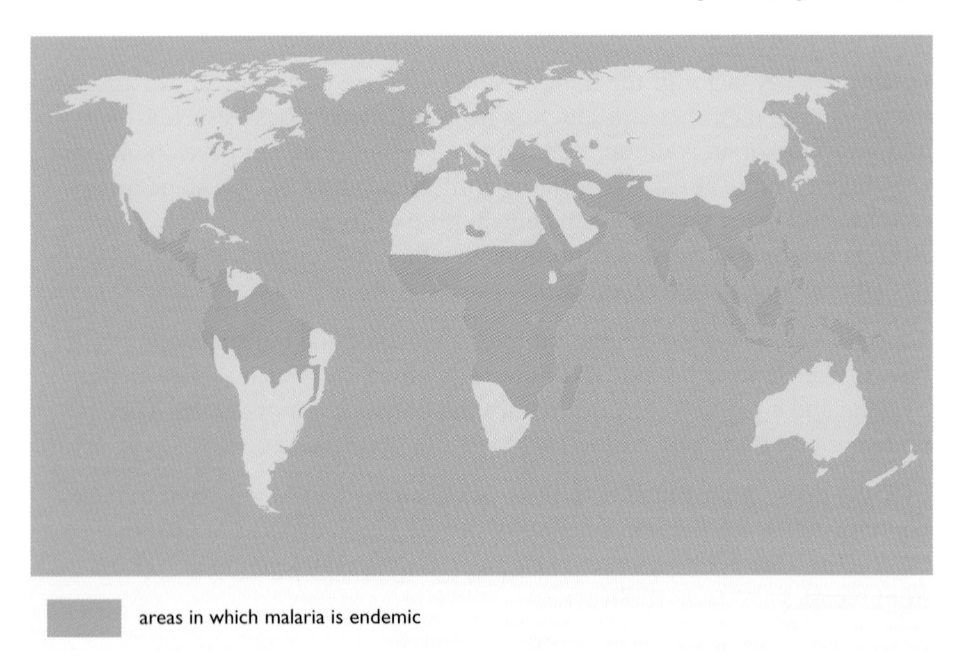

Figure 24.7 *Regions of the world where malaria is endemic.*

Occasionally, a disease that is endemic to one region spreads rapidly to affect large numbers of people in other areas as well. In this case we say that there is an **epidemic**. Note that it is the increase in area of infection, not just numbers, that constitutes an epidemic. If the disease spreads across the entire world, we describe it as **pandemic**. There have been pandemics of 'flu in 1908 (Spanish 'flu), 1958 (Asian 'flu) and 1968 (Hong Kong 'flu).

The 'flu epidemics began in one country as a new strain of the virus appeared. For a period of time, the strain was endemic to that country. As it spread, it caused an epidemic, then a pandemic.

The expansion in travel exposes many people to diseases that are not normally endemic in their own countries. Potentially, this can increase the spread of diseases from an area where they are currently endemic to other, new areas. Before travelling abroad, you should always take advice from your GP about diseases that are endemic in the area and whether any protective vaccination is necessary.

Important infectious diseases

AIDS

AIDS (**A**cquired **I**mmune **D**eficiency **S**yndrome) is one of the most significant worldwide killers of the moment. It is caused by a virus called the human immunodeficiency virus or **HIV**. Figure 24.8 shows the structure of this virus.

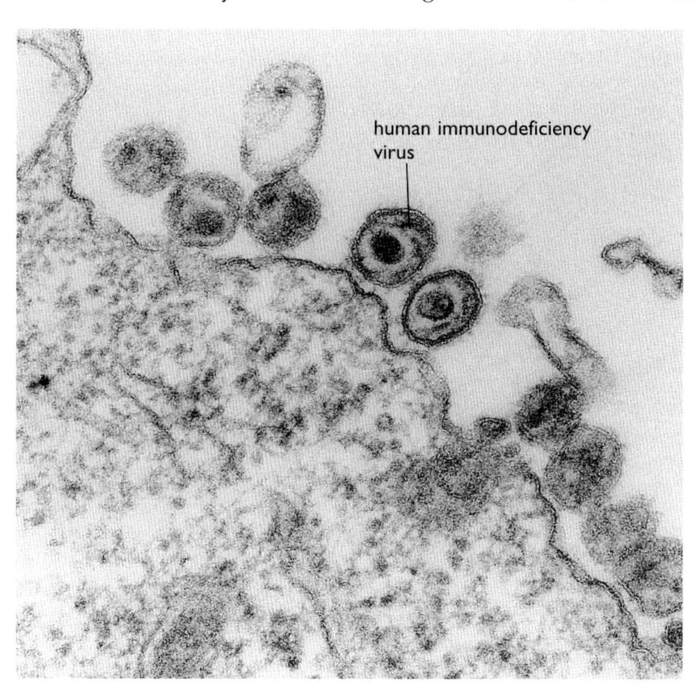

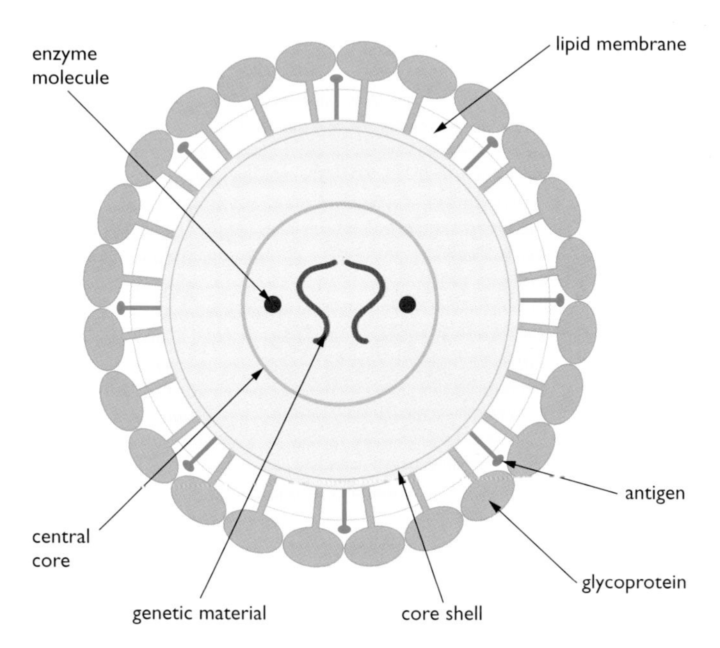

Figure 24.8 *The human immunodeficiency virus (HIV).*

HIV initially infects a type of white blood cell called a helper T-lymphocyte. These cells are necessary if other white blood cells (B-lymphocytes and other T-lymphocytes) are to become active and start fighting infections (see Chapter 25). The course of a typical HIV infection is described below.

1 The genetic material of HIV becomes incorporated in the DNA of the helper T-lymphocyte.
2 When the HIV DNA is activated, it instructs the lymphocyte to make HIV proteins and genetic material.

Part of this immune response to infection by HIV involves the production of antibodies to destroy the virus. At this stage the person is said to be **HIV positive**, because their blood gives a positive result when tested for HIV antibodies.

The period during which the body replaces the lymphocytes as fast as they are destroyed is called the **latency period**. It can last for up to 20 years. The person shows no symptoms of AIDS during this period, but will be highly infective to others.

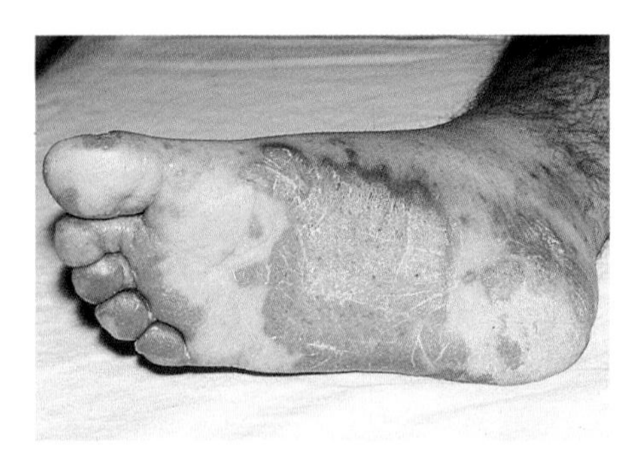

Figure 24.9 *A person suffering from Kaposi's sarcoma.*

The real significance of AIDS is that the cells that are being destroyed are the cells that are needed to help the other lymphocytes destroy the infected cells.

3 Some of the HIV proteins and genetic material are assembled into new viruses.

4 Some of the HIV proteins end up as marker proteins (antigens) on the surface of the cell.

5 These HIV proteins are recognised as foreign proteins.

6 The lymphocyte is destroyed by the immune system.

7 The assembled virus particles escape to infect other lymphocytes.

8 This cycle repeats itself for as long as the body can replace the lymphocytes that have been destroyed.

9 Eventually the body will not be able to replace the lymphocytes at the same rate at which they are being destroyed.

10 The number of free viruses in the blood increases rapidly and HIV may infect other areas of the body, including the brain.

11 The immune system is severely damaged and other disease-causing microorganisms infect the body.

12 Death is often as a result of 'opportunistic' infection by TB and pneumonia, due to the reduced capacity of the immune system that would normally destroy the organisms causing these diseases. Death can also be caused by rare cancers such as Kaposi's sarcoma (Figure 24.9).

Figure 24.10 shows how the levels of virus particles in the blood, HIV antibodies and helper T-lymphocytes change during the course of an AIDS infection.

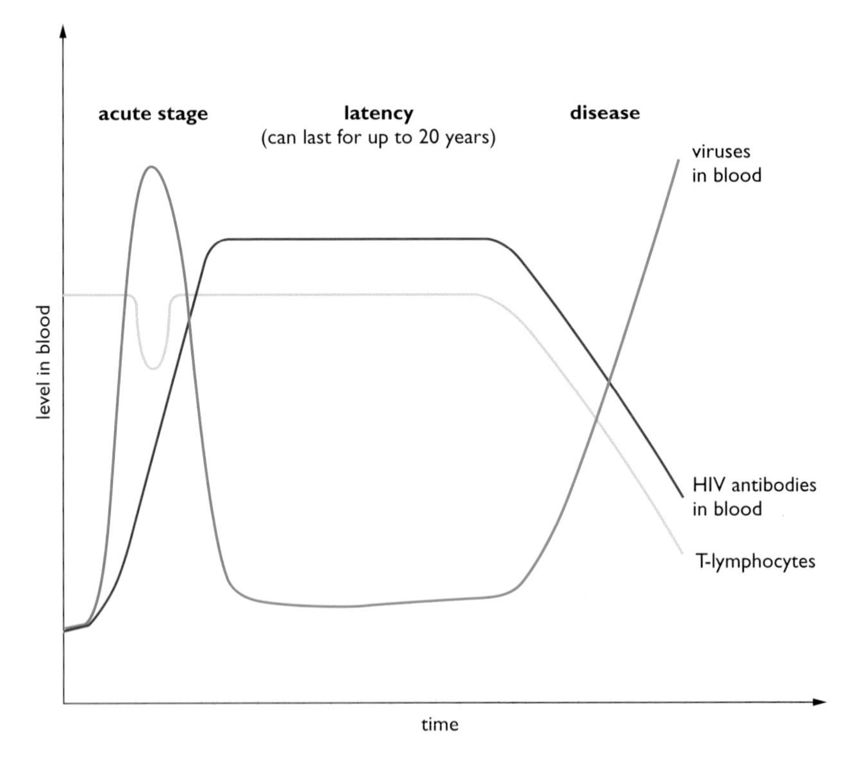

Figure 24.10 *The stages of an AIDS infection.*

AIDS is primarily transmitted by unprotected sexual intercourse, (either homosexual or heterosexual) or by blood-to-blood contact (for example when drug users share an infected needle, or if infected blood is given in a transfusion). AIDS can *not* be transmitted by kissing, sharing utensils (e.g. a cup or a glass), giving blood, skin-to-skin contact or sitting on the same toilet seat as an infected person.

As yet, there is no effective vaccine against HIV, although a number of anti-viral drugs can delay the onset of AIDS. The transmission of AIDS can be controlled, however, by a number of measures. These include:

- use of condoms; although transmission can still occur, the incidence of transmission is greatly reduced
- non-sharing and use of only sterile needles by drug users
- limiting the number of sexual partners; in particular avoiding sex with high risk groups such as prostitutes.

Influenza ('flu)

Influenza is caused by a virus, but this virus does not infect cells of the immune system. The 'flu virus primarily infects cells in the upper airways. Sometimes it does spread down the trachea and into the lungs themselves. 'Flu is transmitted by airborne droplets produced when an infected person coughs or sneezes.

Symptoms of infection include chills and fever, muscular aches as well as loss of appetite and fatigue. These initial symptoms are often followed by a cough, sore throat and runny nose. The fever and aches usually subside after about five days, but the other symptoms last longer and the person may feel weak for about ten days. If the disease does affect the lungs, it causes an acute form of pneumonia that can be fatal within a few days, even in otherwise healthy people.

Influenza can be made worse if the person's breathing system is also infected by bacteria after the 'flu virus. Such infections are called secondary infections. Bacteria causing bronchitis and pneumonia frequently cause secondary infections in young and elderly 'flu sufferers.

There are three strains of the 'flu virus known as the A, B and C strains. Type C produces a much milder form of 'flu than the other two strains. The A and B strains of the virus mutate periodically and produce new strains. This makes everyone susceptible to infection, even if they have had 'flu before. The antibodies they made to the last form of 'flu (see Chapter 25) will be ineffective against the new form. Also, new vaccines must be produced for each new strain, for the same reason.

Mutations alter one or more genes in the virus. Some mutations cause the virus to produce different antigens on its surface.

There is little that can be done to treat the illness, except to ease some of the painful symptoms. Antibiotics are ineffective against viruses so the body must destroy the virus by itself. However, antibiotics may be prescribed to combat secondary bacterial infections. The spread of the disease is controlled to some extent by vaccination, but current 'flu vaccines are only about 60–70% effective.

Malaria

A protozoan called *Plasmodium* causes malaria. Different forms of malaria are caused by different species of *Plasmodium*. The two most significant are

Plasmodium vivax and *Plasmodium falciparum*. The female Anopheles mosquito spreads the parasite when she bites humans to feed on their blood. Figure 24.11 shows the transmission and life cycle of *Plasmodium*.

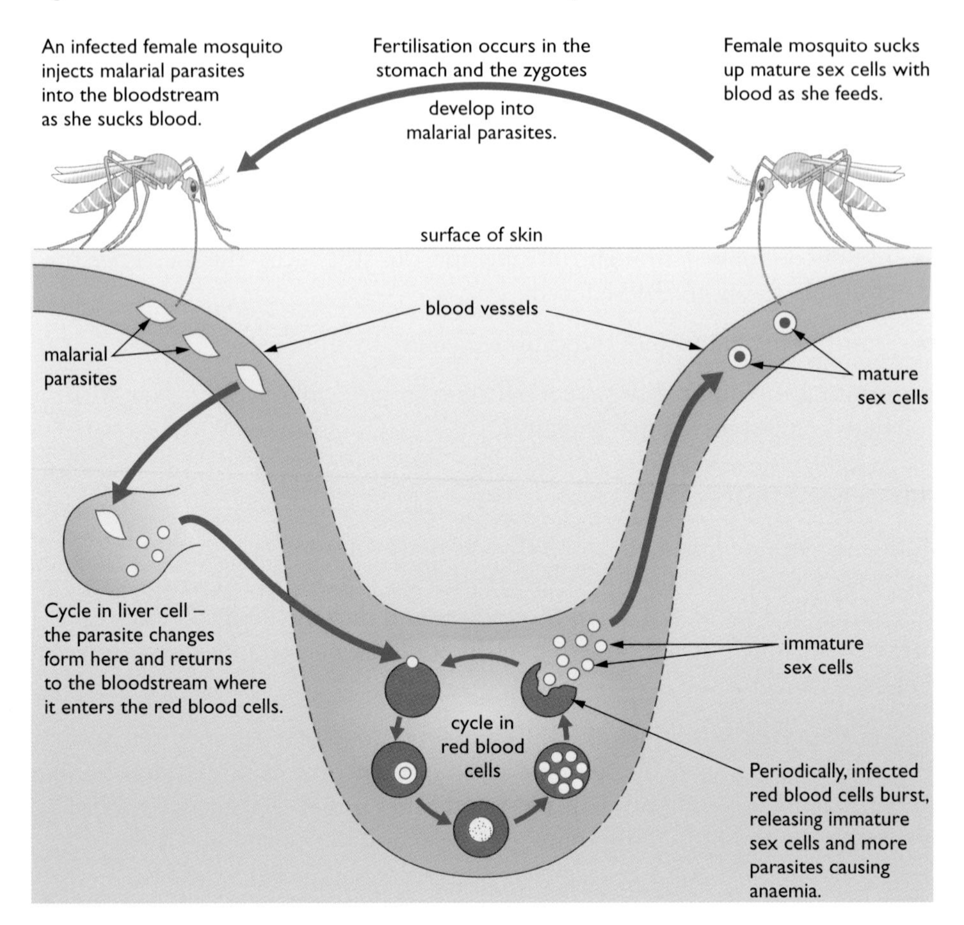

Figure 24.11 *The life cycle of* Plasmodium.

Malaria is endemic in the tropics and sub-tropical areas and infects up to 300 million people per year. This makes it the single most important disease hazard for travellers. The classical symptom of malaria is the fever that is produced when red blood cells are ruptured by *Plasmodium*. The fever develops in three stages:

1 A cold stage, with uncontrollable shivering.

2 A hot stage, in which the body temperature may reach 40.3°C.

3 A sweating stage, in which profuse sweating reduces the temperature to normal again.

Sometimes *Plasmodium* infects the kidneys, liver and brain and these infections are frequently fatal. Malaria is treated with a number of drugs, the most common being chloroquine. However, new strains of *Plasmodium* are becoming resistant to the drug. As yet there is no effective vaccine against malaria, but a number of research groups claim to be making significant progress.

Malaria has been controlled to some extent by use of insect repellents and mosquito netting to prevent mosquito bites, and use of insecticides to reduce local mosquito populations.

Larger organisms as agents of disease

Although microorganisms are responsible for most human diseases, larger organisms also cause disease. Organisms of this kind are **parasites**. They live in or on another organism, called the **host**. The parasite benefits from this association, while the host is harmed. Two examples of organisms that cause disease in this way are tapeworms and roundworms.

Tapeworms are ribbon-like parasitic worms. The adult worm lives in the intestines of humans where it absorbs many of the products of digestion that would normally be absorbed into the bloodstream. Figure 24.12 shows the structure of a tapeworm.

There are three main types of human tapeworm: the beef tapeworm, the pork tapeworm and the fish tapeworm. Humans are the **primary host**, while the cow, pig and fish are **secondary hosts**. The eggs released from the tapeworm develop into larvae in the secondary host. The larvae enter humans when undercooked, infected beef, pork or fish is eaten. Figure 24.13 shows the life cycle of a beef tapeworm.

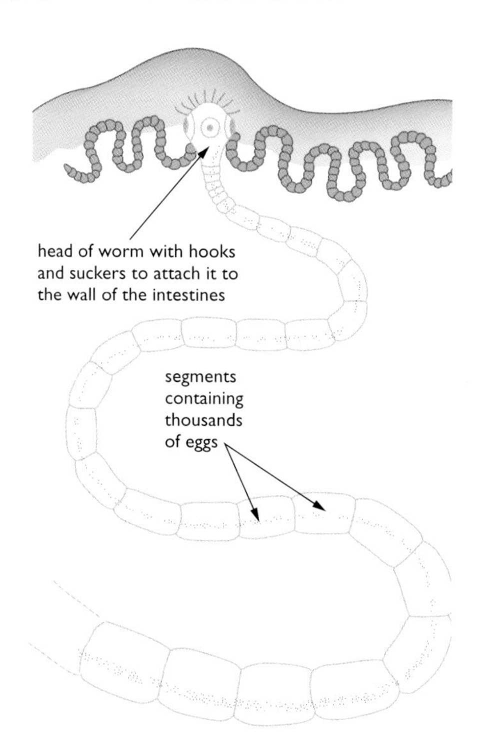

Figure 24.12 *A human tapeworm.*

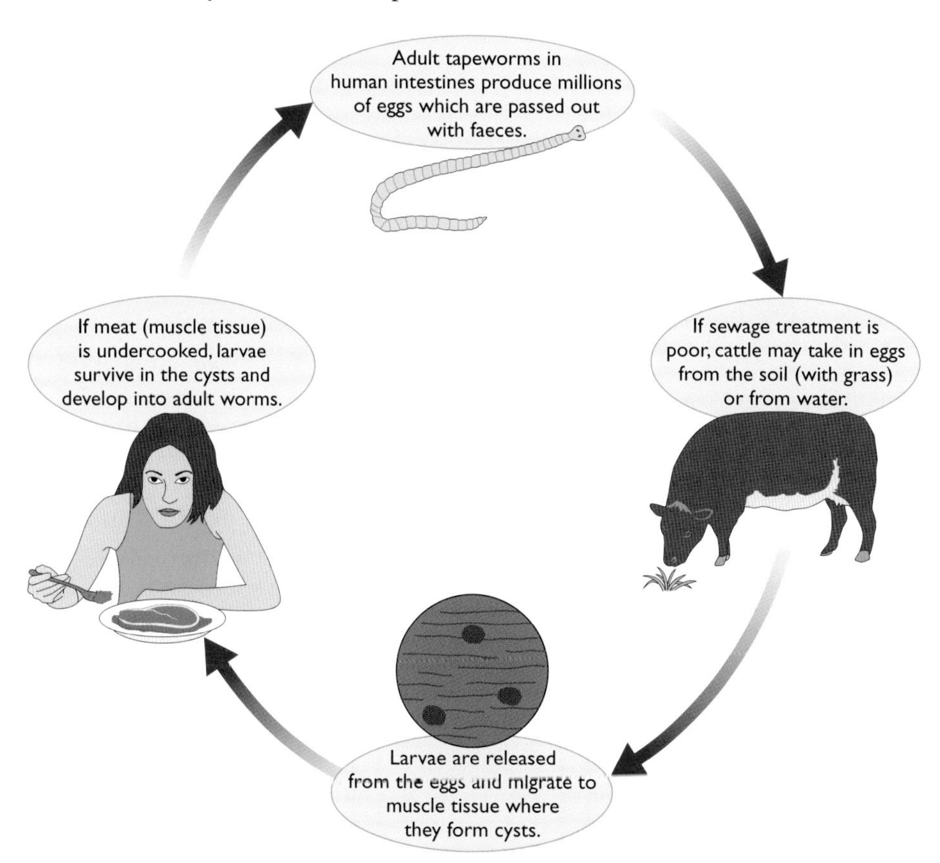

Figure 24.13 *The life cycle of the beef tapeworm.*

Adult beef tapeworms can be up to 9 metres long and release millions of eggs. Despite the size of the worm, people with tapeworms usually show few major symptoms. Tapeworms are rare in more economically developed countries because meat is inspected regularly to check for larvae. Meat is destroyed if it is found to contain larvae and farms producing that meat are inspected. Sewage treatment prevents eggs from reaching cattle and developing into larvae. Proper cooking of meat destroys most of the larvae in their cysts.

1 A dog (usually a puppy) with *Toxocara* in its intestines releases eggs with the faeces. These may contaminate the soil.

2 Children playing with an infested dog, or in soil contaminated with dog faeces, may get eggs on their fingers. If they put their fingers in their mouths they may swallow some worm eggs.

3 The eggs hatch in the intestines and the larvae travel through the tissues to organs such as the brain, lungs, liver and eyes.

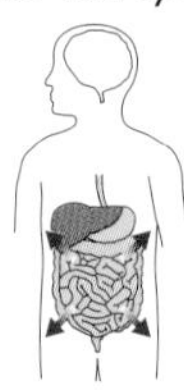

Figure 24.14 *How* Toxocara *is transmitted.*

Roundworms are also called threadworms and there are many different types. One is called *Toxocara* and can have serious consequences if it infects humans. The worm is normally found in the intestines of dogs, especially puppies. Eggs from the worm pass out with the faeces onto soil and grass. Children playing in the same area are at risk of picking up these eggs on their hands. If they then touch their mouths, the eggs may enter the gut. Once in the intestines, the eggs develop into larvae, which migrate through the gut wall and into other tissues. They often find their way to the liver, lungs, brain and eyes. Usually only mild symptoms result, such as a mild fever. However, if a large number of eggs enter, pneumonia, seizures and blindness can result. Figure 24.14 shows how *Toxocara* is transmitted.

The effects caused by *Toxocara* can be treated by drug therapy, but serious cases often require hospital supervision. Prevention of transmission could be relatively simple. All dog owners should have their dogs 'wormed' regularly. There would then be far fewer infected dogs to pass on the eggs. Children, especially young children, should be educated about the dangers of playing with dogs. This would prevent transmission of the eggs.

Plant diseases

Microorganisms can also infect plants and cause disease. Most plant diseases are fungal or viral, but a few bacteria cause diseases in plants and other damage is caused by pests feeding on the plants. The plant diseases that have received most attention are those that have an economic effect. These include diseases of crops such as maize, rice, beans and potatoes, and diseases of plants grown for showing (garden plants and cut flowers).

Viral diseases of plants include those caused by tobacco mosaic virus and tomato mosaic virus. Both infect the leaf tissues of the plant. They reduce the yield by reducing the amount of photosynthesis that can take place, as well as by causing physical damage to the plant.

Rust fungus is a common cause of disease in many garden plants. It infects the leaf tissues and produces red-brown (rust coloured) swellings containing spores. It can cause serious damage as the growth of the spore cases and of other fungal tissues physically disrupts the plant tissue. Rusts of cereals can cause a significant reduction in yield. Rust infection can be treated with a range of fungicide sprays, but it is difficult to prevent infection. Recently, genetic engineering has been used to transfer genes that confer resistance to rust infection to wheat plants. Figure 24.15 compares the leaves of resistant and non-resistant wheat plants, both of which have been exposed to the rust fungus.

Figure 24.15 *Leaves from rust resistant and non-rust resistant wheat plants. Both come from plants that have been exposed to the rust fungus.*

Practising crop rotation can also help to control disease in plants. If the same crop is grown in a field year after year, the pests and parasites of that crop lie dormant in the soil over winter and can re-infect the next year's crop as soon as the seeds germinate. Each year, the extent of infection increases. If the crops are rotated, however, as shown in Figure 24.16, pests and parasites do not build up in the same way. As with diseases in humans, prevention is always better than cure.

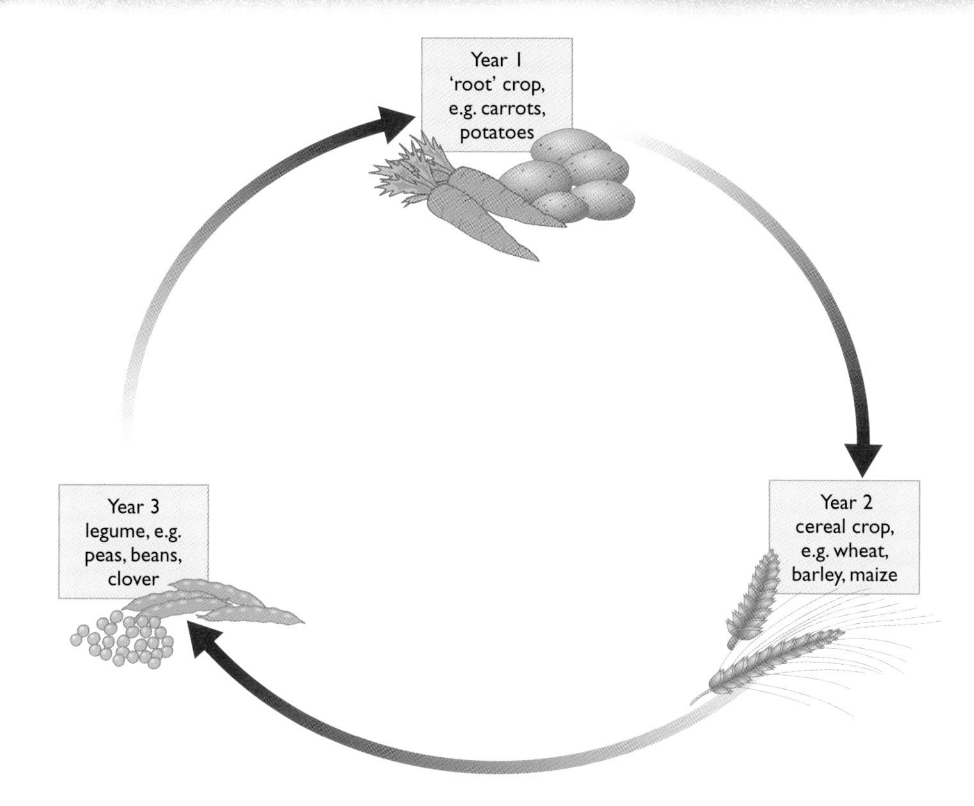

Figure 24.16 *A three-year crop rotation. Changing the crop grown in a field each year prevents the build up of crop-specific pests and parasites in the soil. Including a legume in the rotation helps to maintain the level of nitrates in the soil (see the nitrogen cycle, Chapter 14).*

End of Chapter Checklist

If you haven't got a copy of your specification, read the introduction on page vi.

You will need to be able to do some or all of the following. Check your Awarding Body's specification (syllabus) to find out exactly what you need to know.

- Explain the concept of disease and describe the difference between infectious and non-infectious diseases.
- Explain how different types of microorganism cause disease and give some examples of diseases caused by each type.
- Describe the ways in which disease-causing microorganisms can be transmitted from one person to another.
- Explain how social conditions can influence transmission of disease.
- Understand that the distribution of diseases worldwide can change as endemic diseases spread in epidemics or in pandemics.
- Know the cause, transmission routes, symptoms and treatment of three important diseases – AIDS, influenza and malaria.
- Know that larger organisms can also cause disease and be aware of how tapeworms and *Toxocara* are spread,
- Understand that plants, too, can be diseased.
- Understand the roles of crop rotation and genetic engineering in treating plant diseases.

Questions

More questions on disease can be found at the end of Section F on page 332.

1 What is the difference between an infectious and a non-infectious disease? Give an example of each.

2 AIDS is one of the biggest worldwide killer diseases. It is caused by the human immunodeficiency virus (HIV).

a) Give two ways in which HIV can enter the body.

b) What does being 'HIV positive' mean?

c) The graph shows the changes in the numbers of helper T-lymphocytes during an AIDS infection.

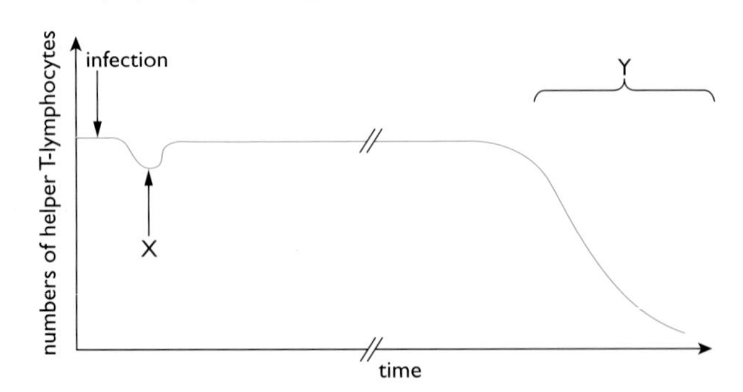

Suggest explanations for the changes in numbers at:

i) the period marked X

ii) the period marked Y.

3 Diseases are sometimes endemic in a certain area of the world.

a) What does endemic mean?

b) What happens when an epidemic of a disease occurs?

c) Give one example of a disease that has undergone an epidemic.

d) Why should you make sure that you are up-to-date with your vaccines before travelling abroad?

4 Respiratory tuberculosis (TB) is a serious disease. It can be transmitted from person to person in airborne droplets. It can also be transmitted to humans from infected cattle in milk. The graph shows the changes in the numbers of deaths from respiratory tuberculosis from 1830 to 1985.

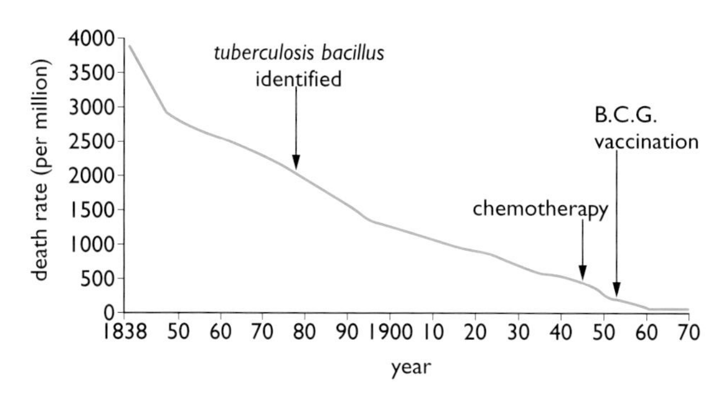

a) Describe the extent to which deaths from TB fell:

i) before vaccination was introduced

ii) after vaccination was introduced.

b) Suggest an explanation for your answers to ***a)***.

5 Diseases of plants are usually caused by viruses or fungi.

a) Name one viral and one fungal disease of plants.

b) What type of disease is rust and how does it damage plants? Explain why rust is such an infectious disease.

c) How can crop rotation help to reduce infection of crop plants?

Chapter 25: Blood and Immunity

Blood isn't just a transport medium. This chapter looks at the important role it plays in defending the body against disease-causing microorganisms.

An organism that causes disease in another organism is called a **pathogen**. Diseases are caused when large numbers of microorganisms invade our bodies and begin to multiply. At this stage we say we have been **infected**. Infection is not just a matter of a few pathogens getting in. They must enter in sufficient numbers to become established and start to multiply.

- When bacteria enter our bodies, they colonise a suitable area and obtain nutrients from the cells or blood in that area. They produce **toxins** (poisons) that make us feel ill.
- When viruses enter, they invade individual cells. Inside living cells, the genetic material of the virus is reproduced by the cell, which also manufactures virus proteins. The cell then assembles these proteins and the genetic material into new viruses. These burst out of the cell and infect more cells.

In either case, an **immune response** is triggered which eventually destroys the microorganisms and we 'get better'. Figure 25.1 shows how the numbers of invading bacteria change during the course of an infection. When viruses infect our bodies, their numbers change in a similar way.

Pathology is the study of diseases. **Pathologists** study the causes, transmission and effects of pathogens on the body.

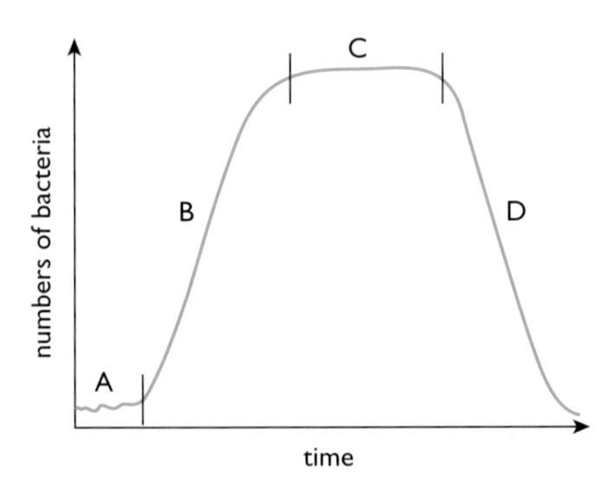

A Bacteria infect the body and find a suitable place to multiply.
B Bacteria multiply rapidly – they have been detected, but there is no response yet. We are ill because of the toxins made by the increasing population of bacteria.
C We are still ill. A lot of toxins have been made, but the body is now killing bacteria as fast as they can reproduce.
D The body is now killing bacteria faster than they can reproduce and we will soon feel 'better'.

Figure 25.1 *The change in numbers of bacteria during an infection.*

Barriers to infection

There are a number of natural barriers that prevent bacteria from entering, or from accumulating in, our bodies. These are illustrated in Figure 25.2.

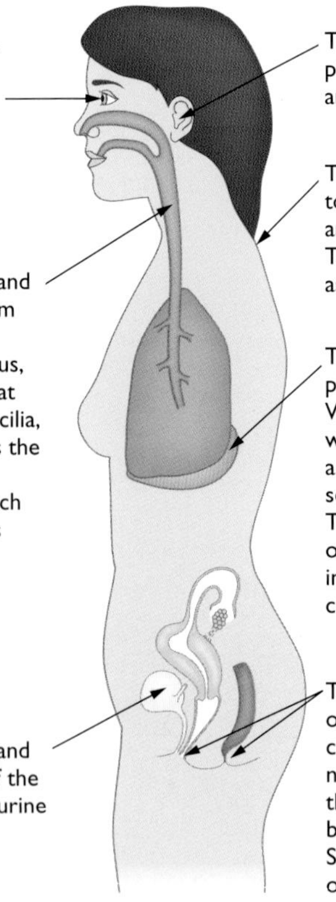

The eyes are protected by tears, which lubricate the conjunctiva and contain a number of anti-bacterial enzymes.

The ears secrete wax, which acts as a physical barrier and has some antiseptic properties.

The skin acts as a physical barrier to the entry of microorganisms and sweat contains a mild antiseptic. The skin is constantly being renewed as cells are lost from the surface.

Cells lining the trachea, bronchi and bronchioles prevent bacteria from accumulating in the alveoli. Some of these cells secrete mucus, which traps bacteria and dust that enters with the air. Others have cilia, which sweep the mucus towards the top of the trachea. Here, it is swallowed and enters the stomach where the hydrochloric acid kills the bacteria.

The lining of the gut provides a physical barrier to infection. We take in millions of microorganisms with the food we eat but most of these are killed by the hydrochloric acid secreted by cells in the stomach lining. Those that are not killed may infect other regions of the gut, but cannot infect the rest of the body unless they can cross the gut lining.

Urine contains a mild antiseptic and also flushes many bacteria out of the urethra (passage through which urine flows to the outside) when we pass urine.

The vagina and anus contain billions of non-harmful bacteria called commensals. They are present in such numbers that they out-compete for the available nutrients any harmful bacteria that may enter. Some also produce antibiotics that kill off harmful bacteria.

Figure 25.2 *The body's natural barriers to infection.*

Babies are born with no commensal bacteria but acquire them within a few days of being born. Sometimes, people lose their natural commensal bacteria and develop serious health problems, as any pathogens that invade these regions have no competitors for nutrients.

Once one of these barriers is pierced, or microorganisms enter in some other way, blood can act in a number of ways to defend the body against disease. Any active response by the body that destroys microorganisms is called an immune response. Some of these immune responses are general, aimed at any microorganism that enters. Others are more specific. There are other responses, that are just aimed at minimising damage and preventing the entry of more microorganisms.

If an area becomes infected following damage to the skin, cells in the area release **histamines**. These chemicals cause the capillaries to swell and to become more 'leaky'. The 'swollen' capillaries allow more blood into the area – it looks red or inflamed. The increased 'leakiness' of the capillaries allows more white blood cells and more antibodies to escape into the infected tissue. Both of these are important in combating microorganisms.

Blood clotting

When we cut ourselves, our blood clots to form a temporary 'plug' that prevents loss of blood and entry of microorganisms. Plugs form in the wall of any blood vessels that have been damaged and in the skin itself. The plug in the skin forms the familiar 'scab'. Underneath the scab, an immune response takes place to destroy any microorganisms that have entered. Scar tissue then forms to heal the wound.

Immune responses

General responses are aimed at any invading microorganism. The most important of these is **phagocytosis**. Some white blood cells can ingest (take in) microorganisms. They do this by forming extensions called **pseudopodia** that enclose the microorganism. This is called phagocytosis and the white blood cells that carry it out are **phagocytes**. Once ingested, the phagocyte encloses the microorganism in a vacuole and digests it. Figure 25.3 shows a phagocyte ingesting a yeast cell.

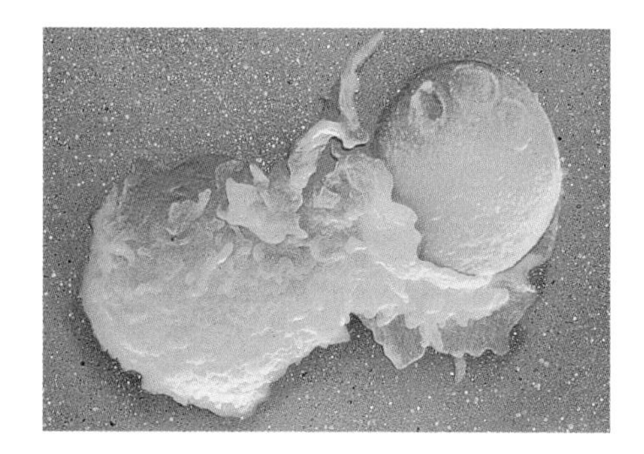

Figure 25.3 *A phagocyte ingesting a yeast cell.*

Specific immune responses are those that our body makes to each individual disease-causing organism. There is one response to the TB bacterium, a different response to the common cold virus, and so on. To make these individual responses, our body has first to be able to recognise the different microorganisms. The cells in our body that can do this are called lymphocytes. There are many kinds of **lymphocytes**, but two are particularly important – B-**lymphocytes** and **T-lymphocytes**.

A phagocyte isn't a particular type of white blood cell. There are several, quite different, types of white blood cell that are phagocytes. The term describes a common function of these cells. Amoeba are also phagocytic!

Microorganisms have individual 'marker chemicals' or **antigens** on their surface. Lymphocytes have **receptor proteins** on their surface that can bind with the antigens on the surface of the microorganisms. Each lymphocyte has slightly different receptor proteins from other lymphocytes and so can bind with different antigens and recognise a different microorganism.

Lymphocytes are white blood cells that are produced from actively dividing stem cells in the bone marrow. B-lymphocytes complete their development in the spleen and the lymph nodes. T-lymphocytes complete their development in the thymus gland – just above the heart.

When a lymphocyte binds with an antigen, it becomes activated and starts to divide rapidly. This will eventually produce millions of the same type of lymphocyte, capable of recognising the same type of microorganism. So far, the story is the same for B-lymphocytes and T-lymphocytes. What happens as the lymphocytes multiply, however, is different.

When B-lymphocytes are activated, most start to produce **antibodies** that are specific to the antigens on the surface of the microorganism. These antibodies bind with the antigens and cause the microorganisms to 'clump' together (Figure 25.5). In this form they are inactive and can be easily killed by phagocytes. Sometimes antibodies cause the cells to burst open.

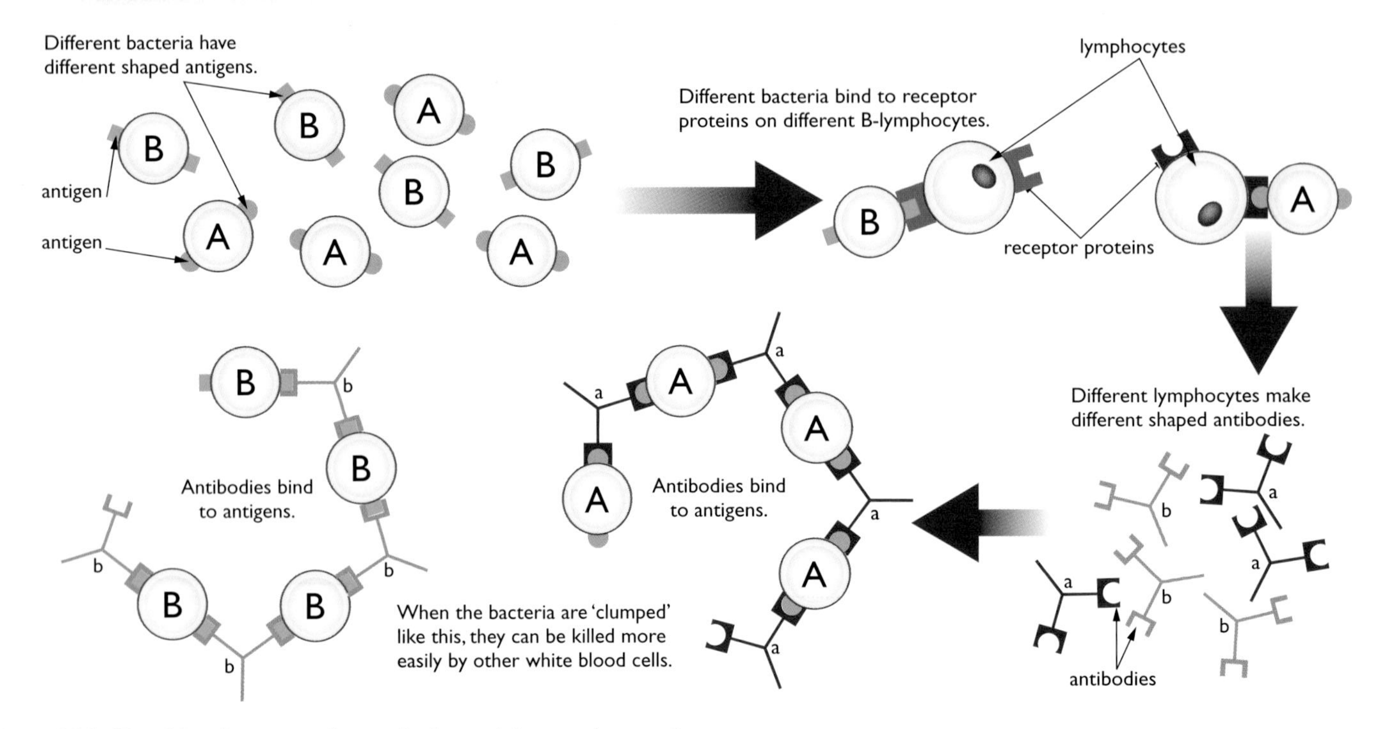

Figure 25.5 *How B-lymphocytes produce antibodies and destroy microorganisms.*

Antibodies are 'Y' shaped protein molecules that bind to a specific protein (antigen) on the surface of a microorganism.

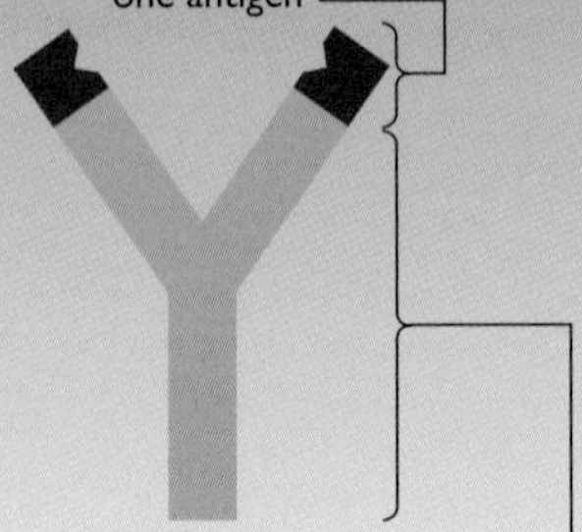

this part of the molecule is specific to one antigen

this part of the molecule is the same in all antibodies

Figure 25.4 *The structure of an antibody molecule.*

It is the *memory cells* that remain in the blood for long periods of time, *not* the antibodies.

Some of the activated B-lymphocytes do not get involved in killing microorganisms at this stage. Instead, they develop into **memory cells**. Memory cells make us **immune** to a disease. These cells remain in the blood for many years, in some cases a lifetime. If the same microorganism re-infects, the memory B-lymphocytes start to reproduce and produce antibodies. This secondary immune response is much faster and more effective than the first (primary) response. The antibody level quickly becomes much higher and the microorganisms are killed before they have had time to multiply to the level where they would cause disease. Figure 25.6 illustrates this.

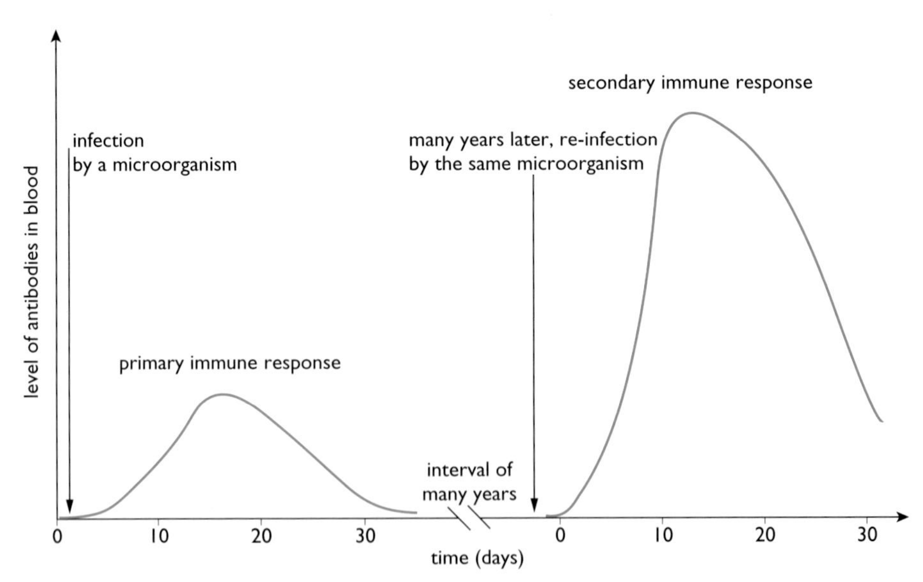

Figure 25.6 *A graph showing the levels of antibody production in the primary and secondary immune responses. Many years may pass after the primary response before we are re-infected and need to make a secondary response.*

T-lymphocytes do not destroy microorganisms in the same way – they do not make antibodies. In fact, the main type of T-lymphocyte actually destroys our own body cells. If this doesn't immediately make any kind of sense, remember that cancer cells are our own body cells 'gone wrong'. Viruses also enter our cells and reproduce inside them – here antibodies cannot reach them.

T-lymphocytes are able to recognise virus-infected cells (and sometimes cancerous cells as well) because of tell tale marker antigens on the surface of the cell. The T-lymphocytes target these cells and kill them in one of two ways:

- Some T-lymphocytes release chemicals that 'punch a hole' in the body cell. All the contents leak out, the cell dies and with it, the viruses inside.
- Other T-lymphocytes press a 'self-destruct button' in the cell. They activate a 'programmed cell death' process that is part of the genetic code of every cell.

Figure 25.7 summarises how T-lymphocytes destroy virus-infected cells.

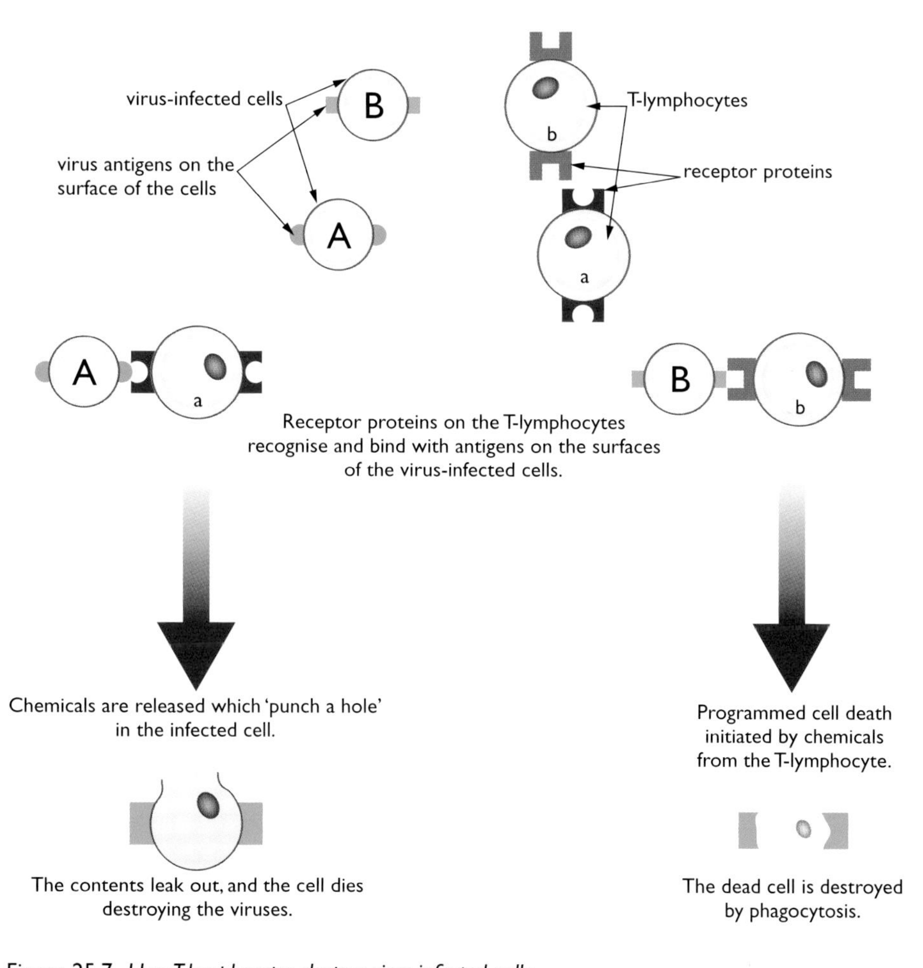

Figure 25.7 *How T-lymphocytes destroy virus-infected cells.*

Our immune systems are at their most active in childhood and adolescence. This is the period of our lives when we meet most diseases for the first time. Although we retain an active immune system throughout adulthood, it is less effective at combating new diseases than it was in our teens.

As with B-lymphocytes, some of the T-lymphocytes become memory cells and remain in the body for many years. They can also mount a speedy secondary immune response if the same microorganism re-infects us. Again, the memory T-lymphocytes are the basis of our immunity to disease.

The process of becoming immune to a disease is called **immunisation**. The responses described so far are **natural active** responses made by *our own bodies* to make us immune to a disease. There are other types of immunity in addition to these.

Natural passive immunity

Passive immunity is a means of acquiring immunity without having to actually produce an immune response ourselves. Normally this occurs only twice in our lives, both times when we are very young:

- We receive antibodies across the placenta before we are born. The antibodies cross the placenta from the mother's blood in increasing amounts throughout pregnancy.
- We also receive antibodies from our mothers in colostrum and breast milk. Colostrum is the liquid produced for the first few days after birth, before the mother produces breast milk proper. Both contain antibodies, but colostrum is a particularly rich source.

'Breast is best' is a slogan used to encourage natural breast-feeding rather than bottle-feeding of babies. One reason why breast is best is that only breast milk can give immunity because of the antibodies it contains.

Passive immunity is short lived. Because babies haven't made the antibodies themselves, the levels drop once breast-feeding stops. However, this passive immunity protects newborn babies for a vital few months while their own immune systems are developing.

Artificial active immunity – vaccination

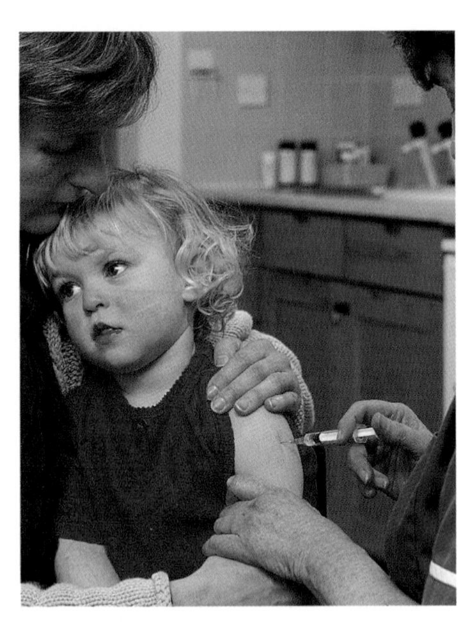

Figure 25.8 *A vaccine is injected to stimulate an immune response to a particular disease-causing organism.*

Vaccination is a way of giving nature a helping hand (Figure 25.8). We can be made immune to a disease without actually contracting (having) the disease itself. A person is injected with some 'agent' that carries the same antigens as a specific disease-causing microorganism. Lymphocytes 'recognise' the antigens and multiply exactly as if that microorganism had entered the bloodstream. They form memory cells and make us immune to the disease. The actual vaccine that is injected may be:

- an **attenuated** (weakened) strain of the actual microorganism (e.g. polio, TB (tuberculosis) and measles vaccines)
- dead microorganisms (e.g. whooping cough and typhoid fever vaccines)
- modified toxins of the bacteria (e.g. the toxins of the tetanus and diphtheria bacteria)
- just the antigens (e.g. the influenza vaccine)
- harmless bacteria, genetically engineered to carry the antigens of a different disease-causing microorganism (e.g. the hepatitis B vaccine).

Different diseases are endemic in different parts of the world. Travellers should seek advice from their doctor before visiting foreign countries about any vaccinations that may be necessary.

ideas evidence

Edward Jenner performed the first vaccination in 1796. At that time, smallpox was a major killer in England. Jenner was an English country doctor and had noticed that people who caught cowpox (a similar, but much milder disease) hardly ever caught smallpox. Jenner vaccinated a small boy called Edward Phipps to show that cowpox could protect against smallpox (Figure 25.9). He first infected the boy with cowpox and allowed him two months to recover from the disease. He then infected him with pus from a smallpox sufferer.

Figure 25.9 *A statue of Edward Jenner infecting Edward Phipps with a vaccine against smallpox.*

Edward Phipps did not contract smallpox and we now understand exactly why – and so should you. The antigens on the cowpox virus are the same as those on the smallpox virus. They stimulate exactly the same immune response and the memory cells formed give immunity against smallpox. The word *vaccination* comes from the Latin word *vacca*, meaning a cow.

Vaccination has reduced the incidence of many serious diseases and has virtually eradicated smallpox from the planet. However, there may be risks with vaccination. If there is a microorganism in the vaccine, it may recover the ability to become infective. In the 1950s, a batch of polio vaccine was not properly prepared, with the result that many children actually contracted polio, instead of being protected against it. Today, the risks are better understood and procedures are much more effective.

Recently there has been concern that the triple vaccine MMR (for mumps, measles and rubella) may cause autism in a small number of cases. The position is far from clear, but the most recent research suggests that there is no link between MMR and autism. However, the problems have raised many questions about vaccination.

Sometimes, vaccination against a microorganism is virtually impossible because the microorganism mutates rapidly with some of the mutations resulting in different surface antigens. This means that any memory cells the body has made will not recognise the new form of the microorganism. A new vaccine must be produced that will stimulate the production of new memory cells. This is why we get so many colds. The common cold virus has a very high mutation rate so by the time a vaccine has been produced, the virus has mutated again.

The influenza ('flu) virus also mutates regularly, but not as fast as the common cold virus. There is enough time to produce a 'flu vaccine that will be effective before the virus mutates again.

Artificial passive immunity

Sometimes people need help immediately so it is not appropriate to give them a vaccination that would make them immune to a disease sometime in the future. In these cases a person can be injected with the actual antibodies to help to boost their own immune response. This is often done in the case of rabies, although there is also a vaccination against the disease.

Louis Pasteur and rabies

ideas
evidence

Louis Pasteur is most commonly remembered for his work on prevention of spoilage of foods, and for the process that bears his name – Pasteurisation. However, he also did important work in the field of vaccination and developed a vaccine against rabies. Rabies is a viral disease, common in animals, that can be transmitted to humans through a bite. Most cases of rabies in humans are transmitted from rabid dogs (dogs with rabies). Figure 25.10 shows the distribution of rabies throughout the world, and the animals mainly responsible for the transmission of rabies in the different areas.

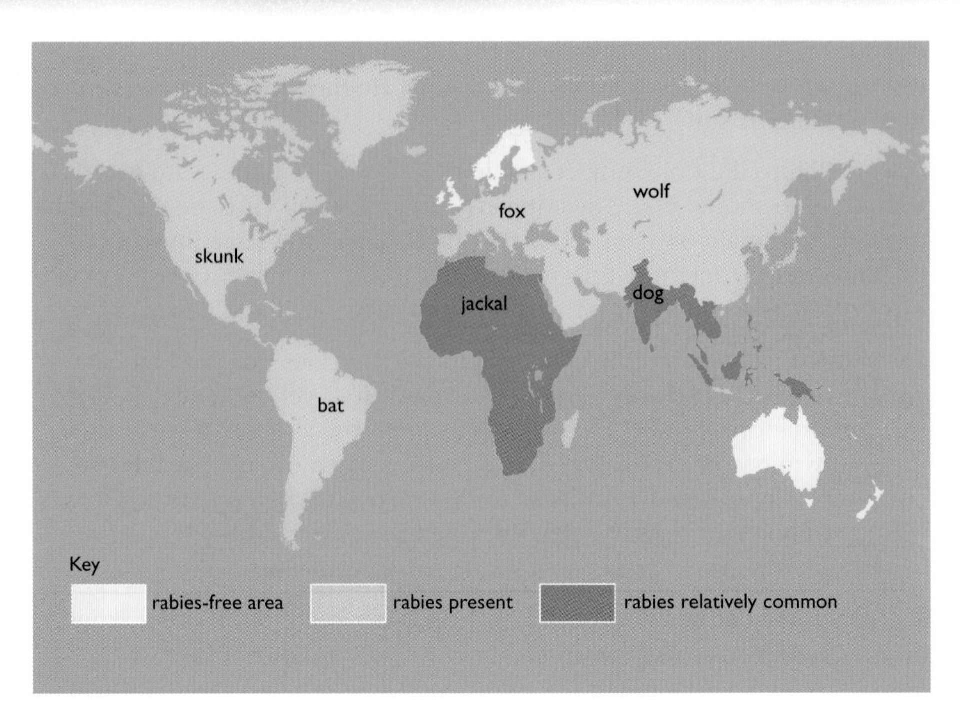

Figure 25.10 *The distribution of rabies worldwide and the animals mainly responsible for its transmission.*

The virulence of a microorganism is its ability to infect and cause disease.

Pasteur had read about Jenner's work and had found similar cases of mild forms of a disease protecting against a more severe form. More importantly, he developed a method of making bacteria lose their virulence. Pasteur found that by keeping cultures of bacteria for a long time – by taking samples and culturing the samples and then taking more samples and culturing them and so on – the bacteria lost some of their virulence. He was able to do this for anthrax and rabies – the first rabies vaccine had been produced. The vaccine had extremely unpleasant side effects and was far from safe, but his work represented a significant breakthrough in immunology. Today's anti-rabies vaccines are much safer and have fewer side effects.

Monoclonal antibodies

When microorganisms enter our bodies, it is common for more than one B-lymphocyte to be able to bind to the antigens on its surface and so become activated. This is because the receptor proteins on the different B-lymphocytes bind to different parts of the antigen. The result is that the different B-lymphocytes multiply and manufacture their antibodies. This is all to the good as it means that more than one type of antibody is actively engaged in destroying the microorganism. Mixtures of antibodies produced by several types of B-lymphocytes are called **polyclonal antibodies**.

***Mono*clonal antibodies** all come from the *same* types of B-lymphocyte. They are all identical and all bind to the same part of an antigen. Monoclonal antibodies can be produced in the laboratory. First, an animal is stimulated to produce antibodies to a particular antigen. Then, the different types of B-lymphocytes that have become activated are separated and cultured. The antibodies produced by each different type are isolated. These are monoclonal antibodies as each was produced by just one type of B-lymphocyte.

A major problem in the production of monoclonal antibodies is that B-lymphocytes do not survive and multiply very well outside the body. To get around this, they are fused with tumour cells, which do survive and multiply well outside the body. The procedure is summarised in the flow chart in Figure 25.11.

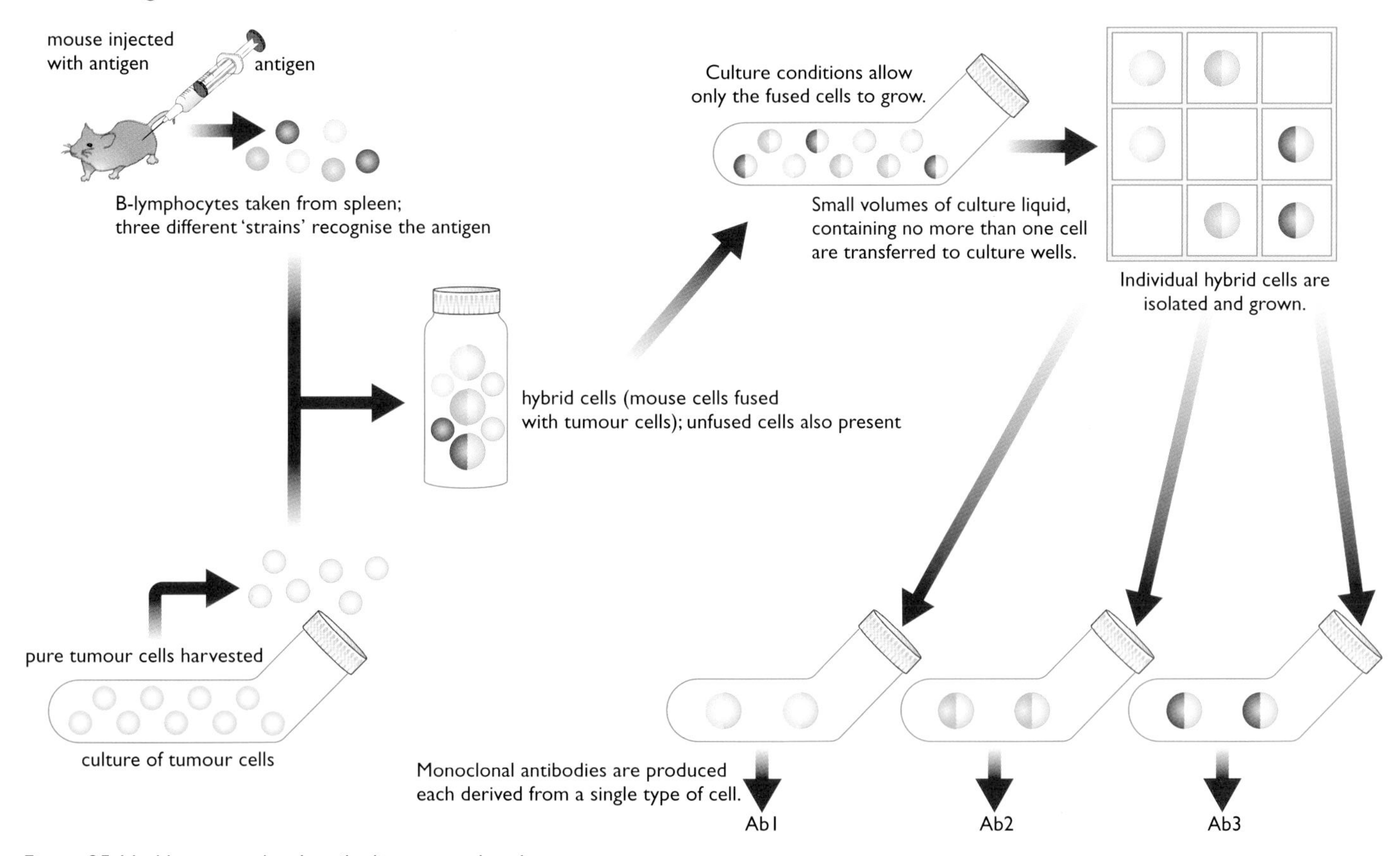

Figure 25.11 *How monoclonal antibodies are produced.*

Individual monoclonal antibodies can distinguish precisely between antigens that only differ very slightly. Because of the precision of their binding, monoclonal antibodies are used in several ways:

- Detecting some cancers, such as lymphomas (cancers of the lymphatic system) – monoclonal antibodies are produced using the cancer cell antigens. Other samples can then be tested to see if similar cancerous antigens are present.
- Treating some cancers – research is underway to try to bind monoclonal antibodies to anti-cancer drugs. At the moment, chemotherapy for many cancers affects not just the cancer cells – it targets *any* cells that are dividing. This includes bone marrow (producing blood cells) as well as many organ linings and hair follicles. Binding the anti-cancer drugs to monoclonal antibodies would mean that only the cancer cells would be targeted. Only a small amount of the drug would need to be used to deliver an effective dose to the cancer cells.
- Pregnancy testing – monoclonal antibodies are made that will only bind with a hormone produced by the embryo. A sample of urine is tested and if a reaction occurs (indicated by a colour change) the embryo hormone must be present and the woman must be pregnant.

- Detecting some viruses – this works in essentially the same way as detecting cancer cells. It is important in screening blood samples for the presence of viruses like HIV.

Organ transplantation

Immunosuppressive drugs 'damp down' our immune responses. They reduce the risk of rejection of a transplanted organ, but they also reduce our ability to fight disease.

Each of us has our own unique set of antigens on our cells. B- and T-lymphocytes will not bind to these because they are 'self' antigens. The antigens on bacteria that infect us are 'non-self' antigens, so at least one B- or T-lymphocyte will recognise them and bind to them. The same is true with organs transplanted from other people. The antigens on the cells of the organ are non-self antigens. The lymphocytes bind with them and set about destroying these 'foreign' cells. This is commonly called **rejection.**

In looking for an donor organ, it is important to find one with antigens on the cells that match those of the patient as closely as possible. Because our antigens are determined by our genes, those of family members are often a good match. Those of identical twins are particularly closely matched. Finding an organ with similar antigens to the ones on the person needing the transplant is called **tissue typing**. A close match of antigens, together with the use of **immunosuppressive drugs**, considerably reduces the risk of rejection.

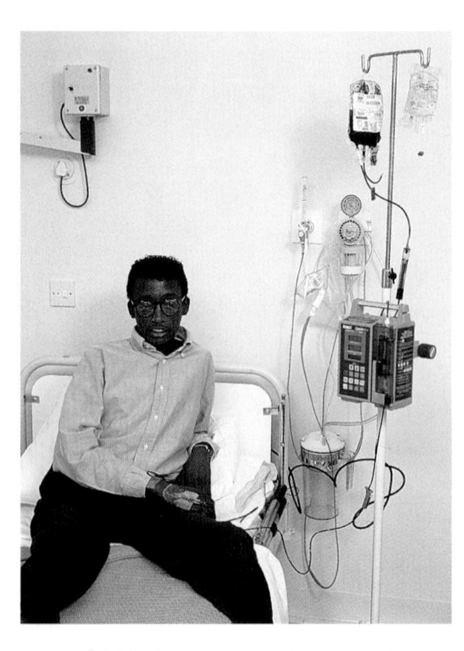

Figure 25.12 *A patient receiving a transfusion of blood.*

Blood groups and transfusions

Successful blood transfusions are only possible because of our knowledge of blood grouping (Figure 25.12). Blood grouping is a kind of tissue typing – blood of an inappropriate group will be 'rejected'.

Your blood group depends on the antigens on the surface of your red blood cells. Two of these are important in blood grouping. They are called the A and B antigens. The four possible blood groups are based on the presence of these antibodies and are called A, B, AB and O.

Besides the antigens on the red blood cells, each person also has antibodies in the plasma. These antibodies will destroy red blood cells with a particular antigen by making them **agglutinate** (clump together). Antibody 'a' agglutinates any red blood cells with the A antigen. Antibody 'b' agglutinates any red blood cells with the B antigen. Table 25.1 gives details of the different blood groups. As a general principle, remember that a person never carries the antibodies that would react with the antigens on their own red blood cells.

These antibodies differ from the antibodies we make against disease-causing organisms in one important way. They are present in our blood *all the time*, so we do not need to be exposed to the antigen to make the antibody.

Blood group	Antigens on red cells	Antibodies in plasma
A	A	b
B	B	a
AB	AB	neither
O	neither	ab

Table 25.1: *Blood groups and their antigens and antibodies.*

It could be fatal for a doctor to give a blood transfusion of type A to a person with type B blood. The person receiving the blood has 'a' antibodies in their plasma that would make the red cells of the transfused blood agglutinate. Blood would clot inside the blood vessels and block them. The safety of a transfusion depends on the *antigens* on the red cells of the donated blood and the *antibodies* in the plasma of the person receiving the blood. If these can react (e.g. antigen A and antibody 'a'), then the transfusion will be unsafe. Table 25.2 shows safe and unsafe transfusions.

Blood group of donor		**Blood group of recipient (antibodies present shown in brackets)**			
Group	**Antigen**	**A (b)**	**B (a)**	**AB (neither)**	**O (a +b)**
A	A	✓	✗	✓	✗
B	B	✗	✓	✓	✗
AB	A + B	✗	✗	✓	✗
O	either	✓	✓	✓	✓

✓ = safe transfusion ✗ = unsafe transfusion

Table 25.2: *Blood transfusions and blood groups.*

Blood group O is sometimes called the universal donor because it can be given to any other blood group. Because there are no antigens on the surface of the red blood cells, there can be no reaction with any antibodies in the plasma of the person receiving the blood. Similarly, blood group AB is the universal recipient. Because there are no antibodies in the plasma, antigens in the donated blood cannot cause a reaction.

End of Chapter Checklist

If you haven't got a copy of your specification, read the introduction on page vi.

You will need to be able to do some or all of the following. Check your Awarding Body's specification (syllabus) to find out exactly what you need to know.

- Describe how bacteria and viruses cause disease.
- List the natural barriers to infection that exist.
- Describe the general defence mechanisms of blood.
- Describe the roles of B- and T-lymphocytes in the natural active immune response.
- Explain how we become immune to a disease, including the nature of the secondary immune response.
- Explain the nature of natural passive immunity, artificial passive immunity and artificial active immunity (vaccination).
- Understand the importance of the work of Edward Jenner in pioneering vaccination, and that of Louis Pasteur in developing techniques of producing non-virulent strains of microorganisms for use in vaccines.
- Understand the nature of organ rejection and how the risk can be minimised by accurate tissue typing and the use of immunosuppressive drugs.
- Understand the importance of monoclonal antibodies and be able to describe how they are produced.
- Explain the biological basis for the ABO blood grouping system in humans and its importance in predicting safe and unsafe blood transfusions.

Questions

More questions on blood and immunity can be found at the end of Section F on page 332.

1 Blood has several immune functions. Some of them are general, others are specific.

a) Name two general immune functions of blood.

b) Describe the role of T-lymphocytes in combating a viral infection.

c) Describe how memory T-lymphocytes are formed and explain their importance in giving lasting immunity to a disease.

2 The diagram shows the levels of antibodies produced by a person during an initial infection and later, when re-infected by the same microorganism.

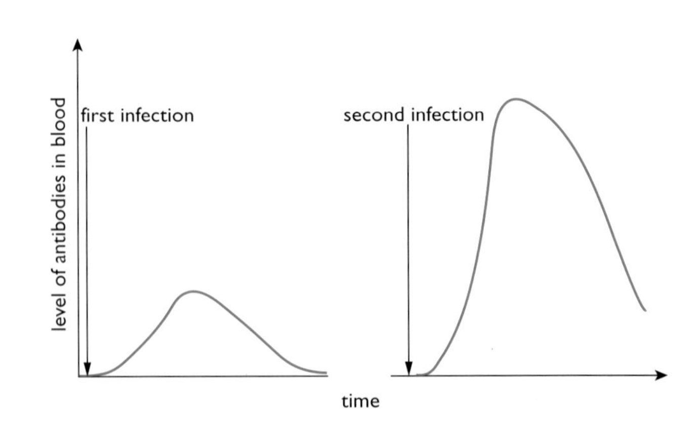

a) Give the names of the initial and subsequent responses by the body to infection.

b) Describe the differences between the two responses in relation to level, speed and duration.

c) Explain how the initial response is produced.

3 One of the problems of organ transplantation is organ rejection. This can be largely overcome by accurate tissue typing and the use of immunosuppressive drugs.

a) Explain what is meant by:

i) organ rejection

ii) tissue typing.

b) Why do transplants between people who are related (e.g. father and son), have a lower rejection rate than those between unrelated people?

c) Explain:

i) an advantage of using immunosuppressive drugs

ii) a drawback to using immunosuppressive drugs.

4 The results of a blood-grouping test are shown in the diagram.

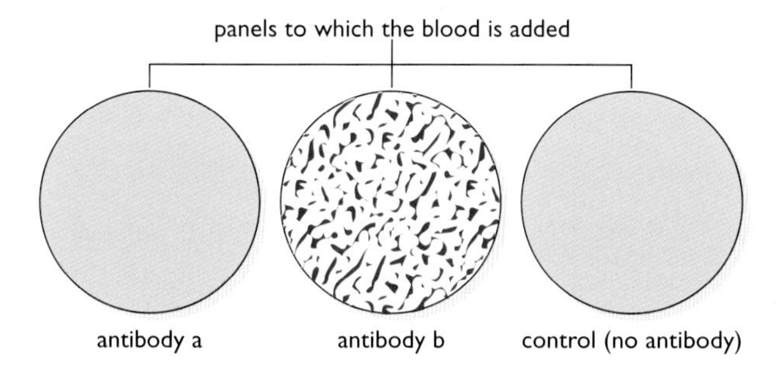

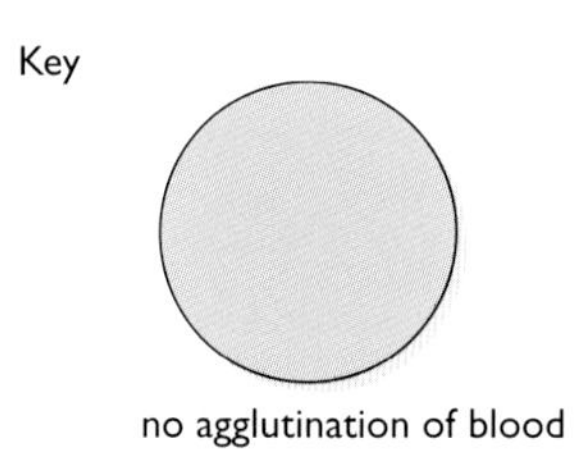

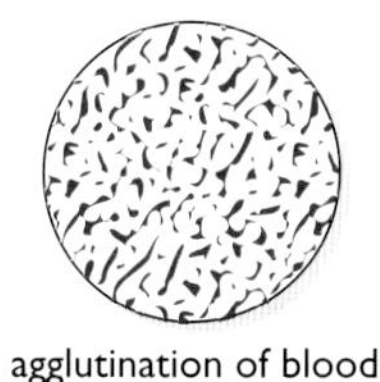

a) What is the blood group of the person being tested? Explain your answer.

b) Explain why:

i) blood group AB is called the universal recipient

ii) blood group O is called the universal donor.

5 Vaccination has been important in reducing the incidence of many infectious diseases.

a) Explain why vaccination has been able to reduce the incidence of infectious diseases.

b) Describe the work of Jenner and Pasteur in developing vaccines.

Chapter 26: Treatment and Prevention of Disease

In previous chapters we have seen how microorganisms cause infectious disease and how our bodies respond. This chapter looks at ways of helping the body to fight infection and prevent attack by pathogens.

Many people just associate the term chemotherapy with cancer treatment, but it has a wider meaning.

All aspirins are derived from the chemical salicylic acid, which is found in the willow tree (*Salix* sp.). Can you see where the name of the chemical comes from? This ancient drug may have a new role in preventing heart disease and strokes as it reduces clot formation inside arteries. One heart surgeon recently said, 'Every man over 50 should take aspirin every day'. Not everyone would agree, but aspirin *is* frequently prescribed to stroke victims.

Figure 26.1 *Aspirin is one of the oldest known drugs.*

Sometimes our bodies cannot cope easily with an infection – they need a helping hand. Under these circumstances, a doctor may prescribe a range of medicines. Some just alleviate symptoms, others kill or slow down the reproduction of the disease-causing organisms. The use of drugs to treat disease is called **chemotherapy**.

One problem with chemotherapy is that natural selection favours resistant mutants (see Chapter 19). This can lead to strains of bacteria evolving that are resistant to a particular drug. Another problem is that drugs can provoke allergic reactions in many people.

Prevention is always better than cure. It is always better to avoid the disease than have to rely on the body's immune response and chemotherapy to combat it. We can minimise the risk in several ways:

- Vaccinations can make us immune to specific diseases.
- Personal and public hygiene measures minimise contact with disease-causing organisms.

Chemotherapy

The most commonly used drugs are aspirin, paracetamol, codeine and other similar drugs that can be bought over the counter at a pharmacy. These drugs are all palliatives – they relieve certain symptoms but do not treat the cause.

Some drugs kill microorganisms or slow down their reproduction. These include the following:

- **Sulphonamides** kill bacteria. They are all synthetic drugs and were widely used before antibiotics were discovered. Their use is now mainly restricted to treatment of infections of the urinary system.
- **Antibiotics** kill a range of microorganisms, but not viruses. The first antibiotics were extracted from microorganisms. An increasing number are now synthetic or semi-synthetic.
- **Anti-viral drugs** (occasionally called anti*viotics*) disrupt the processes by which viruses reproduce inside cells. To date, no anti-viral drug has been developed that is completely effective. The problem arises because viruses reproduce *inside* our cells.

Antibiotics

The discovery of penicillin

ideas evidence

In 1928, in St Mary's Hospital in London, a research microbiologist noticed that one of his cultures of bacteria had become contaminated. A mould was also growing on the culture. He looked more carefully and noticed that there were no bacteria around the area where the mould was growing. It probably looked something like the Petri dish shown in Figure 26.2a.

(a)

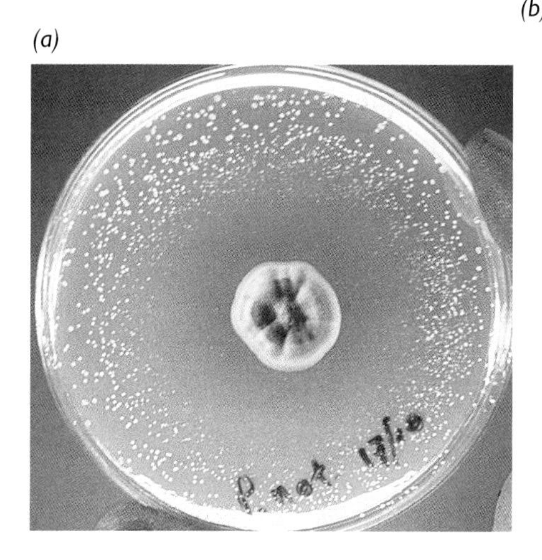

(b)

Figure 26.2 *(a) A Petri dish with a fungus and bacterial colonies growing on the agar. The fungus is secreting an antibiotic that is inhibiting the growth of the bacteria. (b) Alexander Fleming (1881–1955).*

He then deliberately repeated the contamination of a culture – with the same results. He reasoned that the fungus must be secreting a substance that inhibited the growth of bacteria in some way. The researcher's name was Alexander Fleming and he named the substance penicillin – after the mould *Penicillium* that contaminated his cultures. The first antibiotic had been discovered.

Fleming never extracted the pure penicillin. This was left to other researchers – Howard Florey and Ernst Chain. They isolated and purified penicillin and were able to confirm that penicillin did have an antibacterial effect. Florey and Chain also began to develop techniques to scale up the production of penicillin.

Penicillin was the first antibiotic to be discovered. When it was first used to treat bacterial diseases, there were spectacular results. The bacteria had no resistance to the drug and almost 'miracle cures' seemed possible. However, as penicillin was used more and more frequently, mutations in some bacteria gave them resistance to it. Natural selection allowed these bacteria to survive and reproduce. As these resistant strains became more widespread, so the effect of penicillin diminished.

Other antibiotics were discovered and used. These included drugs like streptomycin, chloramphenicol, aureomycin and many others. The pattern was the same with these drugs. Initially they produced spectacular results, then resistant strains appeared, became widespread and the effect diminished.

As people became more aware of antibiotics, they expected (and often demanded) to be prescribed them for even the most trivial infections. This led to over-prescription of the drugs and rapid emergence of resistant strains. Doctors began to prescribe antibiotics in combination for greater effect. Bacteria then evolved that were resistant to several antibiotics – so called 'super-bugs'.

Today, doctors are more aware of this problem and prescribe antibiotics more judiciously. Also, new antibiotics are synthesised in laboratories. There does not need to be a 'search' through a range of microorganisms to see if any produce an antibiotic so far undiscovered.

Centuries before antibiotics were discovered, people living in the countryside knew that rubbing certain moulds into wounds stopped them from becoming infected. They were, unknowingly, administering the antibiotic penicillin that was being produced by the mould.

ideas
evidence

Bacteria can 'genetically engineer' themselves! They can swap plasmids – the tiny circular pieces of DNA that are used in genetic engineering to transfer genes (see Chapter 23). If bacteria swap plasmids that carry resistance to different antibiotics, they can end up with 'multiple resistance'.

How antibiotics act

Bactericidal antibiotics actually kill bacteria, while **Bacteriostatic** antibiotics just stop them from reproducing. **Broad spectrum** antibiotics act against a range of bacteria, while others are more limited in their target. Table 26.1 shows some of the ways in which antibiotics act.

Example	Mode of action	Effect	Type of antibiotic
penicillin	interferes with manufacture of bacterial cell wall	weakens cell wall; water enters by osmosis and bursts the cell (osmotic lysis)	bactericidal
nalidixic acid	interferes with DNA replication	bacteria are not killed, but cell division is impossible so they cannot multiply	bacteriostatic
tetracycline	interferes with protein synthesis	no enzymes can be made to control the reactions in the bacterial cells	bactericidal

Table 26.1: *Comparison of different types of antibiotics.*

Figure 26.3 shows how penicillin acts on bacterial cells. It is particularly effective against bacteria when they are dividing as it interferes with the manufacture of the new cell walls of the daughter cells.

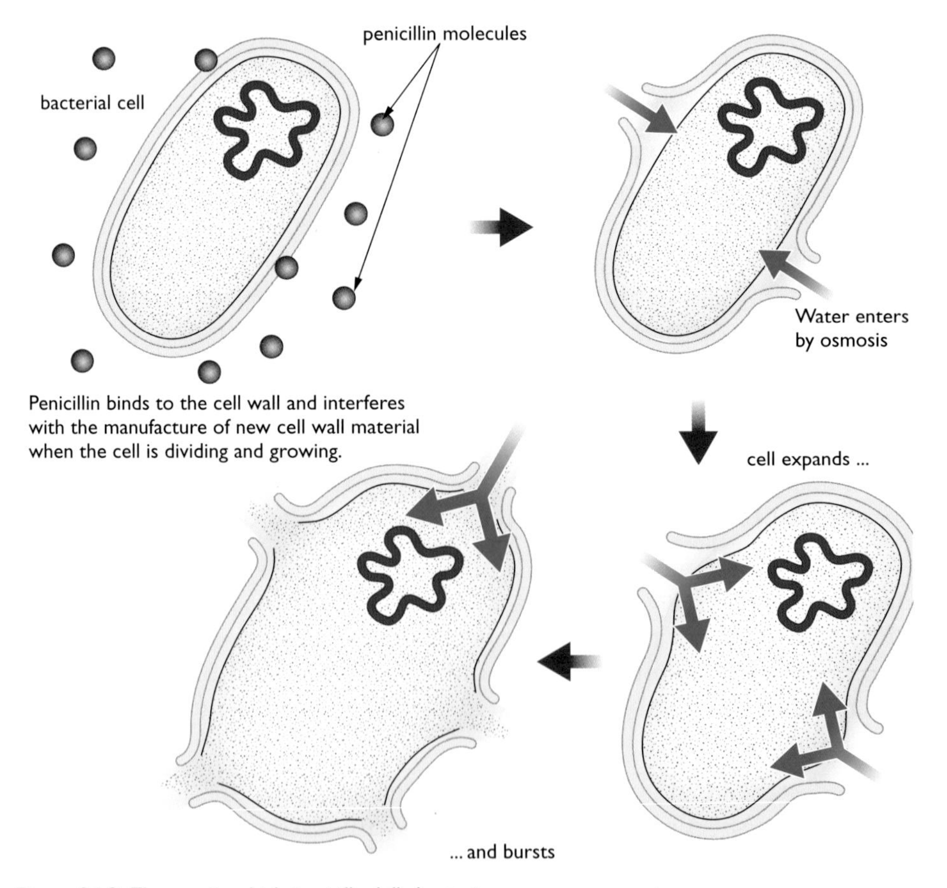

Figure 26.3 *The way in which penicillin kills bacteria.*

Antibiotics have no effect at all on viruses for the following reasons:

- Viruses have no cell walls – so antibiotics like penicillin can have no effect.
- Viral DNA only replicates inside other cells once the virus has infected the cell.
- Viruses do not synthesise their own proteins; they instruct the cell they infect to synthesise their proteins.
- Viruses are only active *inside* other living cells and are protected by the cell as most antibiotics cannot enter human cells.

ideas evidence

Producing penicillin

Penicillin is secreted by the fungus *Penicillium*. Different strains of the fungus can produce slightly different penicillins, that can be active against different bacteria. New strains of *Penicillium* are constantly 'screened' for antibiotic production. Any strain that appears to produce a new type of penicillin with significant anti-bacterial activity is then cultured for further studies to see if the fungus is suitable for bulk production of penicillin (see Chapter 22).

Biochemists can now produce 'designer' penicillins. This involves altering a basic penicillin molecule to give it a slightly different shape. It is done in one of two ways:

- Altering the 'diet' of the fungus during production – this can modify the side chain on the penicillin molecule (Figure 26.4).
- Some strains of the fungus produce penicillin with no side chain at all. A side chain can be created chemically that will bind with the desired cell wall material, and then added to the penicillin molecule.

These two techniques allow production of thousands of different, semi-synthetic penicillins. Each will have a slightly different shape and so have different antibiotic activity.

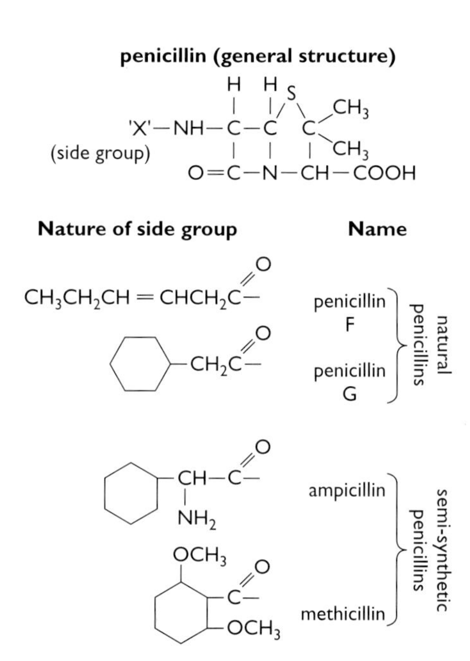

Figure 26.4 *Different side chains give a penicillin molecule different properties.*

Minimising the risk of infection

We have looked at the ways in which our bodies react to infection and at some of the drugs that can be prescribed to give our immune system a 'helping hand'. We can also do a great deal individually and collectively to minimise the risk of infection in the first place. Many of the measures we can carry out are aimed at blocking transmission routes (see Chapter 24).

Preventing transmission in food

We can prevent transmission of food borne microorganisms at various stages. We can prevent microorganisms from getting into the food in the first place, we can treat the food to minimise the rate at which microorganisms multiply in the food (food preservation) and we can cook food properly to kill any microorganisms present.

Preventing microorganisms from entering food

Microorganisms can enter food at several stages: at the point of production, at the point of sale, during storage after production and in the home.

The conditions under which livestock are reared can affect their health and

Figure 26.5 *Contamination of food can be prevented during production by keeping animals in hygenic conditions.*

the food that we get from them. For example, some cheeses are still made from untreated milk. Any harmful bacteria in the milk of the cow (or goat) could therefore find their way into the cheese and from there into us. Poor housing of chickens can allow *Salmonella* to spread rapidly among the birds and lead to infection of the eggs.

The microorganisms that cause disease in crop plants rarely have significant effects on humans, but harvested crops such as grain must be stored properly. Pests could enter the store and contaminate the food with excreta and disease-causing microorganisms.

At the point of sale, food must be stored in such a way as to minimise transmission from humans. Packaging of food prevents this, but loose vegetables and fruit are more at risk. 'Display until' and 'best before' dates are important (Figure 26.6). They tell us when the food is likely to become unsafe to eat because of contamination by microorganisms or toxins from microorganisms. After this date, the numbers of microorganisms may have increased to dangerous levels.

Figure 26.6 *'Display until' dates help us to know if foods are likely to be contaminated.*

In the home, food should be stored in such a way as to minimise transmission. For example:

- cooked and raw foods should not be stored together (bacteria in the uncooked food may be transferred to the cooked food)
- foods that have been frozen should not be refrozen after cooking (any bacteria that remain will multiply when the food is re-cooked)
- food should not be left open to the air on a work surface (bacteria in the air and bacteria carried by insects could contaminate the food).

Food preservation

Some methods of food preservation, such as salting and pickling food, have been used for hundreds of years. Irradiation of food is a recently introduced technique that gives many people concerns about its safety. Table 26.2 describes some methods of food preservation.

Technique	Principles and procedures of the technique	Example of foods preserved this way	Possible drawbacks
salting	High salt concentrations make it impossible for bacteria to multiply. Bacterial cells lose water by osmosis and are killed. Some foods are covered in salt, others are soaked in brine (saltwater).	some meats, fish (kippers), vegetables	Some bacteria can withstand very high salt concentrations.
pickling	Foods are bottled in vinegar – a weak, flavoured solution of ethanoic (acetic) acid. The low pH inactivates most microorganisms.	onions, red cabbage, herrings, mayonnaise	It alters the taste of the food.
pasteurisation	The food is heated to between 63 and 65°C (Figure 26.7) for 30 minutes *or* 71.5°C for 15 seconds. Both techniques kill pathogenic bacteria.	milk, cream, ice-cream, fruit juices, beer	It does not kill all bacteria – spoilage of the food is merely delayed. Spores of harmful bacteria may survive.
ultra-heat-treatment (UHT)	Superheated steam at temperatures of 135–160°C is blown through the food for 2 seconds. This kills *all* bacteria *and* spores.	milk	It alters the taste of the milk.
canning	Food is packed in cans, heated to high temperatures, sealed and then reheated to temperatures of 105–160°C. The high temperatures kill bacteria and spores.	beans, soup, tomatoes	Cans can be damaged when cooled and in transit. This can allow bacteria to enter the can and multiply. Beware dented cans!
drying	Drying removes water from the food so bacteria cannot digest and absorb it. Manufacturers dry food by blowing hot air through it.	cereals, grains, some fruits	Water can easily re-enter once the packets are opened. Bacteria can then cause spoilage.
freezing	Foods are cooled rapidly to temperatures of −10°C. Rapid freezing prevents the formation of large ice crystals which could alter the texture and flavour of the food.	many vegetables, meats, prepared meals	none
irradiation	High-energy gamma radiation from sources such as radioactive cobalt or radioactive caesium is passed through the food. All bacteria and spores are killed, although toxins produced by the bacteria still remain. There is no change in the taste of the food.	Some countries permit irradiation of potatoes, onions, shellfish and some fruits.	Some people are concerned that food may be made radioactive or that carcinogens may be formed in the food. At the moment irradiation is not permitted in the UK.

Table 26.2: *Common methods of food preservation.*

Preventing transmission by other routes

The incidence of many infectious diseases has declined over the past 200 years. This has been due to the improvements in education, diet, living standards and standards of public hygiene coupled with medical advances. Improved water treatment and sewage disposal have reduced transmission of waterborne diseases such as cholera and typhoid fever. These diseases are still endemic where human faeces contaminate drinking water.

Figure 26.7 *A pasteurisation plant.*

Improved housing and living conditions in general have reduced transmission through the air and by physical contact. Better education has led to an improved diet, which has increased our ability to resist infection, and better personal hygiene helps prevent the spread of disease. Data on the decline in the incidence of some infectious diseases in the UK over the past

200 years can be found in Figure 26.8 (smallpox) and Figure 24.6 on page 302 (measles and diphtheria). The death rate from diphtheria was rising until the antitoxin was first used. National immunisation has virtually eliminated the disease. The reduction in death rate from measles is almost entirely due to improved health and standards of living, while the immunisation programme has reduced the incidence of the disease. Disease is much less prevalent in more economically developed countries now than at any time in history. However, measles still kills millions of children each year worldwide. There is still a great deal to do.

ideas
evidence

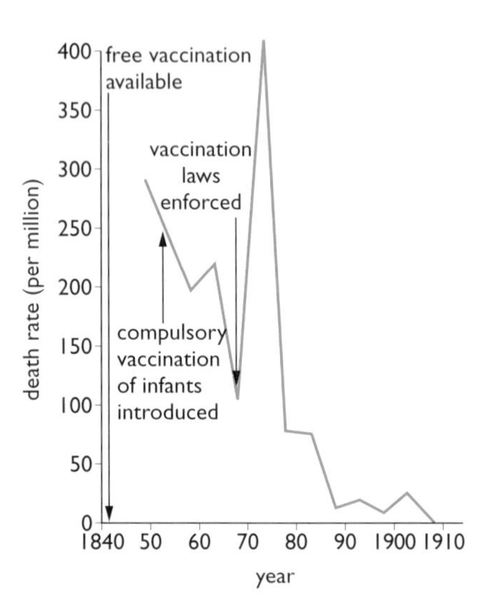

Figure 26.8 *Death rates from smallpox in England and Wales. The decline up to 1850 was largely due to improved living conditions. The sharp rise when vaccination was made compulsory was due to poor technique – many doctors administered the wrong vaccine and used dirty needles.*

It is important to bathe or shower regularly to keep clean. But don't be fooled into thinking that you have got rid of all the bacteria on your skin. There will still be billions left after the shower! You will, however, have got rid of most of the harmful bacteria.

Vaccination programmes have reduced the incidence of many diseases. Vaccinations stimulate our T- and B-lymphocytes to produce memory cells and make us immune to disease (see Chapter 25). It is thought that the smallpox virus now only exists in medical research laboratories as it has been eliminated from the world population by a worldwide vaccination programme.

For such a programme to be effective, not everybody needs to be vaccinated. Once the levels of immune people in a population reach 90%, transmission becomes almost impossible and the disease-causing organism is eliminated from that population. If the level of immunity falls to 85%, then the disease can spread fairly easily among susceptible individuals if it is reintroduced.

Vaccination needs to be sustained in a population to be effective. Many doctors are concerned about the recent decline in the numbers of infants receiving the MMR vaccine. This is because some parents are worried about a possible link between the vaccine and autism (see Chapter 25). The chances of transmission of mumps, measles and rubella are increasing as a result

Education in personal hygiene has also played an important part in reducing the transmission of disease. Simple measures, like sneezing only into a handkerchief, can reduce the spread of microorganisms through the air. Covering cuts and grazes with a plaster or bandage keeps microorganisms out while the wound is healing. Washing our hands after using the toilet and before meals is an important measure in preventing microorganisms from entering our bodies. Just keeping our bodies clean on a daily basis prevents the build-up of microorganisms on our skin.

We now use disinfectants and antiseptics much more than in the past. This also helps to prevent the build-up of microorganisms. A disinfectant is any substance that kills microorganisms and many of the products that we use to clean our work surfaces around the house contain disinfectants. Disinfectants often contain chlorine (e.g. household bleach), which is a powerful oxidising agent and kills all microorganisms.

Antiseptics are mild disinfectants that can be used on the surface of the body. They are less effective than other disinfectants, but do kill or inactivate most harmful bacteria. Their effectiveness often depends on the concentration that is used. Some are bactericidal (kill bacteria) in high concentrations, but only bacteriostatic (prevent reproduction of bacteria) in lower concentrations.

End of Chapter Checklist

If you haven't got a copy of your specification, read the introduction on page vi.

You will need to be able to do some or all of the following. Check your Awarding Body's specification (syllabus) to find out exactly what you need to know.

- Explain the term chemotherapy.
- Describe some of the types of drugs that are used to treat diseases.
- Describe how penicillin was discovered.
- Explain how some antibiotics act.
- Distinguish between bacteriostatic and bactericidal antibiotics.
- Describe how penicillin is produced.
- Explain how the risk of transmission in food can be reduced.
- Explain the principles behind the common methods of food preservation.
- Explain the importance of improved standards of living and personal hygiene on the reduction of the incidence of disease.

Questions

More questions on treatment and prevention of disease can be found at the end of Section F on page 332.

1 The graph shows the changes in the number of reported cases of whooping cough between 1950 and 1970.

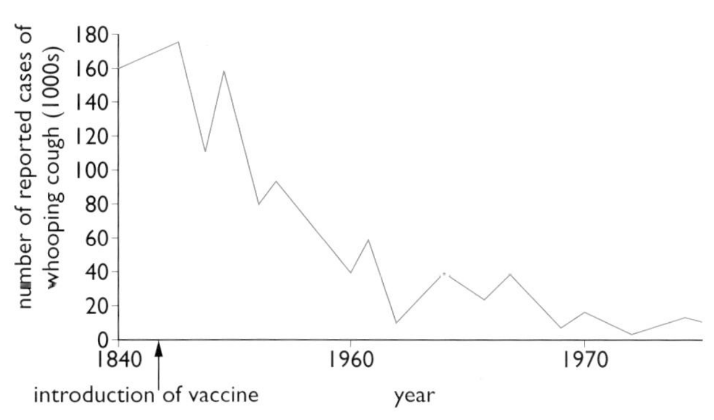

a) Describe the changes in the incidence of whooping cough following the introduction of vaccination in 1952.

b) Explain why you cannot be sure that the increase from 1950 to 1952 was part of a general trend.

c) The numbers of reported cases of whooping cough increased in the 1980s. Suggest one possible reason why.

2 ***a)*** Explain the principles behind the following methods of food preservation:

i) salting

ii) pasteurisation

iii) pickling.

b) Why should food not be consumed after the 'best before' date?

3 Penicillin was the first antibiotic to be discovered in 1928.

a) What is the difference between an antibiotic and an antiseptic?

b) Name two other antibiotics and for each one:

i) state whether it is bacteriostatic or bactericidal

ii) describe how it kills or inactivates bacteria.

c) Explain why antibiotics are ineffective against viruses.

4 Explain the importance of the following in reducing the risk of food poisoning.

a) 'display until' and 'best before' dates

b) cooking food thoroughly

c) not refreezing food that has been frozen once and thawed.

5 Explain fully why doctors are more reluctant to prescribe antibiotics now than they were previously. Your answer should include reference to:

a) appearance of resistant strains

b) natural selection

c) multiple resistance.

End of Section Questions

1 *Escherichia coli* (*E.Coli*) bacteria live in the intestines of humans. Some types of *E. coli* cause food poisoning, others are harmless. The graph shows a growth curve for a population of *E. coli* bacteria grown in a culture at 15°C.

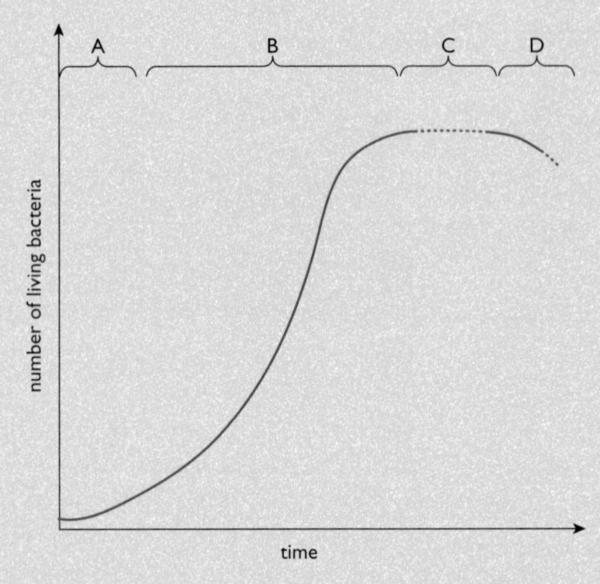

a) *i)* Name the phase labelled A on the graph. *(1 mark)*

ii) Describe what is happening in phases B and C. *(2 marks)*

iii) Give two reasons why the numbers fall in phase D. *(2 marks)*

b) *i)* Copy the graph and sketch curves to represent the growth curve for *E. coli* bacteria grown at 5°C and at 25°C. *(4 marks)*

ii) Explain why each of these curves differs from the one shown. *(2 marks)*

c) Species of *E. coli* that are harmful can sometimes be transmitted to humans in eggs or in foods that contain eggs.

i) Suggest how the risk of transmission of *E. coli* could be reduced at the point of production and in the home. *(2 marks)*

ii) Meringues are made using raw eggs. Why should foods like this be avoided when there is an outbreak of *E. coli* food poisoning? *(1 mark)*

Total 14 marks

2 Rust fungi infect the leaves of many crop plants.

a) Explain two ways in which an infection by rust fungi can reduce crop yield. *(2 marks)*

b) Explain how genetic engineering could help to reduce the damage caused to crop plants by rust fungi. *(1 mark)*

c) Infections of rust fungi can be controlled by spraying the affected crops with a solution of copper sulphate. In an investigation into the effectiveness of this method of control, a field of potatoes was divided into two plots. As soon as the first signs of infection appeared, one plot was sprayed with copper sulphate solution. The other plot was left unsprayed. The relative amount of infection in the two plots was monitored over 50 days. The table shows the results of the investigation.

Day	Relative amount of infection	
	Control (unsprayed)	**Sprayed plants**
0	0	3
10	17	8
20	32	10
30	48	15
40	62	17
50	65	18

i) Plot a graph of this data. *(4 marks)*

ii) What was the relative infection of sprayed and unsprayed plants on day 24? *(2 marks)*

iii) Why was it necessary to have a control group? *(1 mark)*

iv) The investigation was repeated a year later. On day 30, the relative amounts of infection were 45 (unsprayed) and 36 (sprayed). Suggest an explanation for these results. *(2 marks)*

Total 12 marks

3 Microorganisms play an important part in recycling nutrients.

a) Gardeners often take advantage of this and construct compost heaps. They place waste vegetable matter, such as grass cuttings and peel from vegetables, in these heaps. Adding small amounts of soil to the heaps provides the bacteria needed to decompose the vegetable matter. The diagram shows the structure of a typical compost heap.

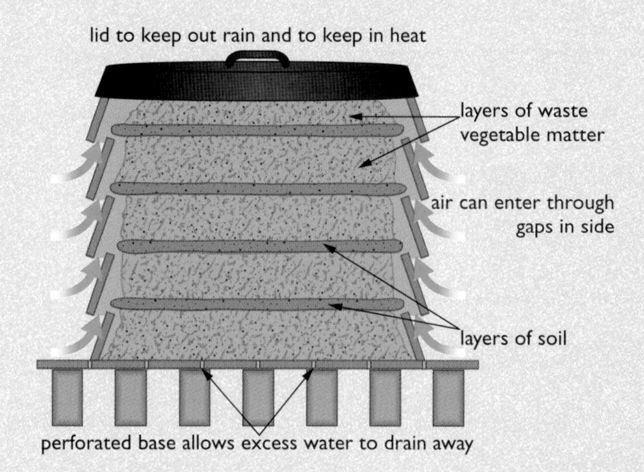

i) Suggest why the heap is constructed with several thin layers of soil and vegetable matter. *(2 marks)*

ii) Explain how the construction of the heap ensures favourable conditions for the microorganisms to decompose the vegetable matter. *(4 marks)*

b) Farmers often use grass in summer to make silage to feed to their cattle over winter. In silage heaps, the grass is fermented by bacteria under anaerobic conditions. The bacteria produce an acid called butyric acid, that preserves the grass. The diagram shows how silage is made.

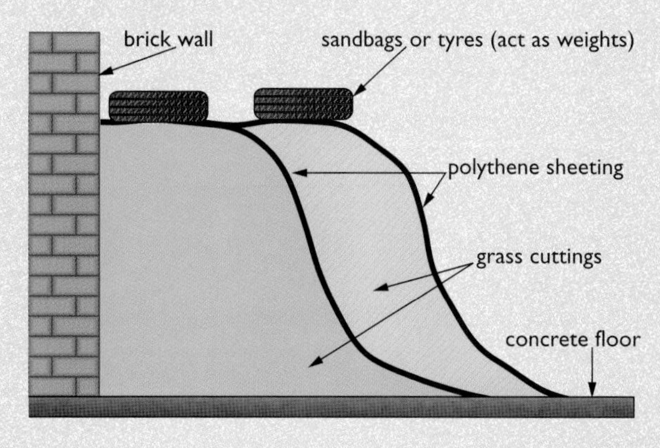

i) Explain how the construction of the silage heap allows favourable conditions for the bacteria to ferment the grass. *(4 marks)*

ii) Which method of preserving human food is most similar to preserving grass by turning it into silage? *(1 mark)*

iii) Explain why this method of food preservation is effective. *(2 marks)*

Total 13 marks

4 The human immunodeficiency virus (HIV) causes AIDS (acquired immune deficiency syndrome). It can be transmitted through sexual intercourse and by other means also. Once in the body, it infects some types of T-lymphocytes. During the latency period of AIDS (which can last up to 20 years) the infected person is HIV positive. The disease cannot be treated using antibiotics and is nearly always fatal. Death is often due to opportunistic infections by TB and pneumonia bacteria.

a) Name two methods, other than sexual intercourse, by which AIDS can be spread. *(2 marks)*

b) What is meant by the 'latency period' of an AIDS infection? *(1 mark)*

c) *i)* What is meant by describing a person as 'HIV positive'? *(1 mark)*

ii) Why can HIV not be treated with antibiotics? *(2 marks)*

d) Why are the infections by TB and pneumonia bacteria described as 'opportunistic' in this case? *(1 mark)*

Total 7 marks

5 Gonorrhoea is a sexually transmitted disease caused by a bacterium. The graph shows changes in the incidence of gonorrhoea from 1925 to 1990.

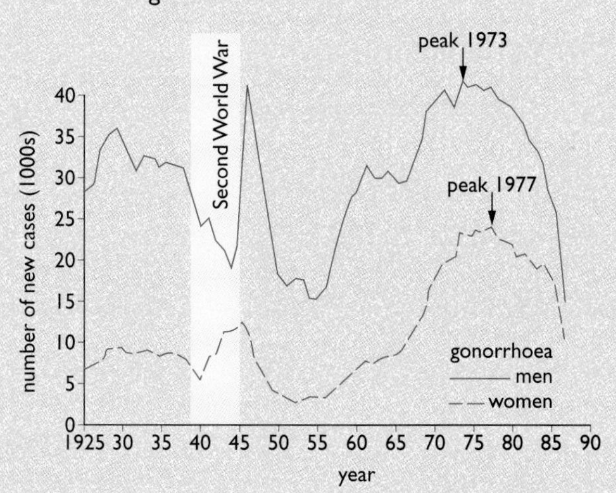

a) Name two other sexually transmitted diseases. *(2 marks)*

b) Describe the general trends in the numbers of cases of gonorrhoea from 1925 to 1990. *(2 marks)*

c) Suggest reasons for the increases in the number of cases of gonorrhoea:

i) during the 1960s

ii) during the early 1970s *(2 marks)*

d) Suggest how the increase in AIDS cases in the 1980s may be linked to the decrease in the incidence of gonorrhoea. *(1 mark)*

e) Gonorrhoea can usually be cured by giving penicillin.

i) Describe how penicillin would kill the bacteria causing gonorrhoea. *(3 marks)*

ii) Suggest why penicillin is not always effective in the treatment of gonorrhoea. *(1 mark)*

Total 11 marks

6 Some genes are transferred from one plant species to another. These genes are called 'jumping genes'. Environmentalists are concerned that genetically modified plants may transfer some of their genes to wild species. Explain their concern about genetically modified plants that:

a) have genes that make them resistant to herbicides (weedkillers) *(2 marks)*

b) have genes that make them resistant to pests *(2 marks)*

c) have genes that increase the yield of the crop they can produce. *(2 marks)*

Total 6 marks

Appendix A: Practical Investigations

Practical investigation (coursework) makes up 20% of your GCSE assessment. It isn't difficult to get full or nearly full marks, but it does take time and patience. If you are prepared to spend that time, it can make a big difference to your final grade.

The next few pages show you how to gain a high score on an investigation. It is important to listen to your teacher's advice on exactly what you need to do in order to get *full* marks. After all, he or she will be marking your piece of work, and will have up-to-date knowledge of what your examiners want.

The investigation we will use as an example is:

Investigate the effect of temperature on the rate of a reaction catalysed by the enzyme trypsin.

You must realise that the version in this book is **incomplete**. You will find it in full on the website supporting the book at www.longman.co.uk/gcse/biology.

How to start

This section shows you how to find background information for the trypsin investigation, but you will need to work in much the same way no matter which investigation you carry out. Use this book and other GCSE textbooks to get you started. You can even use 'A' level textbooks to fill in some details. It might also be worth seeing what you can find on the Examination Boards' websites (try all the Boards – not just your own). There may be material designed for teachers that you could use. You will find web addresses in 'About this book' on page vii. Be wary of any examples of coursework provided by other students on the Internet. Just because they are available on the web, it doesn't mean that they are any good.

Start by reading about enzymes in Chapter 1 of this book. The section on how enzymes work, and how they are affected by factors such as pH and temperature, is particularly relevant. It describes the role of the active site of the enzyme, and why enzymes have an optimum temperature at which they work best.

You can then look up trypsin in Chapter 3. Page 36 tells you that it is a protease, in other words an enzyme that digests proteins into short chains of amino acids. It is made by the pancreas in humans and other mammals.

You now need a substrate for the enzyme to digest. Powdered milk contains a white protein. When you mix the milk powder with water, it forms an opaque suspension. This means that it is cloudy and you cannot see through it because the protein is not very soluble. You can use this as a substrate for the trypsin. If you add a solution of trypsin to the milk suspension in a test tube, it will gradually turn clearer. It never turns completely clear, but eventually you can see through it. We say it is *translucent* (Figure 1b).

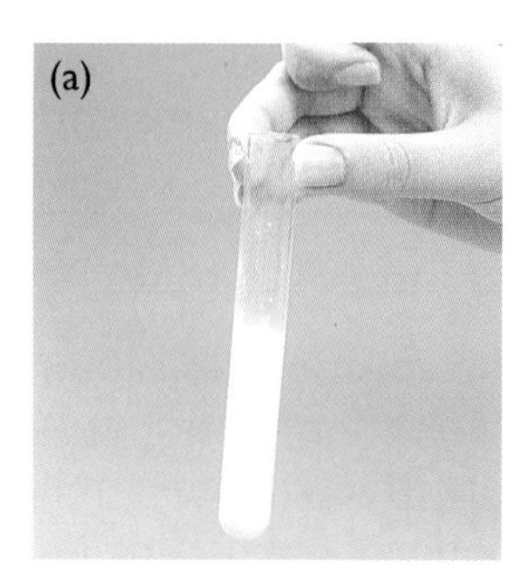

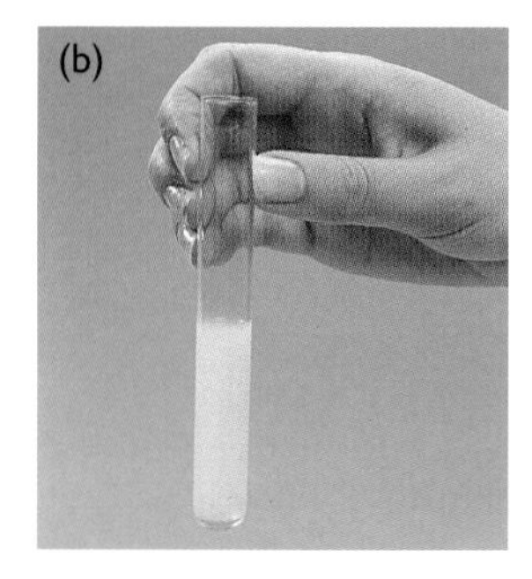

Figure 1 *Here you can see the effect of trypsin on the powdered milk protein* (a) *before treatment with the enzyme and* (b) *after treatment.*

You should now try some preliminary experiments with the milk and trypsin to find out the colour of the digested milk (called the 'end-point colour') and roughly how long it takes for the enzyme to break the protein down. It's no use if this takes hours to happen! You can make a solution that has the same appearance as the end-point by adding an equal volume of 0.1 M hydrochloric acid to a sample of milk. This will break the protein down immediately, without using any enzyme. It is called an 'end-point colour standard'.

Planning

Your teacher will be marking the planning part of your work by matching it to this checklist. The important terms are explained in the following pages.

If you can	Mark awarded
outline a simple procedure	2
plan to collect valid evidence plan the use of suitable equipment or sources of evidence	4
use scientific knowledge and understanding to plan and communicate a procedure, to identify key factors to vary, control or take into account, and to make a prediction where appropriate decide a suitable extent and range of evidence to be collected	6
use detailed scientific knowledge and understanding to plan and communicate an appropriate strategy, taking into account the need to produce precise and reliable evidence, and to justify a prediction when one has been made use relevant information from preliminary work, where appropriate, to inform the plan	8

To score 8 marks, your work must match both the statements in that box, as well as all the other statements for 6, 4 and 2 marks. In other words, for full marks, you need to do everything in the table. It is important to aim for the highest possible mark. Even if you miss it, you can still score well.

You can score odd-numbered marks if your work falls just short of a level. For example, you might score 7 marks if you satisfied the first statement needed for 8 marks, but didn't do any preliminary experiments when some would have been helpful.

You can use the following main headings to help you to get everything in a logical order:

To measure the rate of any reaction, you can either measure how quickly a reactant is used up, or how quickly a product is formed. In this reaction you can measure how quickly the reactant (the milk protein) is used up.

What I am going to do

This investigation aims to find out the effect of temperature on the activity (rate of reaction) of trypsin. This can be done by heating trypsin and its substrate (the powdered milk) to a certain temperature in a water bath, then mixing them and finding out how long the enzyme takes to turn the milk translucent. The procedure is then repeated at other temperatures. The rate of a reaction can be calculated from the time it took to finish.

What I already know

To score 8 marks for planning, not only must everything be based on 'detailed scientific knowledge and understanding', but you have to make it clear how you are going to *use* that knowledge and understanding. List all the relevant things you have found out from books or other sources, and say why you think they might be useful to you. For example:

> I know that enzymes work best at an optimum temperature. For human enzymes this is normally close to body temperature (37°C). Below this temperature, the molecules of enzyme and substrate have less kinetic energy, are moving more slowly and therefore are less likely to collide. Because of this, the rate of reaction is less at low temperatures. At high temperatures (usually above 50°C), enzymes are destroyed or denatured by heat, so the rate again decreases. I will therefore carry out the reaction at a range of temperatures, from room temperature (about 20°C) to a temperature that should cause denaturing (80°C)

It is important to get each solution to the correct temperature before you mix them, otherwise they will start reacting at the wrong temperature. This is called **equilibrating** the solutions to the right temperature. After you have mixed them, you must keep them at this temperature while they react, by leaving them in the water bath.

The key factors to vary, control or take into account

Don't just list the key factors. Explain why you are choosing to control some things and vary others, and why some things don't matter (if that happens to be the case). Your explanations should again 'use detailed scientific knowledge and understanding'. For example:

> I am going to use the same concentration of trypsin in all my experiments. If the concentration of the enzyme were increased, this would mean that the chance of an enzyme molecule colliding with a substrate molecule would be higher, which would affect the rate of reaction. The volumes of trypsin and Marvel must also be constant (5 cm^3 of each). I will take care to measure these volumes as accurately as possible. If either varies, this will upset the final concentration of enzyme and substrate in the mixture.

Remember that none of the examples given are complete. There are other factors that you will need to control.

Preliminary work

Preliminary work involves doing experiments to find out the best conditions for carrying out your investigation. It is essential if you are going to score 8 marks for planning. Describe your preliminary work carefully, and say exactly how it helped you to decide your final plan. Again, wherever possible, explain your choices in terms of 'detailed scientific knowledge and understanding'. For example:

> From my preliminary experiments I found that 5 cm^3 of trypsin solution took less than 10 minutes to digest 5 cm^3 of the milk suspension to the same appearance as the end-point colour standard. This is a reasonable length of time that can be measured by a stopwatch. It will also allow me to carry out a number of readings at each temperature in the time available, so that I could get reliable evidence to test my predictions.

Ask your teacher how many repeat (replicate) experiments he or she expects you to carry out in order to gain full marks. You also need to know how many readings you need to take. In this investigation how many different temperatures will be needed? Most exam boards ask for a minimum of five, so ten-degree intervals between 20°C and 80°C should be OK. However, you may need to do more to find the optimum temperature more precisely.

Producing precise and reliable evidence

'Precise' means that you are measuring things as accurately as possible. Particularly where the quantity you are measuring is small, you should try to measure it using the most accurate equipment you have available. 'Reliable' means that if you repeat your readings, you will get the same results. For example:

> The end-point of the reaction is difficult to judge with accuracy, so I will not attempt to measure the time to less than the nearest second. I will measure the volume of trypsin solution using a graduated pipette. This is accurate to $\pm$ 0.1 cm^3, which is a 2% margin of error. I will repeat the experiment three times at each temperature to check whether my findings are reliable. If my measurements of the reaction time at any temperature don't agree, I will go on repeating the experiment to find out whether I can reject any odd results.

Safety

List all the safety aspects of your plan in detail, explaining why you need to take each precaution. For example:

> Some people are allergic to enzymes, so I will be careful to mop up any spillages and will wash my hands after the practical work.

Doing the experiment

Describe what you are going to do in detail, listing all the apparatus you need. Draw diagrams if they will help your description.

When you have finished describing your method, read it through carefully and ask yourself whether someone else could carry it out successfully if they did it exactly what you have written. You can assume that they know how to use standard apparatus like pipettes and thermometers, but you should stress any unusual points, such as:

> It is important that the enzyme and milk have time to equilibrate to each temperature. To ensure this, I will measure the temperature inside the boiling tubes and keep the solutions at the chosen temperature for three minutes before I mix them together. Afterwards I will put the tube containing the mixture back into the water bath, so that the contents don't cool down while the reaction is taking place.

My prediction

Again, your predictions must be based on 'detailed scientific knowledge and understanding'. For example:

> I predict that temperature will affect the activity of the enzyme trypsin. The trypsin used in this experiment is obtained from bovine (cow) pancreas. Since cows are mammals like humans, the enzyme is likely to have an optimum temperature similar to human trypsin, approximately 40°C. At temperatures below the optimum, the lower kinetic energy of the enzyme and substrate molecules will result in fewer collisions, and a lower rate of reaction. At temperatures above the optimum, the heat will cause the enzyme molecules to denature, so there are fewer of them to catalyse the reaction.

Obtaining evidence

Your teacher will be using this checklist:

If you can	Mark awarded
collect some evidence using a simple and safe procedure	2
collect appropriate evidence that is adequate for the activity record the evidence	4
collect sufficient systematic and accurate evidence and repeat or check when appropriate record clearly and accurately the evidence collected	6
use a procedure with precision and skill to obtain and record an appropriate range of reliable evidence	8

This is a relatively easy section to gain full marks for. You should record your readings or measurements in neatly constructed table(s) that clearly present your findings in a logical order. Tables need a title and the *correct units* must be given. Any observations should also be recorded.

Drawing up a table of results

For example, the 'raw data' of measurements of the time of the reaction at different temperatures, measured in minutes and seconds, might look like this:

Temperature (°C)	Time to completion of reaction (minutes and seconds)		
	(1)	(2)	(3)
20	8.17	7.54	8.31
30	2.28	3.38	3.14
35	2.16	2.12	2.00
40	1.40	1.50	1.30
45	1.26	1.30	1.26
50	1.19	1.20	1.26
55	1.25	1.32	1.25
60	1.42	1.50	1.54
65	2.50	3.20	(7.12) 3.22*
70	7.35	10.12	9.36
80	>30**	>30	>30

* First measurement was anomalous and was repeated to check reliability. It was not used in later calculations of rate.
** Reaction had not taken place by 30 minutes

Table 1: *Time taken for the enzyme trypsin to digest milk protein at different temperatures. Columns show three replicates at each temperature.*

Notice that the times are given as mixed units (minutes and seconds) and will need to be converted to decimal fractions of a minute before they can be used to calculate rates. Calculations like these, from the raw data, are assessed in the next checklist.

Analysing your evidence and drawing conclusions

Your teacher will be using this checklist:

If you can	Mark awarded
state simply what is shown by the evidence	2
use simple diagrams, charts or graphs as a basis for explaining the evidence identify patterns and trends in the evidence	4
construct and use suitable diagrams, charts, graphs (with lines of best fit where appropriate), or use numerical methods to process evidence for a conclusion draw a conclusion consistent with the evidence and explain it using scientific knowledge and understanding	6
use detailed scientific knowledge and understanding to explain a valid conclusion drawn from processed evidence explain the extent to which the conclusion supports the prediction, if one has been made	8

Notice that to gain 6 or 8 marks you don't necessarily have to draw graphs. You can use 'numerical methods' instead. This means doing some reasonably complicated calculations. Although working out a simple average is a 'numerical method', it isn't complicated enough to earn you 6 (or 8) marks.

Calculations

Converting the times in minutes and seconds into time in minutes, then finding a mean time at each temperature, and finally calculating the rate of the reaction, is complicated enough to gain the first point in the 6 marks box.

For example, a time of 2 minutes and 45 seconds is equal to $2\frac{45}{60}$ minutes, or 2.75 minutes. The rate of the reaction is found from this time:

If the trypsin digested 5 cm^3 of milk protein in 2.75 minutes, the rate of reaction in cm^3 per minute is:

volume of milk ÷ time = 5 ÷ 2.75 = 1.82 cm^3/min (or 1.82 $cm^3\ min^{-1}$)

You need to show how you arrived at this formula, not just use it.

Now you can calculate the mean (average) rates at each temperature. It would be best to construct another table for this:

Temperature (°C)	Time (min)			Mean time (min)	Mean rate (cm^3min^{-1})
	(1)	(2)	(3)		
20	8.28	7.90	8.52	8.23	0.61
30	2.47	3.63	3.23	3.11	1.61
35	2.27	2.20	2.00	2.16	2.31
40	1.67	1.83	1.50	1.67	2.99
45	1.43	1.50	1.43	1.45	3.45
50	1.32	1.33	1.43	1.36	3.68
55	1.42	1.53	1.42	1.46	3.42
60	1.70	1.83	1.90	1.81	2.76
65	2.83	3.33	3.37	3.18	1.57
70	7.58	10.20	9.60	9.13	0.55
80	>30	>30	>30	>30	<0.17

Table 2: *Time taken for the enzyme trypsin to digest milk protein at different temperatures. Columns show three replicates at each temperature, mean time and mean rate of reaction.*

Now you should plot a graph of the mean reaction rate against temperature (Figure 2).

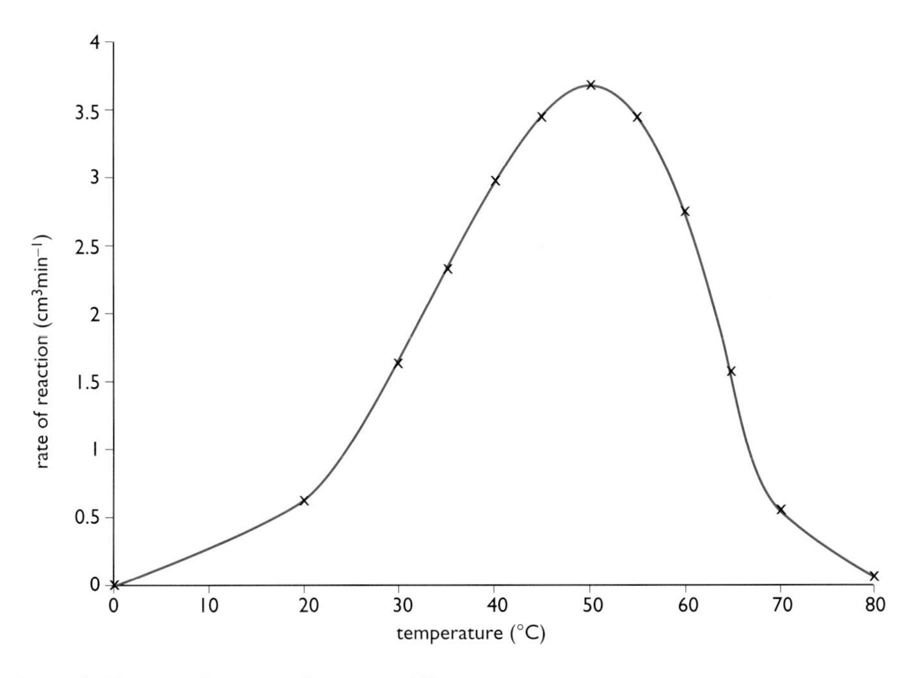

Figure 2 *The rate of reaction of trypsin at different temperatures.*

If you draw a line graph like this, you must connect the points with a 'line of best fit' to fulfil the first point in the 6 marks box. A 'best fit' line is not necessarily straight. In Figure 2 it is a smooth curve passing through the data points. It may also be a smooth curve that lies between each of the points, with some falling on the line, and an equal number either side of it.

Drawing your conclusions

You have to decide whether your results fit your original prediction. Remind the person marking your work exactly what your prediction was. Don't just refer back to your original scientific explanation. Give it again, together with any modifications that your results show to be necessary. For example:

> I predicted that the rate of reaction would increase with temperature, as the kinetic energy of the molecules increased, until the optimum temperature for the enzyme was reached. Above the optimum, the rate of reaction would decrease as the enzyme molecules were denatured. As shown in Figure 2, the results I have obtained fit the general pattern of my prediction. The rate increases from 0.61 $cm^3 min^{-1}$ at 20°C, to a maximum of 3.68 $cm^3 min^{-1}$ at 50°C. It then decreases to 0.55 $cm^3 min^{-1}$ at 70°C. At 80°C, the rate was very slow, and no change in the appearance of the milk protein could be detected after 30 minutes. It is probable that the enzyme had been fully denatured at this temperature, so that the rate of reaction was zero. However, I predicted that the optimum temperature would be about 40°C, and Table 2 and Figure 2 show that it was higher than this, at about 50°C.

It doesn't matter if your prediction doesn't match your results, but you must be able to use your knowledge and understanding to come up with a convincing explanation:

> It is possible that the optimum temperature for trypsin is closer to 50°C. Alternatively, it may be that the enzyme and milk were not held at each temperature for long enough before they were mixed. Denaturation is a time-dependent process, and it may need longer than 3 minutes at 50°C to denature enough of the enzyme to reduce the rate of reaction.

Evaluating your investigation

If you can	Mark awarded
make a relevant comment about the procedure used or the evidence obtained	2
comment on the quality of the evidence, identifying any anomalies comment on the suitability of the procedure and, where appropriate, suggest changes to improve it	4
consider critically the reliability of the evidence and whether it is sufficient to support the conclusion, accounting for any anomalies describe, in detail, further work to provide additional relevant evidence	6

Students often achieve low marks for this last stage. It is worth almost as much as the other three sections, and your answer must be detailed and specific.

Evaluating the experiment

You need to point out any results that appear to be wrong or out of place (anomalous), and try to account for them. For example:

> Although the experiment produced repeatable results (Table 1) some replicate measurements were anomalous, for example the third measurement of the reaction time at 65°C. This was checked by repeating the experiment at this temperature, which led me to reject the first value. I also noticed that the measurements of the rate of reaction became more variable at higher temperatures (Table 2). This could be due to the difficulty in maintaining the water bath at a constant temperature. This is particularly true at high temperatures, when the beaker will cool more quickly. I noticed that however carefully I tried to keep the temperature constant by applying the Bunsen flame or removing it, it still varied by as much as ±2°C.

Improving the experiment

Make sure that the improvements you suggest are detailed and specific to your investigation. You won't get much credit for general comments like 'use more accurate equipment' or 'take more care with measurements'. Some better examples of improvements are:

> The temperature could have been kept more constant by using an electric, thermostatically controlled water bath. These can maintain a temperature to within ±0.5°C.

> The end-point of the reaction was difficult to judge with accuracy, even when comparing it with the end-point colour standard. There are machines called 'colorimeters' that can be used to measure the cloudiness of a suspension. If I had a colorimeter available, I could have compared the cloudiness of the experimental tubes with the cloudiness of the end-point standard.

> In my experimental method I did not control the pH of the reaction mixture. pH affects enzymes, and the pH of the mixture may have changed at different temperatures. I could have controlled the pH by adding a set volume of a buffer solution to the enzyme. Buffer solutions are used to keep the pH constant. I know from my background reading that trypsin works well at neutral to slightly alkaline pH, so a buffer with a pH between 7 and 8 would be suitable.

Extending your experiment

To gain the full 6 marks, you must describe *in detail* how you might extend this investigation to provide additional *relevant* evidence. It's no use, for example, saying that you could repeat this sort of experiment with another enzyme, or find the effect of concentration on the rate of reaction. In this investigation, one problem is that denaturing takes time. It would be a good idea to find out how quickly the trypsin is denatured at high temperatures:

> I could extend this investigation by finding out how quickly trypsin is denatured at high temperatures. From Figure 2, I can see that the rate of reaction is greatly decreased at 65°C, so I would select that temperature to investigate.

I would set up two water baths, one at 65°C and one at a lower temperature where the enzyme works well, such as 40°C. I would place a test tube containing 5 cm^3 of trypsin in the 65°C bath for exactly one minute. I would then remove the tube and place it in the 40°C water bath, along with another tube containing 5cm^3 of Marvel milk. When these tubes were both at 40°C, I would mix their contents, and return the tube containing the mixture to the water bath. I would then measure how long the mixture took to go translucent, as before.

I would repeat this procedure, exposing the trypsin to different periods of time at 65°C, such as ...

You should go on to describe exactly what you would do, what periods of exposure to the high temperature you would choose, and what you predict would happen.

Index